The Paleoarchaic Occupation of the Old River Bed Delta

NUMBER 128

The Paleoarchaic Occupation of the Old River Bed Delta

David B. Madsen, Dave N. Schmitt, and David Page

with contributions by

Charlotte Beck, Daron G. Duke,
George T. Jones, Lisbeth A. Louderback,
Charles G. Oviatt, David Rhode, and D. Craig Young

THE UNIVERSITY OF UTAH PRESS
Salt Lake City

Advisory Board for the University of Utah Anthropological Papers:
James O'Connell and Duncan Metcalfe

19 18 17 16 15   1 2 3 4 5

LIBRARY OF CONGRESS CATALOGING-IN-PUBLICATION DATA

Madsen, David B.
 The paleoarchaic occupation of the Old River Bed delta / David B. Madsen, Dave N. Schmitt and David Page ; with contributions by Charlotte Beck, Daron G. Duke, George T. Jones, Lisbeth A. Louderback, Charles G. Oviatt, David Rhode, and D. Craig Young.
    pages cm. -- (University of utah anthropological paper ; 128)
  Includes bibliographical references and index.
  ISBN 978-1-60781-393-4 (paperback) -- ISBN 978-1-60781-394-1 (ebook)
 1. Limnology--Great Basin. 2. Paleolimnology--Great Basin.
 3. Paleoecology--Great Basin. I. Schmitt, Dave N. II. Page, David (David J.) III. Title.
  QH104.5.G68M34 2015
  577.609792'4--dc23
                                            2014036320

Contents

Figures

Tables

Acknowledgments

The work reported here was supported largely by the U.S. Department of Defense through cultural resource management projects conducted under the direction of Environmental Programs at Dugway Proving Ground and Hill Air Force Base (HAFB). The results of more than 30 such projects, conducted over the course of nearly 20 years, are incorporated in one way or another in this monograph. Project managers for these projects were Kathy Callister and Rachel Quist at Dugway Proving Ground and Jaynie Hirschi at HAFB. Without the enthusiastic support of these individuals the work reported here would never have been accomplished. Supplemental support was also provided by the Utah Geological Survey and by the Desert Research Institute. We are particularly grateful for logistical support provided by personnel from the Utah Test and Training Range facility, HAFB. These include (but are not limited to) Boe Hadley, Tom Larkin, Teresa Neil, and Markus Teters.

During the course of both field and laboratory work associated with the completion of these projects, a large number of individuals provided critical assistance. We wish to extend our sincere appreciation to all these people but in particular wish to thank Jason Fancher, Chris Hall, Jeff Hunt, Kristen Jensen, Monson Shaver, and Geoff Smith. We are particularly indebted to Jack Broughton for help with the identification of fish bones.

Editorial assistance was provided by Reba Rauch, Evelyn Seelinger, and Elisabeth A. Graves, without whom this manuscript would have been filled with innumerable spelling, grammatical, and formatting errors. Those that may remain can be attributed solely to us. Critical reviews that substantially improved the clarity and succinctness of this monograph were provided by Donald Grayson, Kevin Jones, and Marith Reheis. We greatly appreciate their help and note that where problems may occur they likely stem from our failure to follow their advice.

Introduction and Research Perspectives

David B. Madsen, Dave N. Schmitt, and David Page

David B. Madsen, Dave N. Schmitt, and David Page

Background

During the regressive phase of Lake Bonneville, the Old River Bed (ORB) held a river connecting the two major subbasins of the Bonneville basin (Figure 1.1). Beginning sometime around 12,000 ^{14}C BP, the river ran north, draining the southern Sevier basin and emptying into Great Salt Lake. This was a major river, consisting of the combined flow of the modern Sevier and Beaver rivers, many smaller streams, and groundwater draining into and out of the Sevier basin. Moreover, during the Younger Dryas climate period, between ~11,100 and 10,050 ^{14}C BP, this flow was likely enhanced by both increased precipitation and lower temperatures. Sometime after about 8500 ^{14}C BP, water ceased to flow in the ORB, and environmental conditions along the channel began to approach those found at present. During the more than 3,000 years of its existence, however, the river fed a large marsh/wetland system in the ORB delta covering several thousand square kilometers and supported a riverine environment along its length. This 3,000-year interval corresponds almost exactly to the earliest phase of human occupation in the Bonneville basin (e.g., Grayson 1993, 2011 and references therein). Foragers have been drawn to such rich marsh/wetland ecosystems throughout the human history of the Great Basin (Fawcett and Simms 1993; Janetski and Madsen 1990; Madsen 2002; Madsen and Kelly 2010; Raven and Elston 1988), but sites of the Paleoarchaic period are particularly associated with Great Basin wetlands (Beck and Jones 1997; Madsen 2007; Smith 2007; Willig et al. 1988).

Given the alluring physical and temporal context of the ORB, we thought it was quite possible that Paleo-archaic sites were common in the ORB delta, and we initiated archaeological research there in an attempt to identify, describe, and map delta distributaries and intact sites dating to the period. With rare exceptions (e.g., Davis et al. 1969), it has been nearly impossible to map open Paleoarchaic sites in modern Great Basin marsh and lake-margin settings because such sites usually consist of palimpsests of Paleoarchaic and later materials that are not easily separated. Most major wetlands in the Great Basin lie at the end of major river systems, such as the Humboldt, Bear, and Carson rivers, and have been in existence for at least the period of human occupation in the region. As a result of both continuous use of these marshes by foragers and erosional and depositional cycles associated with Holocene climatic shifts, intact Paleoarchaic sites are relatively rare (e.g., Raven 1990; Raven and Elston 1988). The ORB delta differs in that it was forming for only a limited time during the late Pleistocene/early Holocene, and, while erosion has taken place since that time, there has been no movement of the marsh ecosystem back and forth across the landscape, which would result in the disturbance of early sites. Moreover, as warmer and drier climates took hold in the waning years of the early Holocene (e.g., Grayson 2000a; Louderback and Rhode 2009; Madsen et al. 2001; Schmitt, Madsen, and Lupo 2002), the area became relatively unattractive to hunter-gatherers, especially the deflated mudflats of the southern Great Salt Lake Desert, and the impact that later human activities generally had on early sites in most other major wetland systems was probably much more limited in the ORB. In addition, the ORB delta lies on military installation lands that have been closed to the general public for

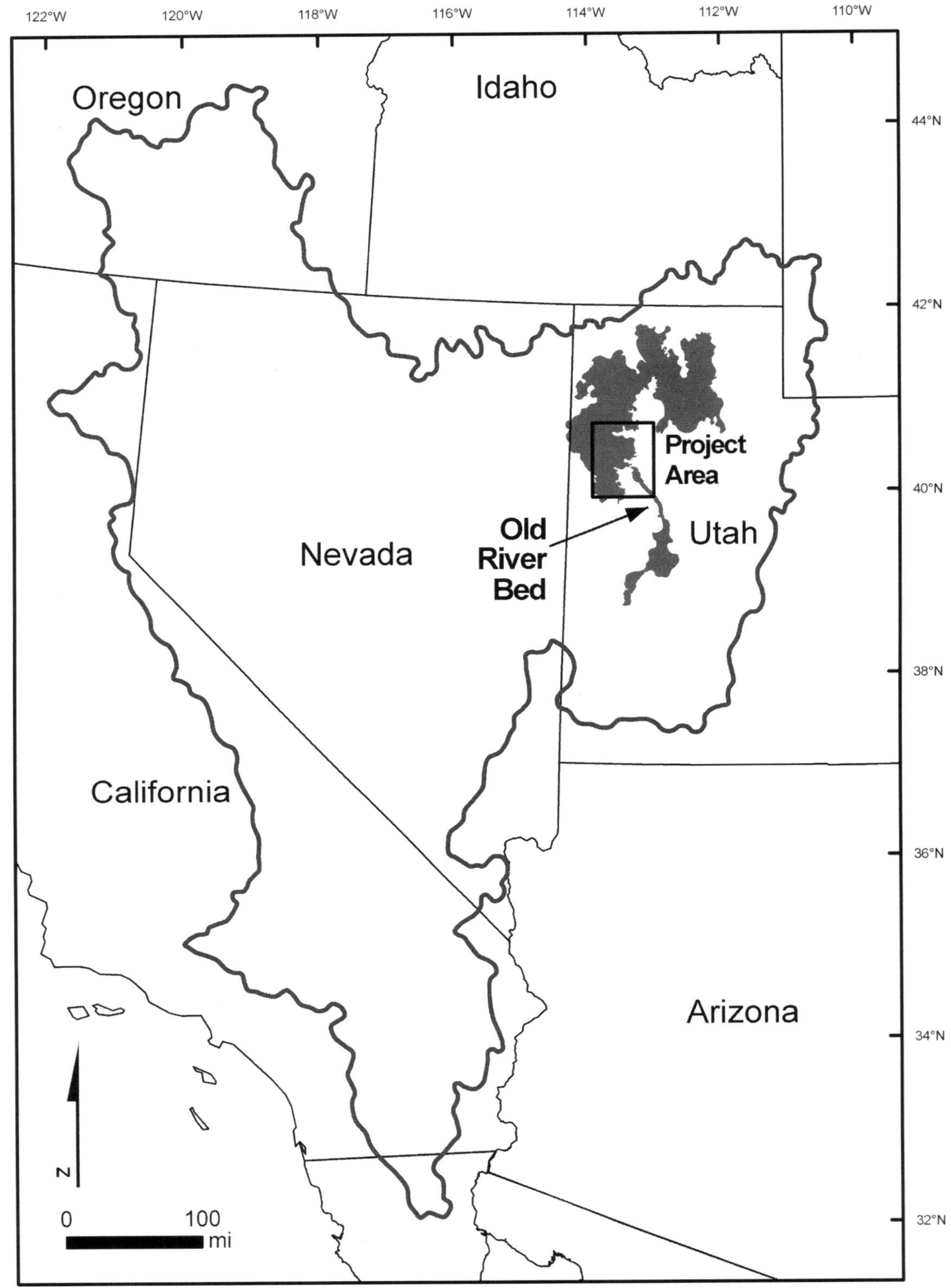

FIGURE 1.1. Position of the Old River Bed river in the eastern Great Basin, joining regressive-phase lakes in the Sevier basin to the south and the Great Salt Lake basin to the north. The lake elevations shown are approximately those of lakes in the basins when the Old River Bed river first began flowing, ~12,000 ^{14}C BP.

more than 60 years, and disturbance by relic hunters has been significantly less than that in more accessible Great Basin valleys.

The results of more than a decade of research in the ORB delta are presented in the pages that follow. Although some aspects of ORB delta geomorphology and Paleoarchaic archaeology have been previously published (Duke and Young 2007; Duke et al. 2007; Oviatt et al. 2003; Rhode et al. 2005; Schmitt, Madsen, Oviatt, and Quist 2007), this monograph presents, for the first time, a comprehensive synthesis of investigations conducted by research teams working at various times in different parts of the delta. We begin by setting the stage with brief discussions on the physical setting and history of investigations, followed by an overview of regional Paleoarchaic archaeology and some specific research domains and hypotheses that served as the impetus for why and how we collected data. Chapter 2 provides an overview of climate and biotic communities during the latest Pleistocene and early Holocene, drawing from regional paleoenvironmental data, and Chapter 3 offers a comprehensive discussion and interpretation of ORB delta geomorphology and chronology. Chapter 4 presents descriptions of sites recorded in the ORB delta, and typological and technological analyses of the lithic tools collected from these sites in the proximal delta follow in Chapter 5. Chapter 6 presents lithic sourcing data and inferences on toolstone acquisition and overall forager mobility. In Chapter 7 we expand on some aspects of tool use, present a gravity model of occupation for large wetland "megapatches" such as the ORB wetlands, and conclude with an integrated summary and interpretation.

SETTING, MODERN ENVIRONMENTS, AND BIOTIC COMMUNITIES

Discussions of regional climates, geology, and Lake Bonneville chronology are available in more detail in Madsen 2000a, Madsen and Schmitt 2005, and Schmitt and Madsen 2005. Here, we provide only a general discussion of the setting in the immediate vicinity of the ORB delta. The delta, in its entirety, lies in Tooele County of western Utah, where it extends north from southeastern Dugway Valley into the southern reaches of the Great Salt Lake Desert (Figure 1.2). Overall, the ORB delta is approximately 55 km in length and encompasses nearly 2,000 km². It consists of a labyrinth of northwest-trending streams that range from broad high-energy channels to meandering low-energy distributaries (see Chapter 3), referred to as "gravel" and "sand" channels, respectively, in earlier reports (e.g., Oviatt et al. 2003; Schmitt, Madsen, Oviatt, and Quist

2007), terms we have discarded here. Most of the delta distributaries lie within the confines of the U.S. Army Dugway Proving Ground (DPG), including approximately 15 km of the ORB main channel (Figure 1.3). A series of channels continue north of the DPG boundary, where they meander across more than 500 km² of deltaic deposits on Utah Test and Training Range lands administered by the U.S. Air Force. A number of streams originating in the Deep Creek Mountains and from Fish Springs on the southwestern margin of the ORB delta contributed to this wetland area, and the size of this great wetlands system exceeds 2,600 km².

East–west tectonic expansion of the Basin and Range Province has created a series of north-to-south-trending fault-block mountains separated by long half-graben valleys filled with Quaternary alluvium. In the Great Salt Lake basin a series of major fault-block mountains and smaller, often isolated, mountains and ridges surround the ORB delta vicinity. These include Camels Back Ridge and the Fish Springs, Simpson, and Dugway ranges to the south; Granite Peak and the Deep Creek Mountains to the west; the Silver Island Mountains to the northwest; Wildcat and Kittycat mountains to the northeast; and the Cedar Mountains to the east. The Deep Creek Mountains reach 3,600 m asl (11,800 ft) and represent the highest mountain range in the ORB vicinity. The Dugway and Simpson ranges are composed primarily of Lower Cambrian limestone formations, whereas Camels Back Ridge consists of an undivided Upper Cambrian limestone of lower Paleozoic age (Hintze 1988). A major igneous intrusive in the Deep Creek Mountains and Granite Peak is composed almost entirely of Precambrian granite (Clark et al. 2008; Hintze 1988). The Cedar Mountains are composed of late Paleozoic sedimentary formations, including silicified sandstones and chert (Clark et al. 2008), overlain by a cap of extruded Tertiary-age igneous rocks that include andesite and rhyolite (Hintze 1975). Additional Tertiary volcanics are scattered along the southern margin of the Great Salt Lake basin, including a major source of obsidian in the Topaz Mountains (e.g., Raymond 1984) and sources of fine-grained andesite and dacite in the Flat Hills of eastern DPG (Page 2008).

Most precipitation in the region is a product of central-northern Pacific storm systems, with minor amounts contributed by storms originating in the Gulfs of California and Mexico. The average yearly precipitation at Dugway, Utah, on the eastern margin of the ORB delta, is 19.9 cm, with snow amounts in the winter months limited to an average of 38.9 cm. The limited contribution of southern monsoons produces a relatively even year-round precipitation regime, with

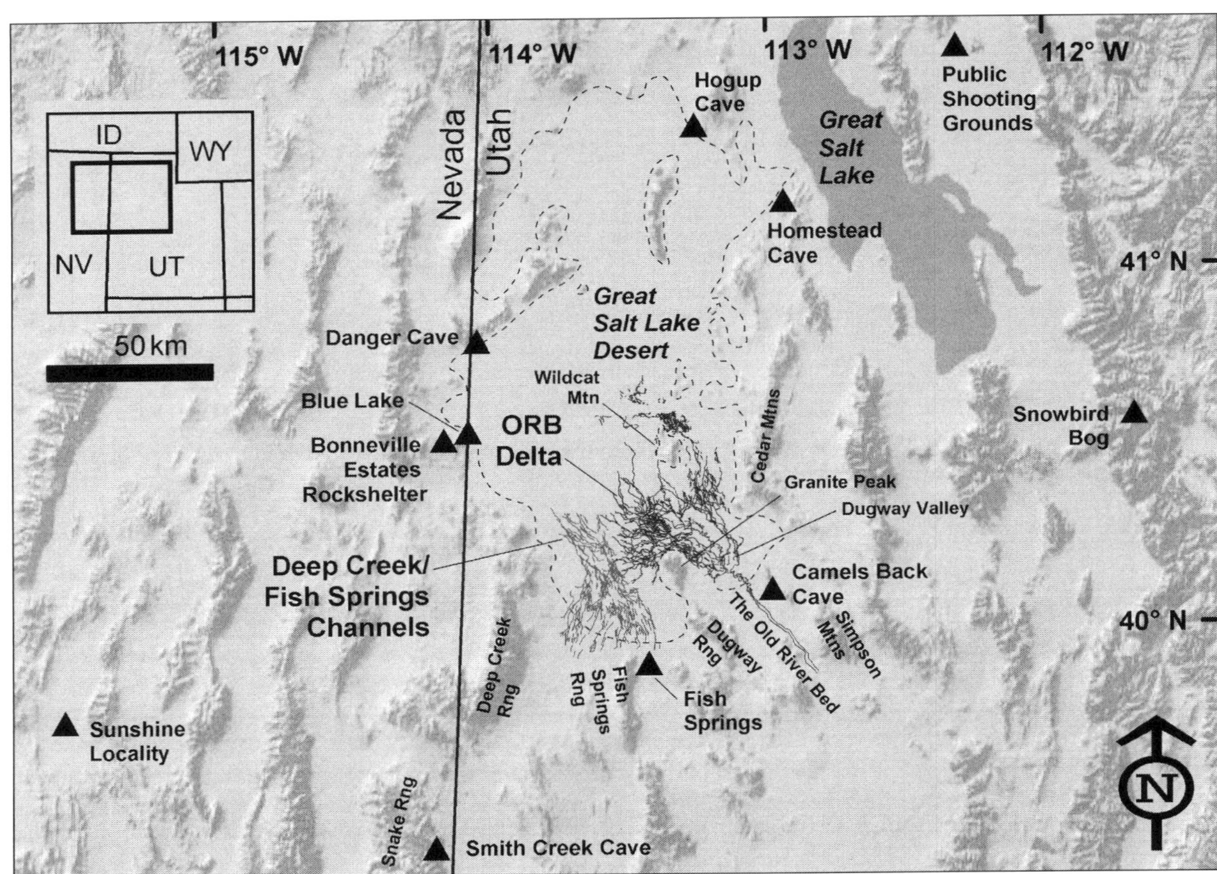

FIGURE 1.2. Satellite image showing visible distributary channels in the Old River Bed (ORB) delta and major surrounding physiographic features (base image from 2006 National Agriculture Imagery Program 1 Meter Orthophotography, http://gis.utah.gov/data/aerial-photography/2006-naip-1-meter-color-orthophotography/).

FIGURE 1.3. View southwest across the Old River Bed incised river valley in southern Dugway Proving Ground. The Dugway Range is in the background.

slightly higher amounts in March through May and relatively smaller amounts in the late summer months. The seasonal temperature variation in the region is often extreme, with the difference between winter and summer average daily maximum temperatures at Dugway being ~31°C. The modern average summer (July–August) high temperature is 34.1°C, and the average winter (December–January) low is 3.9°C.

In addition to the delta itself, the project area contains four primary and distinct landform features: a network of non-ORB stream channels on the southwestern margin of the ORB delta; the central Dugway Valley plain, composed of regressive-phase lacustrine fines (RLF; see Chapter 3) to the south; the vast, often barren mudflats of the southern Great Salt Lake Desert to the north; and a band of dunes that mark the RLF–mudflat transition (Figure 1.2). Stream channels carrying water from the Deep Creek Mountains and from Fish Springs on the southwestern margin of the ORB delta proper were carrying substantial amounts of water during the period the ORB delta formed. Water in these channels contributed to the overall formation and size of what should more properly be considered to be the ORB delta wetlands system, and Paleoarchaic sites are found along these channels as well as along those formed by ORB streamflow.

In the most proximal portion of the delta (the upstream, southernmost portion of the delta at higher elevations), the RLF plain is underlain by silts and clays in a large, low-gradient underflow fan deposited on the bottom of Lake Bonneville after the lake regressed below the ORB threshold and while the ORB river was transporting fine sediment into the lake. These fine-grained lacustrine deposits are overlain by alluvium deposited by the ORB river and by fine sand and silt deposited in aeolian environments during the Holocene. Some portions of the RLF surface support the development of a biological crust on the soils, while others consist of desiccated, sparsely vegetated hardpans that are often scarred by barren tiger stripes (e.g., Ives 1946; Wakelin-King 1999; see Chapter 3). Small aeolian dunes are present in some places on the RLF plain surface. The RLF plain generally slopes down from southeast to northwest in elevations ranging between ca. 1,305 and 1,330 m asl (4,280–4,360 ft) and encompasses more than 300 km² of the proximal ORB delta.

Modern vegetation consists of a xerophytic scrub community containing greasewood (*Sarcobatus vermiculatus*), shadscale (*Atriplex confertifolia*), four-wing saltbush (*A. canescens*), gray molly (*Kochia americana*), halogeton (*Halogeton glomeratus*), snakeweed (*Gutierrezia sorthrae*), Russian thistle (*Salsola* sp.), patches of cheatgrass (*Bromus tectorum*) and pepperweed (*Lepidium* sp.), and an occasional prickly pear (*Opuntia* sp.). The RLF plain contains a number of isolated and widespread low vegetated dunes composed of fine aeolian silt and sand, and in many instances these features are anchored to deltaic channel exposures. The dunes commonly contain most of the species listed above, along with fringe sage (*Artemisia fringata*), Indian ricegrass (*Achnatherum hymenoides*), globe mallow (*Sphaeralcea* sp.), rabbitbrush (*Chrysothamnus* sp.), and occasional clusters of ephedra (*Ephedra* sp.) and rare big sagebrush (*Artemisia tridentata*).

The band of sand and silt dunes at the RLF plain–mudflat interface likely formed sometime during the early middle Holocene, and although a number of periods of stabilization have occurred (cf. Madsen 2000a), they have probably been in place since their formation. They form a ~.5- to 2.3-km-wide arc atop the northernmost edges of the alluvial plain (Figure 3.6) and are at their widest along the northwestern toe of the delta plain north of Granite Peak. Dune crests range from 4 to 13 m above the neighboring mudflats and average about 1,310 m asl (~4,300 ft) in elevation. Like the dunes built atop the RLF plain, dunes at the plain–mudflat transition are often built atop ORB channels/channel margins. They currently contain an array of desert shrubs and grasses that include greasewood, gray molly, snakeweed, Indian ricegrass, halogeton, ephedra, and fringe sage, as well as occasional desert primrose (*Oenothera* sp.) and patches of buckwheat (*Eriogonum nummalare*) and sand dock (*Rumex venosus*).

In the distal portions of the ORB delta (the farthest downstream, northern end of the delta at the lowest elevations), distributaries meander across deflated mudflats in what is now the southern Great Salt Lake Desert. Hundreds of kilometers of channels occur in this area and include ancient streams that continued their course to the north and northwest and others that curved west around the northern end of Granite Peak (Figure 1.2). A number of the channels are topographically inverted, at least in some areas, with the crests of the relatively coarse channel deposits, less affected by wind action, standing as much as 4 m higher than the neighboring mudflats (see Chapter 3). Overall, the mudflats contain the largest surface area of ORB deltaic deposits and the majority of the identified archaeological sites, especially those that date solely to the Paleoarchaic. The mudflats are composed of fine clays and silts, and similar to the RLF plain, they slope gently to the north and northwest in elevations ranging between 1,289 and ~1,305 m asl (4,230–4,280 ft). Although today the area largely consists of an inhospitable, barren landscape

FIGURE 1.4. View from the Old River Bed delta on the mudflats in the southern Great Salt Lake Desert looking southeast. Granite Peak is in the center background.

(Figure 1.4), groundwater supports vegetation in some areas. When present, vegetation consists of iodine bush (*Allenrolfea occidentalis*) (dominant), seepweed (*Suaeda* sp.), and salt grass (*Distichlis spicata*), with occasional patches of greasewood and tamarisk (*Tamarix* sp.) that usually occur within and along the margins of ORB deltaic channels. In the northeastern portions of the distal delta, aeolian dunes at Wild Isle and along the margins of Wildcat and Kittycat mountains support halogeton, greasewood, shadscale, Indian ricegrass, and rabbitbrush.

Modern mammalian fauna in the delta vicinity consists of an array of desert-dwelling species that have been common residents in the area for the past ~8,000 years (Grayson 2000b; Schmitt and Lupo 2005, 2012). Small mammals include Townsend's ground squirrel (*Urocitellus mollis*), Botta's pocket gopher (*Thomomys bottae*), deer mouse (*Peromyscus maniculatus*), desert wood rat (*Neotoma lepida*), chisel-toothed kangaroo rat (*Dipodomys microps*), and long-tailed pocket mouse (*Perognathus longimembris*) (Durrant 1952; Vest 1962; see also Schmitt 2004). Since many of these taxa are fossorial and/or rely on native grasses and forbs for food, they largely occupy RLF and dune deposits in more proximal portions of the delta and are infrequent

(if not absent) in the barren mudflats of the Great Salt Lake Desert. Black-tailed jackrabbits (*Lepus californicus*) are common in open areas of the RLF plain, while more densely vegetated dunes and ephemeral drainages support small populations of cottontail rabbits (*Sylvilagus nuttallii*).

A number of mammalian and avian carnivores that prey on rodents and leporids also occupy the ORB delta area, including coyotes (*Canis latrans*), foxes (*Vulpes vulpes* and *V. macrotis*), bobcats (*Lynx rufus*), golden eagles (*Aquila chrysaetos*), and a variety of hawks and owls. Relatively large numbers of pronghorn (*Antilocapra americana*) range freely across Dugway Valley, and mule deer (*Odocoileus hemionus*) often descend from nearby mountains into the valley bottom, especially during the winter months. The remains of bighorn sheep (*Ovis canadensis*) have been identified in regional late Holocene archaeological deposits (Schmitt and Lupo 2005), and bighorn were likely present historically, but there are no modern records from the immediate ORB delta vicinity. In addition, a number of avian species adapted to aquatic and semiaquatic habitats currently live in areas bracketing the ORB delta (notably Fish Springs to the south), and it is reasonable to suspect that many of these taxa were common, at least season-

TABLE 1.1. History of Research Projects Centered on Proximal Old River Bed (ORB) Delta Archaeology and Geomorphology.

Project	Year(s)	Type(s)[a]	Acres	No. of Sites[b]	Reference(s)[c]
ORB delta/Gilbert shoreline	1999	SR, GE	1,993	14	Madsen et al. 2000; Oviatt 1999; Oviatt and Madsen 2000
West Baker dunes	2000	SR	1,991	11	Schmitt, Madsen, Hunt, Callister, Jensen, and Quist 2002
Lake Öferneet	2001–2002	SR	3,128	43	Schmitt et al. 2003
Test excavation	2003	EX	n/a	2[d]	Madsen et al. 2004
Channel remnants	2005	SR	982	9	Madsen et al. 2006
Southwest channels	2006	SR, GE	633	12	Schmitt, Madsen, Page, and Smith 2007
Southwest channels, pt. 2	2007	SR	625	23[e]	Page et al. 2008
Geomorphic studies	2008	GE, SR	2,627	3[f]	Madsen and Page 2008, 2009
Various, regressive-phase lacustrine fines plain	2009–2011	SR	n/a[g]	86	Page and Schmitt 2010, 2011a, 2011b; Page, Schmitt, Dalldorf, and Wazaney 2012; Page, Schmitt, Rhode, and Sapula 2012; Schmitt and Page 2011; Schmitt et al. 2010; Schmitt et al. 2011

[a] SR = survey and site recordation; GE = backhoe trenching, profiling, and/or other geomorphic investigations; EX = test excavations.
[b] Fully recorded Paleoarchaic/early Holocene sites.
[c] See also Jones, Beck, and Kessler 2003; Oviatt et al. 2003; Page 2008; Rhode et al. 2005; Schmitt, Madsen, Oviatt, and Quist 2007.
[d] Previously recorded sites 42To1683 and 42To1860 (Schmitt et al. 2003).
[e] Four additional sites were observed but were not recorded.
[f] Thirty-three additional sites were observed but were not recorded.
[g] Various block survey projects that incorporated portions of ORB (unnamed) deltaic channels.

ally, when the delta supported slow-moving streams, shallow ponds, and marshlands. Among others, these include the great blue heron (*Ardea herodius*), white-faced ibis (*Plegadis chihi*), snowy egret (*Egretta thula*), long-billed marsh wren (*Cistothorus palustris*), black tern (*Chlidonias niger*), American coot (*Fulica americana*), and a variety of grebes (Podicipedidae), gulls (*Larus* sp.), and ducks and teals (*Anas* sp.) (e.g., Behle et al. 1985; Hayward et al. 1976).

HISTORY OF RESEARCH

To investigate the probability that Paleoarchaic sites occur in the proximal ORB delta, the Directorate of Environmental Programs at DPG and the Utah Geological Survey entered into a cooperative agreement in 1998 to conduct archaeological reconnaissance in and along the margins of what is now the Great Salt Lake Desert mudflats. A large number of development projects, conducted in accordance with the DPG mission, occur annually in this portion of the delta, and these may impact significant archaeological sites, particularly critical and relatively rare sites dating to the Paleoarchaic period. Inventories conducted under this cooperative agreement were initiated in 1999 (Table 1.1) to identify such sites. Our principal research and management goals were to search for, and define the nature of, Paleoarchaic archaeological sites in proximal areas of the ORB delta heavily

impacted by military activities and to map, describe, and evaluate the significance of those sites we could identify. These initial investigations also employed a series of exploratory backhoe- and hand-excavated trenches to examine selected channel remnants in profile and extract organic materials for identification and radiocarbon assay.

These initial surveys (Madsen et al. 2000; Schmitt, Madsen, Hunt, Callister, Jensen, and Quist 2002) encountered a number of Paleoarchaic sites in the delta area that appeared undisturbed by later occupation, and we were confident that others awaited discovery. In 2002 the DPG cooperative agreement with Utah Geological Survey was transferred to the Desert Research Institute, where six additional ORB projects were undertaken over the next seven years (Table 1.1), including additional surveys, test excavations of two sites, and large-scale geomorphic studies that incorporated backhoe trench excavations. Like the earlier projects, reconnaissance included the survey of contiguous blocks, but most of the more recent fieldwork has consisted of linear transects that followed one or more individual deltaic channels (e.g., Page et al. 2008; Schmitt, Madsen, Oviatt, and Quist 2007).

Overall, eight projects in and immediately adjacent the Great Salt Lake basin examined some 12,000 ac in the proximal ORB delta, where 115 Paleoarchaic sites

TABLE 1.2. History of Research Projects in the Distal Old River Bed Delta and Vicinity.

Project	Year(s)	Type(s)[a]	Acres[b]	No. of Sites[c]	Reference[d]
Various Weber State surveys	1991–1996	SR	41,132[e]	25	Arkush and Pitblado 2000
Miscellaneous small surveys	1993–2010	SR	1,307	6	n/a
TS-5 inventory	1998	SR	9,659	57	Carter 1999
TS-5 central area	2000	SR	3,899	31	Carter and Young 2002
South Route inventory	2001	SR	886	9	Duke 2003
Wild Isle testing	2003	EX	n/a	22[f]	Carter et al. 2004
Wildcat 04 inventory	2004	SR	5,037	8	Carter et al. 2005
Wildcat 05 inventory	2005	SR	2,879	60	Young et al. 2006
Wildcat 06 inventory	2006	SR	3,262	44	Duke et al. 2008
Cache and Com sites	2007	EX, GE	n/a	2[f]	Duke 2011
Knolls dunes 09–10 inventory	2009–2010	SR	3,695	25	Duke 2015

[a] SR = survey and site recordation; EX = test excavations; GE = backhoe trenching, profiling, and/or other geomorphic investigations.
[b] As calculated in a geographic information system; some may differ from the initially reported acreage.
[c] Fully recorded Paleoarchaic sites/site components.
[d] See also Duke and Young 2007; Duke et al. 2004; Young 2008.
[e] Much of this area was revisited and subject to close-interval transects in later projects.
[f] Excavations at previously recorded sites.

were identified and recorded in detail and an additional 37 sites were identified and plotted but not formally recorded. Based on content and context, 100 of the 115 formally recorded sites date exclusively to the Paleoarchaic or are dominated by weathered obsidian and fine-grained volcanic flakes and tools that probably reflect Paleoarchaic visits but lack temporally diagnostic artifacts. The remaining sites consist of palimpsests of Paleoarchaic and Archaic/Fremont artifacts in and around Holocene dunes that reflect later occupations. In addition, more recent surveys atop the RLF plain east of Granite Peak have identified additional Paleoarchaic sites/site components associated with deltaic channels (Table 1.1). Overall, these various projects in proximal portions of the ORB wetlands have, as of 2013, identified 238 sites that were occupied by terminal Pleistocene/ early Holocene foragers.

The history of archaeological investigations in the distal ORB delta is provided in Table 1.2. Unlike the investigations to the south in the proximal delta, most of the work in this part of the delta stems from cultural resource inventories in support of various U.S. Air Force projects and primarily consist of large, contiguous block surveys. Initial projects in the distal delta consisted of a series of large-scale, cursory inventories conducted by Weber State University that encountered Paleoarchaic materials along the western margin of Wildcat Mountain and in the dune fields at Wild Isle and South Knolls (Arkush and Pitblado 2000). Over the next several years additional archaeological inventories employing close-interval pedestrian transects were conducted in neighboring areas by Historical Research Associates,

followed by a series of intensive surveys, test excavations, and geomorphic investigations by the Far Western Anthropological Group. Together, these various surveys have examined more than 70,000 ac in the distal ORB delta and identified a total of 265 sites with at least some Paleoarchaic component, 222 of which are defined as wholly or largely Paleoarchaic in nature.

A NOTE ABOUT CHRONOLOGICAL CONTROLS

In the following discussions of the ages of geomorphological features and archaeological sites in the ORB delta, and in our discussions of the chronological relationships of ORB sites to similar sites elsewhere, we rely on radiocarbon age estimates for chronological controls. In so doing we are aware that when calibrated, these age estimates differ, often dramatically, from equivalent calendrical ages. However, with the exception of those dates collected and listed in tables, we have not provided calibration estimates for radiocarbon age estimates discussed in the text. There are two reasons for this. First, with hundreds of textual references to ^{14}C ages, a parenthetical listing of the equivalent calendar age becomes cumbersome; and second and most important, calibration curves have been, and undoubtedly will continue to be, repeatedly refined and are subject to change. This is particularly true for the time period between about 12,000 and 9000 ^{14}C BP. For reference purposes, however, in Table 1.3 we list approximate calibrated ages, derived from Danzeglocke et al. 2012, for the radiocarbon time period covered by the ORB delta.

TABLE 1.3. Radiocarbon Ages and Equivalent Calendar Ages.

^{14}C Age (BP)	Calendar Age (BP)
6500	7400
7000	7800
7500	8400
8000	8900
8500	9500
8600	9500
8700	9600
8800	9800
8900	10,100
9000	10,200
9100	10,200
9200	10,300
9300	10,500
9400	10,600
9500	10,700
9600	11,000
9700	11,200
9800	11,200
9900	11,300
10,000	11,500
10,100	11,700
10,200	11,900
10,300	12,200
10,400	12,300
10,500	12,500
10,600	12,600
10,700	12,700
10,800	12,800
10,900	12,800
11,000	12,900
11,100	13,000
11,200	13,100
11,300	13,200
11,400	13,300
11,500	13,400
12,000	14,000
12,500	14,800
13,000	15,900
13,500	16,500

The Great Basin Paleoarchaic

Throughout much of its ~3,000-year history the ORB delta was occupied by foraging groups who, along with their contemporaries elsewhere in western North America, have come to be known as "Paleoarchaic," although not everyone agrees on the use of the term (e.g., Goebel et al. 2007; Haynes 2007). In-depth treatments of the derivation of the term, together with broad dis-

cussions of the lithic tradition that characterizes these early populations, are available in a variety of sources, and there is no need to repeat those detailed discussions here (e.g., Beck and Jones 1997; Elston 1982; Elston and Zeanah 2002; Goebel et al. 2011; Graf and Schmitt 2007; Jones and Beck 1999; Willig and Aikens 1988; Willig et al. 1988). However, some context for the archaeological research questions we investigated at the ORB delta is necessary. Such a context largely involves the development of the Paleoarchaic concept, the geographic distribution of diagnostic tool classes that characterize the Paleoarchaic, the occupational chronology of these earliest Great Basin peoples, the subsistence practices of Paleoarchaic foragers, their degree of residential mobility, and other controversial topics that remain unsettled.

Concept History

From almost the beginning of archaeological research in the Great Basin, it was recognized that the earliest inhabitants of the region lived along the shores of now-extinct late Pleistocene lakes. Many of the lower shorelines overlooked what had been marsh wetlands formed during the Pleistocene–Holocene transition, and it appeared that occupation of sites on those shorelines was chronologically related to the period when these terminal wetlands formed. In the mid-1930s, for example, Elizabeth and William Campbell began survey work around the paleo-shorelines of pluvial Lake Mohave in the extreme southwestern Great Basin (Campbell and Campbell 1935; Campbell et al. 1937). Along with Ernst Antevs and others, the Campbells were able to show a direct link between occupation of these shoreline sites and a unique artifact complex that included such stemmed points as the "Pinto," "Lake Mohave," and "Silver Lake" varieties. While there has been some debate over the age of these point forms (see the discussion in Chapter 5), the correlation between lacustrine marsh settings and various types of stemmed points has been a steady feature of descriptions of the Great Basin Paleoarchaic.

Contemporaneously, work in several Great Basin cave sites such as Paisley Caves (Cressman 1966; Cressman et al. 1942) and Gypsum Cave (Harrington 1933) was suggesting a possible association between extinct Pleistocene megafauna and early human foragers. The development of any organizational construct that incorporated both the lake-margin sites and the early occupation of the cave sites was slow in coming, however, and when the extinct megafauna–human connection began to be questioned (e.g., Heizer and Baumhoff 1970; Heizer and Berger 1970), pursuit of an integrated understanding of the earliest Great Basin foragers started to lag. There was, on the other hand, a shift to regional studies focused primarily on the definition, chronology, and distribution of particular point forms. Many of these point forms were defined for areas outside the Great Basin, particularly on the Columbia Plateau. By the time Willig and Aikens (1988) published a review of the earliest occupations in western North America, they were able to identify six "phases or horizons," one "pattern," three "complexes," five "traditions," and 12 regional point "styles" that had been defined for the region (see references in Willig and Aikens 1988). Were one to add the patterns, traditions, and complexes that have now come to be seen as defining the entire region inhabited by Paleoarchaic foragers, including those outside the Great Basin, there would be even more than these 27.

The regionally unique lanceolate and stemmed biface forms eventually became known collectively as the "Great Basin Stemmed Series" (Tuohy and Layton 1977), part of the "Western Stemmed Tradition" (Bryan 1980). These phrases largely subsume a number of other similarly named lithic complexes. Here, following Bryan (1980) and Jenkins et al. (2012), we employ the broader "Western Stemmed Tradition" (WST) rather than the more commonly used "Great Basin Stemmed Series" due to the distribution of stemmed bifaces and crescents well beyond the limits of the Great Basin proper, including many areas across western North America (Figure 1.5). WST point types include forms known as "Haskett," "Windust," "Cougar Mountain," "Lind Coulee," "Parman," "Lake Mohave," and "Silver Lake" and other less easily categorized forms (see Chapters 5 and 6). Often included with the WST is a crescent-shaped biface form unique to western North America (Beck and Jones 1997, 2010a; Jones and Beck 1999).

There are a number of other point styles dating to the late Paleoarchaic period that do not fit easily into either the Great Basin Stemmed Series or the Western Stemmed Tradition. These include short-stemmed forms such as "Pinto," "Dugway Stubbies," and several contracting/parallel/convex-stemmed types and even such notched forms as "Butte Valley Corner-notched." These points often co-occur with larger WST stemmed forms on sites dating to the early Holocene and should probably be included as part of the late Western Stemmed lithic tradition. We do so here (see discussion in Chapter 5).

This focus on defining biface forms and lithic traditions was largely due to the absence of buried sites dating to the Paleoarchaic period that might contain a variety of perishable artifacts and subsistence residues. As a result, attempts to understand the lifestyles of the people who manufactured WST tools were limited to extrapolations from the settings in which the sites were found. As the large majority of WST sites were found around the margins of extinct Pleistocene lakes, it was assumed that these people were adapted primarily to wetland environments, and in some areas of the Great Basin, this lacustrine focus came to be known as the "Western Pluvial Lakes Tradition" (Bedwell 1973). Given these wetland settings, as well as an apparent lack of "big-game" kill sites, it was also widely assumed that WST peoples had a rather broad-based subsistence adaptation, somewhat akin to that of the later Great Basin Archaic foragers. Because such an adaptive strategy seemed rather different from that of the fluted point–making peoples east of the Rocky Mountains, a variety of terms such as *proto-Archaic* (Tadlock 1966), *pre-Archaic* (Elston 1982), and *paleo-Archaic* (Willig 1989) began to be used to emphasize the differences between these eastern and western peoples.

Yet the supposed differences between these two eastern and western North American groups have remained largely an assumption, due to the lack of subsistence data. As it became increasingly clear that the subsistence adaptations of contemporary foraging peoples east of the Rockies were much broader than previously thought (e.g., Cannon and Meltzer 2004, 2008), and that at least some of the earliest peoples in western North America were hunting Pleistocene megafauna such as mastodon (Waters et al. 2011), the perceived adaptive differences between the two geographically defined groups have become less apparent. At present, the adaptive implications of the terms *Paleoarchaic* and *Paleoindian* remain largely undemonstrated, although many archaeologists, particularly in the West, continue to think that evidence of adaptive differences will eventually be produced. For now, the two terms remain useful primarily because of the technological differences in stone tool production and forms between groups east and west of the Rocky Mountains (e.g., Beck and Jones 2010a, 2012; Davis et al. 2012). Whether these technological differences imply functional and adaptive differences, reflect different origins and backgrounds, or are merely due to different research histories remains to be discovered.

Distribution

If one uses the distribution of WST "points" and crescents to define the distribution of the western North American Paleoarchaic, then the foraging peoples grouped under that umbrella term ranged from the heights of the eastern Rocky Mountains on the east to the islands off coastal California on the west and from the Baja peninsula on the south to coastal British

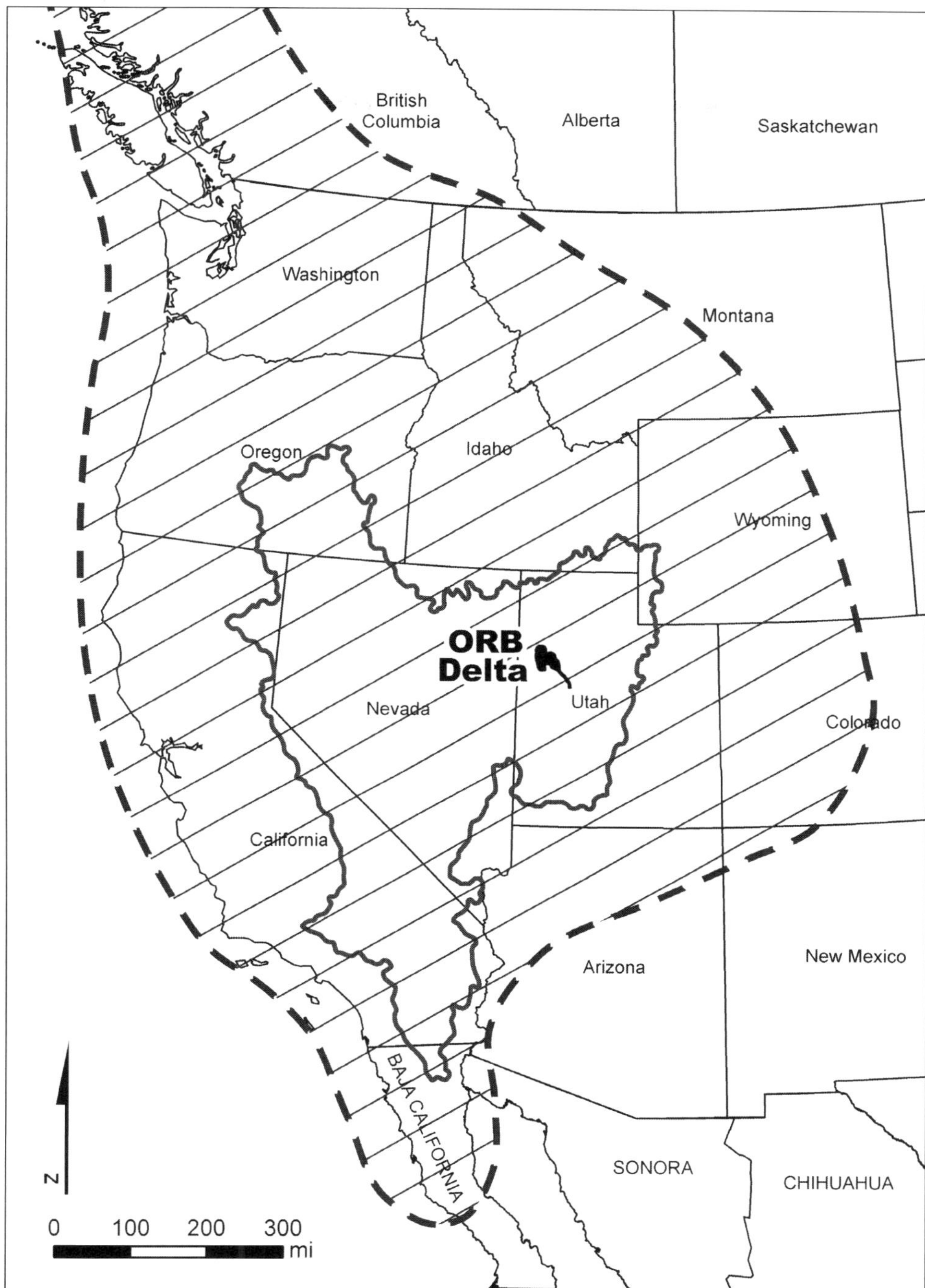

FIGURE 1.5. Distribution of Western Stemmed Tradition diagnostic tool types in western North America. ORB = Old River Bed.

Columbia (Figure 1.5). That area centers on the Great Basin, but the distribution extends well beyond its margins into environmental settings very different from those found in the Great Basin during the Pleistocene–Holocene transition.

WST tool forms, particularly the physically larger varieties of stemmed points, are found as far east as the Middle Park area of the Colorado Front Range on the margin of the Great Plains (Kornfeld and Frison 2000). These include the Haskett and Cougar Mountain varieties as well as a number of other locally named forms that could easily be subsumed under the WST rubric. WST tool forms also occur elsewhere at high elevations in the southern and central Rocky Mountains of Colorado (e.g., Jodry 1999; Pitblado 2003) and Wyoming (e.g., Kornfeld et al. 2001). These occur in relatively low percentages of ~5–15 percent on open sites in the region, and it is difficult to tell whether they represent trade or an actual movement of Paleoarchaic peoples into these high-altitude areas. There are, however, a number of point forms common to the Rocky Mountain corridor, such as "Hell Gap," "Pryor Stemmed," and "Medicine Lodge Creek," that are essentially WST forms and, when combined with tools that have actually been classified as WST, do suggest at least a middle to late Paleoarchaic occupation of the region.

These stemmed point forms extend well into Montana and the southern Canadian Rockies (Carlson and Magne 2007; Frison 1988). Although they have been interpreted to represent a separate adaptive trajectory from that of the Great Basin and Columbian Plateau Paleoarchaic, it seems quite possible that they are related, as suggested by Pitblado (1998) and others. This question will only be resolved by a detailed technological analysis of these various biface forms, since many are so morphologically similar (for example, Hell Gap/Haskett) that the moniker given them seems to be entirely dependent on the background of the analyst. For now we consider the distributional limits of large stemmed biface forms to reflect the distributional limits of the western Paleoarchaic. It should be noted, however, that it is as yet unknown whether these distributional limits represent an expansion of populations making these bifaces or a diffusion of the technology involved.

WST tool forms have been recovered as far south as the central Baja peninsula and the Channel Islands off the southern California coast. These tool forms are associated with coastal adaptations in the region at the Pleistocene–Holocene transition (Reeder et al. 2011 and references therein). A variety of sites on the Channel Islands contain a unique form of stemmed point but also a number of the WST diagnostic crescent bifaces (e.g., Erlandson et al. 2011). On Isla Cedros, off the coast of central Baja (but at the time of occupation connected to the mainland), several sites containing stemmed points "reminiscent of San Dieguito or Western Pluvial Lakes Tradition (WPLT) assemblages (Des Lauriers 2006)" have been identified (Reeder et al. 2011:474).

The extent of WST technology to the northwest is less well known, probably because, as Erlandson and Braje (2011) note, much of the coastal margin dating to around the Pleistocene–Holocene transition has been inundated by Holocene sea level rise. Stemmed point forms dating to the period have been recovered from well-controlled contexts at Paisley Caves in the extreme northwestern Great Basin region of eastern Oregon (e.g., Gilbert et al. 2008; Jenkins 2007) and along the Columbia and Klamath rivers on the Columbia Plateau (e.g., Beck and Jones 2010a; Davis and Schweger 2004). Crescent biface forms are common along coastal California (e.g., Jertberg 1986; Tadlock 1966), but the lithic materials from other contemporary sites along coastal Oregon, Washington, and southern British Columbia tend to be sparse and often nondiagnostic (e.g., Fedje and Mathewes 2005; Fedje et al. 2004). However, stemmed points have been recovered in lag deposits at a number of sites near Vancouver, British Columbia, and have been recovered in stratified contexts at Gaadu Din 1, Gaadu Din 2, and K1 caves dating to between ~10,900 and 10,200 ^{14}C BP (Fedje et al. 2011). At On Your Knees Cave in southeastern Alaska, stemmed points dating to 9900–8800 ^{14}C BP were recovered in the same stratigraphic unit as a human burial (Dixon 2007). Stemmed biface tools have even been recovered from as far north as the North Slope of Alaska, in the Mesa Complex dating to the Pleistocene/Holocene (Bever 2006), although this complex is usually considered to be part of a Plains Paleoindian lithic tradition, like Hell Gap and other stemmed biface forms in the Rocky Mountain cordillera.

Chronology

Although the distribution of WST tool forms is relatively well known and accepted, how that distribution came to be is much more controversial. This is so because it involves questions concerning WST relationships to other North American lithic technologies, their points of origin, their chronology and territorial expansion, and their functional relationships to particular adaptive strategies. As a result, it is difficult to separate an empirical discussion of a Paleoarchaic chronology from more subjective and interpretive ideas about the earliest human inhabitants of North America (see Meltzer 2009 for a thorough review of these interrelated issues).

Briefly, the chronology of Paleoarchaic peoples and their WST lithic technology in western North America was, until recently, largely interpreted in terms of their relationship to the Clovis lithic tradition in eastern North America and to the peoples who employed that technology. As Meltzer (2009) notes, the pervasive assumption has been that Clovis peoples were the first inhabitants of the New World and that all other foraging groups and their associated lithic traditions elsewhere in the Americas derived from those original emigrants. This view has colored interpretations of the age and distinctiveness of possibly contemporaneous non-Clovis lithic technologies, and early hypotheses that there were eastern and western "co-traditions" dating to the Clovis period (e.g., Wormington 1957) have been widely ignored. Put in terms of the western Paleoarchaic, how one interprets age estimates associated with WST tool forms has been largely dependent on whether or not one agrees with the notion that Clovis peoples were the first to enter either North America or the Intermountain Region. That is, the acceptance or rejection of WST age estimates that make them contemporaneous with or even earlier than Clovis has been largely dependent on one's theoretical view.

A pre-Clovis occupation of North America is now generally accepted, and one of the most definitive sites dating to the pre-Clovis/pre-WST era is found in the Great Basin within the distributional area of WST lithic technology. This complex of sites is Paisley Caves in eastern Oregon, where human coprolites have been dated to ~12,300 ^{14}C BP (Gilbert et al. 2008; Jenkins et al. 2012). A second possible pre-Clovis site in western North American is the Manis mastodon site in Washington, dated to ~12,000 ^{14}C BP (Waters et al. 2011), but this remains problematic, as the "bone point" embedded in one of the mastodon ribs has yet to be definitively identified as a tool. In neither case, unfortunately, are diagnostic lithic tools associated with these very early dates. Western stemmed points are, however, securely dated at Paisley Caves to between ~11,300 and 11,100 ^{14}C BP (see below), and the Paleoarchaic period can be definitively said to begin at least by that time period.

Within the Great Basin, it has been difficult to address the issue of Clovis/WST contemporaneity until recently because the majority of sites are open-air sites around pluvial lakes and rivers. As a result, there have actually been rather few age estimates on which to base a reliable conclusion about the beginning of the Paleoarchaic time period. This situation is beginning to change; Beck and Jones (2010a, 2012) and Goebel and Keene (2014) both review the dating of the earliest Paleoarchaic sites in western North America. They come to radically different conclusions, however. Beck and Jones (2010a, 2012), like Bryan before them (e.g., 1980, 1988), suggest that a number of WST sites in the Great Basin and Columbia Plateau date prior to ~10,800 ^{14}C BP and thus are contemporaneous with Clovis sites to the east and southeast dating to ~11,100–10,800 ^{14}C BP (Waters and Stafford 2007). Goebel and Keene review the same data employed by Beck and Jones and, using "a strict regimen of chronological hygiene," conclude that all the purported Clovis-age Paleoarchaic sites have "problematic contexts and/or aberrant dates" (2014:35).

We consider there to be five exceptions to the chronology of Goebel and Keene (Table 1.4; Figure 1.6).

The first of these is the Paisley Caves site in the Great Basin area of southeastern Oregon, where dates on stemmed points come close to passing the "hygiene" requirements of Goebel and Keene (2014). Recent work at Paisley Caves has resulted in the recovery of stemmed points in good association with acceptable radiocarbon dates (Jenkins et al. 2012). One stem base is bracketed by dates of ~11,070 and ~11,340 ^{14}C BP stratigraphically above and below the basal section, while another stem fragment is bracketed by dates of ~10,855 and 11,070 ^{14}C BP. Yet a third stemmed point fragment is bracketed by dates of ~10,050 and 10,965 ^{14}C BP on a human coprolite 35 cm above the fragment and by dates of ~12,140 and 12,260 ^{14}C BP on another human coprolite 15 cm below the point. Based on sedimentation rates between these bracketing dates, Jenkins et al. (2012) estimate its age to be between ~11,135 and 11,600 ^{14}C BP.

Goebel and Keene (2014) also consider, and reject, early Paleoarchaic dates from Smith Creek Cave and Bonneville Estates Rockshelter that we think may be valid. In an earlier essay, using much the same evaluation criteria, Goebel et al. (2007) identify seven "reliable" radiocarbon age estimates from Paleoarchaic deposits in Smith Creek Cave ranging in age from 11,140 to 9940 ^{14}C BP, with the oldest age of 11,140 ± 200 ^{14}C BP derived from hearth charcoal. For Bonneville Estates they similarly list 17 "reliable" Paleoarchaic dates (some of which include averages of multiple dates from the same hearth) ranging from 11,000 to 8900 ^{14}C BP. The two oldest hearths date to 11,040 ± 40 and 10,821 ± 31 ^{14}C BP (average of three age estimates). Despite vetting the reliability of the earliest age estimates from both Smith Creek Cave and Bonneville Estates Rockshelter, Goebel et al. (2007) consider that they need to be "reconfirmed." They accept the remainder, however, and, since only stemmed points were associated with the early hearths from the sites, conclude that "in the western Bonneville basin from 12,850 to 10,650 cal BP [~10,880–9410 ^{14}C BP], humans exclusively made and used Great Basin stemmed points" (2007:158). Graf extends the range even further, concluding that stratum 18b, which contains stemmed points, "accumulated within an 1,100-year span between 13,100 and 12,000 cal BP [~11,200–10,250 ^{14}C BP]" and "likely represents a series of short-term human occupations during that time" (2007:97).

We see no reason to reject the pre-11,000 ^{14}C BP dates, while accepting age estimates of ~10,900 ^{14}C BP. The reasoning of Goebel et al. (2007) seems to be that since stemmed points were not directly associated with the earlier hearths, the possibility remains the hearths were laid down by Clovis-using foraging groups.

TABLE 1.4. Age Estimates Associated with Western Stemmed Tradition (WST) Tools.

Site	Age (^{14}C BP)	Calibrated Age BP (2σ)	Lab No(s).
"Accepted" WST Age Estimates[a]			
Cooper's Ferry	8410 ± 70	9536–9266	Beta-114951
Cooper's Ferry	8710 ± 120	10,157–9522	TO-7346
Bonneville Estates Rockshelter	9435 ± 35	10,749–10,577	Avg. F03.13, stratum 17b
Bonneville Estates Rockshelter	9580 ± 40	11,122–10,741	Beta-207010
Connley Cave 4B	9670 ± 180	11,613–10,503	Gak-2142
Sunshine	9800 ± 60	11,337–11,101	Beta-69782
Sunshine	9840 ± 50	11,380–11,181	Beta-86201
Sunshine	9920 ± 60	11,610–11,217	Beta-86198
Smith Creek Cave	9940 ± 160	12,062–10,871	Tx-1420
Bonneville Estates Rockshelter	9990 ± 50	11,705–11,261	AA-58598
Sentinel Gap	10,180 ± 60	11,754–11,267	Beta-133664
Bonneville Estates Rockshelter	10,030 ± 50	11,763–11,289	Beta-182934
Bonneville Estates Rockshelter	10,050 ± 50	11,817–11,320	Beta-182934
Bonneville Estates Rockshelter	10,090 ± 34	11,906–11,404	Avg. F1.0, stratum 18a
Marmes (floodplain)	10,130 ± 30	12,609–10,794	W-2218
Sentinel Gap	10,130 ± 60	12,026–11,405	Beta-133665
Sentinel Gap	10,160 ± 60	12,072–11,411	Beta-133663
Sentinel Gap	10,180 ± 40	12,043–11,712	Beta-124167
Bonneville Estates Rockshelter	10,250 ± 50	12,367–11,760	Beta-206278
Smith Creek Cave	10,330 ± 190	12,594–11,403	Tx-1638
Bonneville Estates Rockshelter	10,367 ± 33	12,390–12,082	Avg. of 2, stratum 12
Bonneville Estates Rockshelter	10,405 ± 50	12,523–12,081	AA-58593
Smith Creek Cave	10,420 ± 100	12,580–11,944	TO-1173
Bonneville Estates Rockshelter	10,540 ± 40	12,600–12,409	Beta-195047
Smith Creek Cave	10,590 ± 122	12,710–12,102	Avg. of TP8, hearth 3
Connley Cave 4B	10,600 ± 190	12,920–11,827	Gak-2143
Smith Creek Cave	10,647 ± 116	12,784–12,142	Avg. of TP8, hearth 4
Buhl Burial	10,675 ± 95	12,810–12,391	Beta-43055/ETH-7729
Additional Great Basin WST Age Estimates[b]			
Paisley Caves	10,528 ± 30[c]	12,584–12,412	1961-PC-5/18a-10-1
Bonneville Estates	10,690 ± 70[d]	12,754–12,433	AA-58590
Bonneville Estates	10,787 ± 46[d]	12,817–12,571	Avg. F3.15a, stratum 18b
Bonneville Estates	10,821 ± 31[d]	12,847–12,590	Avg. F03.17
Paisley Caves	10,963 ± 30[c]	12,962–12,658	1895-PC-5/16A-23-6A
Bonneville Estates	11,010 ± 40[d]	13,079–12,705	Beta-207009
Smith Creek Cave	11,140 ± 200	13,362–12,638	Tx-1637
Paisley Caves	11,205 ± 40[c]	13,254–12,920	1895-PC-5/16A-24
Paisley Caves	11,354 ± 55[c]	13,347–13,114	1294-PC-5/6D-47-1
Dated Hell Gap/Haskett Points from the Central Rocky Mountains[e]			
Indian Creek, Locus B	9860 ± 70	11,603–11,171	Beta-5119
Lake Minnewanka	9990 ± 50	11,754–11,248	Beta-122723
Indian Creek, Locus B	10,020 ± 110	11,973–11,238	Beta-5118
Casper	10,060 ± 170	12,376–11,191	RL-208
Helen Lookingbill	10,405 ± 95	12,570–11,988	Beta-28877/ETH-4816
Agate Basin, Area 3	10,445 ± 110	12,599–11,987	SI-4430

TABLE 1.4. (cont'd.) Age Estimates Associated with Western Stemmed Tradition (WST) Tools.

Site	Age (^{14}C BP)	Calibrated Age BP (2σ)	Lab No(s).
Dated Coastal WST Sites[f]			
On Your Knees Cave	8760 ± 50	10,114–9555	CAMS-43991
On Your Knees Cave	9150 ± 50	10,485–10,226	CAMS-43989
On Your Knees Cave	9219 ± 50	10,505–10,248	CAMS-43990
On Your Knees Cave	9730 ± 60	11,251–10,800	CAMS-29873
On Your Knees Cave	9880 ± 50	11,592–11,198	CAMS-32038
Gaadu Din 1	9980 ± 30	11,607–11,273	UCIAMS-12386
Cardinalis Creek—DhRn29	10,170 ± 40	12,036–11,649	Beta-226980
Gaadu Din 2	10,220 ± 30	12,074–11,778	UCIAMS-40882
Gaadu Din 2	10,295 ± 25	12,362–11,649	UCIAMS-40882
Cardinalis Creek—DhRn29	10,370 ± 40	12,401–12,072	Beta-241999
Gaadu Din 1	10,550 ± 25	12,602–12,423	UCIAMS-12388
K1	10,592 ± 35[g]	12,683–12,379	EU11B10/EU11B20
Gaadu Din 1	10,615 ± 30	12,665–12,444	UCIAMS-28005
K1	10,735 ± 30[g]	12,970–12,402	EU11CB2a/EU11CB3a
Arlington Springs Burial	10,960 ± 80[h]	13,076–12,223	CAMS-16810

Note: Age estimates with σ >200 years are not included.

[a] From Goebel and Keene 2014.

[b] From Goebel et al. 2007; Jenkins et al. 2012.

[c] Mean of stratigraphic bracketing age estimates over and underlying stemmed point fragment (from Jenkins et al. 2012:Table 1).

[d] Ages considered "reliable" by Goebel et al. (2007) but not included in Goebel and Keene 2014.

[e] From Pitblado 2003.

[f] From Dixon 2007; Fedje et al. 2011; Johnson et al. 2002.

[g] Average of two dates stratigraphically bracketing a stemmed point. Calibrated age reflects combined minimum/maximum of both age estimates.

[h] Calibrated age reflects combined minimum/maximum of both Intercal09 and Marine09 calibrations due to unknown amount of marine resources in the diet.

Parsimony suggests otherwise, however, and it seems reasonable that if the dates themselves are valid and derived from such demonstrably cultural features as hearths, and if only stemmed points were found in the stratigraphic levels from which the hearths derive, then they must be considered to reliably date the early end of the Paleoarchaic occupations at the two sites.

Finally, there are numerous age estimates of between 11,000 and 10,000 ^{14}C BP associated with stemmed points and crescents at coastal Paleoarchaic sites along southern California and Baja. Several of them approach, but do not overlap, the age of Clovis sites (Reeder et al. 2011). At the Arlington Springs site on Santa Rosa Island, however, there is also an age estimate of 10,960 ± 60 ^{14}C BP on bone collagen from a human burial (Reeder et al. 2011). As no Clovis materials have been recovered from the island, the burial appears to be associated with the abundant Paleoarchaic deposits on the north Channel Islands chain. As this early forager was likely consuming marine resources, some marine reservoir effect is probable. Depending on what percentage of her diet consisted of marine or terrestrial

resources, a calibrated age should be slightly older than a comparable terrestrial radiocarbon date and could be anywhere from ~11,200 to 10,300 ^{14}C BP. There is also a bulk collagen accelerator mass spectrometry age estimate of 10,675 ± 95 ^{14}C BP on a paleoarchaic burial from the Snake River Plain in Idaho (Green et al. 1998), but the geological context suggests that Buhl Woman was buried shortly after the Bonneville flood dating to ~14,500 ^{14}C BP (see Chapter 3). Bulk collagen age estimates are also generally thought to be limiting ages (that is, the true age is likely to be older) since the proteins in collagen degrade differentially and select proteins are usually sought for accurate dating (e.g., Johnson et al. 2002). Therefore, select proteins from the Buhl burial would probably date older by an unknown amount than the given age estimate.

Together, these additional age estimates suggest that, at the very least, the beginning of the temporal span for the earliest western Paleoarchaic overlaps that of Clovis (as defined in Waters and Stafford 2007) or, more likely, is contemporaneous with the earliest diagnostically Clovis sites. However, these data are rather

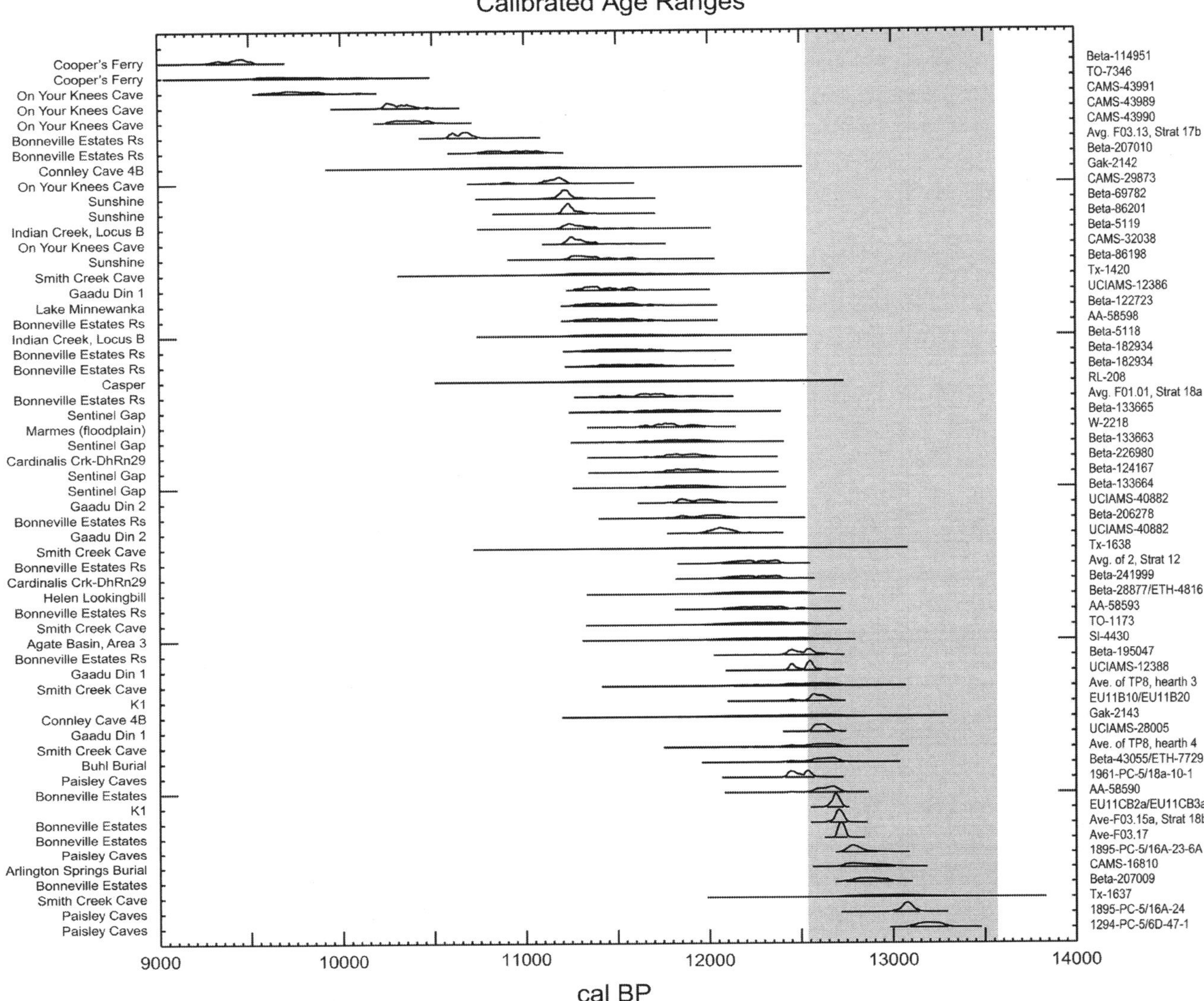

FIGURE 1.6. Chronological data from Table 1.4 displayed graphically (calibrated using Stuiver et al. 2013). The gray bar represents a modified Clovis time period as defined by Waters and Stafford (2007). We have expanded the age range by 200 years on either end of the scale to account for the calibrated age ranges of the youngest (Jakes Bluff, 12,750–12,573 cal BP) and oldest (Blackwater Draw, 13,619–12,670 cal BP) accepted Clovis sites.

limited at present, and many of the age estimates included in Table 1.4 have yet to be vetted in the manner suggested by Goebel and Keene (2014). For now we concur with Davis et al. that

until better chronological dating control on early sites is available, [the] technological evidence of distinctly separate lithic reduction sequences is perhaps the strongest indication for the presence of two contemporaneous cultures or co-traditions during the late Pleistocene–early Holocene period in the far west [2012:53].

There is a somewhat greater consensus that the transition to the later Archaic period occurred ~9500–8500 ^{14}C BP, depending on local conditions and on how one defines the subsistence, mobility, and technological characteristics of the two periods (e.g., Jones and Beck 2014; Madsen 2007). Less well known is whether this transition was gradual (e.g., Jones and Beck 2012; Madsen 2007) or rapid (e.g., Pinson 2007). That is, the chronological end of the Paleoarchaic time span could itself represent an extended period of time, depending on how one defines Paleoarchaic–Archaic differences.

Origins

The pervasive and long-standing argument that the WST of the western Paleoarchaic is related to and derived from the stone tool technology employed by Clovis peoples (e.g., Willig and Aikens 1988) seems to be changing. Both Paleoarchaic lithic specialists (e.g., Beck and Jones 2010a, 2012; Davis et al. 2012) and experts in Paleoindian tool production (e.g., Collins 2014) now suggest that the two lithic traditions are markedly different from one another. Understanding what that difference implies relates directly to the question of Paleoarchaic origins. As Beck and Jones (2010a) and Madsen (2012) note, if the two technological systems are so different that they cannot be directly related, then there are two possible explanations: Either (1) the two traditions were both derived from an earlier, as yet unknown, lithic tradition in North America, or (2) they were derived from separate Old World traditions.

As of 2013, there were three–four sites widely scattered across North America and coastal South America that most archaeologists would agree suggest that much of the New World south of the continental glaciation was occupied by ~12,500 ^{14}C BP. Discussions of these sites and their validity are available elsewhere (e.g., Meltzer 2009), but there is a growing consensus that by this time foraging communities ranged from the Northwest Coast to Florida and from the Northeast south of the glacial margin to the Chilean coast in South America. Assuming that this distribution took some time to develop, a minimum time frame for the initial peopling of the New World is likely to be ~13,500–13,000 ^{14}C BP. That is, it occurred about 2,000–2,500 years before the earliest dated diagnostic lithic complexes of the Paleoindian and Paleoarchaic periods. Whether or not this represents a sufficient time span for the two lithic traditions to have evolved from a common ancestor (or evolved separately in the Old World) is as yet unknown, but it does suggest that any probable common ancestor arrived via a coastal route, as a continental route was closed until after this time period.

Davis et al. (2012), on the other hand, contend that the two lithic traditions most likely represent separate populations that entered North America via different coastal and ice-free corridor routes and, likely, at different time periods. Currently, a continental entry route through a post–Last Glacial Maximum (LGM) corridor between the cordilleran and continental glacial masses is not thought to have been viable before ~13,000–11,500 ^{14}C BP (Clague et al. 2004; Dixon 2013; Munyikwa et al. 2011), so a separate Paleoindian ancestral population must have entered either prior to the LGM or after the ice-free corridor opened. This notion of two separate founding populations, one entering via a coastal route and the other entering via a continental route, gains some support from genetic data (e.g., Kashani et al. 2011; Perego et al. 2008).

Regardless of the origin of Paleoindian peoples, Paleoarchaic groups appear to have originated from Asian source populations via a circum-Pacific coastal route. This idea is of long standing (e.g., Bryan 1986; Fladmark 1979) but has gained traction as a number of sites exhibiting a coastal wetlands adaptation and dating to the Pleistocene–Holocene transition have come to light along the coast of western North America (e.g., Reeder et al. 2011). Erlandson and Braje (2011) review this hypothesis in depth; but, briefly, it holds that sometime during or shortly after the LGM, Upper Paleolithic seafaring peoples who occupied the islands of Japan and the Kuril Archipelago moved along the coast of Kamchatka, the Bering Archipelago, and the coasts of Alaska and British Columbia before settling in along coastal California. These northwest Pacific foragers employed a lithic tradition that involved the production of stemmed bifaces and had a subsistence adaptation that was heavily dependent on fish, shellfish, and waterfowl common to the bays, estuaries, and "kelp forest" of the coastal margin. Beck and Jones (2010a) and Davis et al. (2012) extend this hypothesis by suggesting that these stemmed point–making, wetlands-adapted groups moved inland along major river routes into interior North America, specifically the Great Basin, where they rapidly occupied similar wetland settings around the many pluvial lakes present in the region. This hypothesis remains to be completely tested but, for now, appears to be the most reasonable explanation for the distribution and chronology of a lithic tradition characterized by the production of stemmed bifaces and by a subsistence adaptation focused on lacustrine and riverine ecosystems.

Subsistence Adaptations

The Paleoarchaic foragers of the Great Basin are thought to be characterized by small, mobile foraging groups focused on marsh ecosystems associated with terminal Pleistocene lakes. While a few lithic studies suggest a focus on large-game hunting (e.g., Amick 1997), this conjecture is not supported by the limited available subsistence data (primarily faunal remains) (e.g., Goebel et al. 2011; Grayson 2011). The remains of a mammoth exhibiting possible cut marks and dating to ~11,200 ^{14}C BP were recovered from a moraine-dammed lake at 2,750 m asl in the Wasatch Mountains directly east of the Great Basin (Madsen 2000b). A stemmed point was recovered from these same lacustrine muds, but its association with the mammoth remains is unclear. In the western Great Basin, Paleoarchaic sites contain a variety of small mammals, birds, and fish (e.g., Eiselt 1997; Grayson 1988), as do sites in the northwestern Great Basin (e.g., Bedwell 1973; Connolly 1999; Pinson 1999) and eastern Great Basin (e.g., Goebel et al. 2011; Hockett 2007; Rhode and Louderback 2007). While "large game" were included in the mix, the taxa involved, such as pronghorn, were similar to those taken later during the Archaic.

The large majority of all known Great Basin Paleoarchaic sites were located around the margins of pluvial lakes, but the limited available subsistence data are derived from only a few excavated cave and rockshelter sites dating to the period. How the food resources utilized at these cave sites are related to those used at the much more common open wetlands sites is unclear. The best-documented of these Paleoarchaic cave sites is Bonneville Estates Rockshelter, located less than 50 km from the distal end of the Old River Bed delta distributaries. Paleoarchaic foragers at the site were hunting and collecting a broad mix of animals, including pronghorn, bighorn sheep, mule deer, black bear (*Ursus americanus*), jackrabbit, sage grouse (*Centrocerus urophasianus*), and grasshopper (Caelifera) (Goebel et al. 2011; Hockett 2007). Cactus was a commonly utilized plant, and charred seeds from ricegrass, dropseed sand grass (*Sporobolus* sp.), Great Basin wild rye (*Leymus* cf. *cinereus*), goosefoot (Chenopodiaceae), sunflower (Asteraceae), mustard (Brassicaceae), bulrush (*Scirpus* sp.), and cattail (*Typha* sp.) are common enough in the Paleoarchaic hearths to suggest that they were being utilized (Goebel et al. 2011; Rhode and Louderback 2007). There is no evidence of seed grinding, however, and how these seeds were prepared for consumption is unknown.

It is assumed that the occupation of the large majority of Paleoarchaic sites located in wetland settings involved the procurement of marsh resources including fish, shellfish, waterfowl, and such small mammals as muskrat and beaver. Larger mammals such as deer and antelope commonly winter in such settings as well. Direct evidence of such an adaptive strategy is limited, however. Fecal remains associated with a Paleoarchaic burial from Spirit Cave in the Lahontan Basin dating to ~9400 ^{14}C BP (Kirner et al. 1997) contained small fish bones (probably tui chub [*Gila* sp.] and Tahoe sucker [*Catostomus* sp.]) and *Scirpus* sp. seeds (Eiselt 1997; Napton 1997), which reflect a marsh subsistence focus. The bone chemistry of the Buhl burial from the Snake River Plain also suggests that she was consuming anadromous fish (Green et al. 1998). However, these are the only inland Paleoarchaic sites with direct evidence

of riverine or lacustrine fishing. Subsistence data from Paleoarchaic sites in coastal wetland settings are relatively common by comparison. Forage species include California mussels, Pismo clams (*Tivelas tultorum*), Venus clams (*Chione* spp.), nearshore and kelp forest fishes, sea turtle (*Caretta caretta*), the Guadalupe fur seal (*Arctocephalus townsendi*), and a variety of waterfowl, including goose (Reeder et al. 2011). Similar estuary species were likely also sought along the Northwest Coastal areas, but bear and possibly bison were hunted as well (Fedje et al. 2011).

If the WST tool forms and similar stemmed bifaces found in the upland settings of the central and southern Rocky Mountains do reflect a mid- to late Paleoarchaic presence in the region, then subsistence data from these sites reflect a shift to the hunting of larger alpine game animals during the latter part of the Paleoarchaic period. Haskett/Hell Gap sites dating to after ~10,500 ^{14}C BP commonly contain the remains of bison, as well as the modern pantheon of larger mammals, including bighorn sheep, deer, and, to a lesser extent, elk and antelope, but generally seem to represent a broader-based subsistence focus than do contemporaneous Paleoindian sites on the High Plains (Frison 1988, 1991, 1992). The even later, but still temporally pre-Archaic, "Foothills–Mountain Tradition" shows evidence of increasing diet breadth, with small game becoming a larger part of the diet (Hill 2008). The use of floral resources is less well known, but subsistence in these upland areas may have focused on rhizomes, tubers, and other upland plants.

Mobility

As the vast majority of Paleoarchaic sites in the Great Basin are open sites characterized by lithic scatters and little else, much of the recent Paleoarchaic research in the region has focused on the mobility and settlement systems of these foraging peoples. What is meant by *mobility* is often not defined, and the hypothetical movement of people is based primarily on the identification of the geological sources of the toolstone (mostly obsidian) represented in the open lithic scatters (e.g., Beck et al. 2002; Graf 2001; Jones, Beck, Jones, and Hughes 2003; Jones et al. 2012). Based on the distance to various toolstone source areas from a particular site or set of sites, the size of the territorial toolstone "conveyance zone" can be determined, with larger conveyance zones often thought to reflect higher mobility and smaller zones to result from lower mobility (see the discussion in Jones, Beck, Jones, and Hughes 2003). Yet a conveyance zone only reflects the *range* of a foraging group (that is, the size of the territory it moves around in on a yearly basis), not its *mobility* (Kelly 1992). Did a particular group move long distances infrequently within a particular territory or short distances often, or did part of the group (men) move long distances often while the remainder (women, children, and the elderly) moved less frequently? We would argue further that even estimating the range of a foraging group using toolstone data is suspect because the impact of economically and socially engendered trade networks on toolstone distribution is usually unknown (Newlander 2012).

Despite these criticisms, estimates of Great Basin Paleoarchaic mobility continue to be made, primarily on the basis of toolstone distributions and secondarily on toolstone reduction sequences (e.g., Jones et al. 2012). Yet the sizes of these conveyance zones seem to be getting smaller as more tools from more sites are sourced to a particular geological outcrop. Jones, Beck, Jones, and Hughes originally identified five obsidian conveyance zones within the Great Basin, generally with a north–south orientation reflecting the underlying morphology of mountain ranges and valleys (Figure 1.7A), and interpreted these zones to be "coterminous with the foraging territories of Paleoarchaic foragers" (2003:32). As Madsen (2007) points out, it is not clear whether these distributions reflect the movement of Paleoarchaic groups as a whole or that of male hunters. Madsen (2007) uses an optimal foraging model of Paleoarchaic subsistence strategy proposed by Elston and Zeanah (2002) to suggest an alternative explanation for these toolstone distributions (Figure 1.7B). In that model Elston and Zeanah (2002) contend that male hunting parties were likely coming and going from marsh-ecosystem base camps occupied by women and the rest of the foraging group. Thus, the toolstone distributional data may reflect the long-distance and frequent movement of Paleoarchaic *men*, not that of Paleoarchaic groups as a whole. Madsen (2007), using patch choice models, further contends that the toolstone distributional patterns may be much more variable through space and time than implied by the conveyance zone models of Jones, Beck, Jones, and Hughes (2003; also Jones et al. 2012). If Great Basin Paleoarchaic foragers were focused on marsh and wetland resources, then likely the size of local marsh habitats determined overall mobility. Where such wetland ecosystems were larger, as in the eastern and western Great Basin, foraging groups likely moved less often. Where they were smaller, as in the central Great Basin, moves were probably more frequent.

At the time the models of Jones, Beck, Jones, and Hughes (2003) and Madsen (2007) were proposed, it was not yet possible to distinguish between the two models with the information at hand. However, since then a number of studies have increased the toolstone

source database significantly, and as the number of sourced tools from more and more sites has increased, it has become evident that at least some Great Basin toolstone conveyance zones are smaller than previously thought. Using source data from surface sites in a region of the northwestern Great Basin centered on the Black Rock Desert, Smith (2010) suggests that conveyance zones in the area were much smaller than those envisioned by Jones, Beck, Jones, and Hughes (2003), albeit still larger than later Archaic period zones (Figure 1.7C). Similarly, Jones et al. (2012), using a much larger database from the eastern Great Basin, conclude that conveyance zones were likely smaller there as well (Figure 1.7D). Based on ethnographic data, Newlander (2012) suggests that such zones were likely even smaller. In short, as more and better toolstone source data become available, the size of identifiable conveyance zones, and by inference the degree of residential mobility, is being reduced.

That said, however, there is evidence that in some areas of the Great Basin Paleoarchaic residential mobility patterns were substantially higher than those of later Archaic era foraging groups (e.g., Jones et al. 2012; Smith 2011). This may be a product of the differential sizes of local wetland ecosystems, as Madsen (2007) suggests, since data from the Bonneville basin indicate to Jones et al. (2012) that the size of an occupied marsh area is related to the variance of toolstone sources found within it.

Smith et al. (2013) test the models of Madsen (2007) and Jones et al. (2012) by examining Paleoarchaic toolstone use in the paleo-wetlands around pluvial Lake Parman, one of the smallest of the terminal lakes formed in the Great Basin during the Pleistocene–Holocene transition. They expected, and find, that the use of local toolstone in this small northwestern Great Basin wetland area was much reduced in comparison to percentages found in the ORB delta. They conclude that this demonstrates a pattern of shorter residence times and higher mobility by the Paleoarchaic foragers visiting the Lake Parman wetlands compared with the longer residence times and reduced mobility suggested for the ORB wetlands. Smith et al. go on to suggest that

> the relationship between technological provisioning and occupation span suggested by work at these two areas corresponds reasonably well with patch choice models…which Madsen (2007) applied…to predict that larger resource patches (i.e., wetlands) should have fostered longer overall occupations [2013:4187].

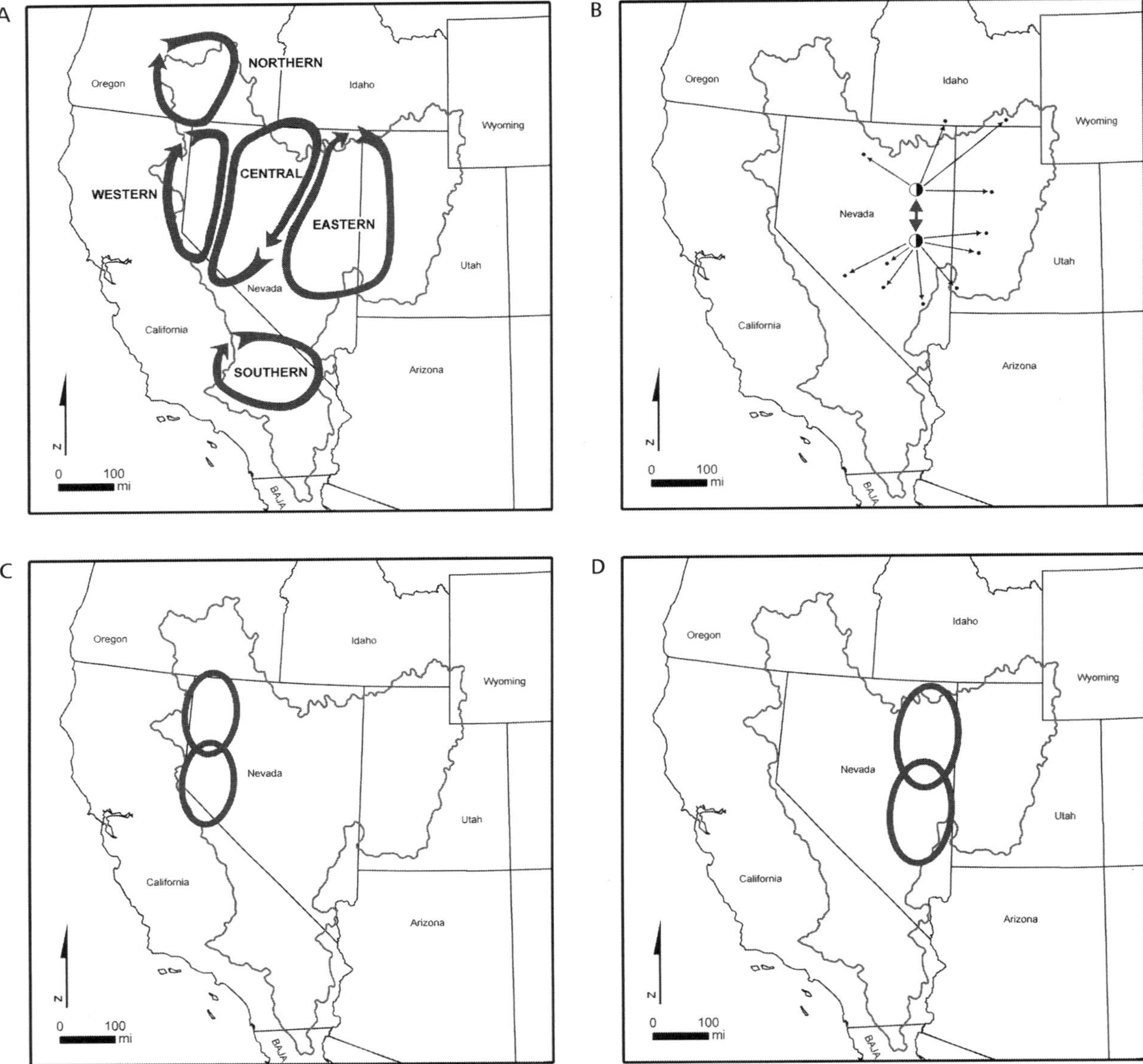

FIGURE 1.7. (*A*) Obsidian "conveyance zones" in the Great Basin as defined by Jones, Beck, Jones, and Hughes (2003); (*B*) alternate explanation of the eastern conveyance zone as defined by Madsen (2007) that involves exchange between separate Paleoarchaic groups; (*C*) western Great Basin conveyance zone as modified by Smith (2010), subdivided into two overlapping zones; (*D*) eastern Great Basin conveyance zone as modified by Jones et al. (2012), subdivided into two overlapping zones.

What remains largely unknown is how social interactions and trade networks impacted the distribution of toolstone during the Paleoarchaic period. Beck and Jones (2011) suggest that formal trade networks were probably missing but, like Madsen (2007), feel that some informal exchange occurred at "social events" where separate groups congregated. They suggest further that this type of exchange increased through time. Newlander (2012), on the other hand, suggests that if modern foragers are any guide, the distribution of toolstone during the Paleoarchaic period was largely a product of formal exchange networks embedded in larger social interaction spheres.

CONCEPTS, HYPOTHESES, AND THE OLD RIVER BED DELTA

The general hypothesis under which we conducted our research in the Old River Bed delta is that the people who lived in the delta between ~11,500 and 8500 ^{14}C BP were likely descended from coastal foragers who, after the initial settling of the New World, appear to have moved rather quickly up major river valleys into lacustrine and riverine wetland settings in the Great Basin that were not very different from the coastal wetlands to which they were adapted. By ~12,500 ^{14}C BP. Paleoarchaic foragers were living in the Great Basin, but as yet there is no evidence that they had developed the distinctive Western Stemmed lithic tradition that became diagnostic of these groups a millennium or so later. By at least 11,100 ^{14}C BP, however, these diagnostic tool forms are found widely scattered across western North America, and the WST is apparently contemporary with the technologically much different Clovis lithic tradition found farther to the east.

These early Paleoarchaic Great Basin peoples appear to have had a relatively broad-based subsistence focus, relying heavily on wetland resources such as fish, waterfowl, rodents, and marsh rhizomes but including larger game from upland areas. Settlement patterns may not have been radically different from those of some protohistoric Great Basin foragers, with the main family group spending most of its time in or near wetland ecosystems and smaller hunting and collecting parties traveling to shorter-term camps in other habitats. If toolstone sources are reliable guides, however, Paleoarchaic foraging territories were larger than those of later Great Basin peoples. The degree of mobility within those foraging territories is unclear and may have been variable from place to place. Where marsh systems were large and productive, movement of the main group to other locations away from those marshes was likely reduced, but where wetland habitats were smaller and less productive, the number and distance of moves were likely larger.

As the population of western Paleoarchaic groups grew, they occupied an increasingly larger territory. As they did so, moving later in the Paleoarchaic period into the Rocky Mountains and other areas where marsh resources were limited, their resource base naturally changed as well. Although their subsistence focus in these areas remained a relatively broad-based "archaic" type of adaptation, the diet included increased amounts of both large and small game, and the plant resources they depended on were locally available upland species. Similarly, as the Great Basin wetland systems diminished in both size and number during the Pleistocene–Holocene transition, the subsistence focus, settlement arrangements, and mobility patterns of later Paleoarchaic groups began to differentiate. Some groups shifted to Holocene "Archaic" patterns earlier than others, depending on the productivity of local wetlands.

ONLINE DATA

Supporting data for several of the following chapters have been placed online in order to limit the length of this monograph and to increase its readability. These data, including geomorphic trench descriptions, site descriptions, artifact photographs, and so on, can be found by linking to the University of Utah Press Web site (http://www.uofupress.com). In the "Books" section, under the title of this monograph—"The Paleoarchaic Occupation of the Old River Bed Delta"—are a series of links to Supplemental Digital Material listed by chapter. In these chapters the initial reference to these online data takes, for example, the following form: "Chapter 3, Supplemental Digital Material."

This supplemental material includes the following links:
Chapter 3:
Trench Descriptions
Individual Channel Location Images
Chapter 5:
Additional Artifact Images
Chapter 6:
Source Assignment Tables
Results of X-Ray Fluorescence and Portable X-Ray
    Fluorescence Analysis

2

Bonneville Basin Environments during the Pleistocene–Holocene Transition

David Rhode and Lisbeth A. Louderback

To place the Old River Bed (ORB) delta archaeological record into a regional environmental context, we summarize paleoenvironmental information from the Bonneville basin for the period between ~12,700 ^{14}C BP, when Lake Bonneville declined from its Provo shoreline (Godsey et al. 2011), and about ~7500 ^{14}C BP, by which time the delta's streams and wetlands had desiccated. Paleoenvironmental data for this period come from a wide variety of sources, including lake reconstructions, sediment cores from wetlands, cave and rockshelter deposits, open-air fossil deposits, glacial sequences, and rodent middens. This region saw significant changes in surface-water availability, changing distributions and abundances of plant and animal species, and other alterations of the landscape. For the people who made part of their living in the ORB delta, these environmental changes impacted the locations and availability of key resources useful or necessary for survival, affecting subsistence strategies and settlement positioning.

The Bonneville Basin in the Bølling-Allerød Interstadial

Pleistocene Lake Bonneville had begun declining from the Provo shoreline by about ~12,700 ^{14}C BP (Benson et al. 2011; Godsey et al. 2011), at or near the start of the Bølling-Allerød interstadial (see Figure 3.2: Bonneville sequence). Miller et al. (2013) suggest that the lake began declining somewhat earlier, around 13,300 ^{14}C BP. The Earth had by this time warmed considerably from its lows of the Last Glacial Maximum, and with the resurgence of Atlantic meridional ocean circulation, Northern Hemisphere temperatures rose rapidly (Clark et al. 2012; Denton et al. 2010; Shakun et al. 2012). These warmer temperatures, coupled with changes in atmospheric circulation, led to drier conditions in much of the Great Basin, though summer temperatures remained cooler than today (Benson et al. 2013; Wigand and Rhode 2002). Protracted drought beginning about this time affected other parts of the North American Southwest as well (Polyak et al. 2012). Lake Bonneville very quickly declined from the Provo level at ~1,480 m to <1,300 m, near or perhaps below the level of the present-day Great Salt Lake (Oviatt et al. 2005). Benson et al. (2011) suggest that the lake level drop was rapid, to low levels by ~12,500 ^{14}C BP. The lake appears to have dropped to the sill dividing the Great Salt Lake basin from the Sevier basin to the southeast by ~12,600 ^{14}C BP (Godsey et al. 2011). Oviatt et al. (1992) consider this steep decline to mark the end of the Bonneville lake cycle and the beginning of the Great Salt Lake phase.

The lake's retreat opened up a vast area of former lake bed to the formation of extensive marshes and playa-margin plant communities. Lake Gunnison, located in the Sevier basin to the south (see Chapter 3 for description and map location), spilled northward to form the ORB and fed the flow of streams out onto the exposed mudflats of the southeast Great Salt Lake Desert (Gilbert 1890; Oviatt 1988; see Chapter 3). In the ORB delta area, channel incision may have begun on the flats as early as ~11,600 ^{14}C BP (see Chapter 3), and wetlands that created organic-rich "black mat" deposits began to form by ~11,000 ^{14}C BP (Oviatt et al. 2003). Marshes formed elsewhere on the basin floor, as at Fish Springs (Godsey et al. 2005) and possibly Blue

Lake (Louderback and Rhode 2009), when these areas became exposed (see Figure 1.2 in Chapter 1 for the locations of these and other localities mentioned here).

On valley margins and surrounding piedmonts in the central Bonneville basin, a mosaic of limber pine (*Pinus flexilis*) woodlands mixed with sagebrush (*Artemisia tridentata*) steppe grew to altitudes at least as low as 1,475 m, with limber pine and other subalpine conifers persisting in lowlands until ~11,800 ^{14}C BP. After this time, sagebrush and shadscale (*Atriplex confertifolia*) came to dominate the lowlands (Rhode 2000a; Rhode and Madsen 1995). The presence of these dry-but-cold-adapted lowland woodland and steppe vegetation communities before 11,800 ^{14}C BP suggests that average July temperatures were on the order of ~6°C cooler than today (Rhode and Madsen 1995). Engelmann spruce (*Picea engelmannii*) and occasionally subalpine fir (*Abies lasiocarpa*) grew in upland canyons in nearby mountains, accompanied by limber pine and, in southern ranges, bristlecone pine (*Pinus longaeva*) (Rhode 2000a; Thompson 1990). Regional pollen records show abundant sagebrush and conifer pollen before ~11,000 ^{14}C BP (Beiswenger 1991; Bright 1966; Louderback and Rhode 2009; Madsen and Currey 1979; Thompson 1992). The Great Salt Lake Core C pollen record (Spencer et al. 1984) suggests a significant drop in the abundance of conifers and a rise in desert shrubs after ~12,400 ^{14}C BP, but this record may be subject to dating problems owing to the influx and reworking of older marl sediments and possible radiocarbon reservoir of the lake (Benson et al. 2011; Godsey et al. 2011; McGee et al. 2012). Glaciers had retreated markedly in the Wasatch Range and Uinta Mountains (Laabs et al. 2009; Laabs et al. 2011; Munroe 2001, 2005; Munroe et al. 2006), but ice persisted in some upland settings (e.g., Little Cottonwood Canyon [Madsen and Currey 1979]).

The mammalian fauna living in and around the Bonneville basin at this time included such large Pleistocene beasts as the Columbian mammoth (*Mammuthus columbi*), giant bear (*Arctodus simus*), mastodon (*Mammut americanum*), horse (*Equus* spp.), yesterday's camel (*Camelops hesternus*), large-headed llama (*Hemiauchenia macrocephala*), ground sloth (*Megalonyx jeffersoni*, *Paramylodon harlani*), helmeted musk ox (*Bootherium bombifrons*), shrub ox (*Euceratherium collinum*), mountain goat (*Oreamnos harringtoni*), long-horned bison (*Bison latifrons*), four-horned antelope (*Capromeryx minor*), flat-headed peccary (*Platygonus compressus*), dire wolf (*Canis dirus*), short-faced skunk (*Brachyprotoma* sp.), Shuler's pronghorn (*Tetrameryx* sp.), and mountain deer (*Navahoceros* cf. *fricki*) (Gillette

and Madsen 1992, 1993; Heaton 1999; Madsen 2000b; McDonald 2002; Miller 1976, 1987, 2002; Nelson and Madsen 1987). Among the smaller mammals, it is notable that pikas (*Ochotona princeps*), small lagomorphs that live today in the alpine and subalpine zones at the tops of the region's higher mountains, thrived at altitudes down to below ~1,510 m, about 800 m lower than they would be expected today (Rhode 2000a).

Remains of insects and other arthropods are often found in pack rat (*Neotoma* spp.) middens in the Bonneville basin. Insects often have quite specific climatic tolerances, so their presence in a midden can provide useful paleoclimatic information (Elias 1997). Scott Elias estimated terminal Pleistocene climatic parameters using insect remains from a pack rat midden dating to ~11,800 ^{14}C BP in the Lead Mine Hills west of the Great Salt Lake Desert (Bonneville Estates #4A [Rhode 2000a]). This midden sample contains several beetle species that have modern ranges chiefly in the Pacific Northwest and California region of western North America (Figure 2.1). The overlapping temperature and moisture tolerances of these beetles suggest that the mean July temperature was ~20.5°C and mean annual precipitation was ~525–675 mm (S. Elias, personal communication 2001; Elias 2007). By contrast, the nearest weather station (Wendover, Utah) records a present-day mean July temperature of 26.4°C and mean annual precipitation of 123 mm. If Elias's climatic estimates based on insect remains are reliable, the region at ~11,800 ^{14}C BP was substantially wetter with much cooler summers than today. Estimated precipitation 4–5.5 times greater than today at Wendover, dry as Wendover may now be, is surprising given the rapidly declining Lake Bonneville, the presence of cold-but-dry-adapted limber pine and sagebrush steppe vegetation in the lowlands, and the absence of cold-but-wet-adapted montane meadow plants and conifers such as spruce and fir. Nevertheless, cooler conditions and greater effective moisture than today are indicated for the Bølling-Allerød interstadial in the northern Bonneville basin.

Although Lake Bonneville stood at relatively low levels, it stayed sufficiently cold and fresh to support a diverse fish fauna, including bull trout (*Salvelinus confluentus*), cutthroat trout (*Onchorhynchus clarki*), cisco (*Prosopium* spp.), and sculpin (*Cottus* spp.), until at least 11,200 ^{14}C BP (Broughton et al. 2000; Hart et al. 2004; Madsen et al. 2001; Rhode and Madsen 1995). Further recession of the lake after ~11,200 ^{14}C BP led to a steep decline in the deposition of fish remains at Homestead Cave, presumably the result of die-offs owing to warmer waters and/or increasing salinity (Broughton 2000; Broughton et al. 2000; Madsen et al. 2001).

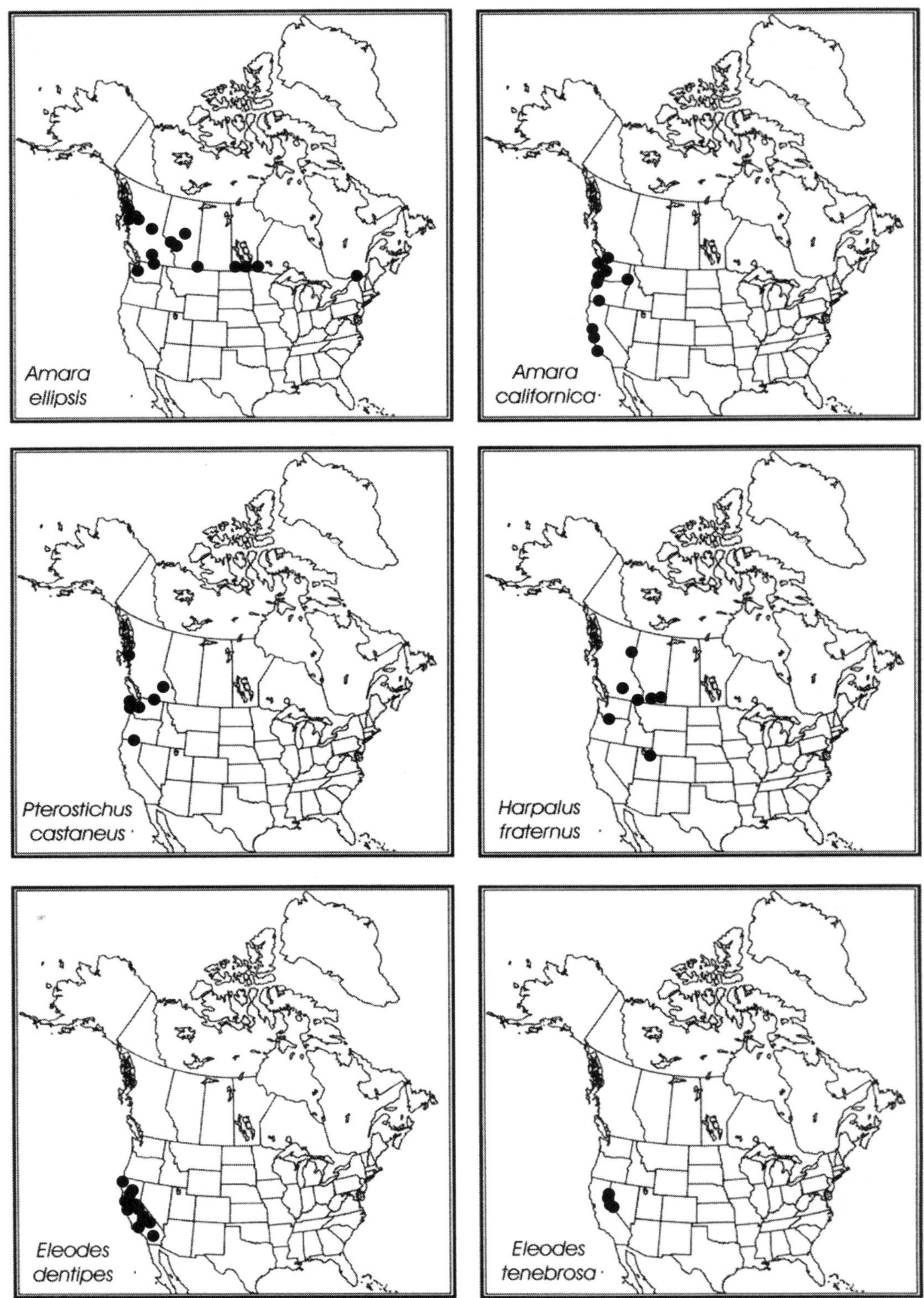

FIGURE 2.1. Modern distributions of beetles identified in a pack rat midden dating to 11,800 ^{14}C BP from the west side of the Great Salt Lake Desert. The presence of these insects in this midden suggests greater annual precipitation and cooler summers than at present (Elias 2007).

TABLE 2.1. $^{87}Sr/^{86}Sr$ Ratios from Mollusks in Old River Bed Delta Channels.

Channel	Easting	Northing	$^{87}Sr/^{86}Sr$	Date (^{14}C BP)	SD	Lab No. (Beta-)	Cal BP (2σ)
Lavender	298777	4449048	.71035	9010	40	221778	9988–9945, 10,057–10,043, 10,246–10,148
Blue	300076	4453212	.70991	9750	40	231555	11,112–11,109, 11,241–11,122
Yellow	296169	4455498	.71010	—			
Yellow	298491	4453615	.71034	10,130	80	221779	11,392–12,057, 11,371–11,366
Yellow	297406	4454186	.71005	—			
Green	298933	4456917	.71017	10,000	40	238669	11,637–11,275, 11,701–11,671
Green	298933	4456917	.71018	10,290	50	238668	11,898–11,827, 12,228–11,953, 12,383–12,249
Black D	294767	4458630	.71000	—			
Gold	290687	4457082	.71030	10,460	60	248474	12,567–12,125

Note: For comparison, the ranges of $^{87}Sr/^{86}Sr$ ratios in carbonates are as follows: (1) Lake Gunnison = .7097–.7105; (2) Great Salt Lake = .7147; (3) Gilbert shoreline = .7119–.7134, Provo level = .7116–.7120, and Bonneville level = .7112–.7118 (Hart et al. 2004). Measurements provided courtesy of Deanna Grinstead and Jay Quade. See Chapter 3 for channel color references.

THE BONNEVILLE BASIN DURING THE YOUNGER DRYAS

The Younger Dryas stadial marks a return to colder Northern Hemisphere temperatures at the close of the last deglaciation, thought to have been triggered by a shutoff of Atlantic meridional ocean circulation and the resulting cooling of the North Atlantic (Alley and Clark 1999; Clark et al. 2012; Shakun and Carlson 2010; Shakun et al. 2012). The Younger Dryas is thought to have begun abruptly at ~10,900 ^{14}C BP and to have ended in ~10,100 ^{14}C BP in the North Atlantic area (Rasmussen et al. 2006; Steffensen et al. 2008), but these age estimates vary somewhat in the literature (e.g., Haynes 2008). The climatic effects of the Younger Dryas in North America and the rest of the world are geographically complex and in many places quite subtle; areas situated close to the North Atlantic cooled dramatically, but regions located farther away from the North Atlantic show much reduced effects on climate, and some areas of North America actually warmed during the Younger Dryas (Shakun and Carlson 2010). Meltzer and Holliday (2010) question whether many Paleoindians would have even noticed the climate changes during the Younger Dryas. In the Bonneville basin, some significant environmental changes did occur that might have made some Paleoindians scratch their heads with interest, puzzlement, or alarm (Goebel et al. 2011).

For one example, at some point Great Salt Lake expanded to cover the Great Salt Lake Desert and form the Gilbert shoreline (Oviatt et al. 1992; Oviatt et al. 2005). Using evidence from the Public Shooting Grounds site north of Great Salt Lake, Oviatt et al. (2005) estimate that the expansion occurred sometime between ~10,500 and 10,000 ^{14}C BP and suggest that the lake stayed at the Gilbert shoreline level only for a few decades and that its level fluctuated between ~1,290 and 1,295 m (see also Currey et al. 1984; Murchison 1989). Subsequently, Oviatt (2014) reconsidered the dating evidence from Public Shooting Grounds and other sites to suggest that the age of the Gilbert rise was closer to ~10,000 ^{14}C BP, at or just after the end of the Younger Dryas. In contrast, Benson et al. (2011) use the sediment record from Blue Lake to argue that the lake stood at the Gilbert shoreline more or less continuously through the Younger Dryas interval, with some fluctuations. However, as Benson et al. note, the Blue Lake core is marked by unconformities, erosion, and sediment reworking during this interval, and thus the core cannot be relied upon to document the persistence of the lake at the Gilbert shoreline through the Younger Dryas. The timing, duration, and extent of the Gilbert rise remain uncertain, though it was probably briefer and less extensive than sometimes thought (Oviatt 2014).

Strontium ratios in carbonate-bearing mollusks from deposits attributed to the Gilbert episode indicate that they lived in a lake environment that contained a mixture of waters from both the Sevier basin and northern Bonneville basin sources (Hart et al. 2004; Quade 2000). In the ORB delta, strontium ratios of mollusk shells from channels (including an *Anodonta* shell from a Black channel) are all consistent with a strictly Sevier basin water source, rather than waters derived from northern sources or a mix of northern and southern sources (Table 2.1). These strontium ratios may indicate continued overflow from Lake Gunnison via the ORB or leakage of groundwater from the Sevier basin through alluvium. In any case, the available strontium data clearly indicate no input of northern water sources into the ORB area through the available record after ~10,500 ^{14}C BP and reinforce the conclusion that Gilbert lake waters did not contribute to or reach the level of the ORB.

A second noticeable difference of the Younger Dryas environment in the Bonneville basin is that, by that time, the latest Pleistocene North American megafauna was already extinct or getting there fast (Faith and Surovell 2009; Grayson 2007; Haynes 2008; Haynes 2009). The latest well-dated megafaunal species in the vicinity of the Bonneville basin is a giant bear from the Huntington locality in the Wasatch Range, dating to the Bølling-Allerød–Younger Dryas transition (Madsen 2000b; Schubert 2010). The demise of the megafauna left the region with a much reduced suite of large herbivores: mountain sheep (*Ovis canadensis*), mule deer (*Odocoileus hemionus*), pronghorn (*Antilocapra americanus*), bison (*Bison bison*), and elk (*Cervus elaphus*). Even these ungulates may have been scarce (Broughton et al. 2008; Grayson 2011:236), though cave and rockshelter deposits dating from the period often contain ample evidence of at least some of them (e.g., Danger Cave, Bonneville Estates Rockshelter, Smith Creek Cave [Bryan 1979; Hockett 2007; Jennings 1957; Mead et al. 1982]). More abundant are the remains of smaller mammals—notably the pygmy cottontail (*Brachylagus idahoensis*), yellow-bellied marmots (*Marmota flaviventris*), bushy-tailed wood rats (*Neotoma cinerea*), northern pocket gophers (*Thomomys talpoides*), and a variety of other rodents—that are uncommon or absent in the desert lowlands today but were plentiful in the Younger Dryas (Grayson 2000a, 2011:187–199; Schmitt and Lupo 2012). The abundance of these montane mammals in lowland settings throughout the Bonneville basin provides strong evidence for a Younger Dryas that was cooler than today. More specifically, the abundance of pygmy rabbit and of sage grouse (*Centrocercus urophasianus*) indicates the prominence of big sagebrush steppe vegetation in the lowlands around Homestead Cave and Bonneville Estates Rockshelter, which today are dominated by desert scrub communities (Grayson 2000a, 2006; Hockett 2007; Livingston 2000; Schmitt and Lupo 2012).

A mixture of desert shrubs and sagebrush-grass steppe grew extensively in the valleys and piedmonts, as both the pack rat midden and faunal records reveal (Grayson 1998, 2000a, 2006; Hockett 2007; Rhode 2000a; Schmitt and Lupo 2012). Shadscale had been an occasional component of the earlier limber pine woodland/sagebrush steppe mosaic; by the beginning of the Younger Dryas it grew much more widely in lowlands, along with other desert shrub taxa such as horsebrush (*Tetradymia* sp.). The pollen record from Swan Lake, well to the north of the ORB in southern Idaho, shows that the Younger Dryas interval began with an abundance of conifers, but these forests were quickly replaced by sagebrush and grass (Bright 1966). Other regional pollen records, such as Blue Lake and Great Salt Lake (Louderback and Rhode 2009; Spencer et al. 1984), are marked by erosional unconformities or a lack of discernible Younger Dryas–age sediments, so they cannot inform on regional Younger Dryas–age vegetation. At Snowbird Bog in the Wasatch Range highlands (Madsen and Currey 1979), the available record suggests that the Younger Dryas interval was cool and dry, characterized by an abundance of sagebrush, alder, and birch and decreased spruce and pine.

THE BONNEVILLE BASIN IN THE EARLY HOLOCENE

The Younger Dryas ended abruptly about ~10,100 ^{14}C BP in the North Atlantic, as meridional ocean circulation kicked back into high gear and tropical ocean waters returned to warm the North Atlantic (Shakun et al. 2012; Steffensen et al. 2008). Summer solar insolation was reaching peak Holocene values (Kutzbach 1981; Kutzbach and Guetter 1986), adding to the heating of higher latitudes and continental interiors. As a result, the intertropical convergence zone moved northward, and global monsoon circulation strengthened. In the American Southwest summer precipitation increased, and temperatures cooled slightly (Miller et al. 2010; Polyak et al. 2004; Polyak et al. 2012); the Pacific Northwest, by contrast, remained significantly drier than at present (Mehringer 1985; Thompson and Anderson 2000). Climate models suggest that higher continental summer temperatures might have resulted in stronger summer monsoonal precipitation in the Great Basin as well (Thompson et al. 1993). However, biotic records in the northern Great Basin, including the Homestead Cave small-mammal record, do not support these models (Grayson 2000a, 2011). Perhaps the lingering presence of continental glacier remnants in subarctic North America (Shakun and Carlson 2010; Wanner et al. 2011) kept northerly continental areas, including the northern Great Basin, sufficiently cool to limit the advance of summer convective storms into higher latitudes.

The decline of Great Salt Lake from the Gilbert shoreline allowed some marshlands that had been drowned to recover in various parts of the Bonneville basin. Some wetlands that lay above the Gilbert shoreline, such as those at the ORB delta, Fish Springs, and Bear River, were not drowned. At Blue Lake, shallow ponds or lagoons ringed with marshes developed about ~10,100 ^{14}C BP; peat deposits began to form by

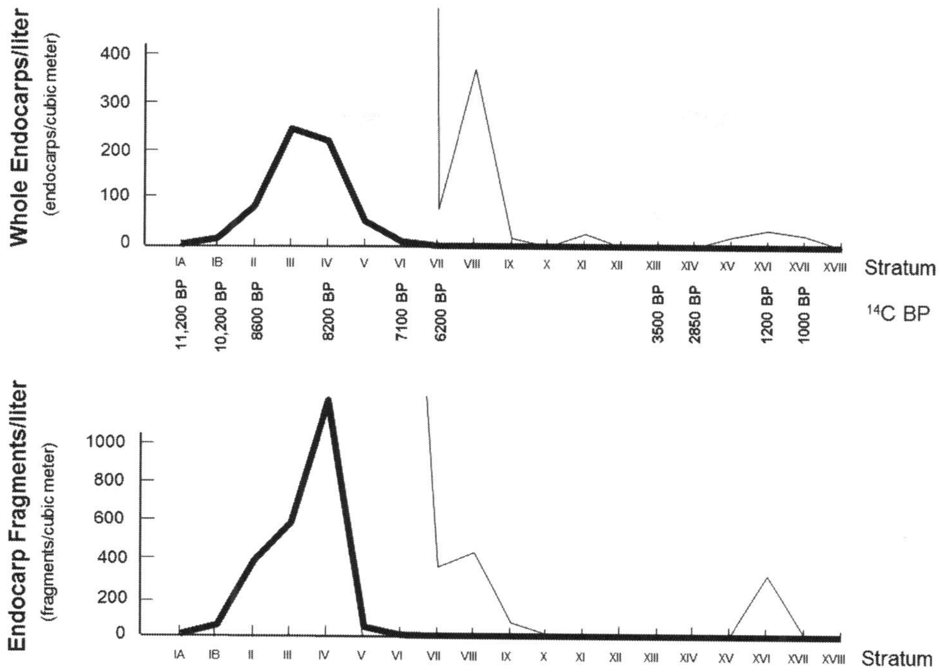

FIGURE 2.2. Stratigraphic distribution of netleaf hackberry endocarps in Homestead Cave. The bold line is the number of whole and fragmentary endocarps per liter; the fine line is a 100× exaggeration to show later Holocene incidences. The vast bulk of hackberry remains occur between 8600 and 8300 ^{14}C BP (from Hunt et al. 2000).

9600 ^{14}C BP (Louderback and Rhode 2009). At the Public Shooting Grounds locality, wetland sediments interpreted to date to ~9700–9300 ^{14}C BP overlie nearshore sediments that had been deposited during the formation of the Gilbert shoreline (Oviatt et al. 2005). Near Danger Cave, slightly organic muds overlying Gilbert beach deposits date to ~9450 ^{14}C BP; post-Gilbert peats date to ~8360 ^{14}C BP (Murchison 1989).

The sparse pack rat midden record from this period indicates that shadscale and sagebrush dominated valley floors (Rhode 2000b), while regional pollen records show that sagebrush steppe decreased dramatically and desert shrubs and grass increased in abundance during the early Holocene (Beiswenger 1991; Bright 1966; Louderback and Rhode 2009; Madsen and Currey 1979; Mehringer 1985). At higher altitudes, a mosaic of sagebrush-grass steppe, mesophilic aspen (*Populus tremuloides*) groves, montane brush communities, and woodlands dominated by Rocky Mountain juniper (*Juniperus scopulorum*) covered the Bonneville basin uplands. Limber pine retreated to protected canyons and mixed with spruce (*Picea engelmannii*) and fir (*Abies concolor* and *A. lasiocarpa*) in higher-elevation subalpine forests. Utah juniper (*Juniperus osteosperma*) and piñon pine (*Pinus monophylla*) apparently did not arrive in the region until the end of the early Holocene (Rhode 2000b; Thompson 1990).

Mesophilic shrubs such as netleaf hackberry (*Celtis reticulata*) and snowberry (*Symphoricarpos* cf. *longiflorus*) grew in rocky lowland settings such as Homestead Cave (Hunt et al. 2000; Rhode 2000b), as they occasionally do in some protected localities at similar altitudes today. The frequency of hackberry endocarps at Homestead Cave declined markedly after ~8000 ^{14}C BP, and they were nearly gone by ~7000 ^{14}C BP (Figure 2.2). Hackberry does not grow at Homestead Cave now, as it requires greater summer moisture than is available there today. It occurs in the nearby Lakeside Range and in higher, better-watered ranges nearby, such as the Stansbury Range.

Hackberry typically occurs in warmer areas where summer precipitation is a major component of the annual rainfall, but it also occurs as far north as the Columbia Plateau, where winter precipitation predominates. It lives mainly on rock outcrops, near springs, and at the bases of cliffs and talus slopes with mesic microhabitats, where its roots can obtain trapped water during the growing season. Localities such as Homestead Cave likely contained enough trapped water in their outcrops and a generally higher water table to support hackberry for about a millennium in the early Holocene. Alternatively, its presence at Homestead Cave may indicate enhanced precipitation during the relatively brief interval between ~9000 and 8000 ^{14}C BP, with maximum

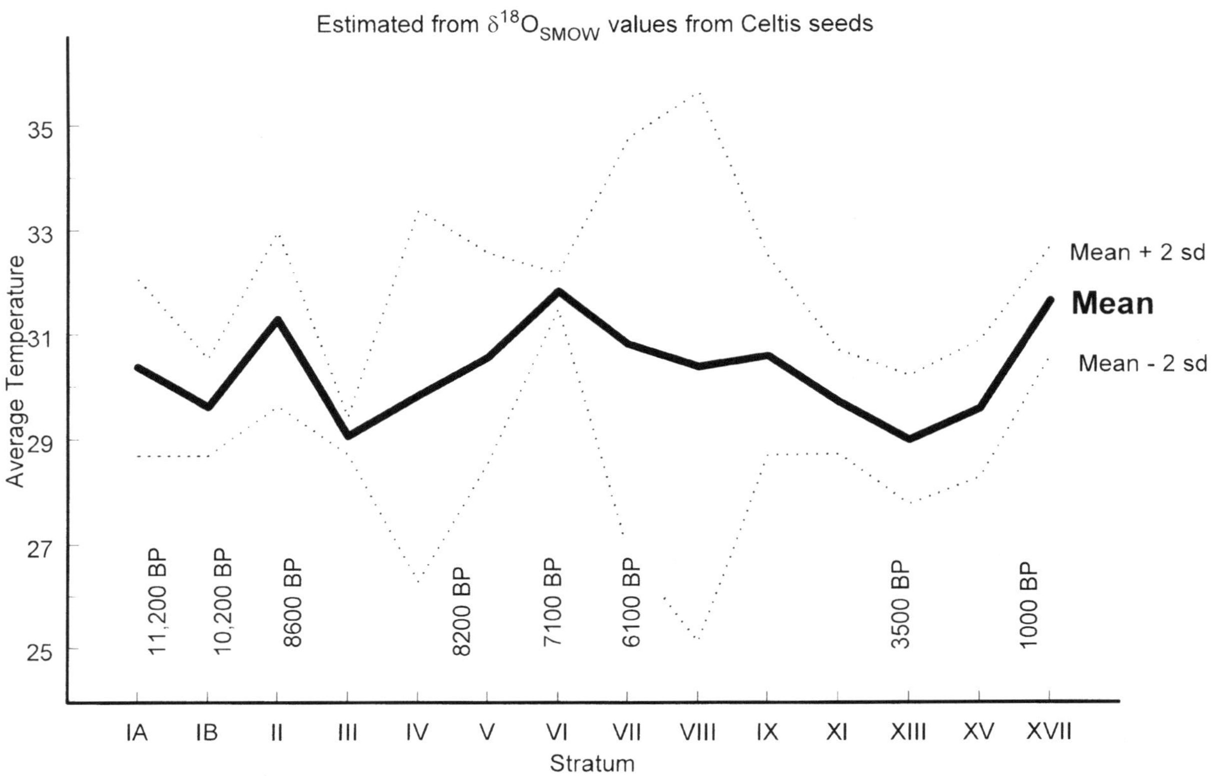

FIGURE 2.3. Average July temperature (°C) estimated from oxygen isotope measurements of hackberry endocarps at Homestead Cave (A. H. Jahren, personal communication March 28, 1998). Dates are ^{14}C age estimates reported by Madsen (2000a). The bold solid line is the average value; the fine dotted lines are two standard deviations. Although the variability is often quite large, growing-season temperatures are significantly greater at ~7100 ^{14}C BP (Stratum VI) than during the early Holocene (Stratum III).

summer solar insolation in the Northern Hemisphere and enhanced monsoon storminess in the Southwest.

The oxygen isotopes in hackberry endocarps have been used to estimate maximum growing-season temperature (Jahren et al. 2001). A. Hope Jahren measured oxygen isotope ratios in a sample of endocarps from Homestead Cave, from separate strata spanning the Holocene, and calculated average July temperatures from those measurements (Figure 2.3). Her results document significant within-stratum variation in estimated growing-season temperature, but they suggest a relatively low growing-season temperature in ~8600 ^{14}C BP, when hackberry attained its peak of abundance at the cave.

THE END OF THE EARLY HOLOCENE

By ~7500 ^{14}C BP, shadscale increasingly dominated valley floors through much of the northern Bonneville basin, though sagebrush habitats apparently persisted in some places (e.g., near Camels Back Cave, east of the ORB [cf. Schmitt and Madsen 2005]). The Great Salt Lake Core C pollen record (Spencer et al. 1984) also documents the continuing decline of conifer woodlands and expansion of xerophytic shrubs in the Bonneville basin beginning sometime after ~9600 ^{14}C BP and completed well before the deposition of Mazama tephra (~6900 ^{14}C BP). The pollen record from Swan Lake (Bright 1966) also reveals a strong increase in the abundance of desert shrubs (cheno-ams) at the expense of conifers, sagebrush, and grass. The pollen record from Grays Lake, southeastern Idaho, indicates warming beginning at ~8500 ^{14}C BP, with a transition from sagebrush steppe to shadscale scrub (Beiswenger 1991). At Snowbird Bog, conifer pollen increased dramatically relative to all other pollen shortly before ~8000 ^{14}C BP, suggesting a rapid transition to a full spruce-pine forest, which Madsen and Currey (1979) attribute to a sharp increase in temperature.

This record of increased drying and a shift from sagebrush steppe to desert shrub communities during the early Holocene is mirrored by the mammalian faunal record from Homestead Cave and other sites (Grayson 1998, 2000a, 2000b, 2006; Schmitt and Lupo 2012; Schmitt, Madsen, and Lupo 2002). Taxa well adapted to relatively cool conditions, such as Ord's kangaroo rat (*Dipodomys ordii*), pygmy rabbit, bushy-tailed wood rat, and yellow-bellied marmot, were abundant before ~10,000 ^{14}C BP but declined after that and were locally extirpated by ~8300 ^{14}C BP. At Homestead Cave, this shift was accompanied by a reduction in waterfowl, but the variety of other avian taxa increased, suggesting that diverse lowland habitats still existed (Livingston 2000). At Camels Back Cave, sagebrush/grass habitats, together with small mammal species including bushy-tailed wood rat, marmot, and white-tailed jackrabbit, began to sharply decline by 9600 ^{14}C BP and disappeared from the record by 8300 ^{14}C BP (Schmitt and Lupo 2005, 2012; Schmitt, Madsen, and Lupo 2002). At Bonneville Estates Rockshelter, the shift began in ~10,000 ^{14}C BP and was essentially complete by ~9000 ^{14}C BP (Schmitt and Lupo 2012).

As the warming and drying trend continued, expansive wetlands dried up. The Blue Lake wetlands were in decline by 8100 ^{14}C BP and largely though not entirely desiccated by ~7000 ^{14}C BP, when marshes dominated by bulrush (*Schoenoplectus* spp.) and sedge (*Carex* spp.) were sharply diminished, replaced by patchy meadows of salt grass (*Distichlis spicata*) and playa margins dotted with iodine bush (*Allenrolfea occidentalis*) (Louderback and Rhode 2009). Other large wetland systems in the region, including the ORB delta, declined or dried up completely between ~8500 and 7500 ^{14}C BP (Kiahtipes 2009; Oviatt et al. 2003; Rhode et al. 2005; Thompson 1992; Wigand and Rhode 2002). It is during this period that dated evidence of human occupation in the Bonneville basin apparently lessens (Louderback et al. 2011), and people who did occupy the area began to intensively process and eat tiny seeds of iodine bush and other small hard-seeded plants at sites such as Danger Cave, Hogup Cave, and Camels Back Cave (Rhode 2008; Rhode and Louderback 2007; Rhode et al. 2006).

Summary

The overall trend in late Pleistocene–early Holocene environmental change in the Bonneville basin is reasonably well known, though many details remain to be worked out. Generally increasing temperatures and overall aridity led to a sequence of dramatic changes in vegetation distribution in the basin lowlands, from subalpine mesic meadow to limber pine/sagebrush woodland/steppe in the Bølling-Allerød, to sagebrush steppe and saltbush desert in the Younger Dryas, and to the increasing dominance of saltbush desert through the early Holocene. Wetlands covered vast areas of the former Bonneville lake bed, but these wetlands diminished in size through the early Holocene until even the largest, such as the ORB, were effectively eliminated; only those wetlands fed by permanent springs (such as Blue Lake and Fish Lake) persisted into the middle Holocene, and even these diminished in size. The Pleistocene megafauna met their demise by or during the early Younger Dryas, resulting in a reduction in faunal diversity; continued warming and drying in the early Holocene led to the replacement of mesic-adapted small mammal taxa with xeric-adapted forms in the lowlands.

How these environmental changes affected human populations living in the Bonneville basin is also becoming better known in broad terms, though here again much is left to be learned. Paleoindian occupations dating as early as ~11,000 ^{14}C BP (see Chapter 1) are known from lower and mid-elevation settings in rockshelters and in open sites on the Lake Bonneville playa and adjacent lowland plains associated primarily with former wetlands and stream courses. The remains from these occupations suggest that people pursued a fairly broad array of animal foods (and likely plants as well), employing a widely mobile land-use strategy focusing on lowland habitats, particularly the wetlands and meadowlands around streams. Through the early Holocene as the environment warmed and desiccated, human occupations were increasingly focused on fewer and smaller high-value wetland habitats. Eventually, by ~8500 ^{14}C BP, a new subsistence strategy involving intensive processing and milling of small seeds from playa-margin plants had emerged in the Bonneville basin. Subsistence pursuits broadened to include a wider variety of mid-level and upland habitats. Hunting technology incorporated notched rather than stemmed projectile points, and the focus shifted toward more small game such as rabbits. In short, from the early to middle Holocene drying trend and the demise of vast wetlands such as the ORB came the spread and integration of the traditional Desert Archaic subsistence and settlement pattern. Whether this shift in strategy was mainly an in situ development or spread primarily from an origin outside the Bonneville basin is not yet known. However, its long-term success as a response to regional aridification appears incontrovertible.

Old River Bed Delta Geomorphology and Chronology

David B. Madsen, Charles G. Oviatt, D. Craig Young, and David Page

The Old River Bed (ORB) delta and associated wetlands occupy a large (~2,600-km^2) area on the floor of the Bonneville basin (Figure 3.1) at the northern end of the ORB, an abandoned river valley eroded into deposits of Lake Bonneville. Gilbert (1890) postulated that during the regressive phase of Lake Bonneville, after the lake had dropped below the topographic threshold between the Sevier basin and the Great Salt Lake basin (the Old River Bed threshold [ORBT]; see Figure 3.1), a shallow lake in the Sevier basin overflowed to the north. The river created by this overflow eroded a meandering, narrow valley, the ORB, in the fine-grained lake sediments on the basin floor. The shallow lake in the Sevier basin has been referred to as Lake Gunnison (Figure 3.1), and studies in that basin (e.g., Mayer et al. 2010) suggest that it may have overflowed through the ORB from about 12,500 to shortly after ~10,000 ^{14}C BP (Figure 3.2). Landforms and deposits at Dugway Proving Ground (DPG) provide a record of environmental change during this time period and into the early Holocene at the distal end of the ORB, hereafter the "ORB delta." Although there is no evidence that the ORB river emptied into a lake at its distal end on DPG during this period, we use the term *delta* to describe the huge area of fluvial wetland landforms and deposits where the ORB river spread out in distributary channels on the flat desert floor.

Our geomorphological studies were initiated to investigate the archaeological resources on DPG, but we have expanded these studies to include examination of the lacustrine and fluvial geomorphology, stratigraphy, and history. This latter work has consisted of mapping the surficial deposits and landforms on aerial photographs and topographic maps, documenting the stratigraphy exposed in backhoe pits and trenches, surveying and mapping archaeological sites as they are related to channel morphology, and dating samples of organics and shells collected from backhoe trenches and outcrops.

FORMATION AND MORPHOLOGY OF THE OLD RIVER BED DELTA

Recent work by Godsey et al. (2011) suggests that Lake Bonneville began to regress from its threshold-controlled Provo shoreline at ~1,480 m asl at ~12,600 ^{14}C BP (Figure 3.2). By ~11,300 ^{14}C BP the falling lake approached modern altitudes at ~1,300 m (Madsen et al. 2001). Lake Bonneville regressed from the Provo shoreline to the altitude of the Old River Bed threshold very rapidly, so that the age of the beginning of regression from the Provo shoreline and the age of the beginning of overflow at the ORBT are within the analytical error ranges of many radiocarbon ages. North of the ORBT the ORB river incised into sediments deposited during the transgressive and deepwater phases of Lake Bonneville to form the ORB valley.

As lake levels continued to drop and the lake margin retreated to the north, the ORB delta also prograded to the north, and the ORB river incised into its older, higher, deltaic deposits. We estimate that shortly before ~11,500 ^{14}C BP. Lake Bonneville had regressed to an altitude where the ORB river was no longer constrained in a narrow valley and it formed bifurcating distributary deltaic channels on the relatively flat floor of the Bonneville lake bed at altitudes of ~1,330 m and lower. Dates on

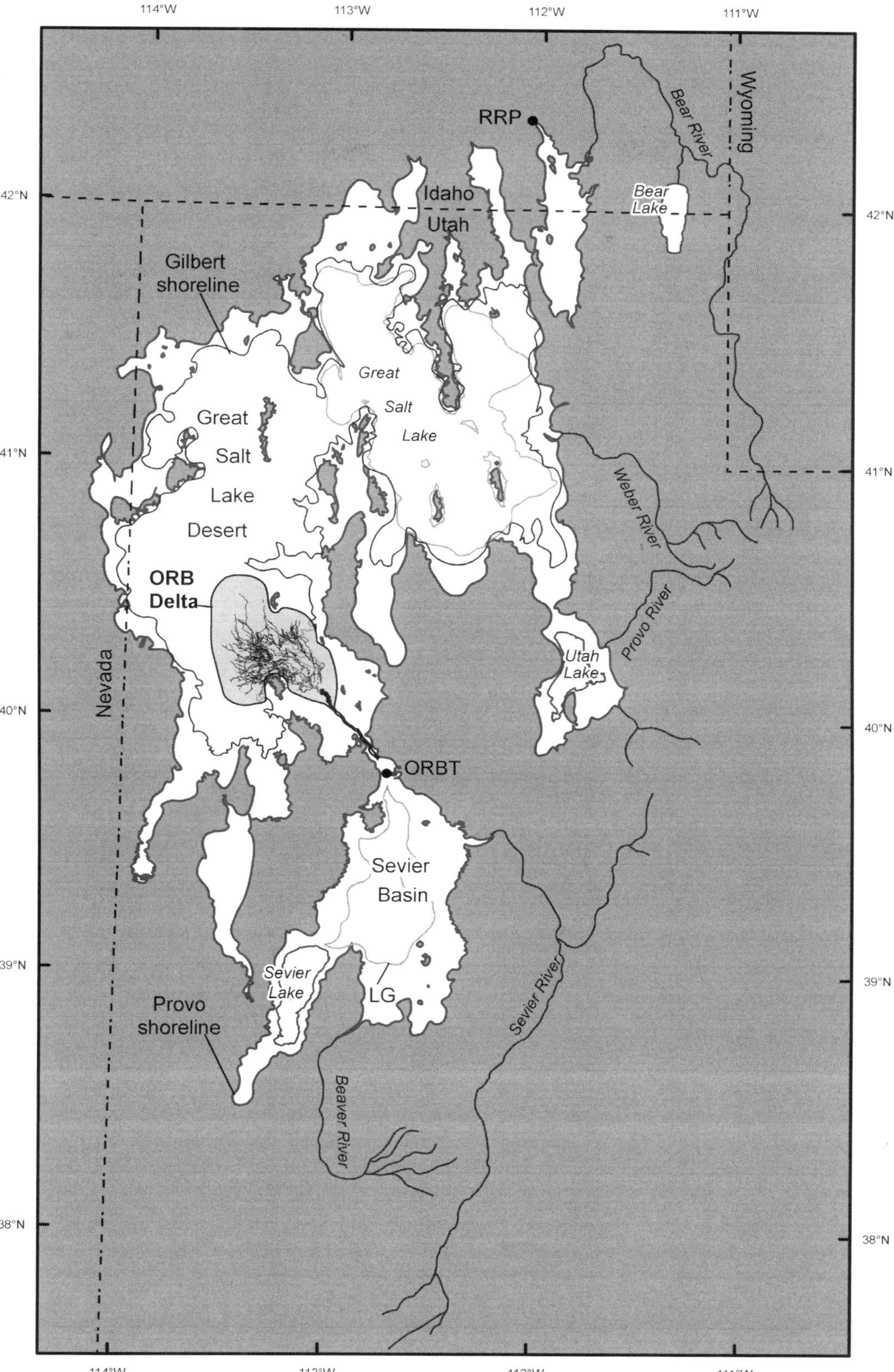

FIGURE 3.1. Map of the Bonneville basin showing the location of the Old River Bed (ORB) delta, the Sevier basin, Great Salt Lake (GSL), and the GSL basin (marked by the Gilbert shoreline, as mapped by Currey [1982]). RRP = Red Rock Pass (the external threshold of Lake Bonneville); ORBT = Old River Bed threshold (low point on the divide between the Sevier and GSL basins); LG = approximate shoreline of Lake Gunnison. The solid line that extends northwestward from the ORBT represents the ORB paleovalley. The position of the Gilbert shoreline is uncertain in the Dugway Proving Ground area, but, as shown partly by the work reported here, it is likely that the Gilbert lake did not rise high enough to reach Dugway (map modified from Oviatt et al. 2003).

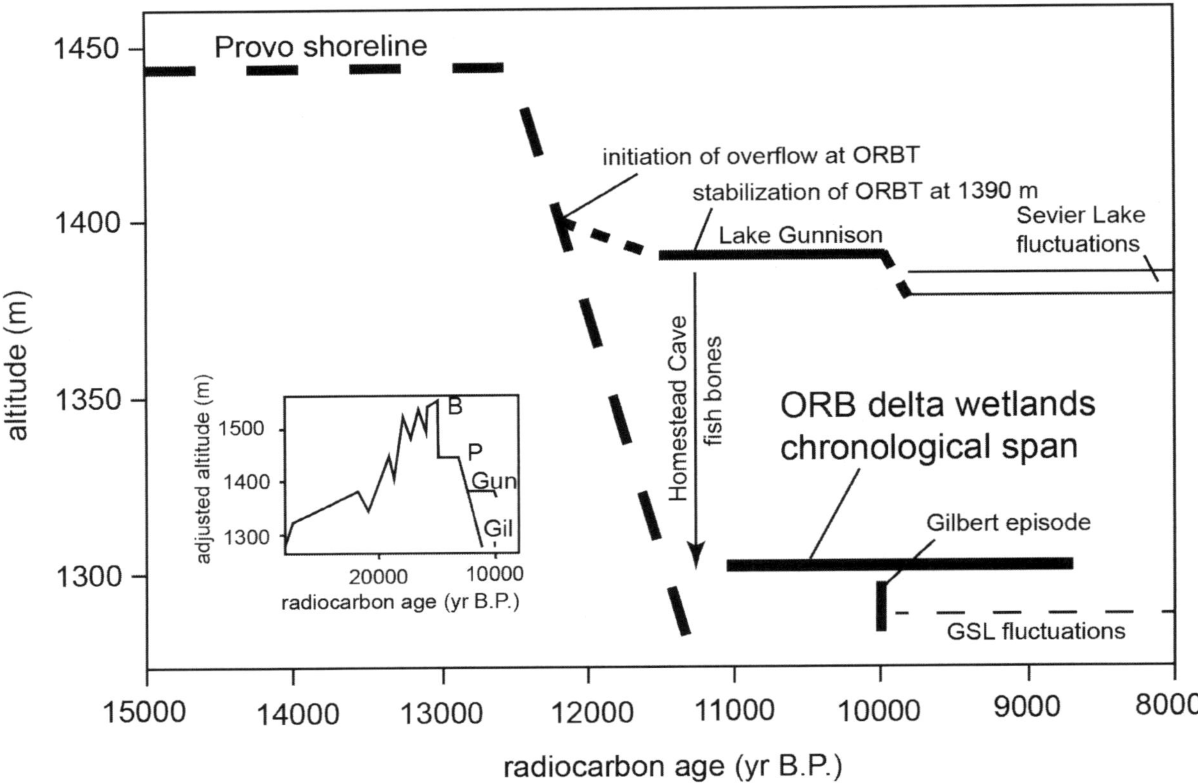

FIGURE 3.2. Chronology of the Old River Bed (ORB) delta wetlands immediately after the regressive phase of Lake Bonneville (modified from Oviatt et al. 2005 and Godsey et al. 2011). The vertical scale for the inset is altitude adjusted for post-Bonneville isostatic rebound using the Currey rebound-adjustment equation (Oviatt et al. 1992). The vertical scale for the main diagram is altitude, not adjusted for isostatic rebound; this scale applies to landforms and events younger than about 12,000 ^{14}C BP. The Provo shoreline is shown on the main diagram at its isostatically adjusted altitude. Lake Bonneville had dropped to the floor of the Sevier Desert at 1,400 m asl soon after it fell from the Provo shoreline. At that point it split into two lakes, one in the Great Salt Lake (GSL) basin and one (Lake Gunnison) in the Sevier basin, and overflow from the Sevier basin caused gradual erosional lowering of the ORB threshold (ORBT). By 11,400 ^{14}C BP the ORBT altitude had stabilized at 1,390 m, and Lake Gunnison continued to overflow northward. Overflow caused incision of the ORB valley into Lake Bonneville sediments. Note that the dated samples that define the ORB delta chronological span were collected above the probable upper limit of the Gilbert episode lake. Post-Gilbert fluctuations of GSL have been below the line labeled "GSL fluctuations." In the inset, B = Bonneville shoreline; P = Provo shoreline; Gun = Gunnison shoreline; Gil = Gilbert episode.

black mat sediments (see below) and on spring marsh peat at the base of a core from Fish Springs (Godsey et al. 2005) indicate that the lake had dropped to elevations below ~1,310 and possibly as low as ~1,290 m by 11,500 ^{14}C BP. We refer to this complex, fan-shaped pattern of meandering, intersecting, and crosscutting distributary channels as the ORB delta (Figure 3.3), although it is important to recognize that the lake margin was far to the north and the lake itself did not control deposition on the delta plain. The ORB delta includes both sheetlike fluvial deposits of mud and sand and numerous channel-fill deposits of sand or sand and gravel. All of the deltaic deposits stratigraphically overlie deposits of Lake Bonneville.

When Lake Bonneville had regressed to altitudes lower than the ORBT and the ORB river had begun to entrench the previously deposited sediments on the floor of Lake Bonneville, the river carried suspended muds and fine sands northward and deposited them offshore in the rapidly regressing lake. These regressive-phase lacustrine fines overlie the Bonneville marl in the Dugway area. These fines are thin (less than several meters in thickness in many places) but probably were originally deposited in a broad fan-shaped form on the floor of the regressing lake. Oviatt et al. (2003:Figure 5) refer to these deposits as "underflow fan" deposits, but we prefer to use the phrase "regressive-phase lacustrine fines" (RLF) in this monograph in order to keep inter-

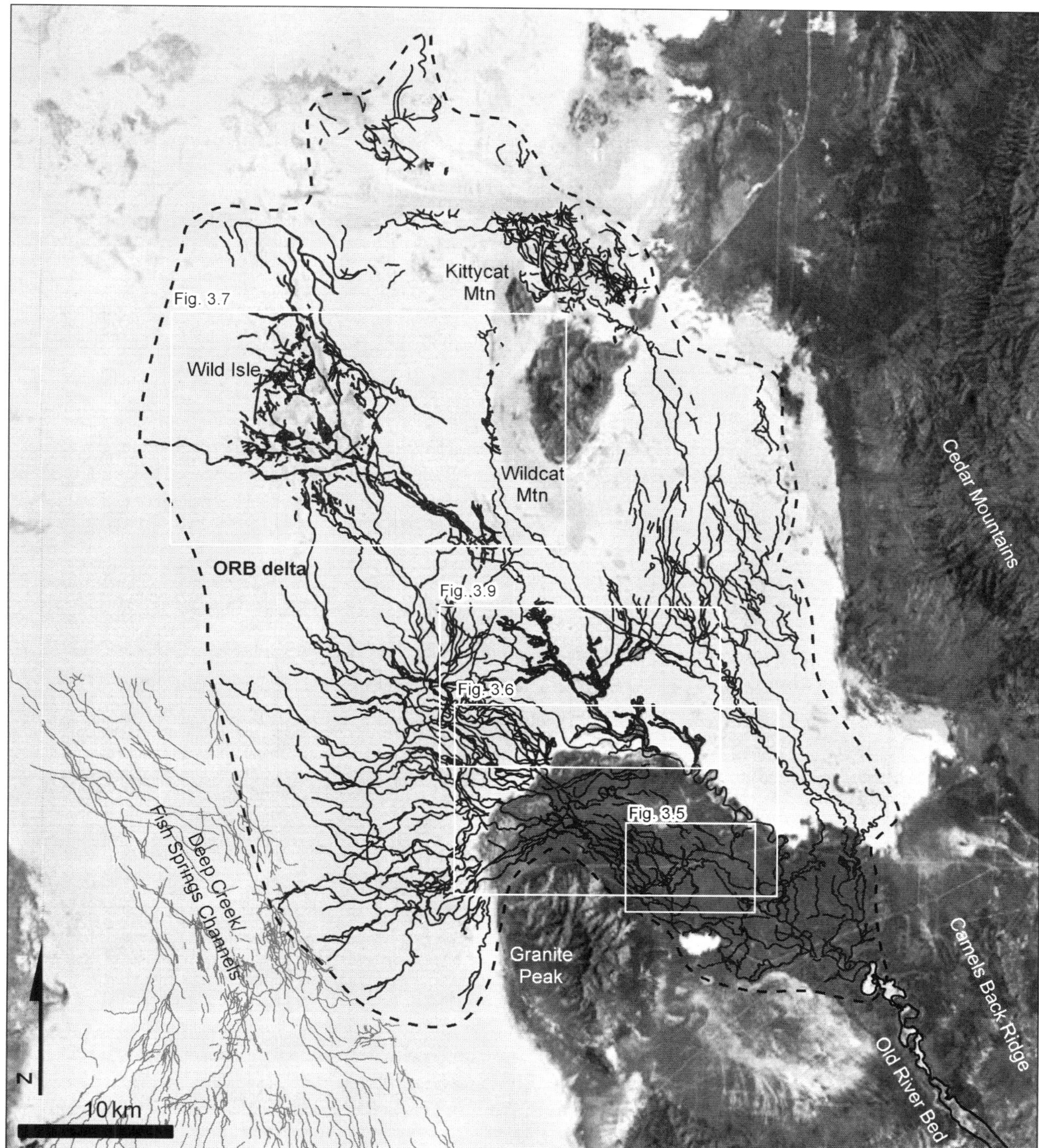

Figure 3.3. Satellite image of the Old River Bed (ORB) delta area showing an outline of the delta and labeling important features (as well as the locations of the images in Figures 3.5–3.7, 3.9).

pretations to a minimum. As the lake regressed, later distributary channels cut through the slightly older RLF deposits. In other areas, deltaic deposits of the ORB delta stratigraphically overlie the RLF (Figure 3.4).

Currey (1982) mapped the Gilbert shoreline on Dugway Proving Ground and identified two Gilbert shoreline points, one just northwest of Granite Peak and the other northeast of Granite Peak (Figure 3.3; Currey's Gilbert shoreline points #20 and #21). However, neither of Currey's points is on a shoreline. Currey's point #21, northwest of Granite Mountain (1,305 m), is actually on Holocene sand dunes, and point #20, northeast of Granite Mountain (1,306 m), is near the upper altitudinal limit of the features we refer to as the

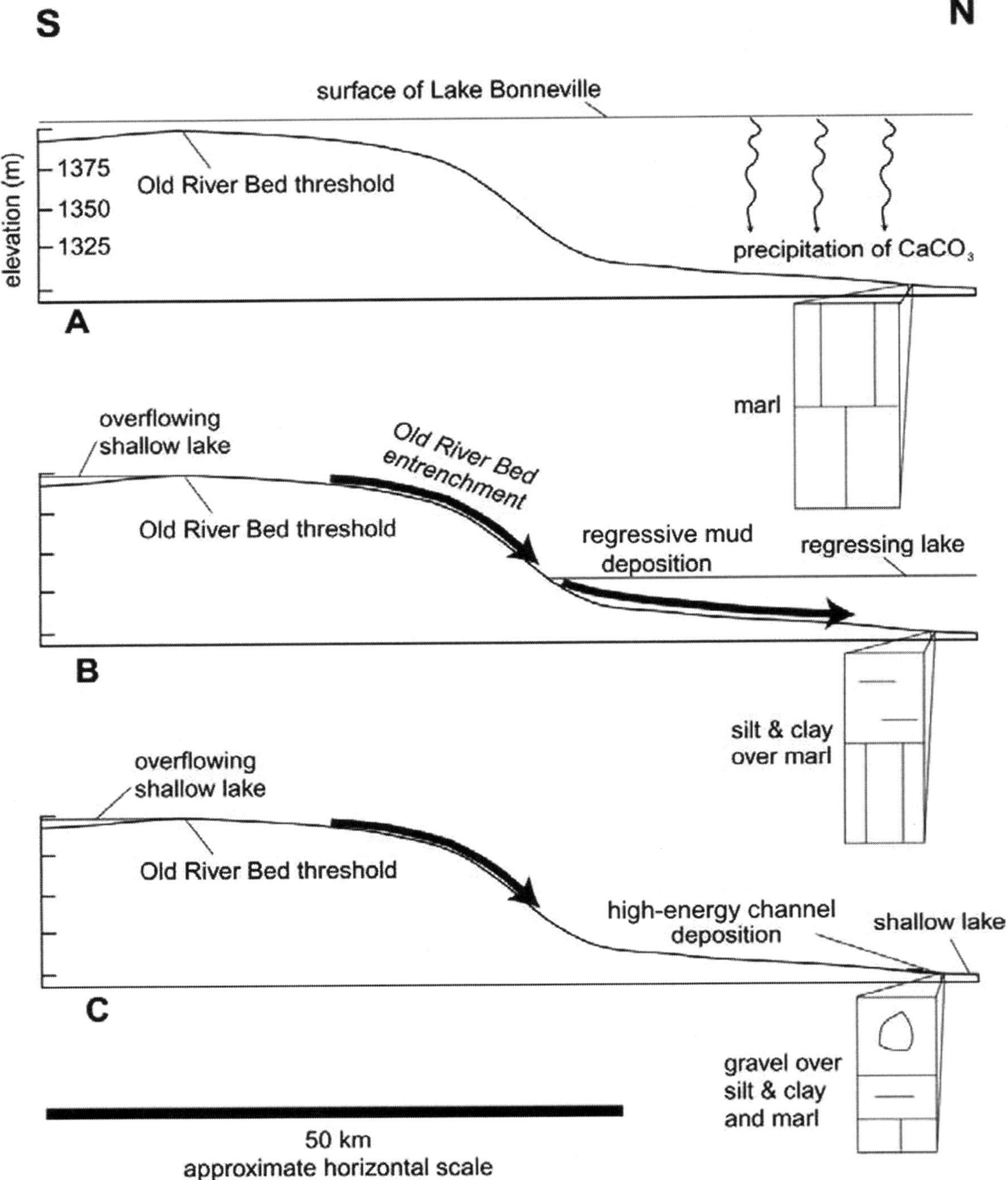

FIGURE 3.4. Schematic diagrams showing changing lake levels during the regression of Lake Bonneville.

high-energy channels. We have found no evidence for a post-Bonneville shoreline or lacustrine deposits at DPG or north of DPG as far as Wild Isle, where altitudes decline to as low as 1,295 m (4,250 ft).

However, the Great Salt Lake (GSL) did rise about 15 m during the event we refer to as the Gilbert episode. The youngest of the organic radiocarbon ages at the base of the Gilbert episode ripple-laminated sand at Public Shooting Grounds (PSG) at the northeast edge of GSL is about 10,100 ^{14}C BP, and a radiocarbon age within the ripple-laminated sand at PSG is about 10,000 ^{14}C BP (Oviatt et al. 2005). Therefore, we interpret the age of the culmination of the Gilbert episode to be about 10,000 ^{14}C BP and think that the episode may have lasted a very short time at its highest altitudes, probably less than a century (within the range of the analytical error of a typical radiocarbon age) (Oviatt et al. 2005). As will be demonstrated below, the fluvial and wetland system at DPG was fully operational before and after the Gilbert episode. Although the Gilbert lake apparently did not reach altitudes high enough to inundate more than the extreme distal DPG delta, it is likely that the climate shift that caused GSL to rise during the Gilbert episode also caused an increase in discharge in the ORB river, and that increase probably affected the fluvial and wetland systems at DPG.

The ORB delta was deposited in fluvial to wetland environments, and the deposits can be categorized as various kinds of channel fills and wetland or marsh sediments. The fluvial channels in the ORB delta were formed in a radiating pattern by freely flowing streams or rivers (see below). We think that this complicated pattern of overlapping and anastomosing channels was created, in part, by the sediment-charged river channels as the river continually shifted in response to changes in sediment load, water discharge, and avulsions in the aggrading fluvial system. Similar processes have been observed in modern rivers (Slingerland and Smith 2004).

The channels are filled with cross-bedded to ripple-laminated gravel or sand deposited by the river, and/or with fine sand and mud, in beds that are draped across the channel margins and channel floors, and these we interpret as slack water deposits. Some wetland or marsh deposits, which consist of sand, mud, or sandy mud, are widespread sheetlike deposits that do not fill channels but cover many square kilometers of the delta area. At all times when the ORB delta was forming, the regional groundwater system was full to capacity, and the water table was above the ground surface. Some of the channels may have been formed by rivers that were generated by surface overflow from the Sevier basin, some may have formed by rivers that were generated by groundwater discharge upstream along the ORB valley (Oviatt et al. 2003), and still other channels on the western margin of the wetlands were formed by streamflow originating in the Deep Creek Mountains. Some of these channels merged with ORB delta channels on the extreme northeastern margin of the distal delta. Extensive wetlands formed in places where the water table exceeded the ground surface over large areas, and river channels probably entered these wetlands at multiple locations. A tremendous volume of sand and mud was delivered to the ORB delta, including its associated marshes, during a relatively short period (~11,500 to ~8800 ^{14}C BP) of the early Holocene.

Lake Gunnison, the shallow overflowing lake in the Sevier Desert, formed as Lake Bonneville dropped below the ORBT. It overflowed at the ORBT because it was fed by two large rivers, the Sevier River and the Beaver River, and by strong groundwater inflow. Lake Gunnison had a small surface area relative to this large inflow. Radiocarbon ages from Lake Gunnison deposits in the Sevier basin range from ~11,400 to ~10,100 ^{14}C BP (Currey 1980; D. S. Kaufman, personal communication 2001; Light 1996; Oviatt 1988; T. W. Stafford and C. G. Oviatt, personal communication 1997). Three radiocarbon ages for *Anodonta* shells collected from spit gravel at Sunstone Knoll near the floor of the Sevier basin at an isostatically rebounded altitude of approximately 1,400 m (Godsey et al. 2011; Simms and Isgreen 1984) average about 12,700 ^{14}C BP, which is within analytical error of the age of the initiation of the regression from the Provo shoreline, less than 50 m higher in altitude. It is likely that overflow began across the ORBT at this time and that the threshold altitude was lowered by erosion to 1,390 m by about 11,400 ^{14}C BP. Apparently, the threshold altitude then stabilized at 1,390 m, and Lake Gunnison overflowed continuously from ~11,400 to sometime after ~10,100 ^{14}C BP.

The oldest radiocarbon age from fluvial and marsh sediments that overlie the Lake Gunnison deposits in the Sevier basin is approximately 9900 ^{14}C BP (Mayer et al. 2010), with a number of other marsh sediment samples dating to ~9500 ^{14}C BP (D. S. Kaufman, personnel communication 2001; Oviatt 1989; Simms and Lindsay 1989), which indicate that Lake Gunnison had dropped below the overflow threshold at least by this time. Three of these four ages are on shell, however, and the chronology and stratigraphy of these wetland deposits remain to be clarified. Between ~10,100 and 9900 ^{14}C BP the source of ORB streamflow is unclear and may have included surface overflow from the Sevier basin, but after ~9900 ^{14}C BP the information currently available indicates that some of the ORB deltaic deposits, including both channel fills and widespread wetland muds, were deposited after Lake Gunnison ceased to overflow and wetlands covered much of the Sevier basin. This basin synthesis has led us to suggest that later fluvial channels, and therefore river flow, in the ORB delta area were fed by groundwater discharge rather than by surface flow from the Sevier basin (Oviatt et al. 2003). There is no local source of water in the Dugway area sufficient to create large rivers, and the only likely source for river water is the Sevier basin.

The modern water table beneath the sheetwash/aeolian surface slopes to the northwest, with an average gradient of .0005 m/m (data from Steiger and Freethy

2001). The water table lies about 5 m beneath the sheetwash/aeolian surface and approximately 1 m below the modern mudflat surface. Along the ORB valley at the southern boundary of DPG, the modern water table lies about 12 m below the floor of the valley (data from Steiger and Freethy 2001). Modern groundwater in the Dugway area is derived largely from the Sevier basin. Groundwater studies in the Dugway area (e.g., Stephens and Sumsion 1978) indicate that more than 40 percent is derived from the Sevier basin via the Old River Bed area, suggesting that even a moderate increase in Sevier basin groundwater could produce renewed surface flow in the ORB valley.

Lake Gunnison, from available observations and radiocarbon ages, apparently overflowed continuously from ~11,400 to after ~10,100 ^{14}C BP, but GSL regressed rapidly during this time period to levels at least as low as modern GSL and then rose briefly during the Gilbert episode (Oviatt et al. 2005). This contrast in water budgets and behavior of the lakes led Oviatt (1988) to suggest different climate controls in the Sevier basin than in the GSL basin: that an enhanced Southwest monsoon affected the southerly Sevier basin but not the northerly GSL basin. Such a hypothesis is supported by climate models (e.g., Mock and Bartlein 1995) indicating early Holocene climatic spatial variability across western North America and by paleobotanical evidence in the Southwest indicating greater-than-present summer precipitation between 12,000 and 9000 ^{14}C BP, consistent with enhanced monsoonal circulation (Davis and Shafer 1992; Spaulding and Graumlich 1986; Weng and Jackson 1999). This hypothesis has not been sufficiently tested to determine its viability, however (but see Jahren et al. 2001). An alternative hypothesis is that the large reservoir of stored groundwater from Lake Bonneville was slow to run out of the system and that groundwater discharge maintained lakes and wetlands during the late Pleistocene and early Holocene, regardless of the regional trends toward dryer and warmer climate. Certainly, early Holocene wetlands were healthy in many places on the floor of the Bonneville basin, including such areas as the ORB delta, the Sevier basin, Tule Valley, Blue Lake, Locomotive Springs, PSG, Jukebox Spring, and Salt Lake Valley.

Mudflats in the northern part of the ORB wetlands extend for many kilometers to the north and west as part of the extensive GSL Desert (Figure 3.3). An abrupt boundary, in many places marked by aeolian sand dunes, separates the groundwater-discharge mudflats from the well-drained fine-grained sheet flow and aeolian deposits on the flat desert floor (Figure 3.3). This interior, southern proximal portion of the ORB delta supports a sparse cover of xerophytic and phreatophytic shrubs (primarily *Atriplex confertifolia* and *Sarcobatus vermiculatus*). The flat desert floor in this portion of the ORB delta is characterized by well-developed vegetation stripes ("desert ripples" or "tiger bush" [Ives 1946; Wakelin-King 1999]), and the surficial deposits on the flat are constantly redistributed by sheet flow and aeolian processes—no ephemeral channels exist.

Trench exposures indicate that this surface layer of reworked aeolian material averages ~1 m thick and obscures subsurface evidence of ORB distributary channels. Their location can, however, be estimated in some locations by differences in vegetation probably related to differences in groundwater flow through the coarser fluvial sediments within the channels (Figure 3.5). Low aeolian dunes are scattered over a broad area and overlie the sheet flow/aeolian deposits. As a result, the ground surface gets closer to the water table toward the north and northwest, and the mudflat margin marks the point where the two surfaces meet. We use the descriptive terms *exposed channel section* and *unexposed channel section* to distinguish these distinctly separate mudflats and sheet flow/aeolian-covered sections of the ORB delta.

Mudflat Deflation

The deflation of the mudflats, which has exhumed the distal end of the distributary channels and truncated the finer sediments in the lower-energy channel forms, is likely related to the proximity of the water table to the surface of the mudflats. Both the ground surface and the water table slope to the north or northwest across the ORB delta, but the ground surface slopes more steeply (.0006 m/m for the ground surface and .0005 m/m for the water table), and the two surfaces intersect. This intersection is marked by the sharp boundary between the mudflats and the flat plain covered with vegetation stripes. This boundary is accentuated by sand dunes and rises in elevation from about 1,300 m directly west of Granite Peak to about 1,313 m on the eastern margin of the unexposed channel section of the ORB. The mudflats are a landscape that is actively undergoing denudation, as shown by the topographic reversal of the distributary channels and by the intricate surface texture evident in aerial photographs. Aerial photos show the surface to be a palimpsest that records a complex geomorphic history following the deposition of Lake Bonneville deposits (the deepwater white marl and the silts and clays of the RLF).

Several processes contribute to the denudation of the mudflat surface (Reynolds et al. 2007). The mud is moistened by near-surface groundwater discharge via capillary action. This probably occurs continuously, but

FIGURE 3.5. Close-up aerial view of channels in the unexposed section of the ORB delta.

the surface dries out during the hot summer when evaporation far exceeds the groundwater discharge rate. This wetting and drying causes the mud to swell and shrink and to pelletize, thus creating loose particles that can easily be entrained by the wind and removed. The larger particles accumulate in dunes at the mudflat margin, and the smaller particles are blown much farther. Salt precipitation in the mud may contribute to the breakup of the mud. On the uneroded section of the ORB delta, on the sheetwash/aeolian surface, the water table is too deep to allow capillary water to rise to the surface and evaporate, so the shadscale and greasewood are able to keep the surface relatively stable.

Hydroaeolian planation (Currey 1990) probably also contributes to the denudation process. Shallow films of water accumulate on the mudflat surface after heavy rains and are agitated by the wind, thus loosening particles that are easily deflated when the surface dries. These water-related processes have accentuated mudflat denudation on the western margin of the ORB delta, where surface elevations are lower, groundwater discharge is higher, and rainfall runoff accumulates. In this area, the distal ends of most distributary channels have been entirely eroded, and channel forms cannot be traced below elevations of about 1,289 m (4,230 ft). A water table close to the surface under the mudflats is apparently an important control on the level to

which deflation erodes the mudflats and is an essential factor in allowing deflation to take place (Reynolds et al. 2007).

The amount of mudflat deflation since the lake deposits were first exposed varies from place to place. Based on the degree of topographic inversion of the largest exposed channel features in the ORB delta (as directly measured by a total station), and assuming that the tops of these channels represent the former surface of the sediments enclosing the channels, at least 4 m of deflation of fine-grained sediments adjacent to the high-energy channels has occurred in upstream locations. We hypothesize that the gravel in the high-energy channels was concentrated in the main channel but interfingered laterally with fluvial sands and muds that were deposited at the same time. Both the channel gravels and the fines accumulated vertically, and channel entrenchment was not involved. Later removal of the fines by deflation left the coarse sands and fine gravels in the channel. Where dunes have protected the fine-grained deposits of the low-energy channels, they are topographically inverted about .4–1.2 m. If these dunes formed shortly after the ORB delta system dried up, then little more than 1 m of the upper mudflats has been removed in the last ~9,000 ^{14}C years. Deflation has not been uniform across the mudflats, however, and in many places even less than a meter of sediment has been removed. In a landform

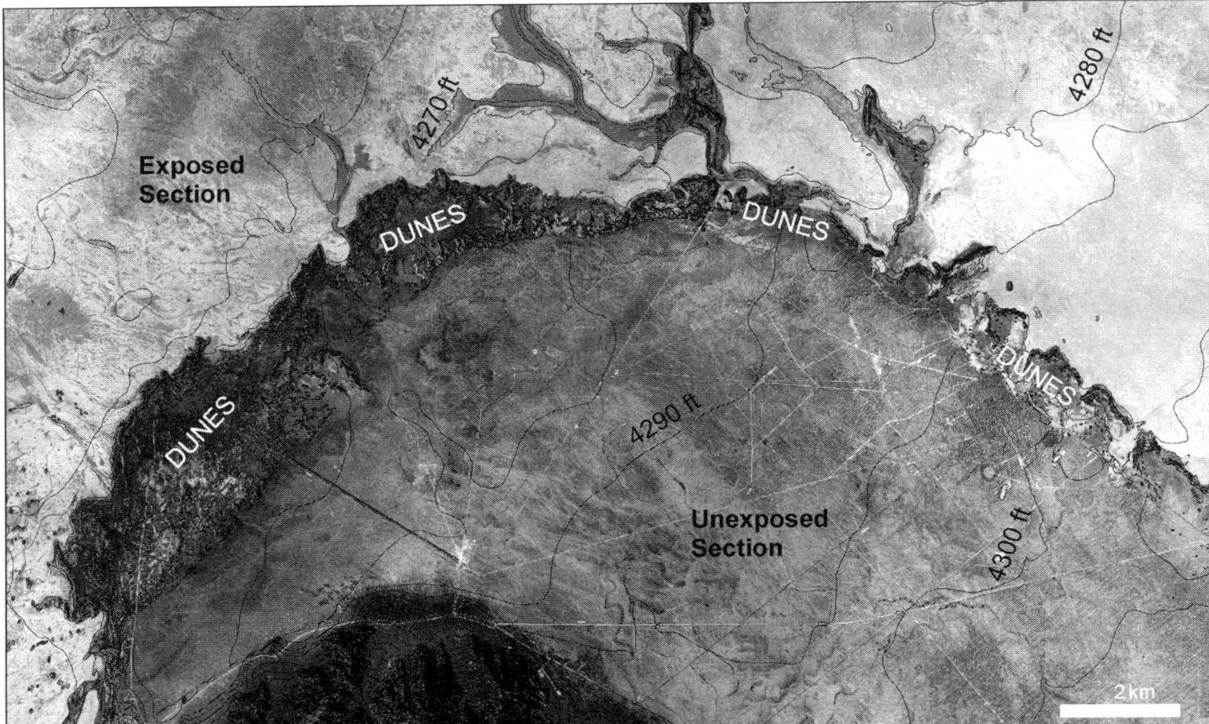

FIGURE 3.6. Satellite image of dunes around the unexposed channel section of the Old River Bed delta.

we informally call "Lake Öferneet," a mudflat area completely surrounded by the coarse sands and gravels of high-energy channels, our elevation measurements indicate that the mudflats are approximately 1 m higher in the interior of the "lake" than they are immediately to the west.

Aeolian Dunes

The boundary between the exposed and unexposed sections of the proximal ORB delta is covered by a .5- to 2.3-km-wide band of aeolian dunes that average ~1,308 m in altitude and stand 12–13 m above the surrounding mudflats (Figure 3.6). The dune band is thickest and highest on the western edge of the boundary. Aeolian dunes also cap areas in the distal reaches of the ORB delta at Wild Isle (Figure 3.7), and expansive active dune systems, such as the West Wildcat Dune (Figure 3.8), extend from the mountain fronts to cover the delta's margins. Extensive dune formation probably started after ~8800 ^{14}C BP when flow in the ORB river finally ended and the ORB delta ceased to form. As we discuss elsewhere in this monograph, while water was flowing through the ORB system, it supplied a network of open flowing distributary channels and groundwater-fed channels. These, in turn, supported a vast wetland/marsh ecosystem that covered most of what is now the GSL Desert, to the west and north of the dune line now

separating the exposed and unexposed sections of the ORB delta, and supported vegetation that protected the fine-grained sediments of the mudflats from deflation. More intensive study of the dune fields postdating the ORB delta will be required to confirm this hypothesis, but certainly dunes were in place by the mid- to late Holocene.

Wild Isle, a small isolated dune field on the mudflats northwest of the main dune band, initially formed as small dunes exposed to wetlands and distributary channels. These small seed dunes eventually coalesced to form a large dune complex on the distal distributary delta. The Wild Isle dune complex developed throughout the Holocene, and at present, deposition and migration continue as individual dunes coalesce along the abandoned channel system. Well-defined arcuate silt dunes are evident on the basin floor west of Wild Isle and southwest of the larger Wildcat dune complex. These isolated dunes are "in motion" (over the long term), with a general direction of migration and sediment transport to the north-northeast. Over time, dune migration brings these dune sediments onto the anchored landform of Wild Isle. Although the dunes are now coalesced, the south-southeast-to-north-northwest trend of individual dunes (evidence of southwest-to-northeast movement) is still apparent in aerial and satellite images.

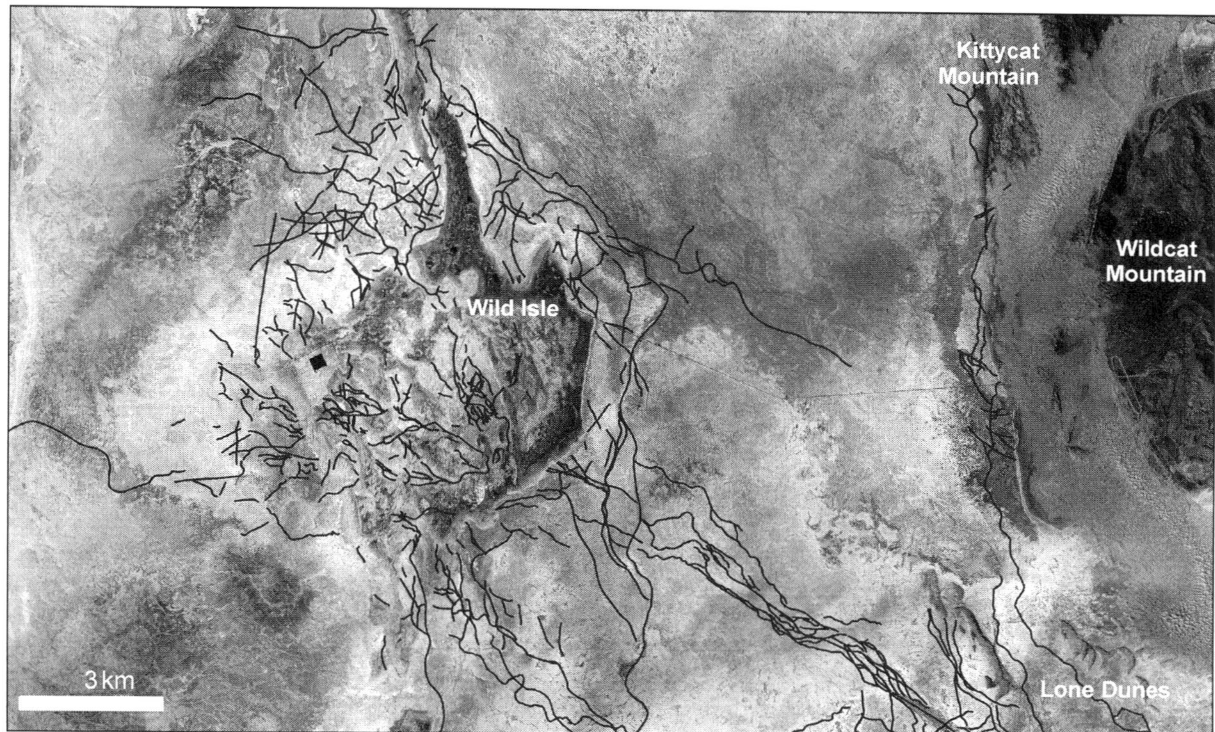

FIGURE 3.7. Close-up aerial view of Wild Isle.

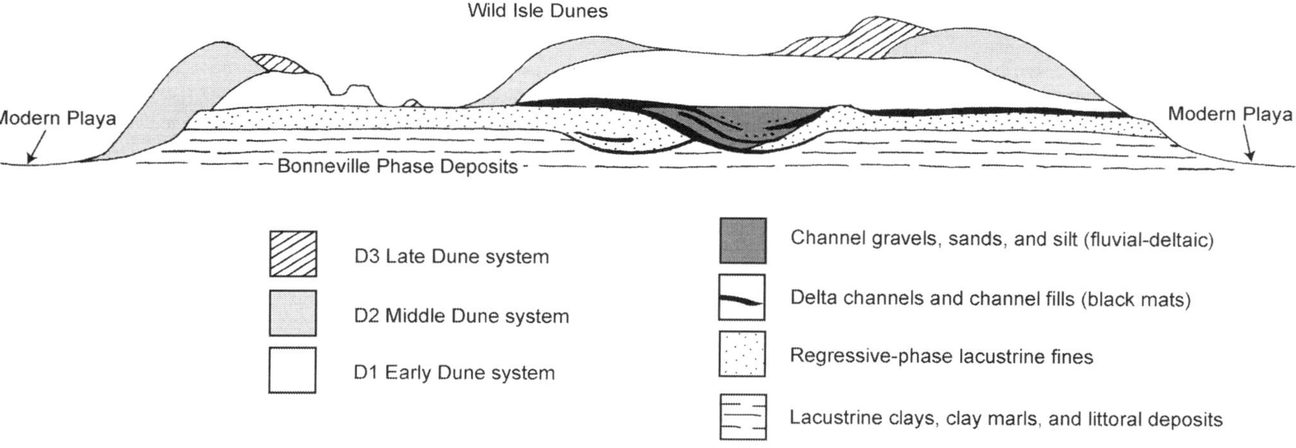

FIGURE 3.8. Schematic diagram of preserved channel settings at Wild Isle.

Archaeological deposits in dunes on both the western and eastern margins of the GSL Desert date to as early as about 4500 ^{14}C BP (Simms et al. 1999) and ~4340 ^{14}C BP (Madsen and Schmitt 2005), with dune formation beginning sometime prior to these ages. Chronologically diagnostic Northern Side-notched projectile points dating to 7500–5500 ^{14}C BP (Holmer 1986; Schmitt and Madsen 2005) on these dunes suggest initial dune-formation dates in at least the early Holocene. Throughout the Archaic to Late Prehistoric periods these dunes were a magnet for human occupation, with foraging likely focused on the collection of Indian ricegrass (*Achnatherum hymenoides*) and other seed crops supported by the moisture held by such dunes (e.g., Madsen and Schmitt 2005). In the expansive West Wildcat dune field a hearth dated to ~3390 ^{14}C BP rests on dune deposits and is now buried due to later dune activity (Carter et al. 2005:98–99). A hearth dated to 240 ^{14}C BP is almost the only evidence of very late use at Wild Isle (Carter and Young 2002:166). As a

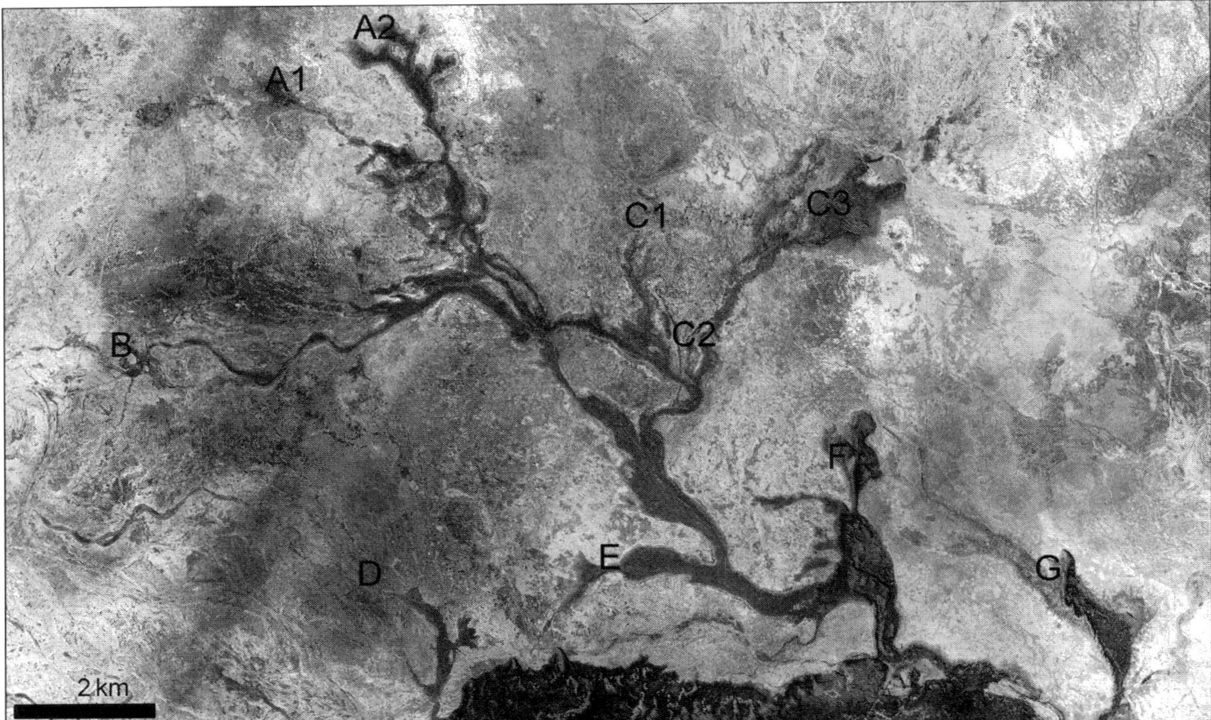

FIGURE 3.9. Close-up aerial view of the high-energy Black channel system, with specific channels labeled.

result of this Holocene occupation of the dunes, artifacts produced after the ORB system was abandoned occasionally occur on and around channels in the exposed channel section of the ORB delta along the fringe of the dune field. These are rarely found away from dune margins, however.

In the vicinity of the Wild Isle dunes, the erosion and deflation of the surrounding playa by wind, water, and ice have lowered the playa and delta surface approximately 1 m since dune accumulation began (Figure 3.8). As a result, the Wild Isle dunes are essentially pedestaled above the modern playa and continue to rest on remnants of the distributary system. Blowouts and openings within the Wild Isle complex expose the old channel system and its associated archaeological record. In most cases, from the proximal ORB to beyond Wild Isle, archaeological components are not eroding from the dunes; rather, they form lag deposits below the dunes. In rare instances they may be well stratified at the contact between the deposits of the ORB and the younger dunes. It is here that bedforms of the distal ORB are preserved.

Distributary Channel Forms

The distributary channels in the ORB delta form a continuum between those made by large high-energy streams or rivers carrying heavy, coarse sediment loads and those made by small low-energy streams carrying much finer-grained sediments. The high-energy channels contain deposits of coarse sand and gravel and, in plan view, are straight to curved and digitate. Some channels have both bulbous and constricted sections along their lengths (Figure 3.9). In transverse cross section, the high-energy channels are topographically inverted, with the crests of the gravel deposits standing 1 to 4 m higher than the surrounding mudflats. They are identified as fluvial in origin by their plan-view form (digitate channel form rather than linear beach form), the composition of the gravel (dominated by volcanic clasts derived from about 50 km to the south along the trend of the ORB), and their longitudinal profiles (they slope gently to the northwest at .0006 m/m, slightly steeper than the mudflat surface). The coarse sand and gravel of the channels is trough cross-bedded, typical of braided stream deposits, and overlies lacustrine mud and marl of Lake Bonneville or organic-rich wetland deposits. The geomorphology and sedimentology of the high-energy channels (braided stream deposits), and their bracketing ages, suggest that they were produced rapidly by bed load streams that debouched onto mudflats and wetlands.

Figure 3.10 illustrates an area of mudflats where at least two, and possibly three, high-energy channels overlap. At this location, channel B overlies, and is therefore

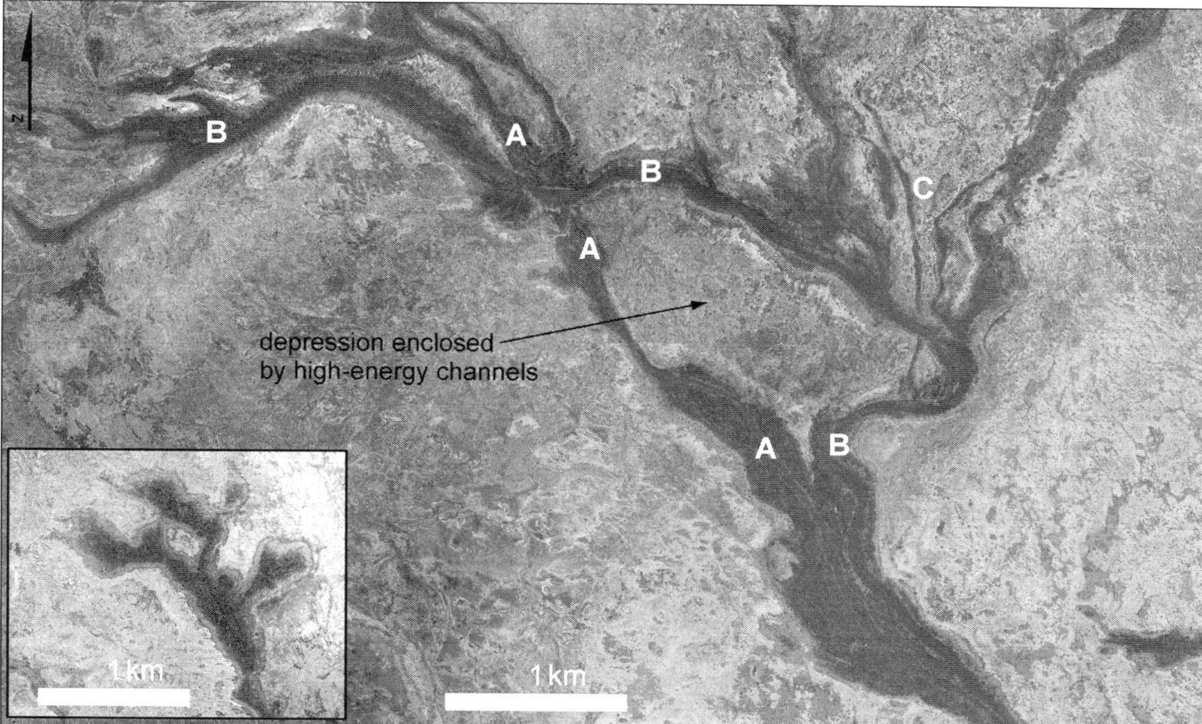

FIGURE 3.10. Vertical aerial photograph of a section of the high-energy Black channels, with specific channels labeled.

younger than, channel A. Channel B is 1.5 to 2 m higher than A where they overlap. Channel C is older than B and could be the same age as A or older. Two radiocarbon samples on a black mat stratigraphically below the northern end of channel A dating to ~11 ^{14}C ka (see below) show that channel A and potentially channels B and C were formed after wetlands had formed in this area.

Low-energy channels are found in the same general area of the mudflats as the gravel channels but are less topographically inverted and are truncated by the erosional mudflat surface. Where they have been protected by dunes, they may stand as much as 1.2 m above the surrounding mudflats, and in other areas they typically stand about .5 m above the mudflats. Many low-energy distributary channels are not easy to identify on the ground because the mudflat surface has been deflated and any fluvial landforms (such as floodplains, natural levees, and point bars) that may have originally existed have been eroded away. In aerial photographs, however, the preserved roots or cores of the low-energy channel deposits exhibit well-developed meander scroll patterns (Figure 3.11). Sediments in these low-energy channels consist of fine to coarse cross-bedded sand and locally include mud that contains organic mats and mollusks.

The high- and low-energy channel forms represent end-member categories, but intermediate forms also exist and actually are the most common channels in the ORB delta. They are straighter and larger in width than the low-energy channels, locally contain some gravel, and are not as topographically inverted as the gravel-armored high-energy channels. Many of these have braided stream morphology similar to the higher-energy channels. We provisionally used the descriptive terms *gravel channels* and *sand channels* in our initial reports (e.g., Oviatt et al. 2003) to distinguish the high- and low-energy channel forms. However, due to the prevalence of intermediate forms, and because these terms also had chronological implications that we have now revised (see below), we here simply refer to individual channels in terms of the relative size of the rivers and streams that formed them. Because some of these rivers and streams were a product of overflow from Lake Gunnison in the Sevier basin, the amount of water in them was determined by inflow from all the drainages that feed the Sevier basin (e.g., the Sevier and Beaver rivers) plus rainfall on the lake, minus evaporation and groundwater losses from the lake. In general, the ratio of input to output in the Lake Gunnison system was quite high during the initial stages of ORB delta formation but low during stages immediately prior to the regression of Lake Gunnison below the ORB threshold and the final drying of the ORB delta ecosystem. However, as we discuss below, there was substantial variation

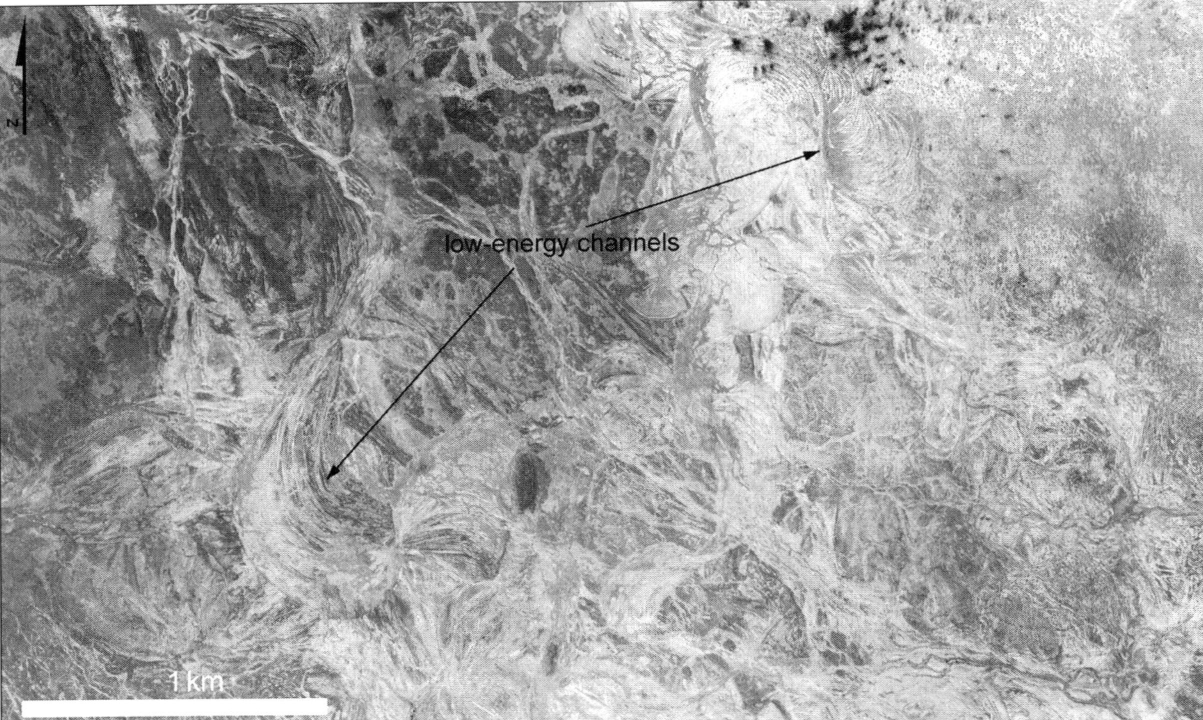

FIGURE 3.11. Vertical aerial photograph of low-energy channels on the mudflats.

in overflow during the approximately 3,000-^{14}C-year period when the ORB delta formed, and high- and low-energy channels were constructed throughout the delta.

GEOMORPHOLOGY OF THE PROXIMAL ORB DISTRIBUTARY CHANNELS

We investigated the geomorphology of the ORB delta episodically over a 10-year period from 1998 to 2008. These investigations essentially took two forms. We mapped distributary channels in the ORB delta through careful examination of aerial photographs, converted channel coordinates into GPS-readable formats, and then followed the channels on the ground from their proximal ends at the interface between the unexposed and exposed channel sections of the ORB delta to their distal ends where they were obscured by mudflat deflation. In the unexposed channel section of the delta, where channels are covered by ~1 m of reworked aeolian deposits, channels were mapped using variations in surface vegetation that reflect differences in subsurface sediments and available groundwater. Once the distributary channels were mapped, we then investigated channel forms, sediment composition, and chronology by cutting backhoe trenches into a variety of channels or, in some cases, by using existing exposures (e.g., in gravel pits). We also sampled sediments exposed on the surface and excavated shallow exploratory pits by hand.

AERIAL PHOTOGRAPHY AND CHANNEL MAPPING

Channels in both the exposed and unexposed channel sections of the ORB delta were initially identified through a combination of aerial photography and satellite imagery. The Esri geographic information system software package ArcMap 9.x and various forms of digital imagery and conventional aerial photographs were used in mapping efforts. These images include 1:40,000-scale, black-and-white U.S. Department of Agriculture Farm Service Agency aerial photographs; 1-m-resolution natural-color National Agriculture Imagery Program orthoimagery from 2006 for Tooele County; 1-m-resolution, black-and-white digital orthophoto mosaics of Tooele County; 30.5-cm-resolution natural-color digital orthophotos from 2002 of U.S. Army Dugway Proving Ground; and various images obtained from the virtual globe program Google Earth. Contrasts between landforms and surface vegetation differ significantly between each source, but by using multiple sources of information at different scales it was possible to trace an extensive network of exposed and unexposed ORB deltaic channels.

A major goal of the channel-mapping exercise was to identify any crosscutting relationships among the channels in order to establish the relative ages of channel formation. Once channels were mapped digitally and data were available in geographic information system format, major, traceable channels were examined on foot and through the use of all-terrain vehicles to verify cross-cutting relationships identified in the aerial and satellite imagery. In selected locations, crosscutting relationships suggested by aerial photography were confirmed by backhoe trenching. This task was made possible by converting the linear channel shapefiles into Trimble-compatible files using Trimble GPS Pathfinder Office 2.90. These files were imported into a real-time differentially corrected Trimble Pro XRS GPS unit and viewed as background files that were traced during pedestrian survey. To date 3,839 linear km (2,385 mi) of channels have been identified and mapped in the exposed and unexposed channel sections of the delta.

During the initial stages of this channel-mapping process it became evident that the extensive meandering and bifurcation of the countless channels in the ORB delta distributary system made it nearly impossible to track individual channels using a single color to represent all the channels in the ORB delta. We therefore began a process of color-coding individual channel systems in order to provide easier visual recognition and to allow comparison of archaeological sites along each channel as a group with other similarly defined groups of relatively different ages. As of this writing we have mapped and color-coded 23 principal exposed channel distributary systems north and west of the dune/mudflat interface separating the exposed and unexposed channel portions of the ORB delta (Figure 3.12). Only a portion of these 23 mapped channels were investigated, however. Channels we studied, either geomorphologically or archaeologically, are listed in Table 3.1.

To some extent, the color-coded lines marking channel routes in Figure 3.12 misrepresent the nature of the ORB distributary system. Most of the channels, particularly the earlier higher-energy channels and many of the intermediate-energy channels, consist of broad, braided channel systems 100 m or more wide, each containing multiple small flow channels and/or meanders within the larger channel margin. All of these systems would be better represented as polygons, as is the case of the high-energy Black channel system shown in Figure 3.12. Representing the channels in such a way, however, is both visually confusing and somewhat misleading, as mudflat deflation and the encroachment of young channels make it difficult to follow the original margins of many channel systems. We therefore stay with the linear representation of most channel systems but note that these lines represent broader channels as illustrated by the Blue and Green channel example in Figure 3.13.

TABLE 3.1. Studied Channels on the Distal Margin of the Old River Bed Delta.

Channel Color Code	Total Mapped Length (km)	General Location
Mango	5.5	Northwest
Mocha	3.9	Northwest
Gold	57.1	Northwest
Black	256.3	North
Limestone	5.6	North
Yellow	57.8	Northwest
Fuchsia	9.2	Northwest
Green	81.4	Northwest
Red	58.3	Northwest
Blue	42.1	Northwest
Lime	17.2	Northwest
Royal Blue	5.6	Northwest
Lavender	66.7	Southwest
Navy	20.2	Southwest
Coral	17.9	Southwest
Orange	52.6	Southwest
Pink	62.7	Southwest
Buff	18.3	Northwest
White	6.6	Northwest
Brown	11.9	Northwest
Light Blue	70.6	North
Seafoam	125.0	Northeast
Rust	53.8	Northeast
Total	1,106.3	

Note: Color terminology is derived from the palette of the software used to produce Figure 3.12.

These exposed channels do not encompass the whole range of distributaries present within the delta, but only the prominent ones that are traceable from the exposed/unexposed margin north of Granite Peak into the distal reaches of the delta. There are many segments of discontinuous channels between mapped channels that have been obscured both by younger channels and by differential mudflat deflation (Figure 3.14). These channels fit within the channel development sequence of the larger deltaic system but were not mapped as part of this research because they are simply impossible to trace for any great distance. We estimate that the total length of these unmapped channels is more than three or four times that of the mapped channels.

Buried channels in the unexposed channel portion of the ORB delta are even less visible than those beyond the dune margin and can only be recognized by faint changes in vegetation such as linear strands of slightly taller greasewood (*Sarcobatus vermiculatus*) or by the presence of pepperweed (*Lepidium latifolium*). Even

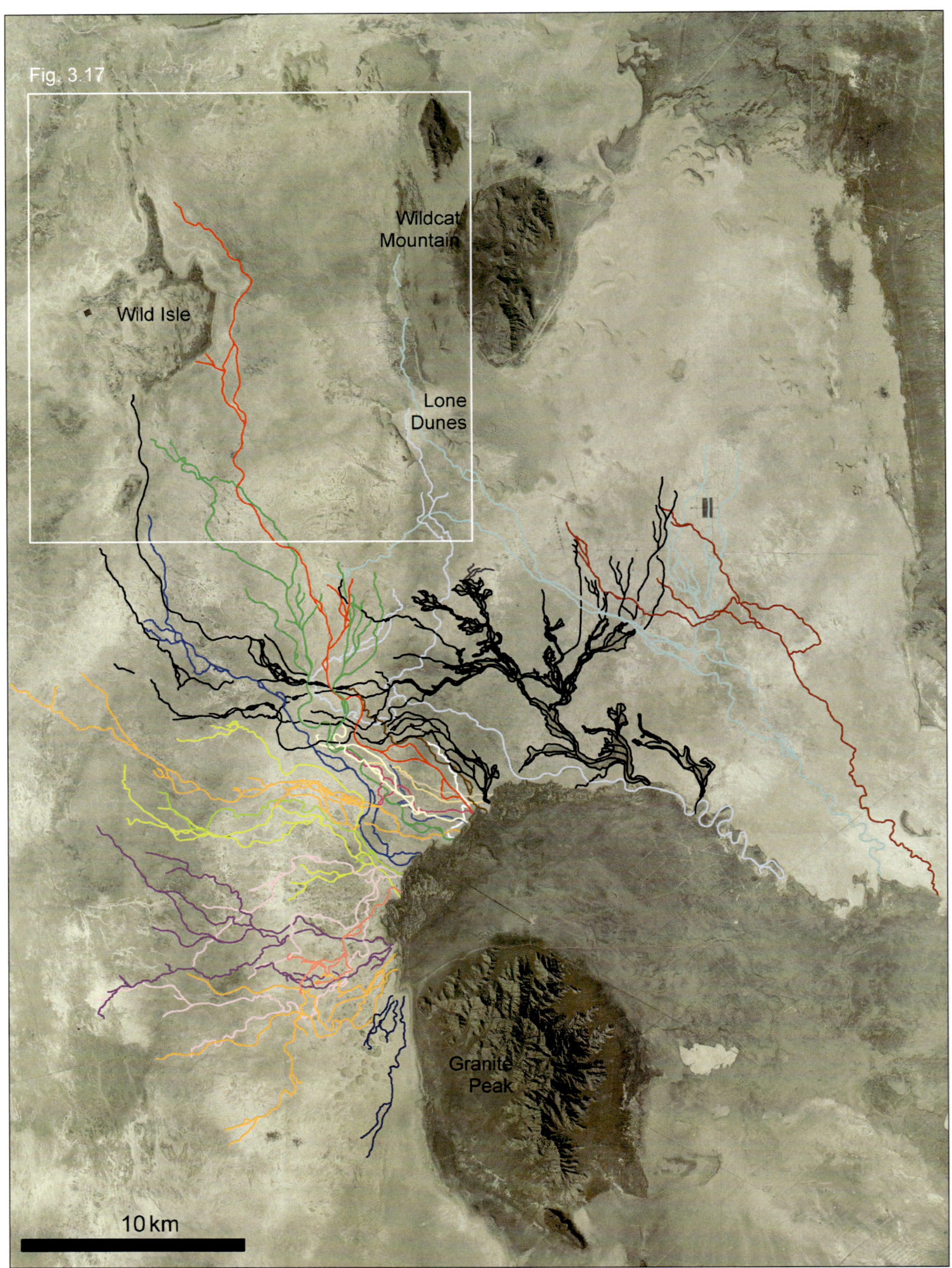

FIGURE 3.12. View of the color-coded channel system in the exposed distal channel section of the Old River Bed delta.

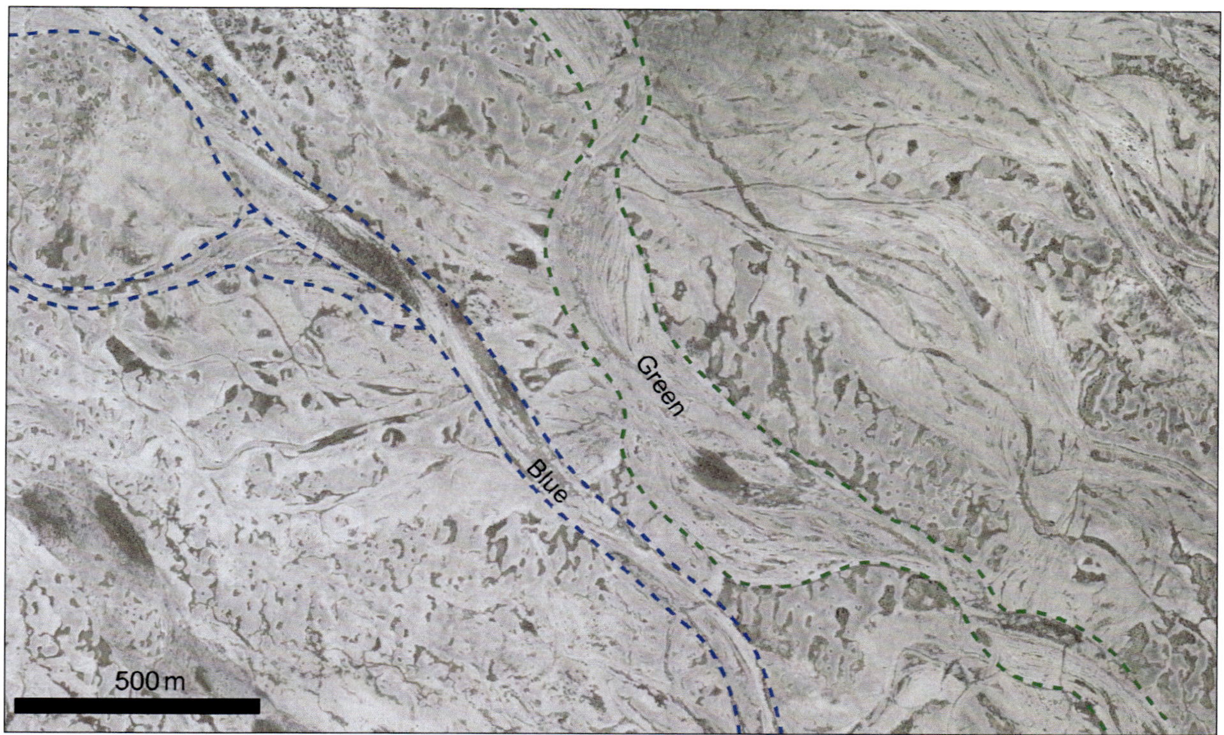

FIGURE 3.13. Close-up aerial view of the Blue and Green channel sections (showing the location of the image in Figure 3.17).

FIGURE 3.14. View of the chaotic background pattern of channels between mapped channels.

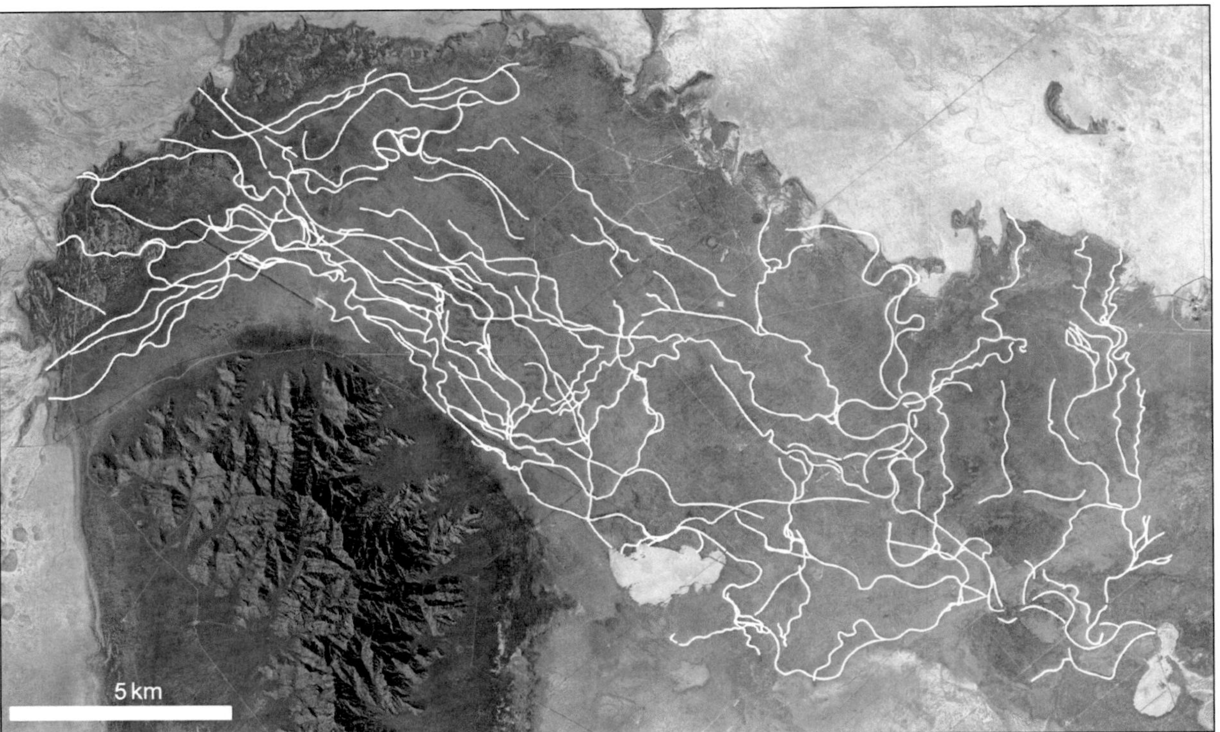

FIGURE 3.15. Probable locations of buried channels in the unexposed section of the ORB delta.

these vegetation differences are difficult to trace for any great length due to facies changes in the channels themselves and to the ~1 m of sheetwash/aeolian silt that covers them. As a result, we made no attempt to color-code individual channel systems within the unexposed channel section of the delta because it is impossible to tell how the small, mappable channel sections relate to one another. We have mapped a total of 442 linear km of buried channels in the unexposed channel section of the ORB delta (Figure 3.15).

We excavated, sampled, and described 32 backhoe trench locations and eight hand-dug or surface sampling localities in the proximal ORB delta (DPG) during five field seasons (Figure 3.16). Sediment descriptions, profile drawings, and photographs of these trench locations are available online (Chapter 3, Supplemental Digital Material; see link information at the end of Chapter 1), while descriptions of the remaining localities are available in Oviatt 1999; Oviatt and Madsen 2000, 2001; and Madsen and Page 2008, 2009.

GEOMORPHOLOGY OF THE DISTAL ORB DISTRIBUTARY CHANNELS

The distal segments of the expansive ORB delta have mostly disappeared. In a few places, however, remnants of the distal delta are preserved beneath Holocene-age dunes. Today, the Wild Isle dunes and portions of the Wildcat dune cap and preserve a small portion of the distal distributary system (Figure 3.17), and because the dunes protect the remnant channels from the intensive deflation affecting the surrounding playa, they provide a unique laboratory in which to study the distal portion of the ORB. The system (i.e., groups of channels, terraces, levees, and splays) is best preserved where channels are buried by early dune accumulations or where local deflation has left channel features as raised topography. In fact, the Holocene-to-modern Wild Isle dunes rest on a pedestal that rises 1 to 2 m above adjacent playa surfaces. It is the subdune pedestal that retains channels and deposits of the former delta.

Archaeological studies in and around Wild Isle have relied on the assumption that wetland resources were once present in the area. Soil horizons or organic-rich deposits (e.g., black mats) of former wetlands are susceptible to surface deflation and scouring; however, where marshes were associated with the channel system, black mats are preserved where sediments that were deposited soon after desiccation of the wetlands are now capping the old channels. The wetland organics can be dated and, therefore, provide clues to the chronology of the distal distributary system and of human use of the delta. Below, we describe discrete channel groups and specific localities where type sections have been studied and sampled.

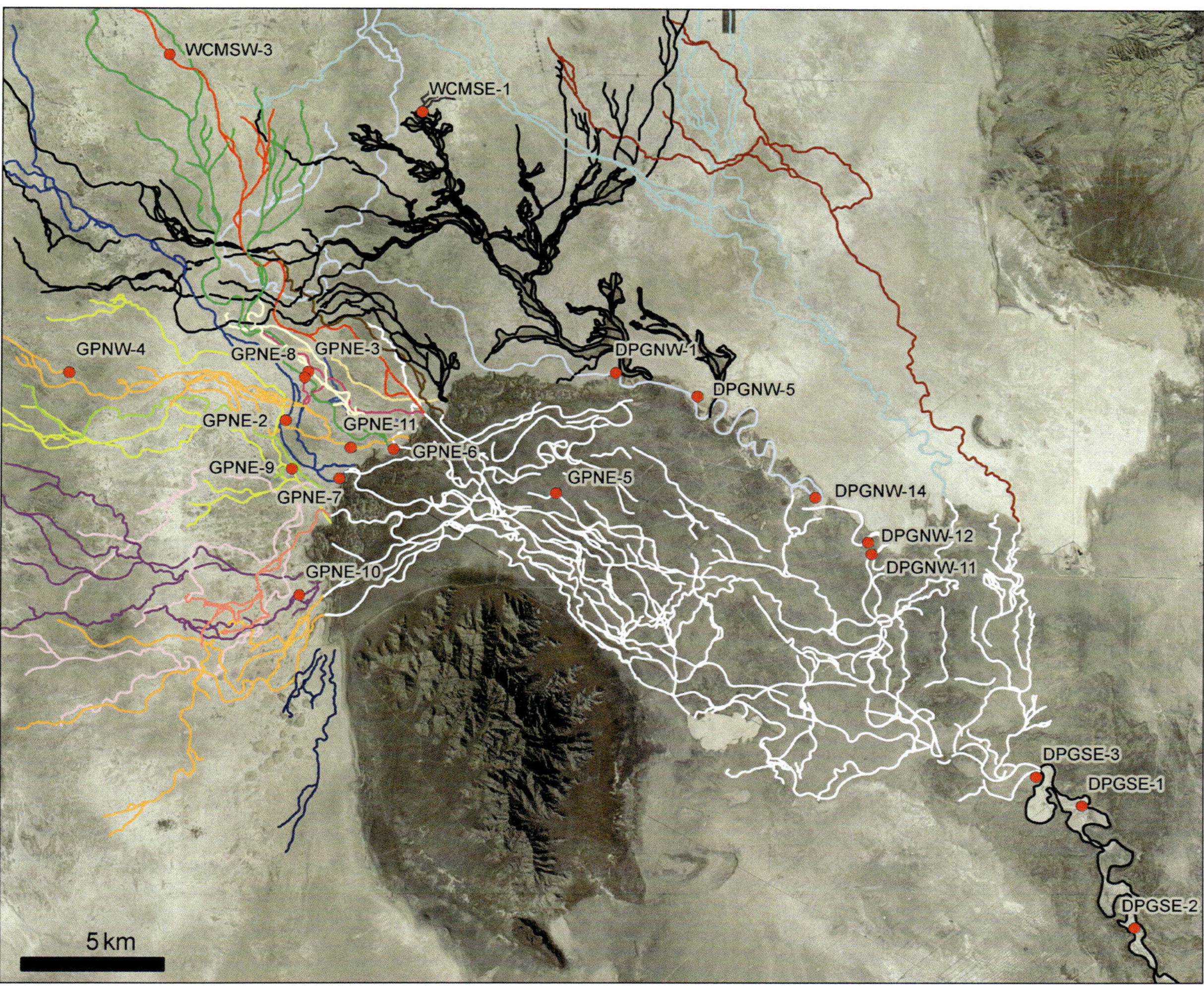

FIGURE 3.16. Trench and surface sampling locations in the proximal Old River Bed delta.

Study Localities

Distal channels or channel segments have been documented at 23 localities within and around the Wild Isle Dune, at Lone Dune, and along the western margin of the Wildcat Dune. A few localities and channels have been studied in detail using backhoe trench exposures. The channels consist of shallow sand-filled swales with locally complex cut-and-fill sequences, ponded facies, splay deposits, braid-like bars and swales, and local terrace sets. Younger, including modern, channels are often captured by older berms and terraces such that older fill sequences acted as channel-bounding levees for current drainage patterns.

Although both the Wild Isle and Lone dunes and the channels preserved beneath them have a general southeast-to-northwest trend, channel directions are independent of Holocene dune orientation, and individual channels can be traced beneath the dunes. Differential erosion, influenced by the presence of capping dunes, has resulted in variable preservation of channel features. Beyond the northern margin of the dunes, channel features have been completely eroded. In several instances, however, remnants of the eroded features are evident due to textural differences in the modern playa surface; this can be easily seen as complex bedforms and meander scrolls in aerial photos (Figure 3.11). Channel features are well preserved in blowouts within the Wild Isle dune complex and in relatively recent exposures along the southern margin of the dunes. Brief descriptions of the sediments and associated age estimates for these localities are available online (Chapter 3, Supplemental Digital Material).

Initial Distal Channels

Although the chronological relationships of the channels preserved on the distal delta are tentative, it is possible, based on stratigraphic position, geomorphic structure, and limited chronological data, to classify

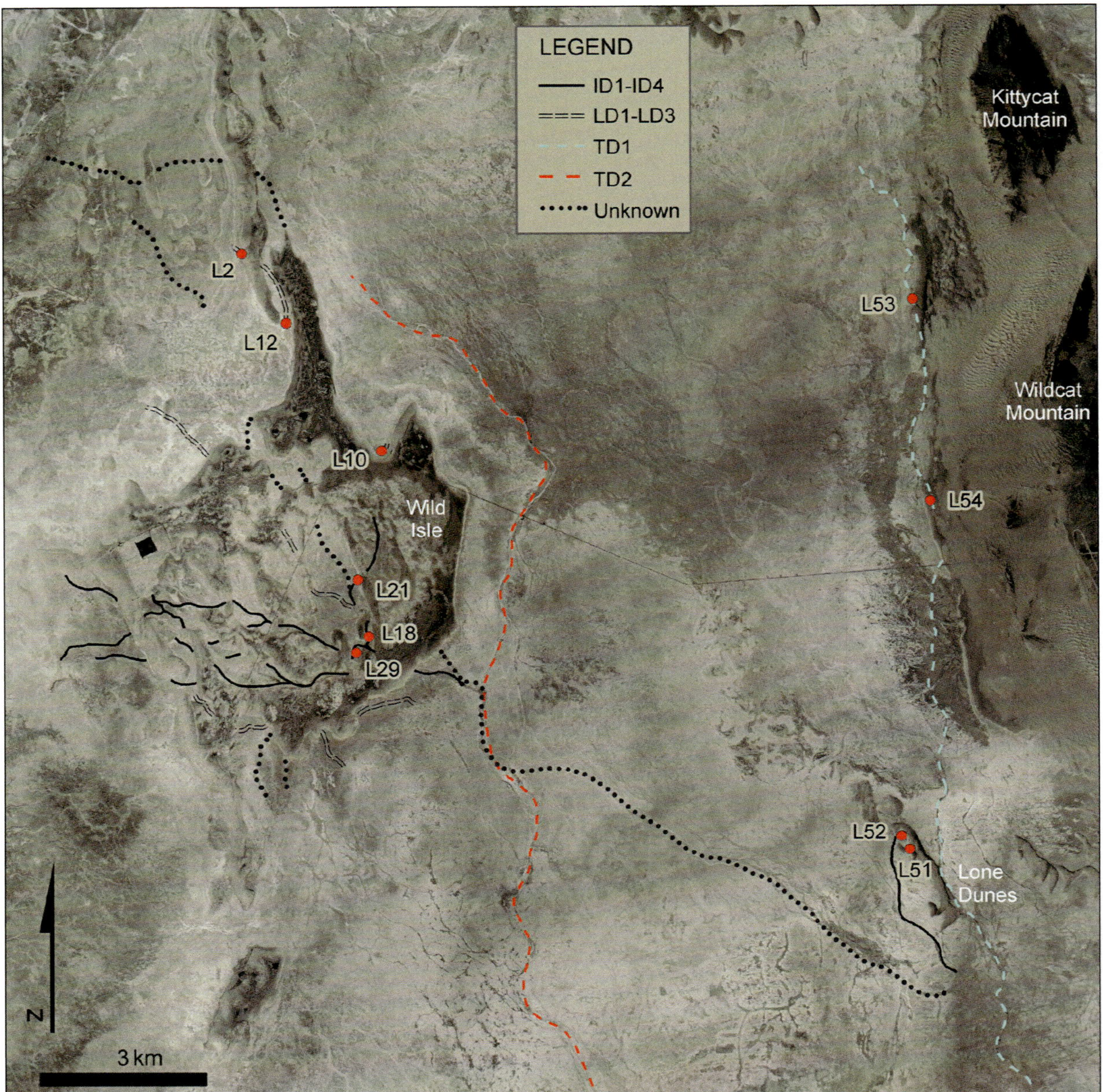

FIGURE 3.17. Distal Old River Bed overview with place-names, study localities, and channel groups. ID = Initial Distal; LD = Late Distal; TD = Terminal Distal.

specific groups of channels. As study of the system continues, it may eventually be possible to associate the distal channels of Wild Isle and elsewhere with their proximal correlates farther up the delta. Here, we divide the Wild Isle channels into Initial Distal, Late Distal, and Terminal Distal delta segments (Figure 3.17).

There are three groups of Initial Distal channels, ID1–ID3, that extend from the southeastern margin of the Wild Isle landform running generally west and north. A fourth set (ID4) in the Initial Delta group is likely associated with early entrenchment preserved

near Lone Dune (Figure 3.17). These channels cut into lake sediments and may be capped by cross- to scallop-bedded alluvial, deltaic, or littoral sands (termed the "red sands" of Wild Isle). ID1 and ID2 channels form a relatively broad set of meanders and minor distributaries under much of the western-central portion of Wild Isle. ID3 channels have a northerly trend and can be traced for the entire south-to-north extent of the Wild Isle landform. These channel sets are likely remnants of the earliest recessional or prograding delta. Unlike later channels, Initial Distal channels cut into lake sediments with well-formed terraces often capped by organic (wetland) soil horizons.

Localities 18 and 29 (Figure 3.17) provide good examples of complex Initial Distal channels in the early delta meander patterns. The localities have overlapping ID1 and ID3 channels and may preserve a prominent shift in channel flows from west to north. Here, the ID1 channel is cut deeply into the drying lake bed. Initial channel entrenchment in the vicinity of Wild Isle (ID1–ID3) and Lone Dune (ID4) may have begun as early as ~11,600 ^{14}C BP, based on the dated terrace at L18. Channels continued to cut into lake clays and formed erosional terraces (occasionally supporting overbank wetlands) until ~10,000 ^{14}C BP. After this time, black mats were generally limited to interchannel bedforms or deposited within distinct fill sequences as channels scoured, stabilized, and dried.

Late Distal Delta Channels

There are three sets of Late Distal delta channels (LD1–LD3) preserved within and around Wild Isle (Figure 3.17). Streamflow was often captured by older features (as at L29), but Late Distal bedforms tend to be stratigraphically separated from the Initial channels by the "red sands." Late channels are cut into the sands and only occasionally cut into deeper lacustrine deposits. Deep cutting is only evident where Late channels were captured by older channel courses or in areas where the channels extend well outside (e.g., north) of Wild Isle. Well-preserved features (e.g., bedforms, levees, splays, and black mats) of the Late channels show clearly defined, sand-filled channels with prominent margins. Levees adjacent to the channels are composed of coarsening-upward sediments (cross-bedded silts to fine sands; LD3 west of Locality 10 is a good example). The levees formed as prograding channels cut into and reworked the "red sands" deposit. Except where the levees are topographically exposed, the contact or transition from channel to levee can be poorly defined. Clasts within the channel fill are typically limited to fine-grained sands and silts, though bedforms (e.g., minor bars) contain medium sands.

The LD1 channel segment cuts through the broad alluvial sand deposit in the central portion of Wild Isle. Several channels capped by the northern reaches of the Wild Isle Dune may be related to LD2 based on similar morphology and the prevalence of reworked "red sands" lining the preserved channel form. The LD3 channel is an east-to-west channel at the southern edge of the Wild Isle landform. Although this southern channel has not been studied in detail, it is clearly a Late channel as it cuts into the "red sands" where the sands are preserved

by the Holocene dune and shows transport of similar material along its course to the west.

Radiocarbon dates from interchannel and Late channel-margin black mats range from ~9900 to ~8900 ^{14}C BP. Although the wetlands formed on infilling channels are present within central Wild Isle and at Lone Dune, they are rare. More common are apparent broad wetlands in northern Wild Isle and a distinct shift to the margin of the Wildcat Dune along an early course of the Terminal Distal ("Seafoam") channel (Figure 3.17).

Terminal Distal Delta Wetlands

A northeastern remnant of the ORB distal distributary system is preserved at and under the western edge of the extensive Wildcat dune field where a prominent Terminal channel (TD1; Figure 3.17), probably related to a channel ("Seafoam") emanating from the proximal region of the ORB system, parallels the margin of the dune landform. The channel cuts across remnant scrolls and meander features of either older channels or, more likely, shallow lacustrine bars and beach remnants.

Radiocarbon dates on a few isolated or interchannel black mats are likely associated with the final stages of the distal delta, the Terminal Distal wetlands. These dates, from ~8700 to ~8300 ^{14}C BP, mark the last pulses of wetland environments. As surface flow stopped reaching the Wild Isle and Wildcat reaches of the delta, wetland productivity would have become increasingly isolated, until replaced by a generally desiccated playa and dune landscape.

INTERPRETIVE CHRONOLOGY AND GEOMORPHOLOGY OF THE ORB DELTA DISTRIBUTARY CHANNELS
Radiocarbon Chronology

The ages of the ORB delta distributary channels are controlled by 85 radiocarbon age estimates on the channels or related features (Table 3.2), 52 of which are from the more proximal portion of the delta on DPG, with the remainder from the more distal delta margin on the Utah Test and Training Range. A great many of these represent samples from the same channel (e.g., there are eight from the Light Blue low-energy channel system; Table 3.2) or from a stratigraphic sequence at a single location that may represent different channels (e.g., there are six from a stratigraphic section within the ORB valley at DPGSE-1; Table 3.2). In some cases, the dates represent multiple samples in a stratigraphic sequence at a single location within a single channel (e.g., DPGNW-11A; Table 3.2). Two of the ages were run on small charcoal fragments redeposited in fluvial

TABLE 3.2. Radiocarbon Age Estimates from the Old River Bed Delta.

Sample No.	Lab No. (Beta-)	Material Dated	δ^{13}C (‰ PDB)	Radiocarbon Age (BP)	Calibrated Age @2σ[a]	East (UTM WGS 84)	North (UTM WGS 8)	Channel	Comment
DPGNW-11A	144506	Plant material	−26.1	10,170 ± 80	12,100–11,400	318818	4450769	Unknown	
DPGNW-11B	144507	Plant material	−24.9	10,060 ± 90	11,970–11,270	318818	4450769	Unknown	
DPGNW-11C	135876	Plant material	−27.2	9660 ± 50	11,200–10,790	318818	4450769	Unknown	
DPGNW-11D	131590	Plant material	−26.1	11,440 ± 50	13,240–13,180	318818	4450769	Unknown	
DPGNW-11E	144508	Plant material	−24.3	10,220 ± 80	12,380–11,510	318818	4450769	Unknown	
DPGNW-12A	144237	Plant material	−21.8	9920 ± 80	11,650–11,200	318687	4451198	Unknown	
DPGNW-T9	282097	Plant material	−23.9	9080 ± 50	10,390–10,180	316747	4444762	Unknown	
DPGNE-T21	282098	Plant material	−18.5	8790 ± 40	10,120–9630	319384	4448189	Unknown	
DPGSE-1A	119848	Plant material	−25.3	11,010 ± 40	13,080–12,710	326196	4441641	Unknown	
DPGSE-1B	123082	Plant material	−26.9	10,180 ± 50	12,060–11,630	326196	4441641	Unknown	
DPGSE-1C	121823	Plant material	−26.3	10,830 ± 60	12,880–12,590	326196	4441641	Unknown	
DPGSE-1E	121824	Charcoal	−25.2	25,180 ± 120	—	326196	4441641	Unknown	Reworked
DPGSE-1F	121825	Plant material	−26.9	9770 ± 50	11,260–11,100	326196	4441641	Unknown	
DPGSE-1G	119850	Plant material	−25.1	9420 ± 120	11,100–10,300	326196	4441641	Unknown	
DPGSE-2B	135251	Plant material	−25.5	9880 ± 40	11,390–11,310	328059	4437155	Unknown	
DPGSE-2E	137105	Plant material	−22.2	8990 ± 110	10,400–9700	328059	4437155	Unknown	
DPGSE-3A	131592	Plant material	−24.7	8850 ± 40	10,160–9740	324603	4442724	Unknown	
DPGSE-3B	135252	Shell (*Anodonta*)	−6.9	9950 ± 90	11,760–11,210	324603	4442724	Unknown	Limiting (≤)
DPGSE-3C	135253	Shell (*Anodonta*)	−6.6	9330 ± 90	10,750–10,250	324603	4442724	Unknown	Limiting (≤)
GPNE-5A	120190	Plant material	−24.2	9520 ± 40	11,080–10,680	307710	4452944	Unknown	
DPGNW-1D	120189	Shell (Lymnaeidae)	−9.5	9710 ± 60	11,240–10,790	309803	4457312	Light Blue	Limiting (≤)
DPGNW-5B	119846	Charcoal	−25.9	14,140 ± 140	17,630–16,860	312665	4456464	Light Blue	Reworked
DPGNW-5C	119847	Plant material	−25.9	9850 ± 90	11,700–10,900	312665	4456464	Light Blue	
DPGNW-5F	124298	Plant material	−14.2	9470 ± 90	11,120–10,510	312665	4456464	Light Blue	
DPGNW-5G	124299	Plant material	−12.4	9450 ± 90	11,100–10,440	312665	4456464	Light Blue	
DPGNW-5H	124300	Plant material	−21.8	9170 ± 80	10,550–10,200	312665	4456464	Light Blue	
DPGNW-5I	124301	Plant material	−15.9	9350 ± 100	11,060–10,250	312665	4456464	Light Blue	
DPGNW-14	131594	Plant material	−24.0	8800 ± 40	10,130–9670	316823	4452807	Light Blue	
GPNE-3	250000	Plant material	−22.0	9200 ± 60	10,510–10,240	299004	4457321	Buff	
GPNE-7A	231555	Plant material	−24.5	9750 ± 40	11,241–11,110	300115	4453471	Blue	
GPNE-7B	238667	Plant material	−25.5	9450 ± 40	11,060–10,570	299326	4453728	Blue	
GPNE-7C	248473	Plant material	−22.7	9530 ± 60	11,110–10,610	299326	4453729	Blue	
GPNE-8A	238669	Plant material	−27.2	10,000 ± 40	11,700–11,280	298869	4457120	Green	
GPNE-8B	238668	Plant material	−27.2	10,290 ± 50	12,380–11,830	298869	4457120	Green	
GPNE-6	234320	Plant material	−26.8	9330 ± 50	10,690–10,300	301998	4454537	Green	
GPNE-9	221779	Plant material	−23.0	10,130 ± 80	12,060–11,370	298426	4453818	Yellow	
GPNE-10	221778	Plant material	−25.9	9010 ± 40	10,250–9950	298712	4449251	Lavender	
GPNE-11	250001	Plant material	N/A	9250 ± 60	10,570–10,260	300509	4454589	Gold	

GPNW-4A	248474	Plant material	−23.6	10,460 ± 60	12,570–12,130	290623	4457285	Gold	
GPNW-4B	249999	Shell (*Gyraulus*)	−9.0	10,720 ± 60	12,750–12,550	290623	4457285	Gold	Limiting (≤)
WCMSE-1A	168834	Shell (unidentified snails)	−10.2	10,590 ± 40	12,500–12,420	302960	4466682	Limestone	Limiting (≤)
WCMSE-1B	251958	Plant material	−13.9	11,020 ± 60	13,090–12,700	302464	4466094	Black A	
WCMSE-1C	251959	Plant material	−15.2	11,050 ± 60	13,110–12,720	303026	4466269	Black A	
WCMSW-3	248475	Plant material	−26.0	9800 ± 60	11,340–11,100	294082	4468753	Red	
WCM-L53-01	267663	Organic sediment	−20.8	8300 ± 50	9440–9130	299765	4483650	Seafoam	Limiting (≤)
WCM-L53-02	267664	Organic sediment	−25.9	10,570 ± 60	12,650–12,240	299765	8365044	Seafoam	Limiting (≤)
WCM-L53-03	267665	Organic sediment	−22.0	8740 ± 50	9900–9560	299765	4483650	Seafoam	Limiting (≤)
WCM-L54-01	206215	Organic sediment	−22.6	8590 ± 80	9880–9440	300138	4479844	Seafoam	Limiting (≤)
WCM-L54-02	234837	Organic sediment	−25.0	9090 ± 60	10,480–10,170	300138	4479844	Seafoam	Limiting (≤)
WCM-L54-03	234838	Organic sediment	−24.3	8910 ± 50	10,200–9700	300138	4479844	Seafoam	Limiting (≤)
WCM-L54-04	234839	Organic sediment	−24.4	8980 ± 50	10,240–9920	300138	4479844	Seafoam	Limiting (≤)
WCM-L54-05	234840	Organic sediment	−23.4	9210 ± 60	11,0550–10,240	300138	4479844	Seafoam	Limiting (≤)
WCM-L54-06	236937	Shell (Lymnaeidae)	−10.4	8890 ± 50	10,190–9780	300138	4479844	Seafoam	Limiting (≤)
WCM-L54-07	236938	Shell (*Physa*)	−10.2	9830 ± 60	11,390–11,160	300138	4479844	Seafoam	Limiting (≤)
WCM-L54-08	236939	Shell (Lymnaeidae)	−11.8	9960 ± 50	11,680–11,240	300138	4479844	Seafoam	Limiting (≤)
WCM-L54-09	237011	Shell (*Physa*)	−10.9	8940 ± 40	10,200–9920	300138	4479844	Seafoam	Limiting (≤)
WCMNW-L10-01	234836	Organic sediment	−22.6	8880 ± 60	10,190–9740	290324	4480874	ID2 (Red?)	Limiting (≤)
WCMNW-L10-02	237010	Shell (unidentified snails)	−5.8	10,570 ± 40	12,620–12,420	290324	4480874	ID2 (Red?)	Limiting (≤)
WCMNW-L2-01	179084	Organic sediment	−14.7	8890 ± 60	10,200–9770	287692	4485305	ID2 (Red?)	Limiting (≤)
WCMNW-L12-01	154760	Shell (unidentified snails)	−10.9	9480 ± 60	11,080–10,570	288434	4483326	ID2 (Red?)	Limiting (≤)
WCMNW-L18-03	263307	Organic sediment	−26.2	9870 ± 50	11,400–11,200	289940	4477165	ID1 (Red?)	Limiting (≤)
WCMNW-L18-04	263308	Organic sediment	−25.9	10,320 ± 60	12,410–11,840	289940	4477165	ID1 (Red?)	Limiting (≤)
WCMNW-L29-01	263307	Organic sediment	−21.9	12,900 ± 60	16,070–15,020	289760	4477180	ID1 (Red?)	Reworked
WCMNW-L29-03	263310	Shell (*Helisoma*)	−8.0	10,980 ± 60	13,070–12,680	289604	4477210	ID1 (Red?)	Limiting (≤)
WCMNW-L29-04	263311	Shell (Lymnaeidae)	−8.1	11,090 ± 60	13,130–12,750	289604	4477210	ID1 (Red?)	Limiting (≤)
WCMNW-L29-05	154761	Shell (unidentified snails)	−9.9	11,080 ± 70	13,130–12,720	290032	4478905	ID1 (Red?)	Surface
WCMNW-L5-01	120199	Shell (unidentified snails)	−9.5	9640 ± 60	11,190–10,780	290014	4479465	ID3 (Red?)	Surface
WCMNW-L18-01	179083	Organic sediment	−19.4	10,000 ± 80	11,820–11,240	289898	4477519	ID3 (Red?)	Limiting (≤)
WCMNW-L18-02	263305	Organic sediment	−15.9	10,220 ± 60	12,140–11,630	289832	4477461	ID3 (Red?)	Limiting (≤)
WCMNW-L18-05	307887	Organic sediment	−18.5	11,580 ± 40	13,580–13,290	289820	4477430	ID3 (Red?)	Limiting (≤)
WCMNW-L21-01	267657	Organic sediment	−26.2	10,140 ± 50	12,040–11,410	289898	4477519	ID3 (Red?)	Limiting (≤)
WCMNW-L21-02	267657	Organic sediment	−22.5	13,550 ± 60	16,920–16,460	289898	4477519	ID3 (Red?)	Reworked
WCMNW-L51-03	267660	Organic sediment	−25.1	10,250 ± 50	12,370–11,760	299658	4476470	ID4	Limiting (≤)
WCMNW-L51-01	267658	Organic sediment	−24.8	9260 ± 50	10,570–10,280	299658	4476470	ID4	Limiting (≤)
WCMNW-L51-02	267659	Organic sediment	−25.1	10,030 ± 50	11,760–11,290	299658	4476470	ID4	Limiting (≤)
WCMNW-L52-01	267661	Organic sediment	−24.1	9680 ± 50	11,220–10,790	299508	4473725	ID4	Limiting (≤)
WCMNW-L52-02	267662	Organic sediment	−23.7	12,430 ± 50	14,980–14,140	299508	4473725	ID4	Reworked

[a] Calibrations to estimated calendar years at 2σ are derived using Calib 6.0 (Reimer et al. 2009).

sediments and were collected and analyzed only to provide limiting ages on the deposition of those sediments. In both cases the ages are anomalously old (~25,200 and ~14,100 ^{14}C BP) and are of little use in establishing the ages of ORB distributary channels.

Thirty-seven of the age estimates are derived from organic matter in black mats directly associated with distributary channels. Nearly all of these were collected from the proximal delta area. This organic matter consists of very small plant fragments that were unidentifiable for the most part. In some cases samples could be tentatively identified as sedges, such as *Schoenoplectus*, or reeds, such as *Phragmites*, but even these samples were too fragmentary for identifications to be conclusive. We therefore simply describe the samples as "plant material" in Table 3.2. We consider these 35 age estimates to be the most reliable of the radiocarbon dates, providing close estimates of periods when there was flowing or open water in the channels.

Thirty-three of the samples are also derived from black mats but are listed in Table 3.2 as "organic sediment" because no individual plant parts could be separated from the bulk sample and directly dated. While we think that these age estimates are, for the most part, derived from plant material compressed into the black mats, it is possible that they contain reworked Lake Bonneville muds containing some organic material. Where this appears to be clearly the case we have noted that in Table 3.2 in the comment column. Although the bulk of the remaining organic sediment dates are consistent with the age estimates on plant material, we cannot discount the possibility that they are contaminated, and we consider them to be limiting ages, with the true age of the sample equal to or younger than the given age estimate.

Fifteen of the age estimates are derived from various genera of mollusks, including the freshwater clams *Anodonta* sp. and *Sphaerium* sp. and the aquatic snails *Gyraulus* sp., *Physa* sp., and *Stagnicola* sp. In most cases, the shells were dated only after careful preparation. For example, a radiocarbon age estimate was obtained on a sample from trench DPGNW-1D only after selecting four or five well-preserved *Stagnicola* sp. snail shells from the coarse sand and crushing them lightly to make sure no grains of sand were trapped in the whorls. The snail shell fragments were then washed briefly in dilute hydrochloric acid (~1 percent), washed in deionized water, dried, and powdered in a mortar and pestle. X-ray diffraction analysis showed the powdered shells to consist of 100 percent aragonite, thus suggesting that the sample was free of secondary $CaCO_3$, which usually appears as the mineral calcite. As a result, young carbon from a secondary source is unlikely.

Despite this careful preparation, the problem of anomalously old age estimates derived from mollusk shells remains. Eight of the 15 dated mollusk samples are paired with age estimates on plant material from the same location and depositional unit. In several cases the paired plant material and shell samples represent the same, or close to the same, age (e.g., dates WCM-L54-[01–09] on the distal ORB delta margin; Table 3.2), but in other cases plant remains and shells produced widely different results (e.g., DPGSE-3; Table 3.2). In all the latter cases, the shell ages are older, sometimes considerably older, than their companion plant material ages. This reflects a problem of the uptake of old carbon or carbonates during shell growth common to the dating of mollusk shells (see Oviatt et al. 2005 for a discussion of the problem in post–Lake Bonneville wetland deposits). We therefore also use age estimates from shell samples as limiting ages, particularly when they represent stand-alone samples. That is, we consider the true age of a shell sample to be younger than, or the same age as, the analyzed age estimate.

In many ways, the 68 remaining age estimates on plant remains and bulk organic sediment samples from channel fills also represent limiting ages. Twenty are samples from channels either in the unexposed channel section of the ORB delta or on the far distal margin that cannot be reliably associated with mapped, color-coded channels. While useful in a general way in estimating the overall developmental chronology of the ORB delta (Figure 3.18), they do not help in estimating the ages of individual channels or the archaeological sites associated with them. Most of the 22 age estimates on plant remains are derived from black mat samples collected either on the surface of a channel filled by sediment or from organic lenses within the sediment package filling the channel. As a result, they do not date the time the channels originally formed but, rather, are related to periods after the main distributary streamflow abandoned those channels and the channels were fed by slack water from the main channel or by groundwater after the channel completely filled with aeolian and/or sheetwash sediment. Thus, in most cases, the age of initial channel formation occurred prior to the age of the oldest dated plant material from an individual channel. These dates do, however, demonstrate a continuing formation of extensive wetlands during these periods (Figure 3.18).

Individual Channel Chronologies

Together with relative age relationships determined by careful examination of air photos to identify where the channels intersect and overlap one another (Figure 3.19), these radiocarbon ages make it possible to define

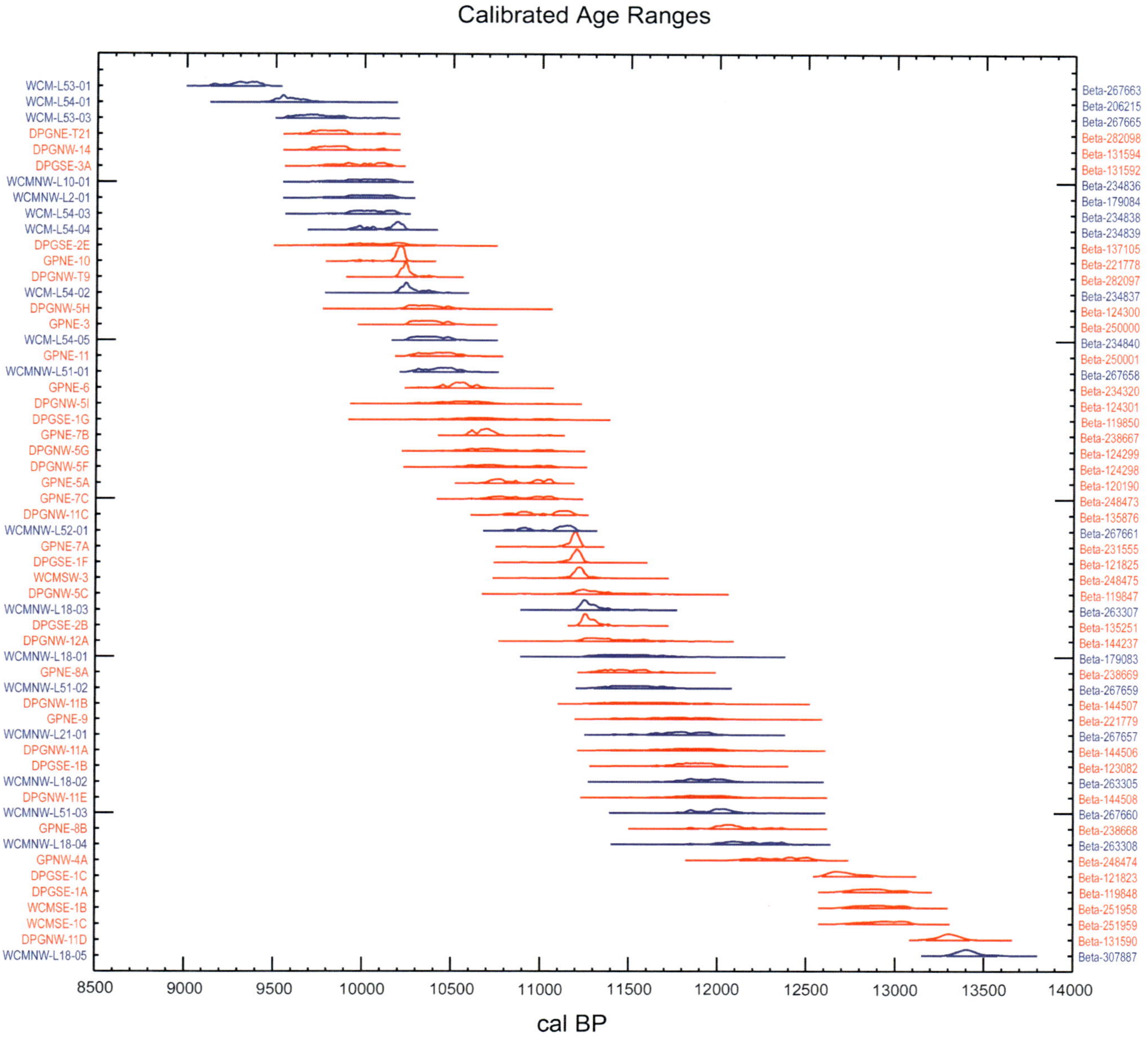

Calibrated Age Ranges

FIGURE 3.18. Linear array of organic radiocarbon calibrated ages (red = plant material; blue = organic sediment).

a chronological sequence of channel formation. By extension, these age relationships provide relatively useful estimates of the ages of the archaeological sites associated with them, although, as we discuss below, there are important caveats in making such associations.

Mango, Mocha, and Gold Channels

The three morphologically similar Mango, Mocha, and Gold channels represent the earliest recognizable channels in the exposed portion of the ORB delta (individual images showing the locations of these and other channels discussed separately below are available in the Chapter 3, Supplemental Digital Material).

The three channels exit the dune fringe surrounding the unexposed channel section of the ORB delta on its northwestern edge (Figure 3.12). They are all inset into the surrounding mudflats, and much of their channel fill has been removed by deflation, with the channel surfaces now lying below the surface of the surrounding mudflats. Neither topographically inverted channel margins nor gravel "islands" are present, and all three channels appear to have been formed by low- to moderate-energy streams. The upper portions of the channels appear to have been sheared off to the level of the surrounding mudflats by various water-related processes and possibly by wind deflation. The initial

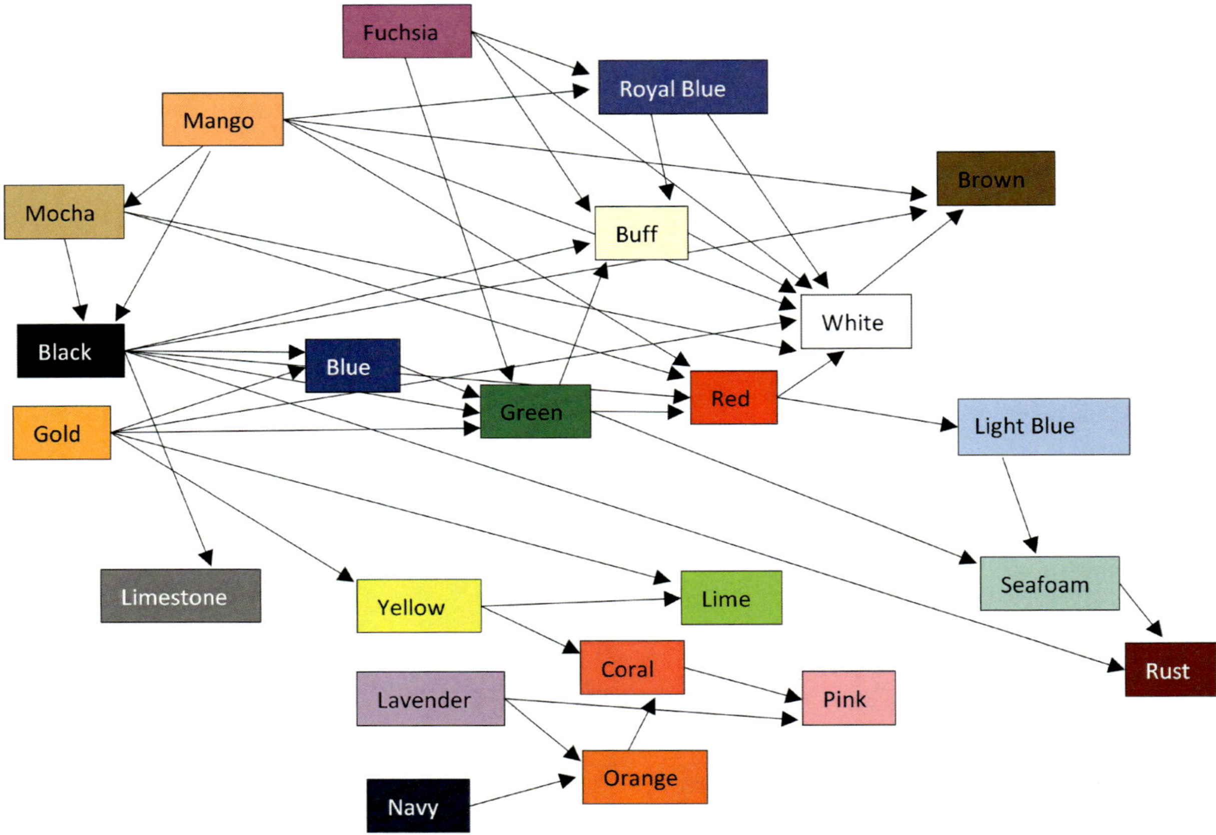

FIGURE 3.19. Relative age sequence of mapped Old River Bed channels.

deflationary episode occurred prior to the formation of Black Channel D, which cuts Mocha and Mango channels. The Mango and Mocha channels are disturbed at their distal ends and are only traceable to elevations of ~1,298 m, but the Gold channel once extended to elevations below 1,290 m.

There are no direct ages on the Mango and Mocha channels, but plant remains in a black mat in the Gold channel date to ~10,460 ^{14}C BP, so the channel must have formed sometime prior to that time. A paired age on gastropod shells (*Gyraulus* sp.) from the same sample dates to ~10,720 ^{14}C BP but must be used as a limiting date, as noted above. A third age estimate of 9250 ^{14}C BP on a black mat extending over the Gold channel fill and onto the surrounding mudflats provides a maximum age for groundwater flow in the channel. The Mango and Mocha channels are cut by Black Channel D and other younger channels. The Gold channel is cut by Yellow, Green, Blue, and numerous other younger channels. Based on the two direct limiting age estimates on the Gold channel, the age of the black mat underlying Black Channel A, and the position of the channels relative to

the Black channels, which may be of Gilbert episode age, we think that the probable age of the channels is ~11,300–10,500 ^{14}C BP.

Black Channels

The Black channels are high-energy channels on the northern margin of the exposed channel section of the ORB delta (Figures 3.10 and 3.13). At least 10 of these topographically inverted, gravel-filled channels can be recognized in the exposed section of the ORB delta, but we label them simply as Black Channels A–G because we have only limiting age estimates on the Black channel system, they likely formed within only a few hundred years, and they cannot be readily distinguished from one another. The bulbous sections of the Black channels, at their ends and along the length of the channels, might have been created when the bed load–charged rivers flowed onto the irregular topography of the wetlands and mudflats. Higher-elevation digitate channel ends may be present under aeolian silt in the unexposed channel portion of the delta.

There is a date of ~11,400 ^{14}C BP on a black mat in

a gravel pit in the unexposed channel portion of the ORB delta (locality DPGNW-11), but whether or not the gravel in the pit is related to one or more of the Black channels is unclear. Black mats in the channel fill above the channel gravels date to ~10,200 ^{14}C BP. There are two dates of ~11,050 and ~11,020 ^{14}C BP on a black mat associated with the northernmost Black Channel A2 distributary that underlies, and thus predates, the formation of the channel and deposition of the channel gravels (locality WCMSE-1). We also have limiting dates of ~11,010 and ~10,830 ^{14}C BP on charred plant material from a black mat overlying coarse fluvial gravels at the end of the inset portion of the ORB channel south of the unexposed channel portion of the ORB delta (locality DPGSE-1). These dates indicate that the regressing Lake Bonneville had dropped below an elevation of ~1,297 m sometime prior to ~11,000 ^{14}C BP. The Green and Blue channels, along with numerous younger channels, crosscut the ends of Black Channels B and D at elevations lower than those of the digitate channel ends (Figure 3.12). That is, they cut the Black channels at elevations below ~1,297 m where the ORB rivers flowed subsequent to the formation of the digitate channel margins. The earliest date of ~10,300 ^{14}C BP on a Green channel black mat at locality GPNE-8 provides a younger limiting age for the formation of these two Black channels.

Together, these age estimates suggest that Black Channels B and D formed between ~11,000 and 10,300 ^{14}C BP. The ages of the other Black channels are unknown but postdate ~11,000 ^{14}C BP. These high-energy channels may be associated with the Gilbert episode climate change. The increase in the ratio of input to output that caused the GSL to rise approximately 15 m during the Gilbert episode may also have caused Lake Gunnison to increase its overflow rate—and therefore to increase the discharge and stream power of the ORB river. If so, the Black channels may have formed during the Gilbert interval or immediately prior, during its transgressive phase. All these Black channels may have formed in a century or less, possibly in only a few decades. This chronology applies only to Black Channels A–E, as we have neither direct nor relative age estimates for Channels F–G, and it is possible that these high-energy channels are related to lake levels at the close of the Lake Bonneville regression.

Limestone Channel

The Limestone channel can only be traced for a short distance on the northern end of the northernmost Black Channel A2, but it meanders on and across the channel, as well as through adjacent areas of Bonneville Lake sediments (Figure 3.12). We obtained an age estimate of ~10,590 ^{14}C BP on *Sphaerium* sp. shells from one of the Limestone channel distributaries, but since the date is on shell it represents a limiting age, and the true age of channel formation may be younger. Since the Limestone channel postdates Black Channel A2, which is likely of Gilbert episode age, the channel was probably formed sometime between ~10,500 and ~10,000 ^{14}C BP.

Yellow and Fuchsia Channels

The Yellow channel exits the dune fringe of the unexposed channel portion of the ORB delta south of the Blue channel (Figure 3.12) and flows to the northwest, where it can be traced to elevations of ~1,292 m. Meanders are more common at its distal end, suggesting that the Yellow channel stream was losing competence and carrying capacity. On the exposed mudflats, at channel locations where heavy sediments were deposited (e.g., point bars on interior channel bends), the Yellow channel is topographically inverted. Coarser gravels in these locations have armored the surface, limiting deflation and forming gravel "islands" on the mudflats. There is a single direct age estimate of ~10,130 ^{14}C BP on a sample collected from the upper fill of the Yellow channel, so it must have formed before that time. It cuts the Gold channel, and is therefore younger, and probably dates to between ~10,500 and ~10,200 ^{14}C BP. The Fuchsia is undated and unsurveyed but may be of a comparable age.

Green Channel

The moderate- to high-energy Green channel flows north of the Yellow channel and cuts a number of older channels, including Black, Gold, and Mocha, and is, in turn, cut by a number of younger channels, including Buff and Red (Figure 3.12). It is morphologically similar to the Yellow channel and runs first west and then northwest into the Wild Isle dune complex. It can be traced to an elevation of ~1,291 m.

There are four direct radiocarbon age estimates of ~10,290, ~10,000, ~9800, and ~9330 ^{14}C BP related to the Green channel. The oldest of these is on plant material associated with *Anodonta* sp. shell and with fish bones, probably Utah chub (*Gila atraria*). These suggest flowing water, and the ~10,290 ^{14}C BP date may be close in age to when the channel actually formed. The ~10,000 ^{14}C BP date is from the same location and may represent prolonged flow along the channel. The two dates are on channel-fill black mats and probably represent groundwater-fed marsh growth within an abandoned channel. However, both dates overlap at the two-sigma range when calibrated, and channel formation and abandonment may have occurred relatively quickly. The ~9800 ^{14}C BP date is on a black mat on the mudflat surface toward the distal end of the ORB delta where the Red channel cuts into and follows along the older Green channel. The sample may thus be related to the Red channel, but it provides a limiting age for the older Green channel. The ~9330 ^{14}C BP date is substantially later but may not even be related to the Green channel. Although the channel from which the sample was taken is within the Green channel margins in air photos, it runs perpendicular to the Green channel flow and may represent a younger, but now obscure, crosscutting channel. The Green channel contains considerable quantities of coarse sand and gravel and was a relatively high-energy distributary stream. However, there are fewer gravel "islands" along its course, so it apparently had less energy than the Yellow channel and may represent diminishing streamflow in the ORB river.

Red Channel

The Red channel is a geomorphically uninvestigated and archaeologically unsurveyed channel that exits the dune fringe of the unexposed channel portion of the ORB delta just south of Black Channel D (Figure 3.12). It is rather enigmatic in that near its proximal end it resembles the Mango and Mocha channels, which it cuts, while at its distal end it resembles the Green and Yellow channels and is characterized by topographically inverted gravel "islands." These enigmatic channel forms may be more apparent than real and may be due to the great difficulty in mapping the ORB distributary channels from aerial photography. As noted above, one of the Red channel distributaries intersects a Green channel distributary and flows along the older channel for some distance toward the distal margin of the ORB delta to elevations below ~1,296 m. We have mapped this as a Red channel, and a ~9800 ^{14}C BP date on a black mat on the channel surface is likely related to Red channel formation, but the channel may more properly be mapped as the "Christmas" channel (i.e., it is both Red and Green), with sediment packages in different areas related to different periods of formation. Unfortunately, most of this channel is located in a closed impact area on DPG and is unavailable for investigation. Certainly, the Red channel postdates the Black channels, and although the single ~9800 ^{14}C BP age estimate may reflect its true age, it may also be considerably older.

Blue (Lime, Royal Blue?) Channels

The Blue channel also exits the dune fringe of the unexposed channel portion of the ORB delta on its northwestern margin (Figure 3.12). It cuts the Black, Gold, and Yellow channels but is cut by few younger channels, as the main distributary flow appears to have shifted to the south after it was formed. There are three direct radiocarbon age estimates of ~9750, ~9530, and ~9450 ^{14}C BP on black mats in the channel fill. All three overlap at the two-sigma range when calibrated. The latter two are from the same profile and come from black mats in a series filling the channel and probably representing groundwater flow within the channel. However, the ~9530 ^{14}C BP sample is from near the bottom of the sequence and may more closely provide a limiting age for when the meander was formed. The older age estimate is from a separate location and appears to represent a different meander within the overall Blue channel. Whether or not it represents the age of actual flow in the channel is unknown, but it suggests that the channel originally formed prior to ~9800 ^{14}C BP.

Geomorphically, the Blue channel contains some coarse sand and gravel, but much less than either the Yellow or Green channel. No gravel "islands" are formed along its course, and it appears to have been a lower-energy stream relative to the Green channel. It can be traced to an elevation of ~1,292 m. The Lime channel is not directly dated but cuts both the Gold and Yellow channels. It has roughly the same sediment load as does the Blue channel and may be approximately the same age. The Royal Blue channel, much farther to the north, cuts Fuchsia and is in turn cut by Buff. While undated and unsurveyed, it also may be close to the Blue channel in age.

Lavender Channel (and Younger Channels, Including Navy, Coral, Orange, and Pink)

The Lavender channel is one of a series of channels formed on the extreme southwestern margin of the ORB delta and exits the delta west of the north end of Granite Peak (Figure 3.12). It and the other younger channels in the area are characterized by a lack of gravels and represent low-energy streamflows with considerably more extensive meander patterns than the older channels. The Lavender channel flows mostly due west and can be traced to an elevation of ~1,293 m. A single age estimate of ~9010 ^{14}C BP is from a black mat sample that is associated with the channel, but whether or not this represents the approximate age of channel formation or is a product of later groundwater flow is unknown. Despite the lack of crosscutting relationships, given its location on the delta and the low-energy streamflow characteristic of this channel, its probable age may be close to the single age estimate. The Navy, Orange, Coral, and Pink channels are very similar and, although undated, likely are of a similar age.

Buff Channel (and Younger Channels, Including White and Brown)

The Buff channel is a low-energy channel that exits the dune fringe on the northwestern margin of the unexposed section of the ORB delta (Figure 3.12). It is a small, shallow channel that appears to have formed quickly and then was abandoned. That is, it has neither the braided channel form nor sequential meanders within a wider inset channel that are characteristic of the larger, older channels. It is late in the relative channel sequence, as it cuts the Green and Fuchsia channels and is only intersected by White, which is then cut by Brown. It can be traced out to the west only to an elevation of ~1,296 m. Thin lenses of coarse sand and small gravel occur at the base of the channel fill, but the overlying deposits consist of very fine sands and mud containing marsh vegetation, snails, and *Anodonta* sp. Four of the clay lenses consist of reddish orange stained clays. The sequence suggests that the narrow channel may have been cut by a single flood event and then gradually filled by very low-energy streamflow. A date of ~9200 ^{14}C BP on vegetation in silty clay deposits immediately overlying basal sands may be close in age to the original channel construction.

Although undated, the Brown channel may be the youngest channel to form on the western margin of the ORB delta, as it cuts the White channel and, by extension, most of the other channels in the area. It formed sometime after ~9200 ^{14}C BP, possibly sometime after ~9000 ^{14}C BP. Sites associated with the channel contain substantial numbers of Pinto points.

Light Blue Channel

The Light Blue channel exits the unexposed channel section of the ORB delta margin on the northeast side of the delta and meanders along its northern margin in and out from under the dune fringe before turning to the northwest (Figure 3.12). It is characterized by relatively low-energy streamflow, and sediments in the channel fill consist of fine to medium sands with little gravel. However, along its distal end where it is exposed and disturbed by deflation, some gravels are evident in small gravel "islands." The Light Blue channel can be traced to an elevation of ~1,295 m near the northern distal margin of the ORB delta, where it meanders close to the Seafoam channel. In places along the dune margin, where it has been protected from deflation and only recently exposed, this sand channel fill is topographically inverted 1–1.5 m. Elsewhere, the channel has been deflated to the level of the surrounding mudflats. Several radiocarbon age estimates are available for this channel, but all were collected from locations on the eastern, proximal end

of the channel. These six dates are ~9850, ~9700, ~9470, ~9450, ~9350, and ~8800 ^{14}C BP, suggesting a prolonged period of channel filling by postflow groundwater-fed marsh systems. However, since the channel winds in and out of the dunes and the adjacent uneroded portion of the ORB delta, it is impossible to be sure that these dates all apply to the same channel system. As a result, we cannot be entirely sure the these dates apply to the channel at its distal end and to associated archaeological sites within and along its margins to the northwest.

Seafoam and Rust Channels

The Seafoam channel exits the uneroded portion of the ORB delta on its far eastern margin and runs northwest. It can be traced to an elevation of ~1,294 m immediately west of Wildcat Mountain, where a black mat associated with the channel has been dated to ~9210, ~8985, and ~8590 ^{14}C BP (Figure 3.12). A bulk sediment date of ~10,570 ^{14}C BP is inconsistent with the other age estimates from the Seafoam channel and is likely contaminated by carbonate. A series of stratigraphically younger channels, possibly related to meanders/braided channels within the main channel, are dated from ~9090 to as late as ~8300 ^{14}C BP by four age estimates on bulk organic sediments. The channel sequence appears, morphologically, to have formed near the end of the ORB delta sequence when flow was much reduced, but as noted above these very latest age estimates may simply be underestimated. Alternatively, these deposits may represent extreme storm events that episodically flooded the ORB distributary system prior to its mid-Holocene disappearance rather than permanent or annual streamflow. The Seafoam channel is characterized by well-developed meander scroll patterning and represents very low-energy streamflow. Sediments in the channel have not been investigated fully, but the channel fill appears to lack the coarse sand and gravel deposits characteristic of older channels.

The Rust channel exits the uneroded portion of the ORB delta on its far eastern margin and runs northwest. It can be traced to an elevation of ~1,300 m south of Wildcat Mountain. Although it has not been investigated on the ground, it appears to cut Seafoam, which is morphologically similar (Figure 3.12).

GEOMORPHIC SUMMARY AND ARCHAEOLOGICAL IMPLICATIONS

Investigation and dating of distributary channel features in the ORB delta indicate that all recognizable channels in the exposed portion of the delta date to ~11,500–8800 ^{14}C BP. High- to medium-energy streamflows that deposited the now topographically inverted gravels and coarse sand were limited to the early portion of this period. Rivers in these channels were likely fed by overflow from Lake Gunnison from ~11,500 ^{14}C BP to before ~9900 ^{14}C BP. Smaller low-energy streams, probably fed by groundwater, continued to support wetland/marsh environments until shortly after ~8800 ^{14}C BP. Limited isotopic data from shells deposited in these channels suggest that this groundwater also originated in the Sevier basin (Chapter 2).

Evidence of earlier streamflow in the ORB delta area is limited primarily to black mats that cannot be directly related to particular channels. A black mat below Black Channel A2 dating to ~11,000 ^{14}C BP overlies crossbedded sands that were apparently associated with low-energy streams, and a number of similar black mats on the distal end of the Red channel near Wild Isle date to between ~11,000 and ~10,500 ^{14}C BP (Daron Duke, personal communication 2013). Several pre-Gilbert channels cut by Gilbert episode high-energy channels, and dating to sometime before ~10,500 ^{14}C BP, are also characterized by comparatively low-energy streamflow, and there is no evidence of large high-energy rivers dating to the pre-Gilbert time period in the exposed portion of the ORB delta. However, coarse fluvial gravels are found below a black mat in the entrenched ORB valley that dates to ~11,000 ^{14}C BP, suggesting that high-energy streamflow was present in at least the proximal, unexposed portion of the delta sometime prior to that date. There is also an age estimate of ~11,400 ^{14}C BP on an organic black mat sample overlying coarse sands and gravels at section DPGNW-11, but we are unsure of this sample's relationship to the coarse distributary sediments. Ages of black mats in finer-grained sediments well up in the section average to ~10,200 ^{14}C BP. Finally, a bulk organic sediment age estimate of ~11,600 ^{14}C BP from channel ID3 in Wild Isle suggests that this channel, and possibly channel ID4, may be related to a Lake Bonneville late regressive phase.

Given the geographic position of these earliest dated channels, it appears that at least some of the pre-Gilbert regressive Bonneville-phase distributaries flowed north from the entrenched ORB valley; continued north through areas now covered by later, topographically inverted Gilbert-age high-energy channels; and then turned west in the area of Lone Dune, continuing to Wild Isle and beyond (Figure 3.20). ORB gravels reworked into bars near the head of the ORB delta (Figure 3.21) suggest relatively high-energy streamflow during the early to middle stages of the Lake Bonneville regression.

During climate change associated with the Gilbert episode, sediments associated with high-energy channels were deposited on top of marsh/wetlands formed in the ORB delta. During the following Holocene fine-grained wetland sediments were deflated away, leaving only the coarser sands and gravels of the topographically inverted high-energy channels. During the period immediately following their deposition, these inverted gravel-filled channels may have provided dry "highways" into the ORB wetlands.

Paleoarchaic sites in the exposed portion of the proximal ORB delta take two basic forms. There are sites formed along channel margins and sites found within channels. Separate categories of these two primary forms also occur. For example, sites on the channel margins can be subdivided into linear sites, which were likely associated with natural levees, and sites within sharp channel meanders on what were likely point bars. The age relationships of the channels described above have different implications for these two basic kinds of sites.

The sites along the channel margins, particularly the point bar category of sites, appear to be related to periods when there was open flowing water in the channels. This relationship, although likely, is not entirely definitive since there may also have been open, slack water–fed or groundwater-fed ponds in the channels even after they were abandoned by distributary streamflow. The archaeological sites within channel margins, on the other hand, obviously postdate open water flow within the channels and could have been occupied only after the channels were abandoned and silted in. Black mat formation within the channels suggests that site occupation within the channels may have been associated with groundwater-fed wetland habitats. Based on these relationships, we can use the channel chronology to identify the approximate time periods when the sites were occupied. A number of sites in the more distal portions of the ORB delta are less clearly associated with channels, although they are on or near isolated black mat sections, suggesting some form of wetland development. On these sites artifacts are often thinly scattered over an extended area, and the sites have less easily defined boundaries.

If the geomorphic sequence defined above is valid, then all sites on the exposed portion of the ORB delta probably date to after ~11,200 ^{14}C BP, with the bulk of them dating to after ~10,500 ^{14}C BP. Sites associated with the Light Blue and Seafoam channels on the northern and eastern margins are even younger, with the youngest age estimates on these channels being ~8900–8700 ^{14}C BP (two slightly younger age estimates of ~8600 and ~8300 ^{14}C BP are likely in error, as the channels/black mats from which the samples were taken are also dated

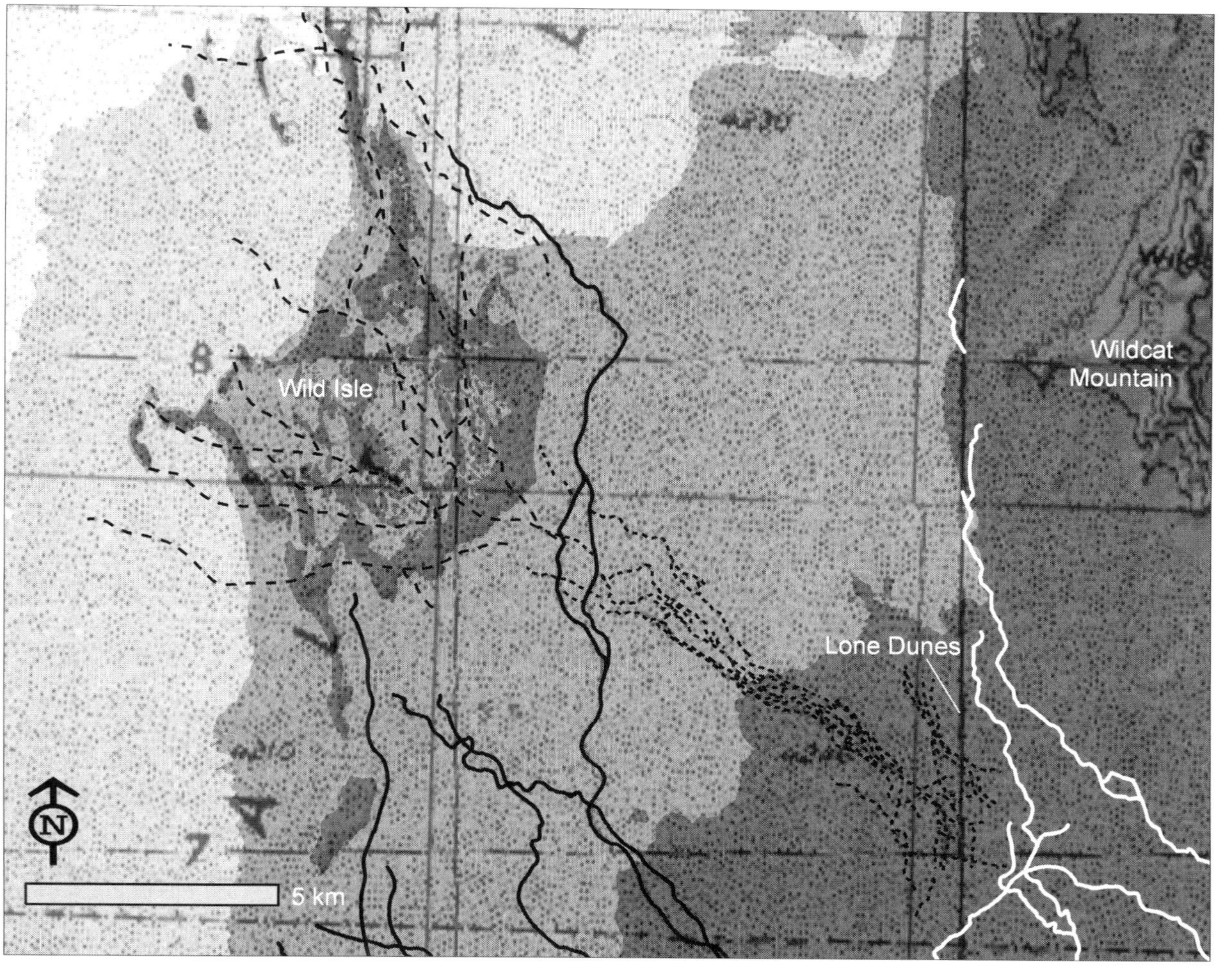

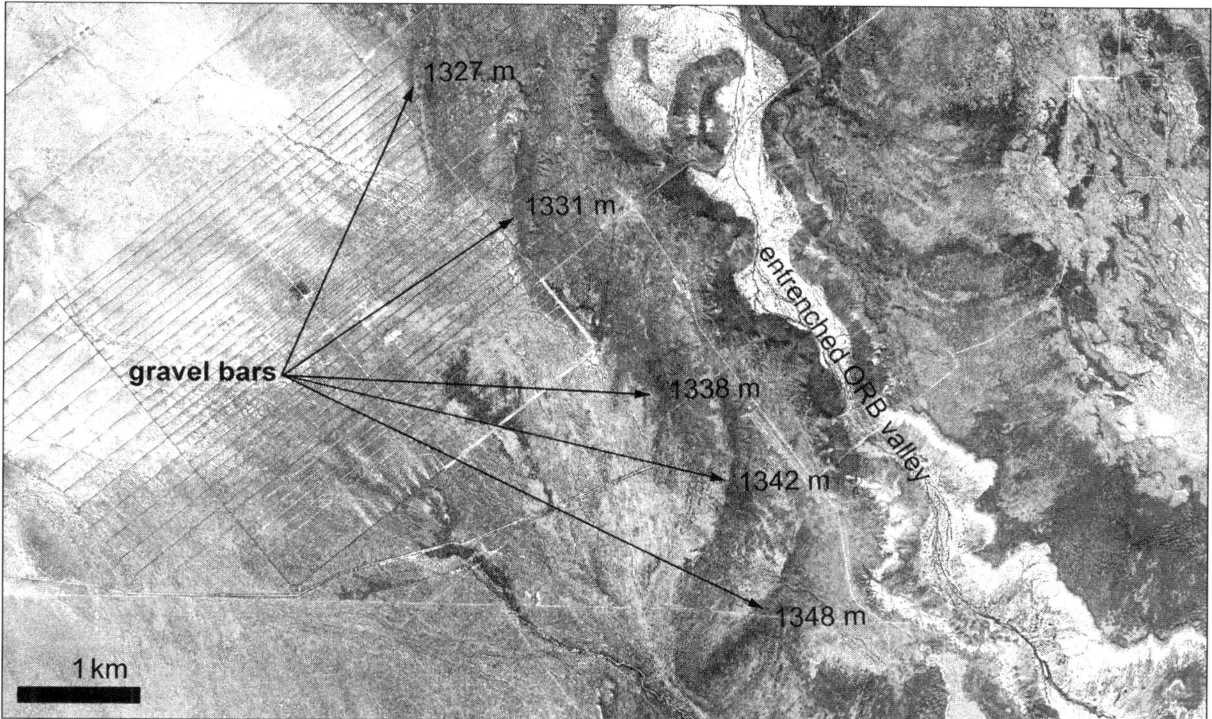

FIGURE 3.20. Mapped channels in the distal portion of the Old River Bed delta.

FIGURE 3.21. Gravel bars at elevations of ~1,327–1,348 m formed by nearshore currents. ORB = Old River Bed.

to ~9210 and ~8980 ^{14}C BP and are cut by a younger channel dated by five radiocarbon age estimates to ~9000–8900 ^{14}C BP). That is, sites with some degree of integrity associated with lower-energy, lower-elevation channels appear to postdate the transgression of GSL during the Gilbert episode.

In sum, the geomorphological sequence suggests that the Paleoarchaic sites identified to date in the ORB delta were deposited sometime after ~11,200 ^{14}C BP. Scattered cultural material on the deflated playa surface north and northwest of the better-defined proxi-mal delta distributary system may be associated with the marsh/wetland that formed in that area at ~11,000 ^{14}C BP. Elsewhere in the delta system our transect surveys indicate that virtually all artifacts are directly associated with fluvial channels, either in the channels or immediately adjacent them, and there are only a few isolated artifacts on the playa surface between them. With the possible exception of a few sites on the distal mudflats dating to ~11,200–10,500 ^{14}C BP, the primary occupation span of the ORB delta appears to be about 1,700 ^{14}C years, from ~10,500 to ~8800 ^{14}C BP.

Descriptions and Classifications of Paleoarchaic Sites in the Proximal Old River Bed Delta

Dave N. Schmitt

Throughout most of its ~3,000-year active history, the Old River Bed (ORB) delta offered prehistoric foragers a vast wetland containing a variety of attractive resources. A few Archaic and later groups inhabited some portions of the delta, but the majority of the sites were created by Paleoarchaic (PA) foragers as middle Holocene desertification and corresponding declines in productivity (e.g., Grayson 2011 and references therein) transformed much of the region into an uninviting landscape. Archaeological and geomorphic investigations clearly indicate that the formation of the delta was a dynamic event; streams and associated marshlands developed in certain areas at certain times, and as the river changed course, some channels dried up, while others maintained groundwater-fed marshes for long periods of time. With this ever-changing environment it is reasonable to assume that the delta's PA human inhabitants were equally dynamic, moving from patch to patch in response to changes in both resource types and abundances. Among other factors, these movements were a consequence of local resource depletion, the blossoming of more fruitful marshlands in distant contexts, freshwater and aquatic fauna in newly formed streams, and recently exposed patches of dry ground for both long-term occupation and short-term use.

These purportedly frequent movements are evident in the ORB archaeological record as the sites are as diverse as they are abundant. More than 230 PA sites have been identified during various sample surveys in proximal portions of the delta. Of these, 115 sites have been recorded in and adjacent what is now the Great Salt Lake Desert mudflats and include massive scatters containing large and diverse artifact assemblages, diffuse scatters dominated by tools, and small, dense clusters containing abundant stone detritus. While variable site sizes, contents, and densities clearly reflect different activities, distinguishing the types and duration of the behaviors that created them can be elusive, especially when the regional record consists exclusively of surface scatters of flaked-stone tools and detritus.

Analyses of PA sites in the proximal delta employ measures of assemblage richness and some middle-range theoretical assumptions in an attempt to "breathe a measure of behavioral life" (Thomas 1988a:393) into this large and complex record. Specifically, tool richness measures are compared with tool-to-debitage ratios in an attempt to classify individual sites and distinguish site-specific activities (Figure 4.1). Long-term (cf. residential) camps are anticipated to be marked by large numbers of artifacts and artifact classes. These areas served as the focal point of processing, fabrication, and preparation activities and should be characterized by technologically rich tool kits and low tool-to-debitage ratios resulting from recurrent stone tool manufacture and maintenance. Although the size of long-term camps will be variable given differences in the extent of favorable living space and group size, material remains are expected to be relatively dense, and some may contain discrete artifact concentrations resulting from intensive manufacture and/or processing activities.

Short-term camps are expected to contain a similarly diverse suite of artifact types but with a higher tool-to-debitage (T:D) ratio, as the by-products of tool fabrication will be fewer in number than those in camps

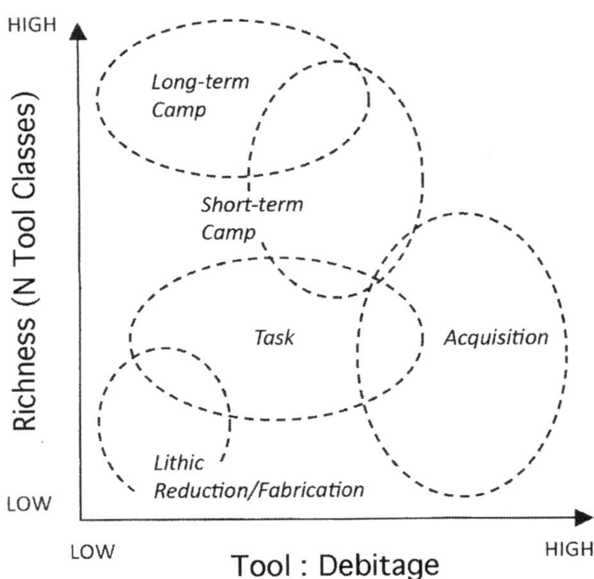

FIGURE 4.1. General model of Old River Bed delta Paleo-archaic site classifications based on the relationship between tool richness and tool:debitage ratios (inspired by Thomas 1988a, 1989).

TABLE 4.1. General Artifact Classes Used in Diversity Measures.

Artifact Class
Beak scraper
Biface (combined types)
Chisel
Concave scraper/notch
Core
Crescent
Drill
Edge-modified/utilized flake[a]
End/side/slug scraper
Graver
Ground stone[b]
Hammerstone
Knife/Cody knife
Projectile point
Uniface

[a] Simple modified/utilized flake tools were recorded during survey and not collected.
[b] Only one ground-stone artifact was observed in a multi-component site (1182; Light Blue channel) containing stemmed and Archaic points.

that are a product of prolonged stays. As in long-term camps, materials in short-term base camps are expected to be densely clustered, and their extent will likely be contingent on group size and context. In rather stark contrast, sites that largely served as lithic-reduction loci should possess very low T:D ratios and small tool assemblages containing few classes (Figure 4.1). Sites that served largely or exclusively as reduction loci are anticipated to consist of small artifact clusters, and the most convincing of these will contain cores, broken bifaces, and/or edge-modified flakes associated with abundant lithic detritus.

Sites that contain widely scattered artifacts dominated by tools are considered to represent patches where activities centered on resource acquisition. In some instances tools were resharpened and new tools were made, but it is projected that acquisition loci will largely consist of complete and fragmentary tools that were lost or discarded during procurement forays. While the number of tools and especially tool types will depend on exactly what type of resource(s) was being pursued, harvesting loci will likely contain abundant edged tools, and hunting areas will be dominated by complete and broken projectiles. The size of procurement sites will be contingent upon the size of the patch that was being exploited, and they will likely occur in any context that provided food resources, construction materials, or any

other utilitarian goods, including active channels, channel margins, and groundwater-fed marshes.

Finally, a number of sites probably represent overlapping acquisition, reduction, and processing activities. These general task sites (Figure 4.1) are expected to be variable in size, content, and context and may overlap with other site types. For example, a site that served as a resource-acquisition and lithic tool–fabrication locus may contain a variety of tools classes and a low to moderate T:D ratio, while a location used for resource harvesting, occasional tool maintenance, and initial resource processing may resemble a short-term camp, although artifact densities are expected to be lower.

The 15 tool classes used to measure assemblage richness in the proximal delta sites are presented in Table 4.1. Included are projectile points (which may have served other functions), drills, crescents, finished knives, and gravers (see Chapter 5). Scrapers were segregated into three functional types based on differences in use-edge morphology (i.e., linear convex [side, end], concave [notch, "spokeshave"], and pointed [beak]), and all bifaces were combined into one general class. Use edges of combination tools were counted as two classes if no individual types were identified. For example, a site containing bifaces, projectile points, and a side scraper/graver was identified as having four tool types, while a site with projectile points, a side scraper, and a side scraper/graver was tallied as having three functional classes.

ANALYTIC CONSTRAINTS AND STRENGTHS

Before proceeding with site descriptions and functional ascriptions, a few interpretive limitations and strengths warrant discussion. With respect to the former, it is well known that most archaeological deposits have been (and continue to be) impacted by burrowing animals, erosion, and/or a host of other postdepositional processes, and sites in the ORB delta certainly prove no exception. In fact, the duration and severity of erosion and deflationary processes at these sites are significantly greater than at other sites in the region. This is especially the case for sites on the barren Great Salt Lake Desert mudflats. Deflationary processes have removed massive volumes of sand and silt from the central and distal delta (see Chapter 3), and in most instances the associated archaeological materials have dropped 1–3 m below their primary context. During and after this vertical displacement the artifacts were also subject to horizontal movement resulting from innumerable freeze–thaw, high wind, and sheetwashing cycles. Although most of these movements were likely limited, some materials have been displaced many meters from their point of origin, while other, smaller artifacts (notably pressure flakes and flake fragments) have been completely removed.

A second concern is the reoccupation of sites by PA foragers during the delta's active history. Some sites likely represented favorable locations, and whether it was several months or several hundred years later, people returning to these areas deposited additional artifacts. As a result, I have attempted to identify PA palimpsests by incorporating data on site setting and size and overall artifact abundances and spatial distributions in addition to assessing the relationship between artifact richness and T:D ratios. Despite this attempt, it is likely that some multiple occupation loci went unrecognized.

An additional and related limitation centers on postdepositional artifact weathering (see Chapter 5) and the identification of simple flake tools. Whether flakes were detached from a multidirectional core or a large biface (see Schmitt, Madsen, Oviatt, and Quist 2007:Figure 6.7), the use of flakes for expedient cutting and scraping tasks was probably very common, and these artifacts likely represented a fundamental part of the tool kit for men and women foraging in the ORB wetlands. Although utilized/retouched flake tools have been identified in delta sites, the evidence for use-wear on a number of volcanic artifacts has doubtless been obliterated by long-term particle abrasion and chemical weathering, and many of these tools went unrecognized.

Despite these shortcomings, PA sites in the proximal delta largely represent discrete scatters separated by several hundred meters of artifact-barren flats and appear to represent separate use episodes. Moreover, and with few exceptions, each of the recorded sites was subject to a detailed examination that included the classification and quantification of every artifact. Because there is little (if any) potential for buried culture-bearing deposits in most contexts (see Madsen et al. 2004), ORB site classifications and comparisons employ comprehensive information and lack the sample size effects that pester most archaeological interpretations.

SURVEY METHODS

The various project areas in the proximal delta were investigated by teams of archaeologists most often composed of a four-member crew. Survey tracts were completely covered by parallel transects and included pedestrian reconnaissance and survey from ATVs at low speeds (2–4 mph) on the mudflats; given the white salt efflorescence background on the barren mudflats on which the black fine-grained volcanic (FGV) and obsidian artifacts lay, it is likely that all sites were identified. Regardless of technique, the spacing interval varied between about 15 (pedestrian) and 30 m (ATV) depending on the amount of vegetation cover and context. When surface artifacts were encountered, the area in the immediate vicinity of the materials was examined via pedestrian survey at closely spaced (~2-m) intervals. Cultural materials were defined as a site or as an isolated find based on this detailed examination. Sites were carefully examined to determine the spatial extent of the artifact scatter, as well as to identify all tools and cultural features evident on the surface. The site boundaries were identified, and orange plastic datum stakes were driven into the approximate middle of each site. Site locations were determined by taking and recording a GPS (Universal Transverse Mercator) reading at each of the datum marker posts; North American Datum of 1927 was the primary datum until 2007, and World Geodetic System 1984 was used in each of the subsequent projects. When sites were formally recorded they were evaluated in compliance with the National Historic Preservation Act, the Archaeological Resources Protection Act, and Army Regulation 200-4.

Relevant data on the local ecological setting, geomorphology, degree and kind of disturbance, diagnostic artifacts, and so on were recorded on Intermountain Antiquities Computer System site forms especially modified for use at Dugway Proving Ground by the cultural resources manager of Environmental Programs. Provenience and descriptive information were recorded for isolated finds on special forms created for that purpose by Dugway Proving Ground cultural resource management personnel. A detailed map of each site was constructed by taking and recording GPS readings at formed artifacts, cultural features, and site boundaries. Individual flakes in diffuse scatters were characterized (e.g., decortication, secondary interior) and mapped with a portable GPS unit. Discrete concentrations of flakes and/or lithic tools were identified as loci. These loci were described (including tallies of flakes by reduction stage), boundaries were plotted, and, when present, any formed tools within a locus were individually mapped and often collected. Where pertinent, a series of GPS readings were taken on, or along the margins of, primary geomorphic features (e.g., dune margins, channels) within and immediately adjacent site boundaries.

Each recorded site was photographed from two or more angles, and specific features were photographed where relevant. All identifiable artifacts were pin-flagged prior to both photography and mapping to ensure that site dimensions and activity areas were accurately recorded. In most instances all tools and debitage were characterized and mapped, save for at some large and dense sites where debitage was tallied by type and only formal tools were plotted. Diagnostic artifacts were mapped and collected, and a sample of lithic tools was collected (or all were collected), from the majority of the formally recorded sites. A GPS reading was taken at the location of all collected artifacts, where each was assigned a specimen number and recorded as a Universal Transverse Mercator grid point. Tools that were characterized in the field and not collected (notably fragmentary bifaces and edge-modified/utilized flakes) were also assigned specimen numbers and plotted.

Survey Parcel and Site Locations

Eight survey projects in the proximal delta in and along the margin of the Great Salt Lake Desert mudflats examined approximately 12,000 ac and identified 115 PA sites that were fully recorded (see Table 1.1). The initial investigations consisted of a series of contiguous block surveys in the northeast delta and along its eastern margin at the regressive-phase lacustrine fines (RLF) plain–mudflat interface (Madsen et al. 2000) and linear surveys that included a swath of a Black distributary along the southwestern edge of Lake Öferneet (Schmitt, Madsen, Hunt, Callister, Jensen, and Quist 2002) and segments of Black Channels D and E (Schmitt et al.

2003). Since most of the initial surveys centered on prominent, topographically inverted distributaries, an exploratory rectangular block (134.2 ac) was investigated on the flats north of the Black channels (see below) to identify the nature of PA materials and their possible association with ancient streams. The area consists of extensively deflated low-energy channel meander remnants, including stream deposits that were cut by (i.e., predate) the formation of the Black channel, and intervening barren mudflats. Although survey identified more than 200 lithic artifacts in the tract (Schmitt et al. 2003), no discernable sites were discovered, and we suspect that this random and often diffuse scattering of materials is the result of postdepositional transport (e.g., sheetwashing and wind movement) over the past ~10,000 ^{14}C years. It is possible that materials were deposited by later foragers pursuing resources in a distant groundwater-fed patch, but the overall context and the results of other investigations in the delta (see below) suggest that these artifacts were once clustered along low-energy, pre-Black channels and may represent some of the earliest human occupations in the proximal delta.

Beginning with the 2005 field season, reconnaissance centered on various channel segments in more westerly portions of the delta by following one or more individual distributaries. These linear transects ranged from about 100 to 200 m in width and were designed to examine various channel segments in their entirety along with 20–25 m of the bordering mudflat surfaces. Surveys included segments of Red, Blue, and Green channels in the central delta; Yellow and Lime channels to the west; and portions of the Lavender channel near the delta's southwestern margin (Madsen and Page 2008; Madsen et al. 2006; Page et al. 2008; Schmitt, Madsen, Page, and Smith 2007). Together, these linear projects in the western delta identified and recorded 48 sites dating to the PA.

While crossing the mudflats between identifiable western ORB channels innumerable times over a five-year span, we did not locate any sites that were not directly on or immediately adjacent deltaic channels. In an attempt to formally examine this purported site–channel relationship in the region, we initiated two linear, essentially north–south surveys in 2007 that crosscut numerous east-to-west-flowing channels. The eastern transect was placed midway between where the channels emerge from the underflow fan margin and the western/northwestern end of the channels where they disappear into the eroded mudflat surface (Figure 4.2). Four PA sites were identified during survey of this 150-m-wide, 9-km-long transect, and in 2008 the transect was expanded to a width of 450 m, where 12 new

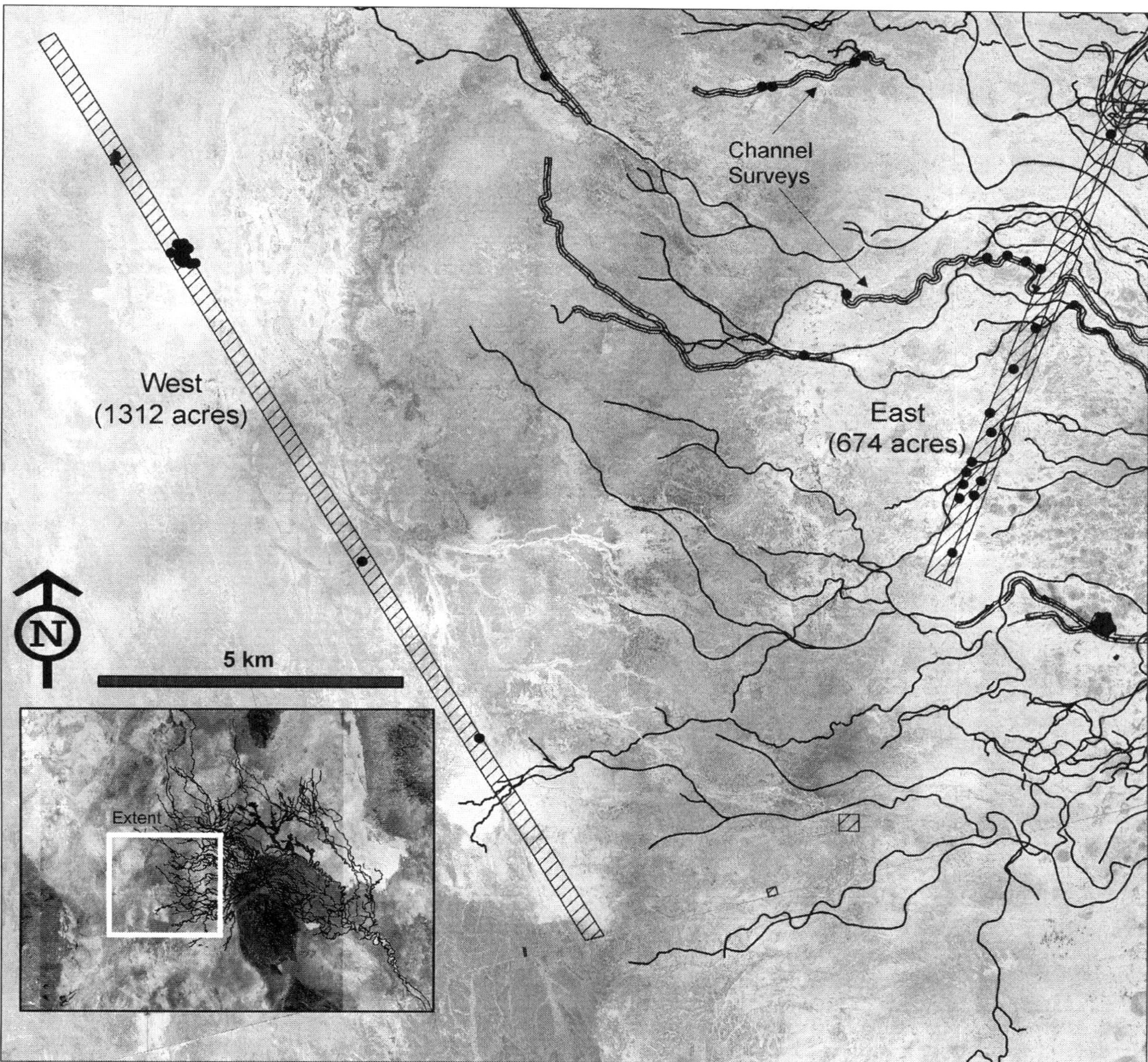

FIGURE 4.2. Location of two linear survey tracts in the western Old River Bed delta.

sites were identified (Madsen and Page 2008). All 16 sites were found directly on or immediately adjacent either formally named channels or unnamed dissected channel sections; no sites were located on the mudflats between channels.

Survey of a second transect was undertaken on the mudflats west of where any channels can be recognized and traced from satellite imagery (Figure 4.2). This western transect was 17.7 km long by 300 m wide and oriented northwest–southeast. Four PA sites were identified during survey of this parcel, and, again, each was discovered in direct association with scatters of pea gravels that represent eroded channel remnants. Occasional isolated flakes and tools were also identified, but they are widely scattered (i.e., one item was encountered every ~200–300 m during transect) and seem to be either associated with channel ends or between closely adjoining distributaries.

Together with the results of linear and block investigations to the east, these survey transects indicate that human activities centered on channels and channel margins, and in most instances it was because these areas provided dry ground. In fact, just as afternoon shade draws human activities in extreme desert environments (e.g., Bartram et al. 1991; O'Connell 1987; O'Connell et al. 1991), elevated patches of dry ground were doubtless the most attractive and favored locations for Paleoarchaic peoples inhabiting the delta. Clearly, foragers lost and discarded tools while hunting and harvesting resources in shallow streams and brackish marshes, but high ground was certainly preferred, indeed necessary, for processing resources, preparing and eating meals, leisure, and sleeping. As such, it is not surprising that the majority of sites are on topographically inverted channel segments, inner bends of oxbows, and natural levee deposits. In most instances these sites postdate channel flow, and foragers likely used these elevated patches of dry ground while extracting resources from nearby flowing channels and groundwater-supported marshes. Conversely, when levees, oxbow point bars, or other patches of favorable ground were available along the margins of active streams, foragers may have inhabited these locations and created a record that is older than the age estimates from radiocarbon assay of organics deposited at the end of channel flow. Overall, and while the primary context of a number of sites remains ambiguous, it appears that approximately 90 percent of the PA sites occur directly in channels that represent the occupation of partially exhumed dry ground. A few others likely represent within-channel foraging activities during streamflow, and there are at least six sites largely restricted to levees or point bars that appear to have served as elevated ground overlooking active streams. Discussions on the chronological implications of ORB site locations are presented below.

SITE DESCRIPTIONS AND CLASSIFICATIONS

Various archaeological surveys in the proximal delta in and immediately adjacent to the Great Salt Lake Desert basin identified 115 sites that contain materials dating to the PA. A number of additional Paleoarchaic sites/components are reported in more southern reaches of Dugway Valley in association with ORB deltaic channels atop the RLF plain (see Table 1.1). Since these sites may contain shallowly buried artifacts, lack radiocarbon age estimates, and often contain later materials within or immediately adjacent their boundaries (e.g., Page, Schmitt, Dalldorf, and Wazaney 2012; Schmitt et al. 2010), they are not discussed here. However, a large number of artifacts, notably stemmed and Pinto points, have been collected, and the results of geochemical sourcing are presented in Chapter 6.

Of the sample of 115 sites, 100 are exclusively PA in age, and the following discussions focus on these early visits to the proximal delta; an additional 265 PA sites/site components have been recorded in the distal reaches of the delta, 222 of which are considered wholly or predominantly PA in age (see Chapter 1, Table 1.2 and references therein). Sites are classified and described by channel, and the channels are presented oldest to youngest based on the chronological sequence of their formation. In a few instances channel/site aggregates

are presented together based on similarities in stream formational ages and/or proximity. Since all of the sites are in Tooele County of western Utah, the state ("42") and county ("To") designators are omitted, and sites are identified by the last numbers of the Smithsonian trinomial.

Black Channel Sites

The Black channel complex consists of a series of broad, gravel-filled, and topographically inverted delta distributaries with clusters of artifacts typically scattered along their crests and margins. Although the age of Black Channels F–G are unknown, radiocarbon age estimates in the A–E complex suggest that they were flowing from ~11,000 to 10,300 ^{14}C BP (see Chapter 3). The most complex sites in the proximal ORB delta are associated with the Black channel distributaries along the northern and especially eastern margins of the delta near the dune–mudflat transition (Madsen et al. 2000; Schmitt et al. 2003). More precisely, most sites sitting atop the Black channels must necessarily postdate stream formation, given the high-energy environment that created the Black channel complex. In most cases the Black channels formed a barrier along which the stream(s) that created the later channels flowed, and they were apparently used by human foragers as high ground overlooking the later channels (e.g., Light Blue) and their associated resources. The only dry areas were probably partially exhumed segments of Black channels and natural levees formed along the subsequent, low-energy stream margins, and it is these areas where human activities appear to have been concentrated.

Archaeological reconnaissance in various portions of the Black channel complex identified 33 sites containing PA materials and three additional sites comprising PA artifacts intermixed with materials dating to the Archaic and/or Fremont periods (Table 4.2). The PA sites range in size from ~100,000 to 250 m^2 and average about 100 artifacts per site, and like on most sites in the proximal delta, items tend to be diffusely scattered, as the mean number of artifacts per 100 m^2 is less than one (Table 4.3). Differences in artifact types and abundances and the extent of their distributions clearly suggest that the sites represent a range of activities. Included are campsites containing abundant debitage associated with large numbers of tools and tool types, lithic-reduction loci marked by limited tool classes and abundant flaked-stone detritus, and large, diffuse artifact scatters with high T:D ratios that likely manifest resource-acquisition/processing locations.

Based on artifact abundances, diversity, and overall contexts, five Black channel sites appear to have served

TABLE 4.2. Selected Attributes and Contents of Sites Associated with the Black Channel in the Proximal Old River Bed Delta.

Site No. (42To-)	Projectile Points[a]	Area (m²)	No. of Tool Classes[b]	No. of Tools	Pieces of Debitage	Tool:Debitage Ratio	Artifact Density[c]
1356	S	3,162	1	1	33	.03	1.07
1368	S	9,257	3	15	20	.75	.38
1369	S, E	12,177	3	17	~200	.09	1.78
1370	S	3,981	2	8	~250	.03	6.48
1371	S, E	43,933	7	103	~200	.52	.69
1385	—	250	1	1	~50	.02	20.40
1666	S, A	11,914	3	13	~60	.22	.61
1667	—	1,022	0	0	20	.00	1.96
1668	S	4,203	4	7	20	.35	.64
1669	S, E	1,816	1	3	7	.43	.55
1670	—	5,465	2	2	29	.07	.57
1672	S, E, A	8,048	3	8	27	.30	.43
1673	S	3,489	1	2	20	.10	.63
1680	—	751	2	4	11	.36	2.00
1681	S, E	619	2	6	10	.60	2.58
1682	S	308	3	6[d]	6	1.00	3.90
1684	S, E	11,808	6	18	~50	.36	.58
1685	S, E, A	33,233	9	37	~300	.12	1.01
1686	S, E	19,109	8	69	~500	.14	2.98
1687	S, E	1,157	2	10	8	1.25	1.56
1688	S, E	16,121	6	38	~225	.17	1.63
1858	E	8,898	3	3	81	.04	.94
1859	S, E	1,793	2	6	17	.35	1.28
1860	S	4,242	5	5	~100	.05	2.48
1861	S, E	25,837	6	18	~50	.36	.26
1862	S, E	13,543	4	18	~70	.26	.65
1872	S, E	41,792	9	68	~100	.68	.40
1873	S, E	16,829	4	26	45	.58	.42
1874	S, E	4,744	4	6	16	.38	.46
1875	S, E	10,932	3	19	56	.34	.69
1876	S, E	15,432	4	9	7	1.29	.10
1877	M	1,516	5	6	9	.67	.99
1878	S	2,043	3	9	19	.47	1.37
1920	S, E	24,001	8	44	~70	.63	.47
1923	S	1,083	6	11	3	3.67	1.29
1924	S, E	99,462	6	167	~150	1.11	.32
Total (mean)		463,970	(3.8)	783	~2,839	(.49)	(1.79)

[a] S = Great Basin Stemmed and lanceolate; E = Early Holocene stemmed, Stubby, and Pinto; A = Archaic dart and/or arrow; M = Great Basin/Early Holocene stemmed.
[b] General functional classes/types (see Table 4.1). Some tallies include artifact classes that were identified in the field and not collected.
[c] Number of artifacts per 100 m².
[d] Includes three fragments of a late-stage production biface that were refitted.

as base camps. Site 1686 (Figure 4.3) is a ~100-x-360-m scatter along the top and northern edge of Black Channel B. It contains some 500 pieces of debitage and 69 tools (Table 4.2), which include gravers, unifaces, and a number and variety of projectile points, scrapers, and bifaces. Approximately 800 m downstream, 1688 consists of a linear cluster of flaked-stone tools and ~225 pieces of detritus dominated by core-reduction flakes. The associated tool kit reflects an array of processing and fabrication tasks; it contains a number of projectile

Channel	No. of Sites[a]	Area (m²) Range	Area (m²) Mean	Mean No. of Tools	Mean Pieces of Debitage	Mean Tool:Debitage Ratio	Mean Artifact Density[b]
Black	33	250–99,462	12,447	22	74	.30	.77
Yellow	14	1,504–22,713	7,595	20	44	.44	.84
Green	7	861–90,972	29,251	37	~135	.27	.59
Blue	10	1,329–52,087	18,619	36	~122	.29	.85
Lavender	9	803–40,757	10,292	27	~184	.15	2.04
Light Blue	18	11–83,483	17,123	8	~96	.08	.64

Note: Channels with fewer than three recorded sites are not included (see Tables 4.6 and 4.9).
[a] Does not include palimpsest sites containing Archaic darts and/or arrowpoints.
[b] Number of artifacts per 100 m².

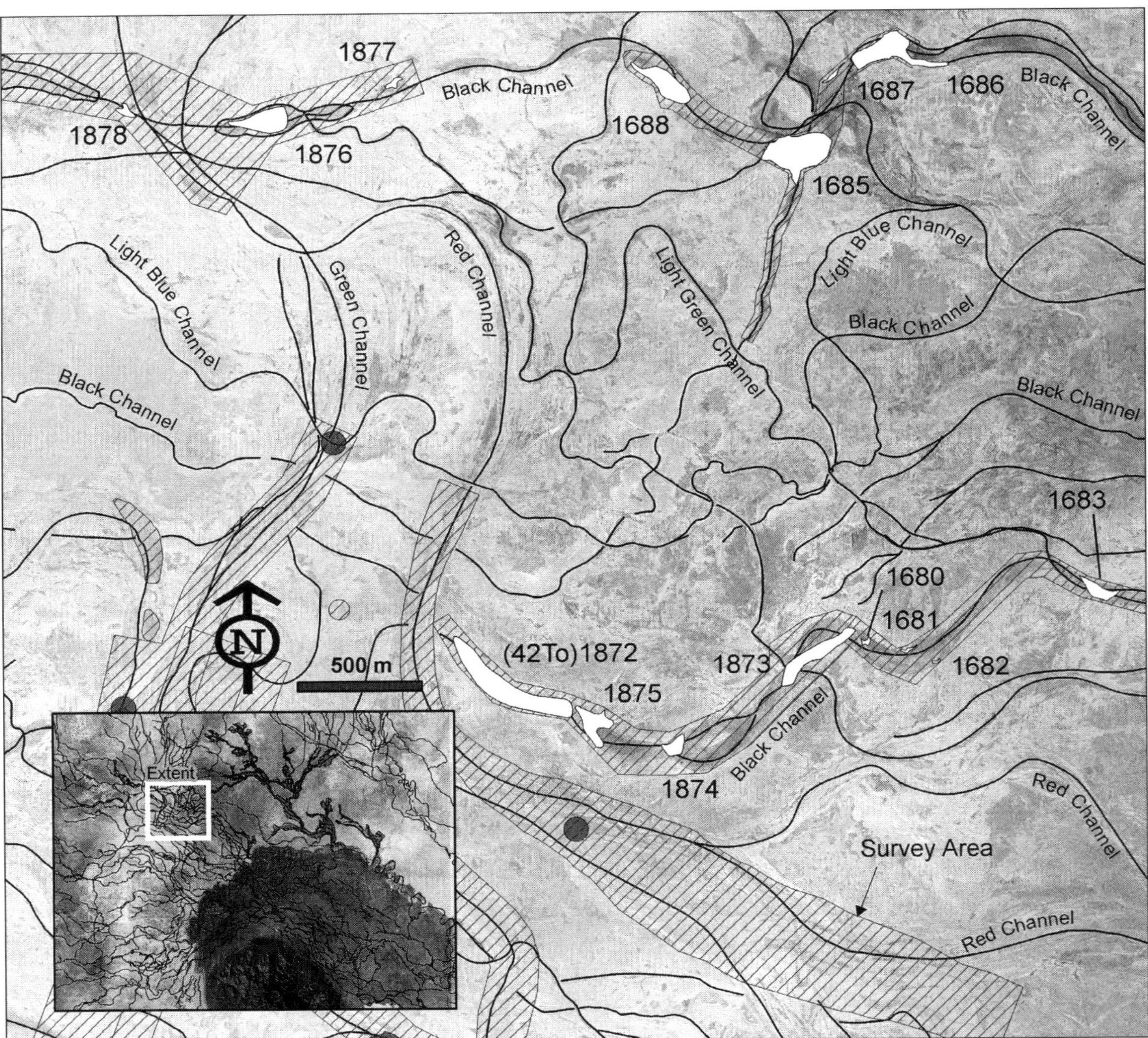

FIGURE 4.3. Location of survey areas and recorded sites (*labeled*) associated with the Black and Light Green channels. The shaded circles mark the location of additional unrecorded sites.

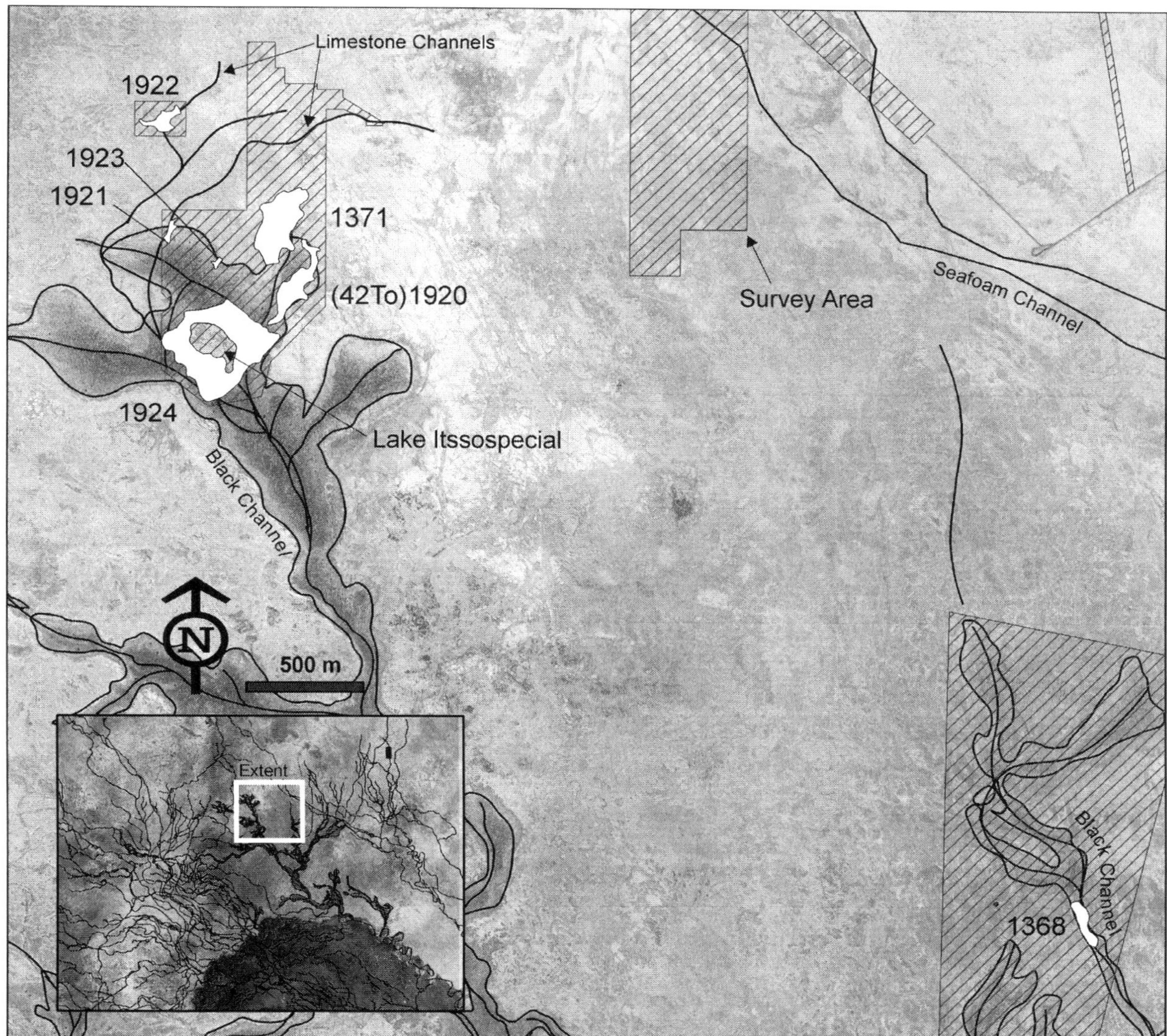

FIGURE 4.4. Location of survey areas and recorded sites (*labeled*) associated with the Black and Limestone channels.

points, bifaces, and scrapers (including a combination scraping-graving tool), along with a crescent and a chisel.

Neighboring sites 1920 and 1371 represent camps along the top and eastern margin of a broad segment of Black Channel A2 immediately north and east of Lake Itssospecial (Figure 4.4). In addition to numerous PA points, both contain a variety of tool classes (e.g., bifaces, cores, scrapers, and gravers [and a drill and chisel at 1920]) with moderate T:D ratios (Figure 4.5). At both sites diffuse scatters of flakes and tools extend east well out onto the mudflats and likely reflect task/procurement areas in an adjacent shallow-water marsh, with the main occupation area being on more elevated portions of the channel (Schmitt et al. 2003). Site 1371 is almost twice as large as 1920, it contains some 200 flakes, and

its sizable tool assemblage (*n* = 103; Table 4.2) is one of the largest in sites recorded in the proximal ORB delta.

Site 1872 contains a large and diverse tool kit and a moderate T:D ratio and also appears to have served as a camp. It is a large, linear (575-×-94-m) scatter of lithic tools and detritus that predominantly occur along the southwestern margin of Black Channel D. Materials are concentrated in some areas, and it contains more tool classes than any other site in the Black channel complex. Among others, tools include a large number and array of PA points/point fragments, bifaces, scrapers, and expedient flake tools, along with a chisel and a Cody knife (see Chapter 5). As at sites 1920 and 1371, artifacts extend out onto the mudflats, where the channel inhabitants may have lost and discarded tools while harvesting adjoining marsh resources.

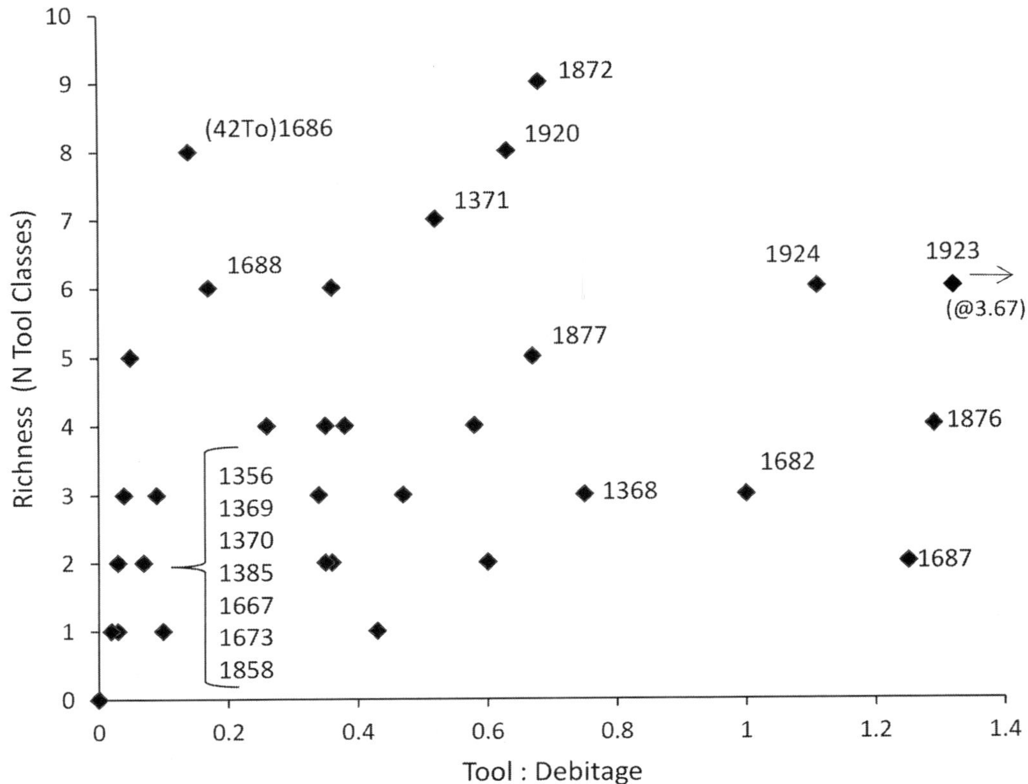

FIGURE 4.5. Scatterplot of the relationship between the number of tool classes and tool:debitage ratios for sites associated with the Black channels. The labeled plots represent the primary sites discussed in the text.

At least seven sites contain few tool types, are dominated by debitage, and appear to have largely served as lithic-reduction loci. Three of these sites—1356, 1673, and 1858—are relatively large and diffuse scatters that may represent a series of brief reduction episodes dispersed by postdepositional processes. Sites 1356 and 1673 each contain one artifact class (stemmed points), while the 1858 tool kit consists of a stemmed point, beak scraper, and core associated with 81 flakes. Assemblages that reflect the most persuasive evidence for tool manufacture and maintenance occur at a site we term "Öferneet Falls" (1370) and three other nearby sites (1369, 1385, and 1667). Site 1369 is on the edge of a broad channel segment along the margin of Lake Öferneet (Figure 4.6). The center of the site is a relatively dense concentration of artifacts (two–three items per square meter) surrounded by a diffuse scatter that extends down onto the Lake Öferneet playa margin. Approximately 200 flakes occur and include large numbers of FGV flakes detached from early- to middle-stage bifaces, and the tools consist of one combination scraper/graver, 10 stemmed points/point fragments, and six fragmentary bifaces. Öferneet Falls contains four Western Stemmed

Tradition (WST) points, four broken bifaces, and ~250 pieces of debitage, and 1385 is a small, dense cluster of approximately 50 flakes with a single production biface in association. Site 1667 is a small scatter of 20 obsidian core-/early-stage biface-reduction flakes containing no tools. Flaking stages at 1385 reflect obsidian and FGV core-reduction and biface-thinning tasks, while Öferneet Falls is dominated by orange/brown chert shatter and core-reduction detritus that may reflect the use of local deltaic stream cobbles. Both Öferneet Falls and 1385 are atop elevated surfaces of topographically inverted channel deposits 2–3 m above the surrounding mudflats, and they contain the highest artifact densities of the 36 sites recorded in the Black channel complex (Table 4.2).

Based on content and context, sites 1368, 1687, and 1876 appear to represent task sites where foragers stationed to procure, and perhaps at times process, wetland resources. Although variable in size, each contains high T:D ratios (Table 4.2; Figure 4.5) and represents diffuse, linear scatters atop and along the margins of inverted channel deposits. Site 1368 consists of two projectile point fragments (including a stemmed point

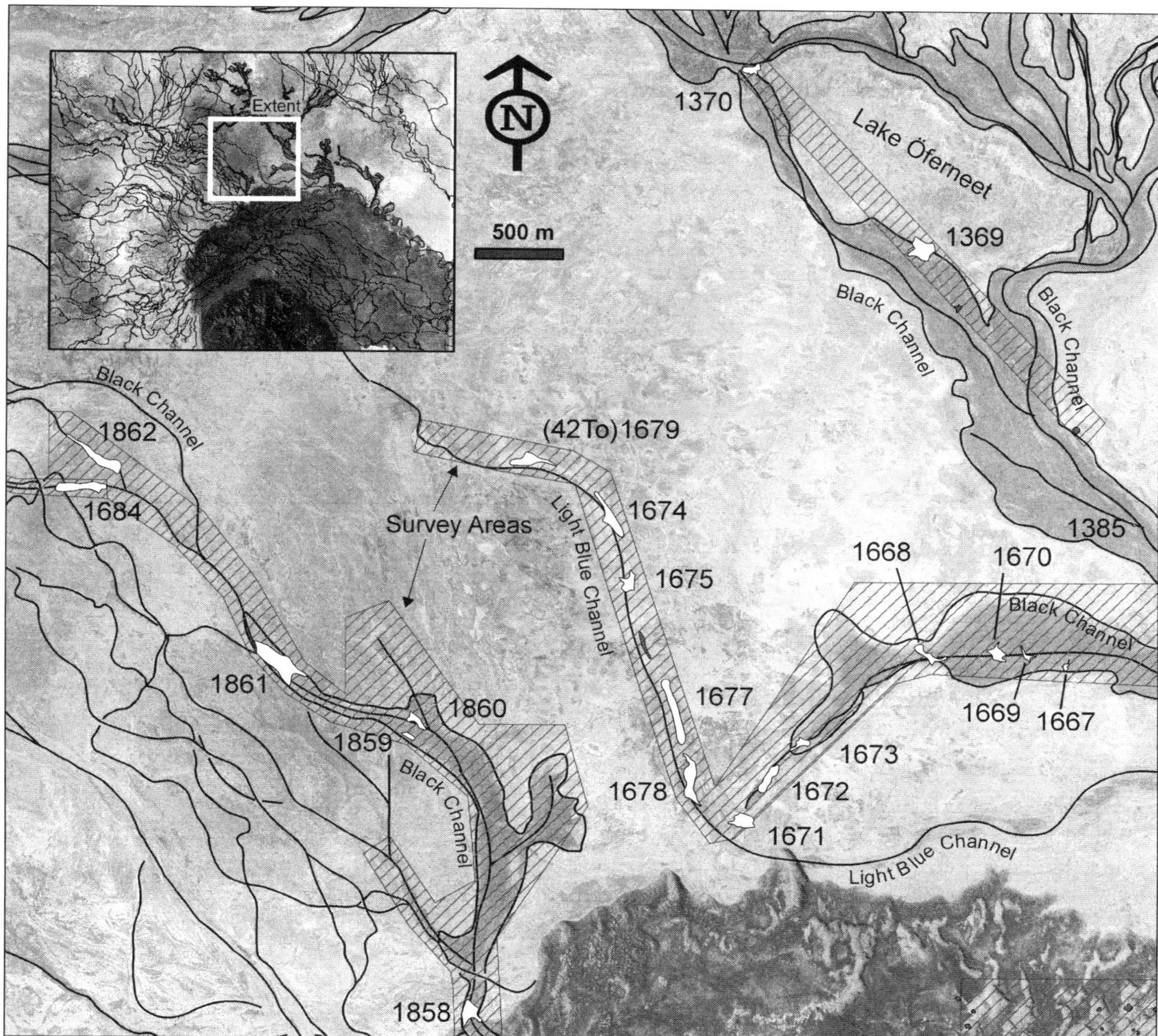

FIGURE 4.6. Location of survey areas and recorded sites (*labeled*) associated with the Black and Light Blue channels.

blade), an end scraper, eight bifaces, four edge-modified flakes, and 20 pieces of debitage. It is near the end of an exhumed Black channel distributary that once formed a peninsula jutting well into the marsh. Site 1876 is a diffuse scatter of seven flakes and nine tools (stemmed points, bifaces, one uniface, and a scraper/graver) atop Black Channel B, and 1687 is a thin and relatively small ($1,157$-m²) scatter of 10 tools and eight flakes along the northern channel margin upstream of 1876. Eight of the 10 tools at 1687 are projectiles (including Stubby [$n = 2$], Lake Mohave [$n = 1$], and Silver Lake [$n = 3$] types), which suggests that the site may have served as a hunting location on more than one occasion.

Three small artifact clusters containing moderately diverse tool kits and high T:D ratios may represent resource-processing stations. Site 1682 (Table 4.2; Fig-

ure 4.3) is a small, linear scatter containing six flakes and three tool classes on the margin of a slightly topographically inverted channel segment. The tools consist of two stemmed point blades, an end/side scraper, and three (refitted) pieces of a late-stage production biface/knife that measures 15 cm in length. Site 1877 is a ~$1,500$-m² scatter of nine flakes and six tools. Materials are atop an east-to-west-trending "island" of Black channel deposits anchored by gravels. The small yet diverse tool kit contains a projectile point fragment, an edge-modified flake, two bifaces, one side scraper, and an FGV chisel. In the northerly reaches of Black Channel A2, 1923 consists of a $1,083$-m² artifact cluster atop exhumed stream deposits containing three flakes and 11 tools representing six different types; it contains the highest T:D ratio of sites in the proximal ORB delta (Table 4.2; Figure

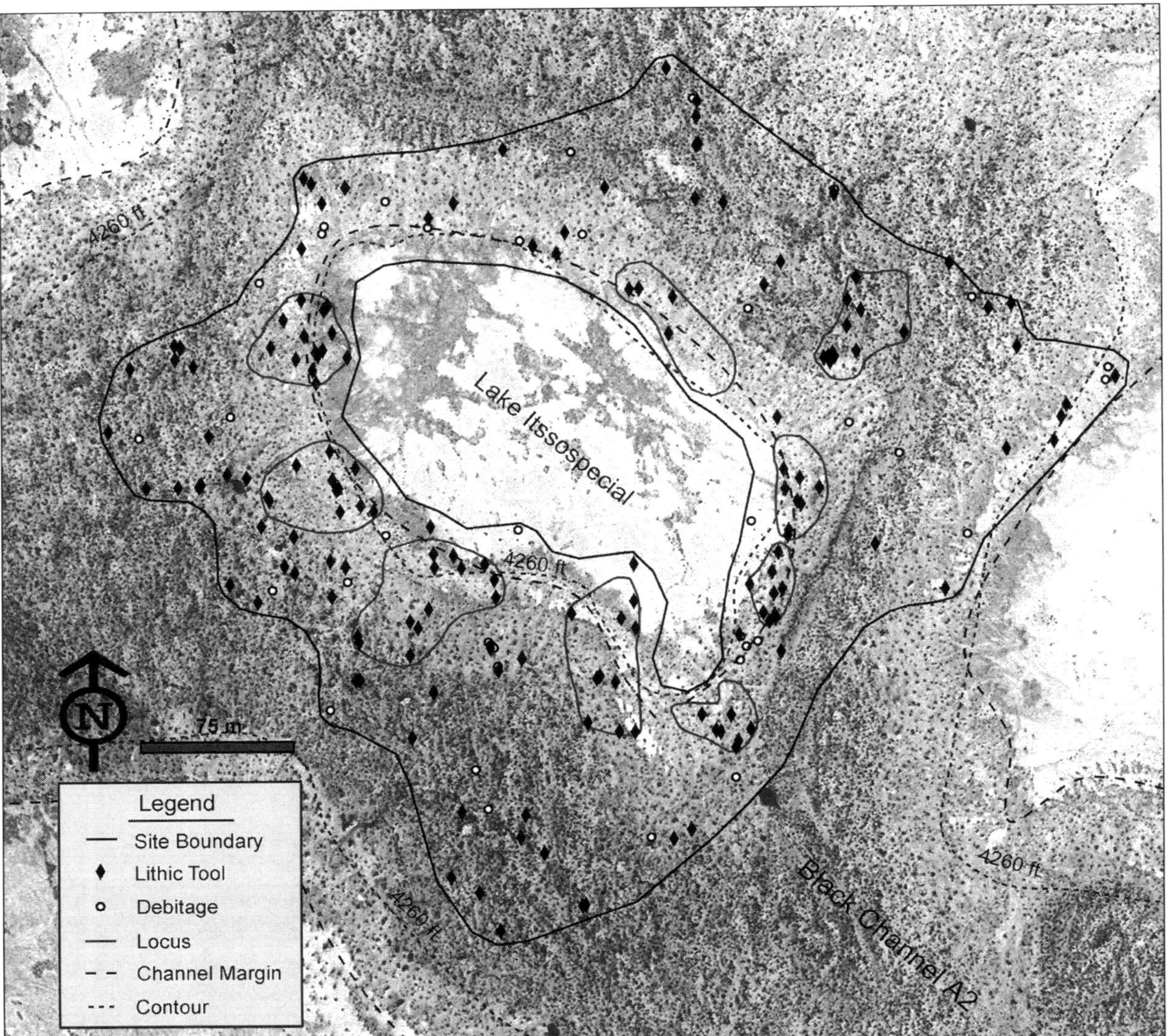

FIGURE 4.7. Plan map of Krispy Kreme Vista (42To1924) on the shores of Lake Itssospecial. Only tools are plotted in the artifact concentration loci.

4.5). Included are (one each) a WST point blade, biface, core, beak scraper, and end/side scraper and six utilized/modified flakes.

One of the more unique sites in the ORB delta is Krispy Kreme Vista (1924), near the northern end of Black Channel A2. As its name suggests, it is a donut-shaped site on the crest and interior margin of an exhumed channel complex that encloses the small Lake Itssospecial basin (Figures 4.4 and 4.7). Cultural materials occur in a continuous scatter around the basin and include a number and variety of projectile points, scrapers, bifaces, drills, unifaces, and simple flake tools and a crescent. Artifact types and densities suggest that the site largely witnessed resource acquisition and processing, but the wealth of tools and presence of nine

artifact concentrations probably reflect multiple visits that included some prolonged stays.

The remaining sites associated with Black channels appear to have largely served as short-term task sites that witnessed varying procurement, processing, and/or tool-production activities. Sites 1669, 1670, 1680, 1681, and 1857 are all relatively small, nondescript scatters containing a few flakes and tools representing one–two classes (Table 4.2). In the western reaches of the Black channel system, sites 1684 (Figure 4.6) and 1873 (Figure 4.3) are large, linear scatters containing 45–50 flakes. Each is dominated by stemmed points and point fragments ($n = 12$, 67 percent of the tool kit, and $n = 18$, 69 percent of the tool kit, respectively), and they appear to have largely served as hunting areas. However,

TABLE 4.4. Selected Attributes and Contents of Sites Associated with the Yellow Channel in the Proximal Old River Bed Delta.

Site No. (42To-)	Projectile Points[a]	Area (m²)	No. of Tool Classes[b]	No. of Tools	Pieces of Debitage	Tool:Debitage Ratio	Artifact Density[c]
2950	—	2,523	2	3	39	.08	1.66
2951	S, E	6,935	3	38	54	.70	1.33
2952	S, E	6,269	3	20	42	.48	.99
2953	S, E	3,941	3	12	22	.55	.86
2954	S	1,712	4	7	10	.70	.99
3219	S, E	13,535	9	69	115	.60	1.36
3220	S	1,504	3	6	10	.60	1.06
3221	S, E	8,617	3	8	65	.12	.85
3222	S, E	2,110	5	9	22	.41	1.47
3223	S	22,713	4	19	98	.18	.52
3224	S	13,362	5	29	33	.88	.46
3225	S, E	2,952	4	13	15	.87	.95
3226	S, E	17,964	4	41	40	1.03	.45
3521	—	2,187	2	2	55	.04	2.61
Total (mean)		106,324	(3.9)	276	620	(.52)	(1.11)

[a] S = Great Basin Stemmed and lanceolate; E = Early Holocene stemmed, Stubby, and Pinto.
[b] General functional classes/types (see Table 4.1). Some tallies include artifact classes that were identified in the field and not collected.
[c] Number of artifacts per 100 m².

additional tool classes at these sites reflect fabrication and processing tasks; site 1684 contains a biface, an obsidian blade core, two scrapers, and two combination scraper/gravers, and 1873 contains a graver, a side scraper, and six bifaces. Site 1861 consists of a diffuse artifact scatter with a diverse array of projectile points and cutting and scraping tools and likely witnessed resource acquisition and processing. Finally, 1860 is an often dense cluster of approximately 100 flakes and five tools representing five different classes. The wealth of debitage suggests that lithic reduction/fabrication was paramount, but the small yet diverse tool kit includes a biface, stemmed point blade, graver, beak scraper, and side scraper, and it appears that processing activities may have also occurred.

Yellow Channel Sites

The Yellow channel exits the RLF plain north of Granite Peak, where it splits into three primary distributaries that flow northwest across what is now the southern Great Salt Lake Desert mudflats. It contains gravels and coarse sands indicative of a relatively high-energy flow and is topographically inverted in most areas, including elevated point bars on interior channel bends where heavy sediment loads were deposited. Based on its relationship to neighboring dated channels, the channel probably formed in 10,500–10,200 ^{14}C BP (see Chapter 3). Archaeological surveys of Yellow channel distributaries identified a total of 14 sites that appear to

date solely to the PA (Table 4.4; Figure 4.8). On average these sites are small in size, contain small assemblages of lithic debitage, and relative to other channel aggregates in the proximal delta, possess high T:D ratios (Tables 4.3 and 4.4). Most of the sites associated with the Yellow channel contain two–four tool types (Figure 4.9) and seem to manifest task sites centered on stone tool production, resource procurement, and/or resource processing. Site 3219, however, possesses a large and diverse tool kit associated with more than 100 pieces of debitage and appears to represent a short-term camp. It is a large (and at times dense) scatter around an exhumed gravel-bar "island" associated with the channel (Figure 4.10). The scatter extends up the lower slopes of the gravel island, but most materials lie within the confines of the channel and were likely deposited on a patch of exhumed channel deposits that postdates its flow. Overall, the site contains nearly equal numbers of FGV and obsidian debitage (~60 flakes each), with the FGV dominated by biface-thinning flakes and the obsidian dominated by core-reduction debris. Sixty-nine tools were observed and include a variety of projectile points and bifaces, a crescent and a drill, and a number and variety of scrapers and simple flake tools.

Two Yellow channel sites (2950 and 3521) are dominated by debitage, contain two tool classes (Figure 4.9; Table 4.4), and appear to represent brief use episodes that centered on tool manufacture and maintenance. Site 2950 consists of a relatively small (~2,500-m²)

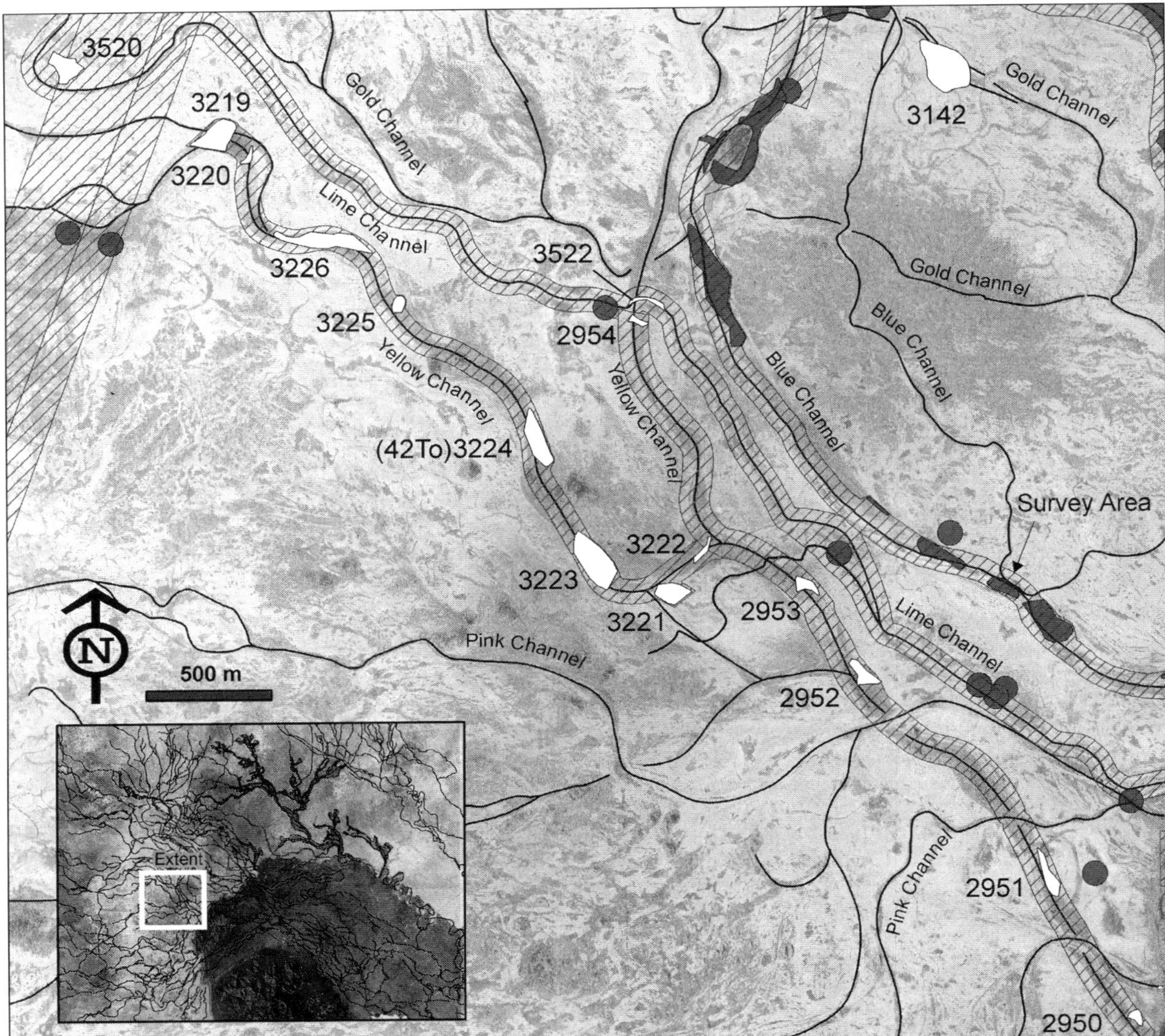

FIGURE 4.8. Location of survey areas and recorded sites (*labeled*) associated with the Yellow, Lime, and Gold channels. The shaded circles mark the location of additional unrecorded sites.

artifact scatter on the main distributary channel where about 20 FGV and obsidian flakes occur on both channel margins. The debitage consists of relatively equal numbers of core-reduction and biface-thinning debris, and the associated tools include an edge-modified flake and broken biface. Site 3521 is a similar small cluster of debitage and tools wholly confined within a Yellow channel distributary, and it appears that the site was occupied after the channel ceased flowing. A small concentration of ~20 FGV interior flakes is located near the west end of the site, with a thin veneer of flakes and tools (one obsidian core and a fragmentary biface) extending upstream. Although site 3221 also contains abundant lithic detritus and a low T:D ratio (Figure 4.9), its overall context and content suggest that it rep-

resents resource-acquisition activities in addition to tool manufacture. Specifically, it is in a shallow pond-like depression along the channel margin, materials are more sparsely scattered over a larger area, and it contains five stemmed points and three bifacial tools.

Five sites along the Yellow channel are considered to largely represent resource-extraction loci. Site 2951 is a linear scatter of 54 flakes, 20 bifaces, seven flake tools, and 11 projectile points that represent a variety of types (see Chapter 5). Artifacts are most abundant in a ca. 30-m-diameter area near the center of the site, with a few flakes and tools occurring on the mudflats to the northeast and others extending downstream along the center of the topographically inverted channel segment. Site 2954 consists of a ~75-×-30-m scatter of 10

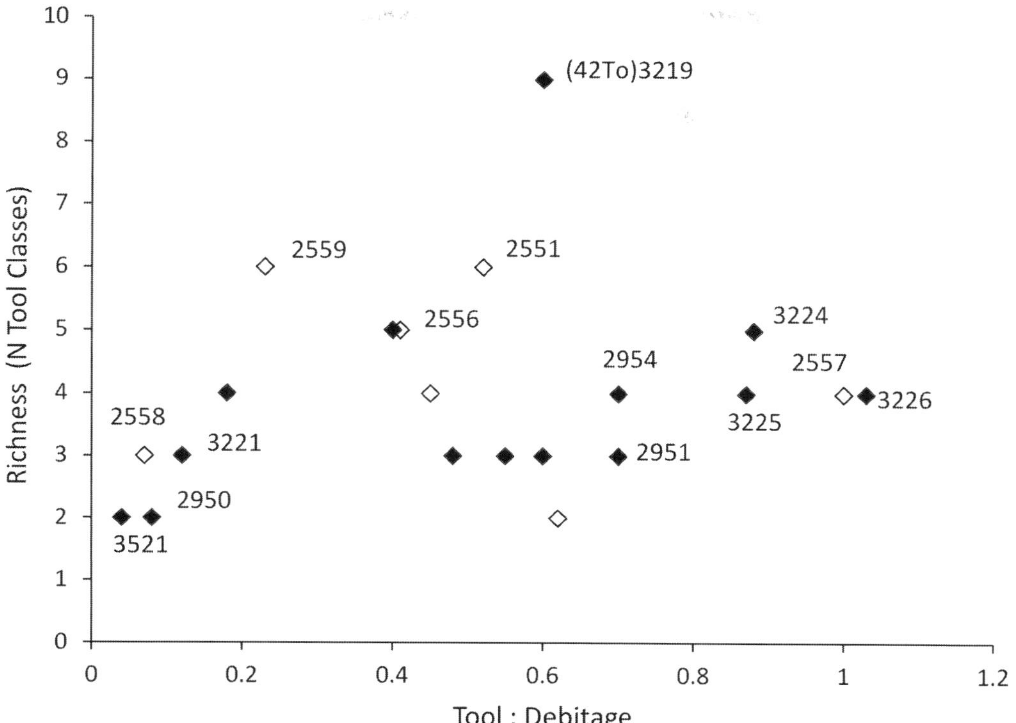

FIGURE 4.9. Scatterplot of the relationship between the number of tool classes and tool:debitage ratios for sites associated with the Yellow (*shaded*) and Green (*open*) channels. The labeled plots represent the primary sites discussed in the text.

flakes and a small yet diverse tool kit containing four projectile points, a scraper, one bifacially modified flake tool, and an obsidian biface. Because some materials are associated with the Yellow channel and others parallel a crosscutting distributary, the site may actually be associated with either channel; regardless, artifact types and distributions tend to signal resource-acquisition activities in a small patch. Three sparse, linear lithic scatters contained wholly within the channel/channel margins are also classified as procurement sites. Site 3224 (Figure 4.11) consists of an array of FGV and obsidian flakes and tools along a Yellow channel segment slightly deflated (~10 cm) below the surrounding mudflats. It contains 11 bifaces, three WST point blades, three scrapers, a crescent, and a variety of flake tools associated with 33 widely scattered pieces of debitage. Although smaller in size, 3225 is also in the bottom of a deflated channel segment and possesses a similarly high T:D ratio (Figure 4.9). Artifacts are dominated by FGV and include 15 flakes and 13 tools, including projectile points, bifaces, scrapers, and utilized flakes. Downstream to the west, 3226 is a diffuse scatter of flakes and tools more than 400 m in length and is bounded on its up- and downstream ends by inverted gravel-bar islands. The debitage includes 21 core-reduction flakes and 18 biface-thinning

flakes, and the tool kit consists of stemmed points, bifaces/biface fragments, scrapers, a crescent, and numerous utilized/edge-modified flakes.

The remaining sites associated with the Yellow distributaries largely contain three–four artifact classes, have low to moderate T:D ratios, and appear to represent various task loci. Sites 2953 and 3220 are both relatively small lithic scatters containing three tool types each and a thin veneer of debitage (Table 4.4), and 3223 is a large, diffuse scatter of 98 flakes associated with 19 tools, which include a few bifaces, gravers, and edge-modified flakes; the small projectile point assemblage includes Cougar Mountain and Pinto types, and it appears that 3223 witnessed at least two occupations.

Green Channel Sites

The Green channel flows northwest from the RLF–mudflat interface approximately 2.2 km east of the Yellow channel, where it splits into a series of somewhat linear distributaries that continue their northwesterly courses for more than 12 km. The Green channel contains abundant small gravels and coarse sand, and it appears to be the result of moderate to high energy flow. Radiocarbon assay of plant remains associated with *Anodonta* shell and fish bone returned a date of

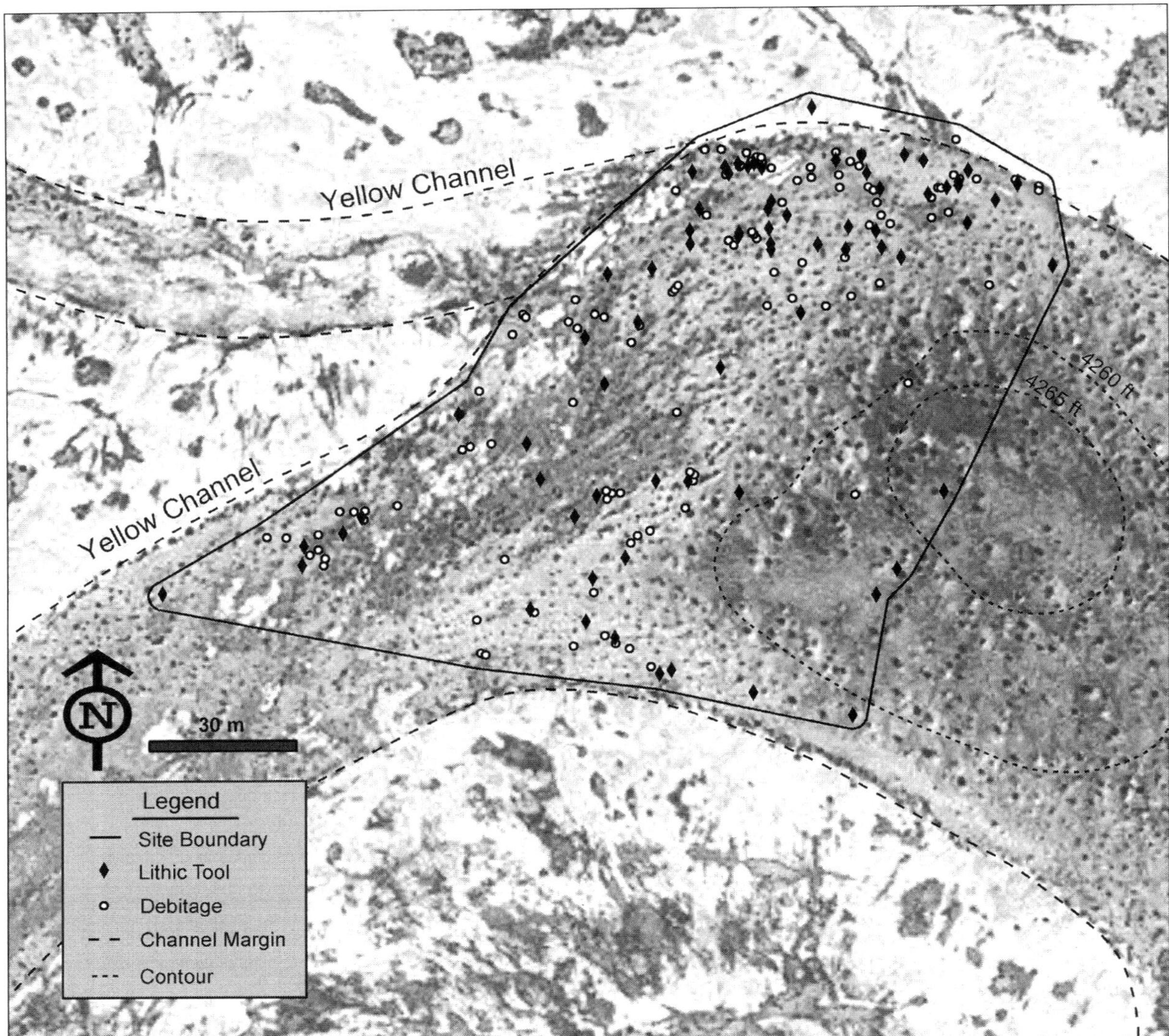

FIGURE 4.10. Plan map of 42To3219 on the Yellow channel.

10,290 ^{14}C BP, which appears to mark channel flow (see Chapter 3). Seven PA sites were discovered while surveying a linear (~200-m-×-10.2-km) segment of the channel (Table 4.5; Figure 4.12). Overall, they tend to be large in size and contain relatively large numbers of flakes and tools (see Table 4.3). Three sites—2551, 2556, and 2559—contain abundant and diverse artifact assemblages with low to moderate T:D ratios (Table 4.5; Figure 4.9) and are classified as camps. Site 2551 is a linear scatter covering some 47,700 m² within the margins of a Green distributary where a number of areas in the center of the channel have deflated below the surrounding mudflats. The site is 50–100 cm above the mudflat surface, where it is armored by pea gravel. Artifacts are most abundant in a ~350-×-60-m crescent-shaped scatter within a broad curve in the channel, with materials

extending upstream and especially downstream from this main concentration (Figure 4.13). Approximately 120 flakes and 62 tools were observed, the latter including bifaces, scrapers, edge-modified/utilized flakes, a split obsidian cobble, a crescent, and a large number of stemmed points.

Site 2556 is a ~22,000-m² scatter of lithic tools and approximately 125 pieces of detritus containing obsidian, FGV, and some chert. Materials occur in deflated areas between a series of migrating channel meanders and, hence, tend to form linear stringers of artifacts (Madsen et al. 2006). Debitage types indicate that biface reduction was common, and the 51 tools found in association include projectile points, gravers, bifaces, and scrapers that reflect a variety of fabrication and processing tasks. Finally, site 2559 consists of an extensive

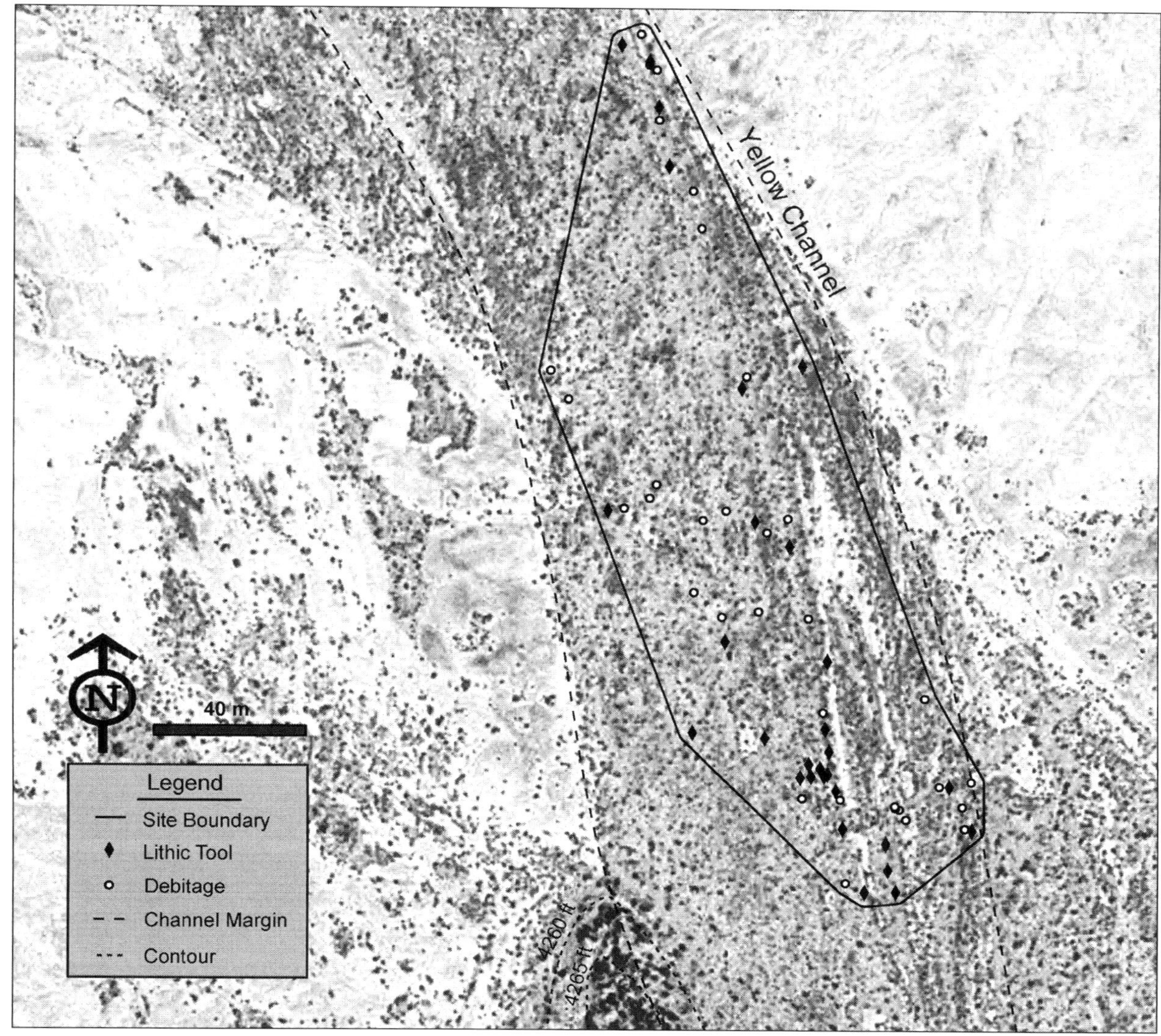

FIGURE 4.11. Plan map of 42To3224 on the Yellow channel.

TABLE 4.5. Selected Attributes and Contents of Sites Associated with the Green Channel in the Proximal Old River Bed Delta.

Site No. (42To-)	Projectile Points[a]	Area (m²)	No. of Tool Classes[b]	No. of Tools	Pieces of Debitage	Tool:Debitage Ratio	Artifact Density[c]
2551	S, E	47,757	6	62	122	.51	.38
2552	S	30,749	4	13	122	.11	.45
2553	S, E	3,146	2	8	13	.62	.67
2556	S, E	22,070	5	51	~125	.41	.80
2557	S, E	861	4	8	8	1.00	1.86
2558	S, E	9,204	3	10	134	.07	1.63
2559	S, E	90,972	6	104	423	.24	.61
Total (mean)		204,759	(4.3)	256	~947	(.42)	(.91)

[a] S = Great Basin Stemmed and lanceolate; E = Early Holocene stemmed, Stubby, and Pinto.
[b] General functional classes/types (see Table 4.1). Some tallies include artifact classes that were identified in the field and not collected.
[c] Number of artifacts per 100 m².

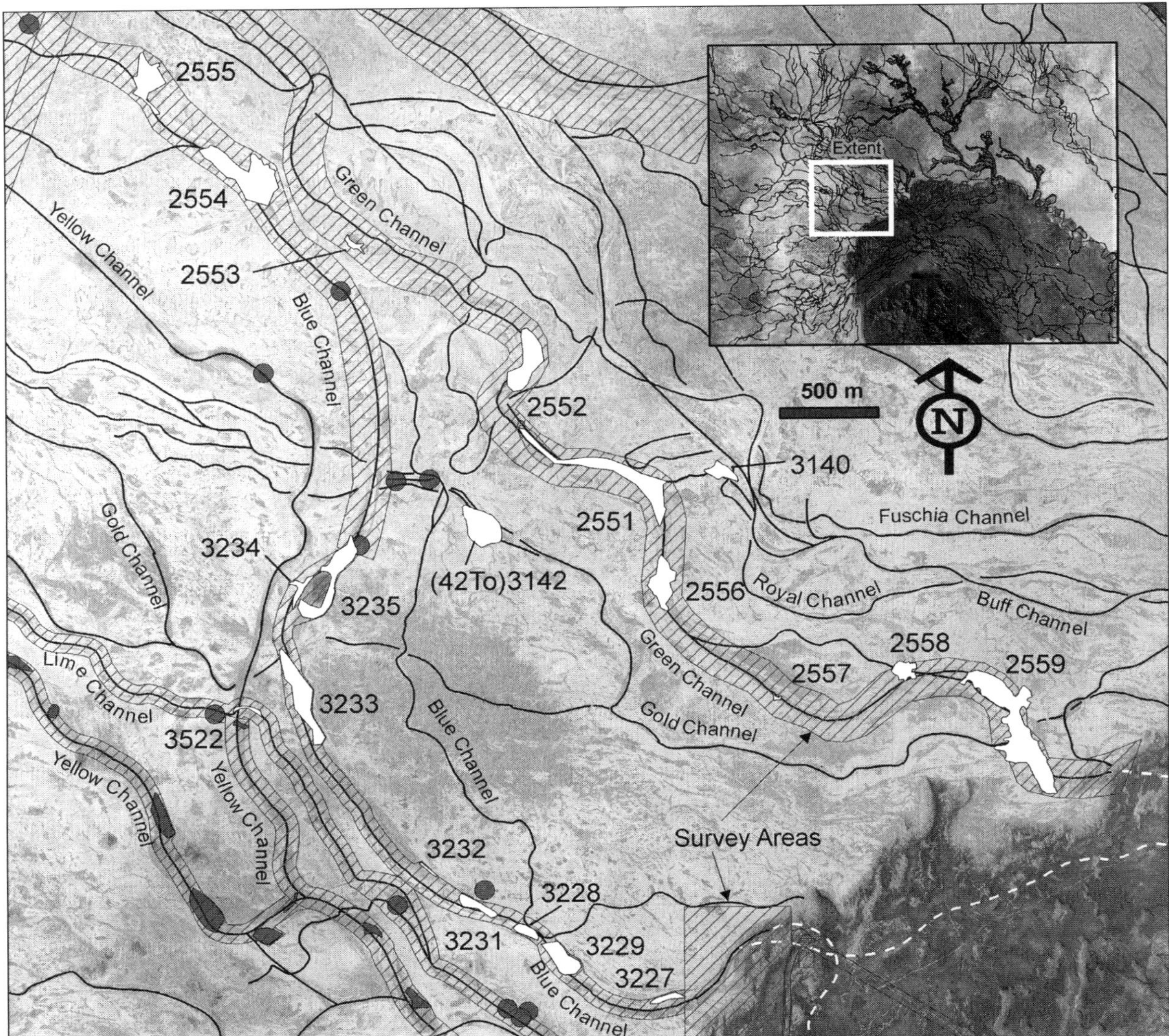

FIGURE 4.12. Location of survey areas and recorded sites (*labeled*) associated with the Green, Blue, Lime, and Buff/Royal Blue channels. The shaded circles mark the location of additional unrecorded sites.

scatter of 104 lithic tools and more than 400 flakes along a broad S-shaped curve of the Green channel (Table 4.5; Figure 4.12). The site covers more than 22 ac and is in excess of 800 m in length. In a few areas artifacts extend well out onto the mudflat surface, where foragers probably exploited neighboring marsh resources. Several concentrations of lithic debris occur and appear to mark tool-manufacture and -maintenance areas (Madsen et al. 2006), and artifacts are also abundant between stream meanders on what may have once been natural levees parallel to channel flow. The context of these materials and the presence of numerous WST and Early Holocene projectile point forms (see Chapter 5) likely reflect overlapping camps that include an early occupation associated with channel formation. The 2559 debitage assemblage contains large numbers of obsid-

ian core-reduction flakes and FGV biface-thinning/manufacture debris, and the associated tools consist of 29 bifaces, 31 edge-modified flakes, two scrapers, and four gravers.

Site 2557 is a small, diffuse scatter of eight pieces of debitage and eight tools down the southern edge of a slightly exhumed Green channel distributary. A small concentration of five artifacts occurs in one area, but the remainder of the flakes and tools are strung along the channel in a linear fashion and likely represent items lost and/or discarded by people working in a small patch. Conversely, 2558 consists of a larger and denser artifact scatter dominated by obsidian debitage and served predominantly as a lithic-reduction locus. Artifacts are distributed in and between a younger channel that is deflated to the level of the surrounding mudflats, a

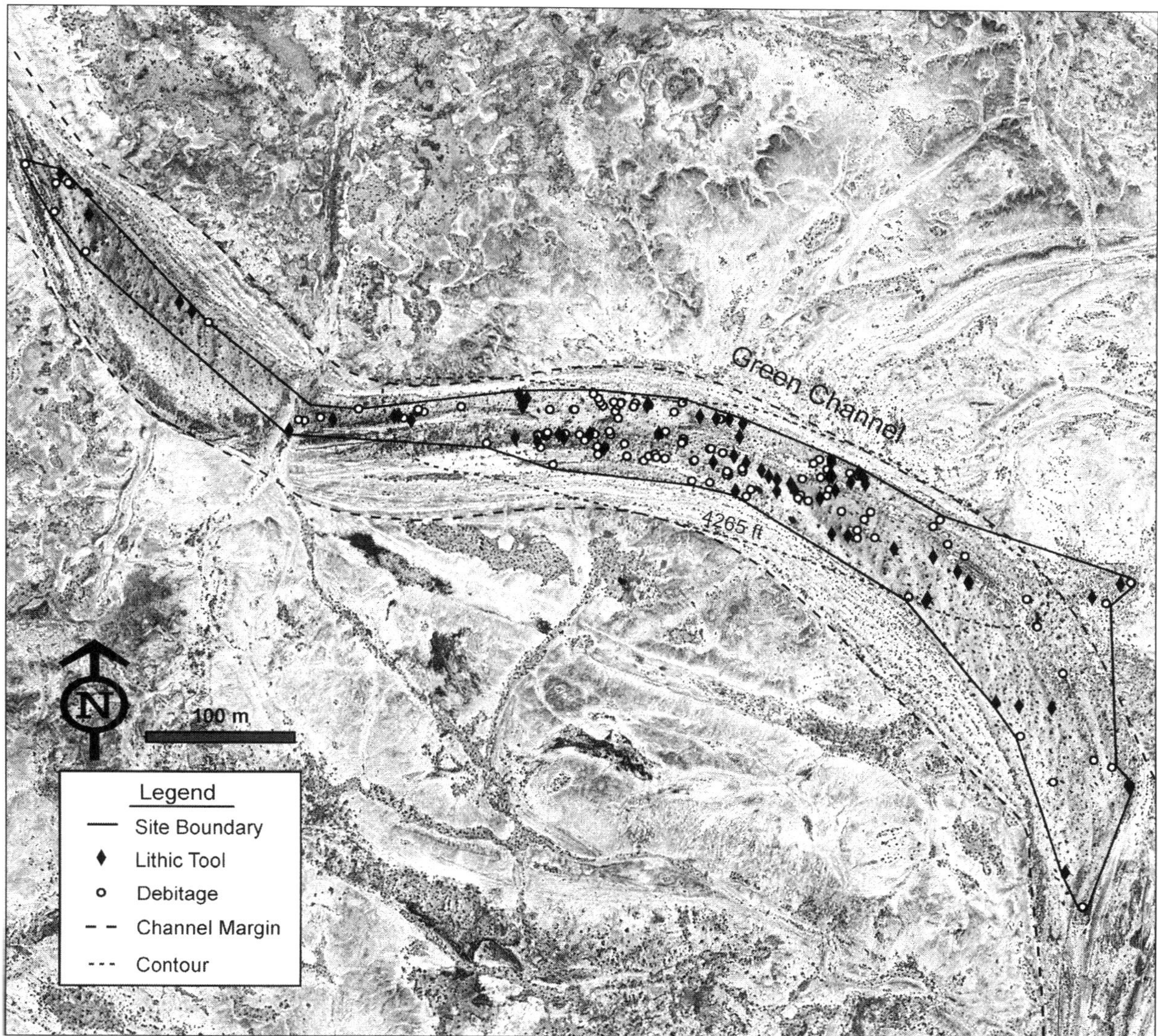

FIGURE 4.13. Plan map of 42To2551 on the Green channel.

slightly inverted segment of the Green channel, and an associated low silt dune. The tools consist of an edge-modified flake and two fragmentary bifaces, as well as seven stemmed points/point fragments that may reflect the hunting of game in or immediately adjacent the site. Finally, 2552 is a large, diffuse artifact scatter dominated by debitage (Table 4.5) on exhumed deposits armored by pea gravels, and 2553 consists of a diffuse, crescent-shaped array of 13 lithic flakes and eight tools along the western margin of a broad distributary segment.

Blue and Lime Channel Sites

The Blue channel (Figure 4.12) consists primarily of two roughly parallel streams (a main terminal channel and an associated distributary) that converge and diverge on a northwesterly course from the RLF–mudflat transition. The deposits contain some small gravel and

coarse sand, but less than that in the Yellow and Green channels, and it appears to have been a lower-energy stream. Three radiocarbon age estimates from black mat deposits in two separate locations span from ~9750 to 9450 ^{14}C BP (see Chapter 3), and it appears that the channel originally formed prior to ~9800 ^{14}C BP. To the immediate west, the Lime channel emerges from the RLF, where it parallels the Blue channel for ~3 km and then bends on a westerly course. Its sediment load is very similar to that in the Blue channel and is considered to be about the same age. Ten PA sites were identified while surveying two linear segments of the Blue channel (total length ~6.5 km [Page et al. 2008]), and two sites were recorded along the Lime channel during a linear (11.3-km) survey of the channel from where it emerges from the underflow fan margin to well out onto the deflated mudflats (Madsen and Page 2008).

TABLE 4.6. Selected Attributes and Contents of Sites Associated with the Blue and Lime Channels in the Proximal Old River Bed Delta.

Site No. (42To-)	Projectile Points[a]	Area (m²)	No. of Tool Classes[b]	No. of Tools	Pieces of Debitage	Tool:Debitage Ratio	Artifact Density[c]
Blue Channel							
2554	S, E	52,087	5	51	~100	.51	.29
2555	S, E	23,335	4	29	28	1.04	.24
3227	—	4,548	3	6	24	.25	.66
3228	S, E	5,756	6	17	24	.71	.71
3229	S, E	17,849	5	23	~500	.05	2.93
3231	S, E	8,838	4	8	59	.14	.76
3232	—	1,329	3	5	13	.39	1.35
3233	S, E	35,282	7	48	112	.43	.45
3234	S, E	5,840	6	53	87	.61	2.40
3235	S, E	31,322	10	123	270	.45	1.26
Total (mean)		186,186	(5.3)	363	~1,217	(.46)	(1.11)
Lime Channel							
3520	S, E	9,722	7	48	71	.68	1.22
3522	S, E	2,489	5	10	23	.43	1.33
Total (mean)		12,211	(6.0)	58	94	(.56)	(1.28)

[a] S = Great Basin Stemmed and lanceolate; E = Early Holocene stemmed, Stubby, and Pinto.
[b] General functional classes/types (see Table 4.1). Some tallies include artifact classes that were identified in the field and not collected.
[c] Number of artifacts per 100 m².

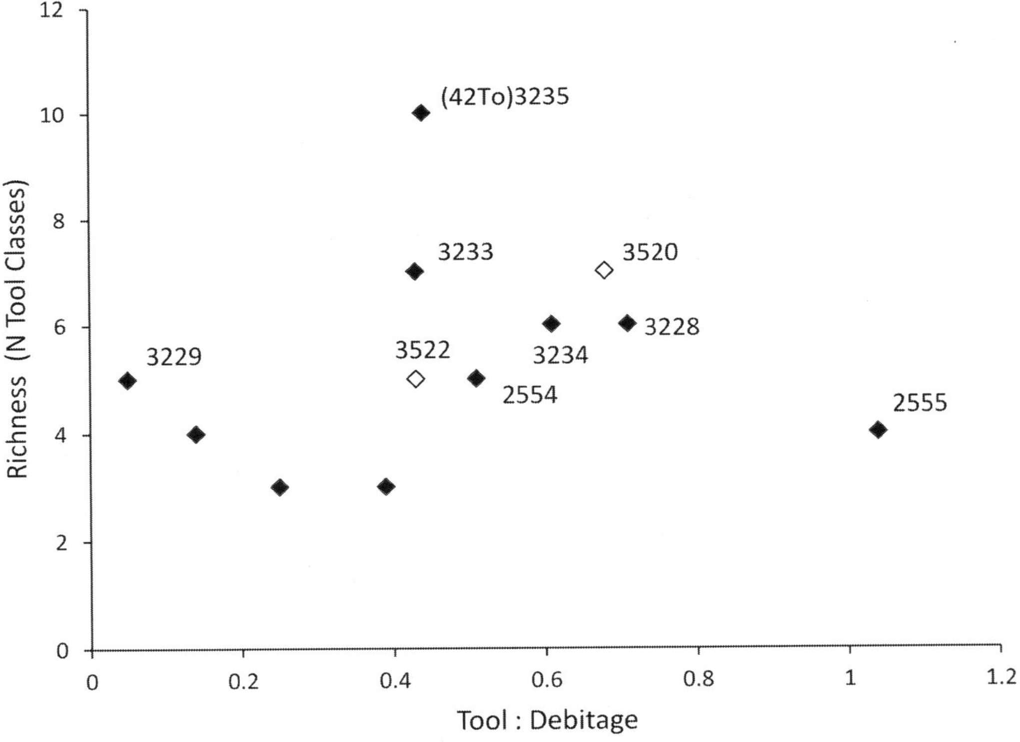

FIGURE 4.14. Scatterplot of the relationship between the number of tool classes and tool:debitage ratios for sites associated with the Blue (*shaded*) and Lime (*open*) channels. The labeled plots represent the primary sites discussed in the text.

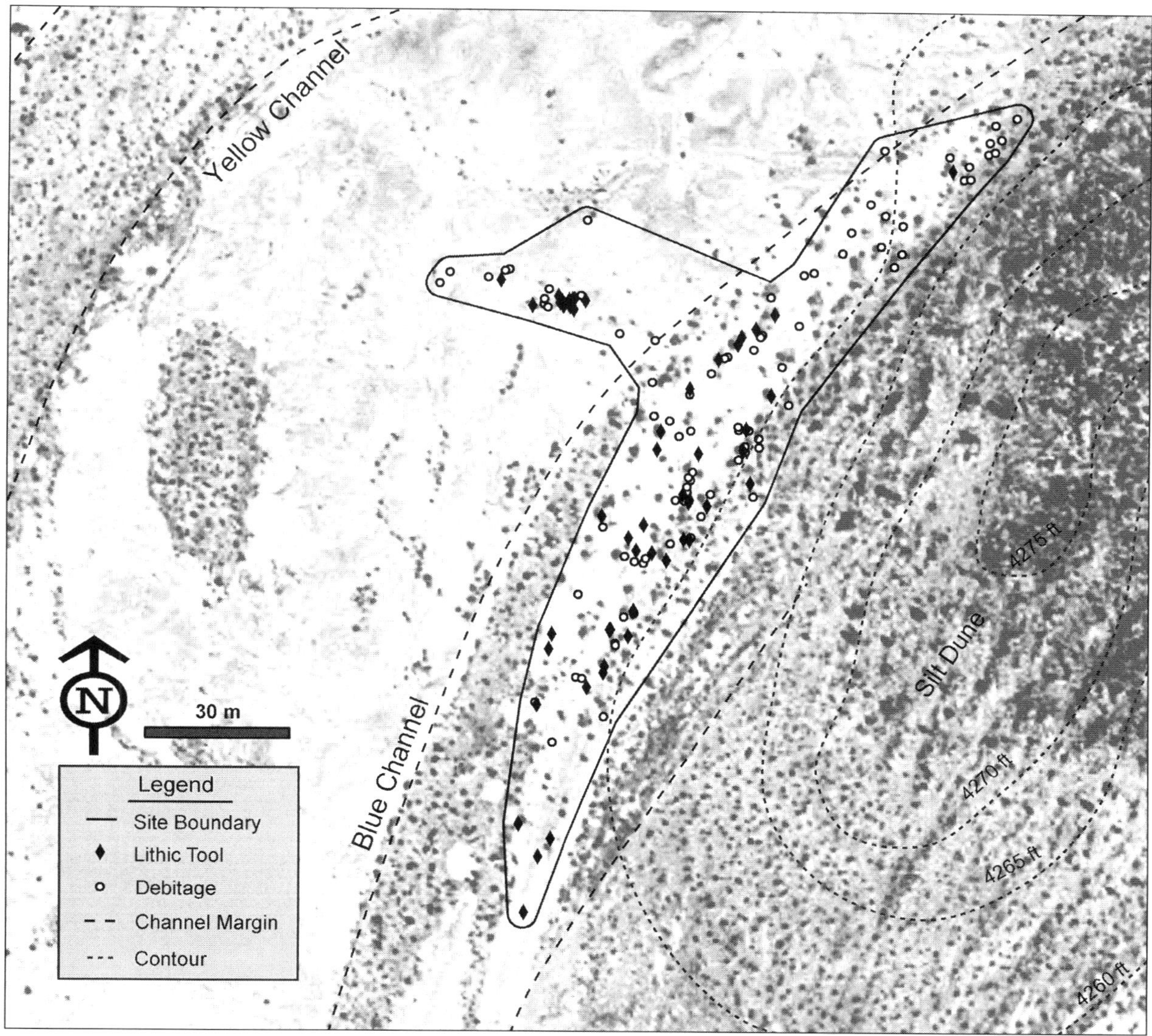

FIGURE 4.15. Plan map of 42To3234 on the Blue channel.

Sites associated with the Blue channel represent a diverse array of activities and appear to represent both brief and prolonged stays. They tend to be large in size, and most sites contain abundant tools and tool classes (Tables 4.3 and 4.6). Neighboring sites 3233, 3234, and 3235 contain abundant artifacts and a variety of tool classes (Figure 4.14) and are classified as base camps. Site 3233 is a long, linear array of obsidian (dominant) and FGV tools and debitage along a Blue channel distributary. This channel is immediately south and west of the main terminal channel, and its setting suggests that the site may be older than most other sites associated with the main Blue channel (Page et al. 2008). Flaking stages include some core-reduction detritus and abundant ($n = 74$) biface-thinning debris. Forty-eight tools were observed and include bifaces, WST

stemmed points, three cores, 12 edge-modified flakes, and five scrapers representing four different types. Immediately downstream, 3234 (Figures 4.12 and 4.15) is a small, and in places dense, artifact cluster along an older meander of the Blue channel that appears to have served as a short-term camp on a couple of occasions. It may be associated with a natural levee either formed when water was flowing in the terminal Blue channel or deposited in association with the older channel itself. Regardless, the site context suggests that it is older than sites wholly contained within the terminal Blue channel, notably neighboring site 3235. Debitage types reflect core reduction through late-stage biface thinning, and the associated tools ($n = 53$) consist of beak, end, and slug scrapers; a host of bifaces and utilized/edge-modified flakes; an obsidian core; and 26 projectile points that include Cougar Mountain, Haskett, Silver Lake, Stubby, and Pinto types. Adjoining site 3235 ("Big Blue") consists of a large and often dense lithic scatter wholly contained within the terminal meander of the Blue channel. A large number of artifacts occur along the western margin of the channel at the interface between the older and terminal channels, and it appears that foragers may have been occupying a stream bank/levee in this area. A large dune atop the channel encroaches the southwestern portion of the site, and it is likely that some buried culture-bearing deposits are present in this area. Approximately 275 flakes occur at 3235 and signal core reduction through the manufacture/maintenance of mid- to late-stage bifaces. Its large and diverse tool kit includes a knife, drill, and graver; four cores; and a large number and variety of bifaces, scrapers, expedient flake tools, and stemmed points.

At least three Blue channel sites are considered to represent acquisition loci, although they are variable in size and likely reflect size differences in associated patches. Site 2554 (Table 4.6; Figure 4.12) consists of a large scatter ($52,087 \, m^2$) of FGV and obsidian tools and detritus on a slightly topographically inverted channel segment armored by fine pea gravel. The scatter extends some 100 m to the east of the main site area onto the mudflat surface, where foragers likely exploited adjoining marsh resources. Approximately 100 flakes were observed along with 51 tools that include bifaces, modified flakes, three scrapers, one knife, and numerous projectile points dominated by Stubby and Expanding and Square stem types. Downstream (~northwest) of 2554, site 2555 consists of several diffuse stringers of stone tools and detritus along a series of Blue channel segments that formed as the stream was migrating to the north. Artifacts occur in deflated areas between the slightly elevated channel deposits, and in a few areas they extend north/northeast out onto the neighboring mudflats. Lithic tools include two obsidian scrapers, nine bifacially modified flakes, six biface fragments, and 12 projectile points that largely represent Early Holocene types. Site 3228 is a sparse scatter of 24 flakes and 17 tools within the main (terminal) channel of the Blue channel distributary. All the observed materials are within the channel margins and likely were deposited after primary channel flow when the streambed supported a brackish wetland. Most ($n = 20$) of the debitage consists of biface-thinning debris, and the associated tool kit is dominated by WST points/point fragments ($n = 10$) along with a few bifaces, scrapers, a bifacially modified flake, and an obsidian core.

Site 3229 is a rather unique lithic scatter containing abundant debitage and 23 tools that include projectile points, bifaces, scrapers, a uniface, and expedient flake tools. It is a linear distribution of artifacts contained within the margins of the terminal Blue channel. Lithic materials (notably obsidian biface-thinning flakes), although generally widespread, are concentrated near its north end. Here, and across most other site areas, the biface-thinning debris includes both large flakes detached from early-stage bifaces and small detritus from late-stage production/maintenance. Based on the types and distribution of materials, the site appears to have served as a resource-acquisition/processing area that witnessed extensive lithic-reduction activities, or it functioned as a short-term camp where stone tool manufacture was common.

The three remaining sites consist of thin artifact scatters containing few tool classes and appear to manifest task areas. Site 3227 consists of a diffuse veneer of 24 flakes and six tools, and 3231 contains 59 flakes and eight tools scattered across some 8,800 m^2 of partially inverted stream deposits. Near the center of 3231 artifacts are generally ~2–4 m apart, with diffuse scatters running up- and downstream from this area. Site 3232 is a small linear scatter of 13 FGV flakes associated with two bifaces, a scraper, and two edge-modified flakes. Materials are most abundant along the outer (eastern) main Blue channel margin where it meets the mudflats, which may reflect resource-harvesting/processing activities in a small wetland patch.

Paleoarchaic sites associated with the Lime channel include a short-term camp (3520) and a task site (3522) that appears to have centered on resource acquisition (Table 4.6; Figure 4.14). Site 3520 (Figures 4.8 and 4.16) is a relatively dense cluster of 71 flakes and 48 tools that represent at least seven types. It is on the inner side of an oxbow that was migrating to the south when it was abandoned. Artifacts appear to have been deposited on what would have been a point bar on that inner bend, but the area is now deflated to the elevation of the surrounding mudflats. At the approximate center of the site there is a ca. 25-m-diameter area containing numerous flakes and tools, with lithic materials feathering out from this central location another 80–100 m across the bar. The 3520 tool kit contains stemmed points, bifaces, unifaces, gravers, a variety of scrapers, and a tabular quartzite manuport. Site 3522 consists of a linear array of lithic tools and flakes along the outer margin of the Lime channel where it crosscuts a segment of the more gravel-filled Yellow channel distributary. Overall, the site is nearly 150 m long, but only about 20 m wide at maximum, and is mostly restricted to the channel fill. Given its context and small yet diverse tool kit—five stemmed points/point fragments, two slug scrapers, one crescent, one

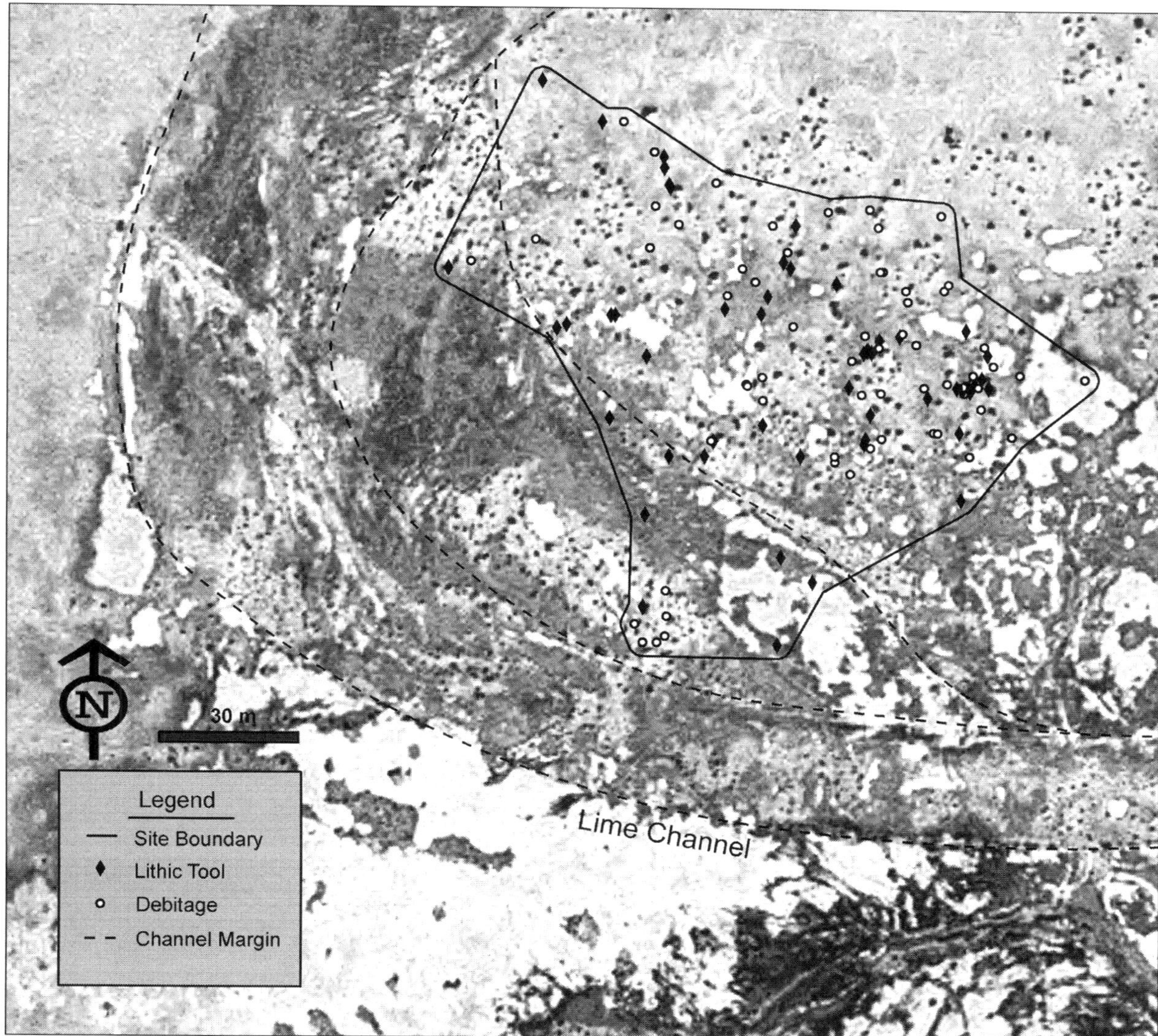

FIGURE 4.16. Plan map of 42To3520 on the Lime channel.

uniface, and a scraping-graving tool—the site was probably created by foragers procuring marsh resources in
a small patch, and the presence of Haskett and Pinto
points probably reflects two brief visits.

Lavender Channel Sites

The Lavender channel complex (Figure 4.17) is a series
of curving streambeds that flow west along the extreme
southwestern margin of the ORB delta. Given the lack
of gravels and often pronounced meander patterns,
the Lavender channels were doubtless formed by low-
energy streamflow beginning at approximately 9100
^{14}C BP. A total of 11 archaeological sites was recorded
in association with the channel during reconnaissance
of approximately 7.5 linear km encompassing a series of
distributary segments. Two of the sites (2947 and 2948)

are rather large palimpsests of PA and younger materials (Table 4.7), and the remaining nine sites contain
stemmed points and/or extensively weathered obsidian
and FGV artifacts and are considered to be wholly PA in
age. These sites tend to possess large numbers of tools
and low T:D ratios, most are small in size, and together
they have the highest artifact densities of any channel
aggregate in the proximal ORB delta (see Table 4.3).

Three of the sites contain hundreds of flakes and
diverse tool assemblages (Table 4.7; Figure 4.18) that
likely reflect prolonged occupations where a variety of
fabrication and processing tasks were performed. One
of these—2945—is a massive artifact scatter containing
39 tools and some 700 flakes. Materials occur on a deflated mudflat surface between and on migratory distributaries associated with the channel, and although

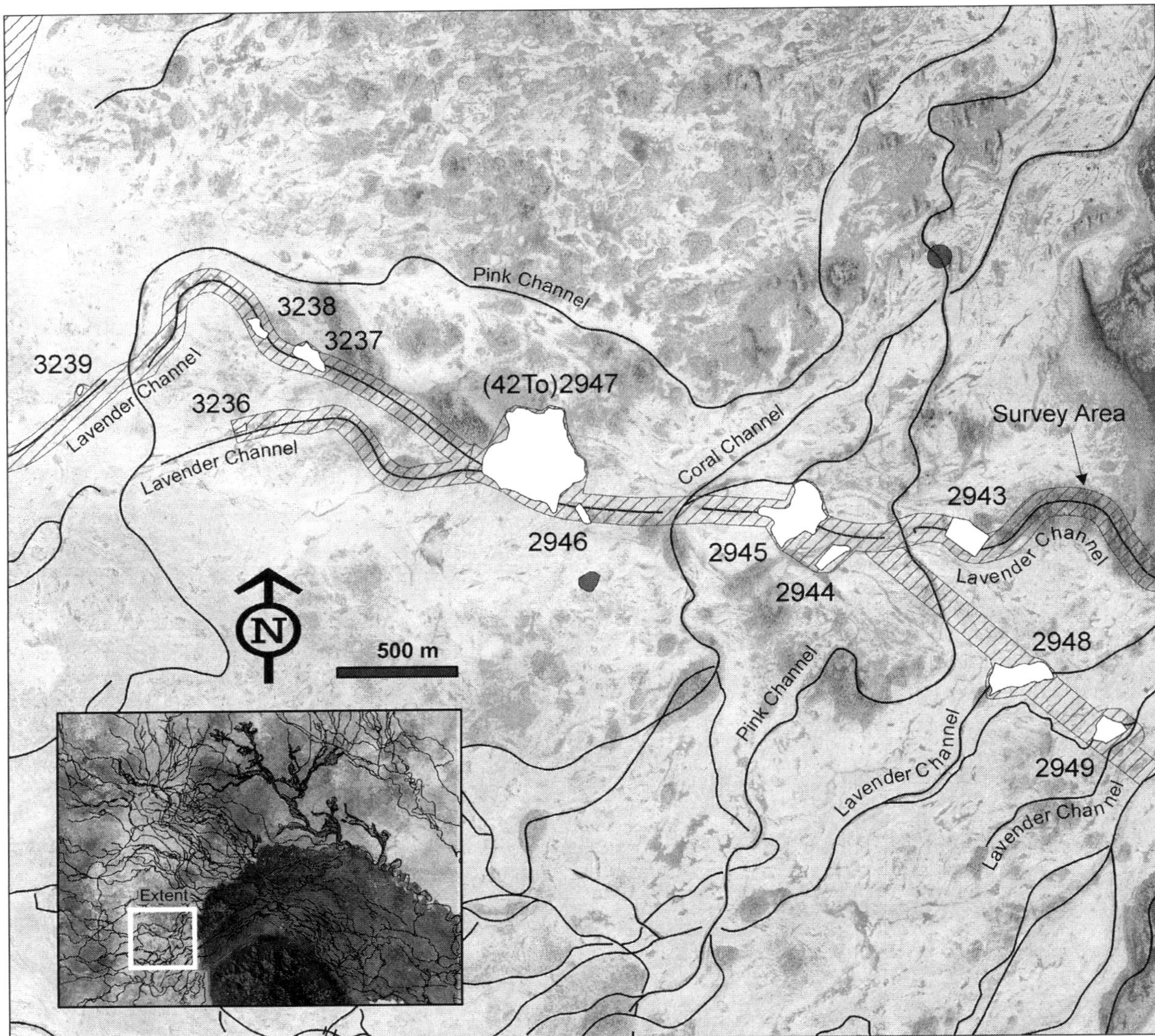

FIGURE 4.17. Location of survey areas and recorded sites (*labeled*) associated with the Lavender channel. The shaded circles mark the location of additional unrecorded sites.

generally widespread, two discrete artifact clusters were observed. Although these concentrations currently occur in the low areas between the topographically inverted channel segments, it appears that they were originally located on the channel margins. Tools include numerous stemmed points, scrapers, bifaces, and edge-modified flakes and a drill and a graver, and the debitage is dominated by biface-thinning flakes along with shatter and decortication debris that reflects some primary reduction.

Neighboring sites 3237 and 3238 (Figure 4.17) contain large and technologically rich assemblages and possess the highest artifact densities of sites recorded in the Lavender channel system (Table 4.7). Both are wholly contained within the main channel distributary,

and, while currently deflated, they likely represent the occupation of dry ground on partially exhumed stream deposits that postdate channel flow. Site 3237 contains a large concentration of flakes and tools in the southwest portion of the site, where artifact densities reach 10–15 items per square meter in some areas. Overall, debitage types reflect both core reduction and the manufacture of middle- to late-stage bifaces, and tools include slug scrapers, end scrapers, bifaces, gravers, a drill, an FGV core, and numerous stemmed points that include seven Silver Lake types. Site 3238 is a smaller camp immediately downstream containing three artifact concentrations with a veneer of flakes and tools scattered in between (Page et al. 2008). The tools consist of unifaces, bifaces, cores, modified flakes, WST projectile points,

TABLE 4.7. Selected Attributes and Contents of Sites Associated with the Lavender Channel in the Proximal Old River Bed Delta.

Site No. (42To-)	Projectile Points[a]	Area (m²)	No. of Tool Classes[b]	No. of Tools	Pieces of Debitage	Tool:Debitage Ratio	Artifact Density[c]
2943	S, E	17,820	3	13	23	.57	.20
2944	S	6,866	4	8	37	.22	.66
2945	S, E	40,757	7	39	~700	.06	1.81
2946	S, E	2,830	4	11	57	.19	2.40
2947	S, E, A	121,234	4	37	~1,200	.03	1.02
2948	S, E, A	23,085	3	21	189	.11	.91
2949	S, E	10,192	3	35	51	.69	.84
3236	E	1,005	4	7	25	.28	3.18
3237	S, E	8,459	7	78	~500	.16	6.83
3238	S, E	3,892	7	45	~250	.18	7.58
3239	—	803	4	4	9	.44	1.62
Total (mean)		236,943	(4.5)	298	~3,041	(.27)	(2.46)

[a] S = Great Basin Stemmed and lanceolate; E = Early Holocene stemmed, Stubby, and Pinto; A = Archaic dart and/or arrow.
[b] General functional classes/types (see Table 4.1). Some tallies include artifact classes that were identified in the field and not collected.
[c] Number of artifacts per 100 m².

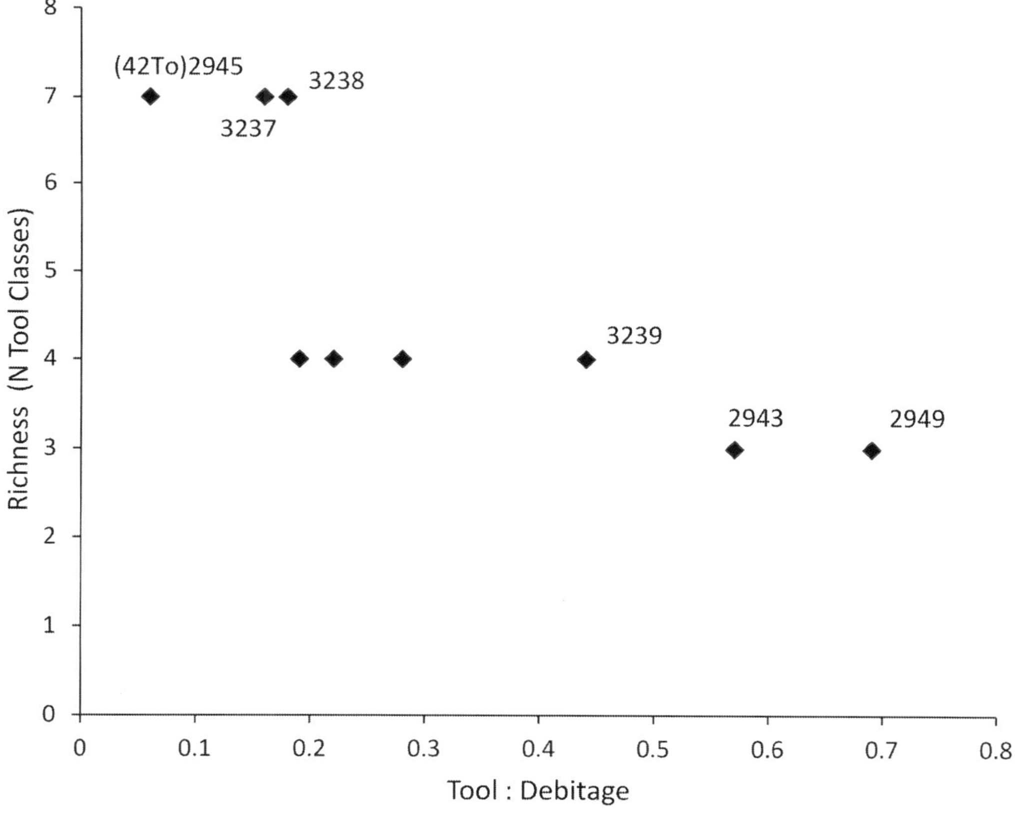

FIGURE 4.18. Scatterplot of the relationship between the number of tool classes and tool:debitage ratios for sites associated with the Lavender channel. The labeled plots represent the primary sites discussed in the text.

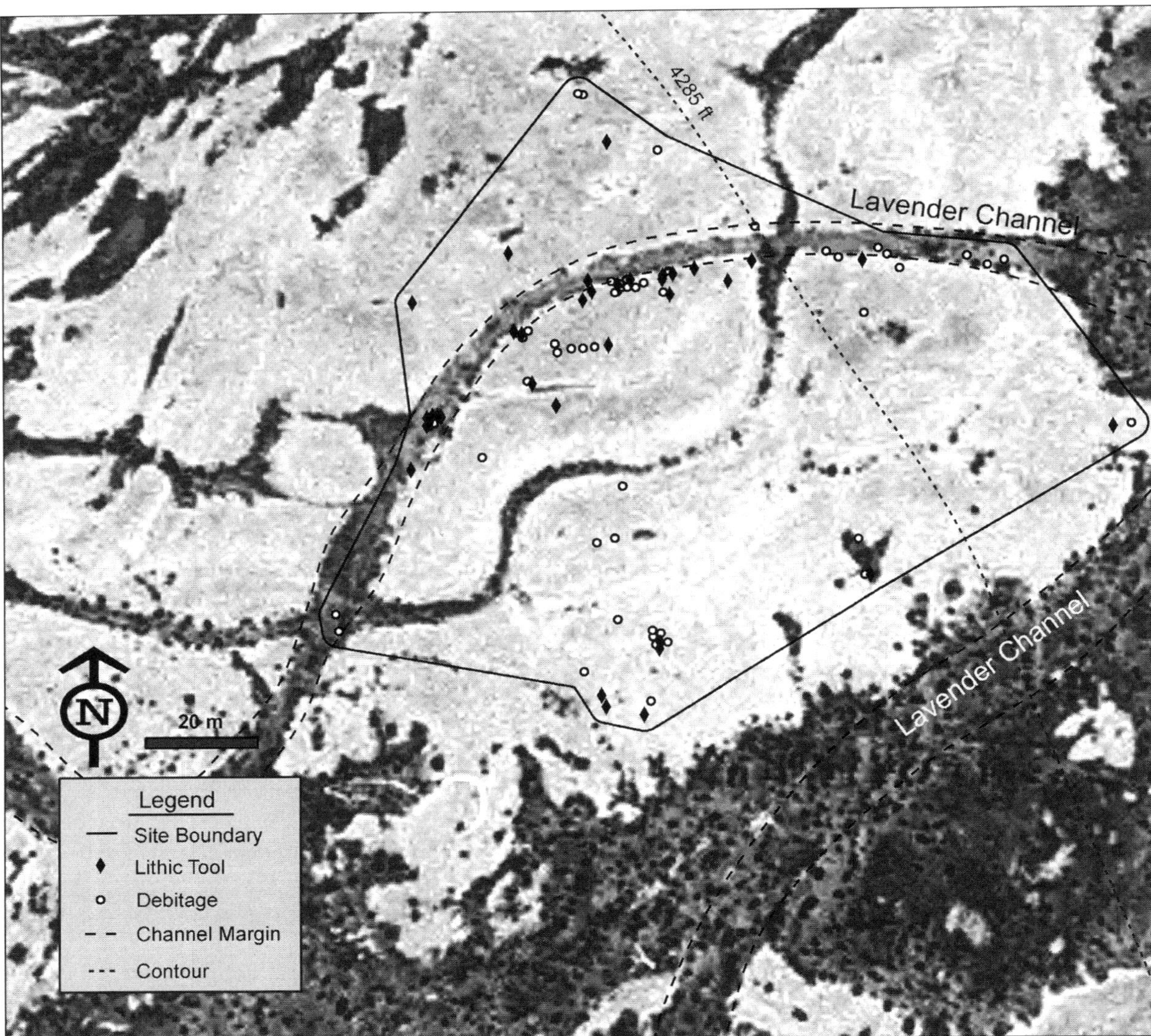

FIGURE 4.19. Plan map of 42To2949 associated with the Lavender channel.

and a host of scrapers, and the ~250 pieces of debitage reflect core reduction through late-stage biface manufacture and maintenance.

Sites 2943 and 2949 consist of relatively large scatters of sparsely distributed artifacts containing high T:D ratios (Table 4.7; Figure 4.18) and seem to represent resource-acquisition areas. Site 2943 is on the inner bend of a channel meander that is grading to the south. A few flakes were observed atop the channel, but most of the artifacts are scattered across the neighboring eroded mudflat surface (Schmitt, Madsen, Page, and Smith 2007). The debitage ($n = 23$) largely consists of biface-thinning debris, and the associated tool kit contains three stemmed points, four fragmentary bifaces, and six modified/utilized flakes. Site 2949 is a smaller and denser artifact scatter along a segment of a narrow Lavender channel distributary. Some cultural materials occur along the partially exhumed stream margin (Figure 4.19), and others extend out from the channel onto the neighboring flats, where foragers may have deposited artifacts while procuring marsh resources. Tools include nine stemmed points/point fragments, two beak scrapers, 14 bifaces, and 10 simple flake tools that largely exhibit bifacial modification. Based on content and context site 3239 also represents a task site, but it may have functioned as a resource-processing area rather than an acquisition locus. It is a small (803-m^2) discrete scatter on the bank of a Lavender distributary; it contains only nine flakes; and the small yet diverse tool kit consists of a Stubby point, an end scraper, a side scraper with a graver bit, and an FGV crescent.

Three additional Lavender channel sites contain

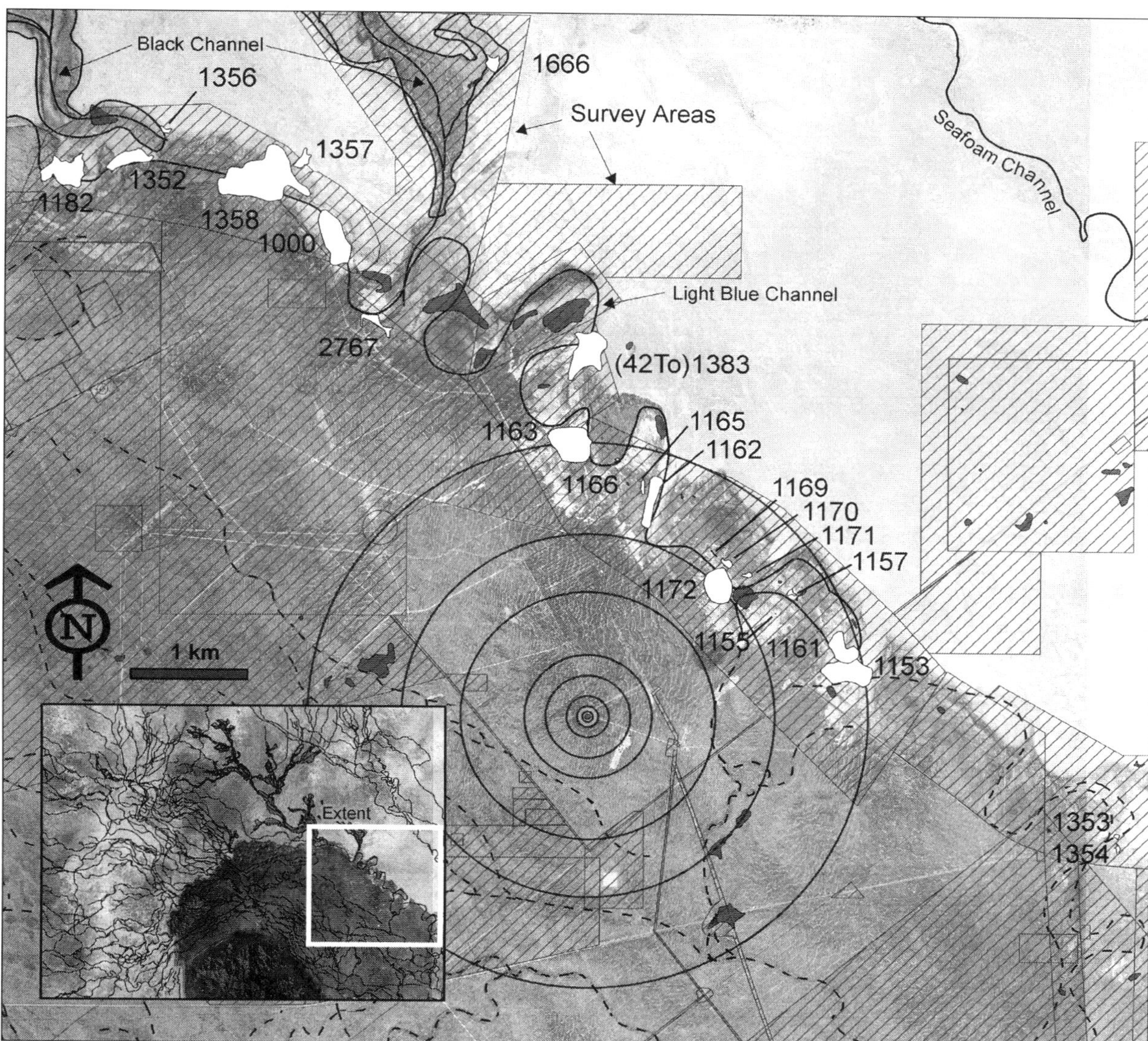

FIGURE 4.20. Location of survey areas and recorded sites (*labeled*) associated with the Light Blue and Black (*upper left*) channels along the Old River Bed eastern margin. The shaded locations along the channels and vicinity mark recorded sites dating to the Archaic/Fremont periods and palimpsests containing Paleoarchaic and younger materials. The concentric circles are part of a modern military grid.

four tool classes and a thin veneer of lithic detritus and likely represent short-term task sites. Site 2944 consists of 37 flakes and eight tools (Table 4.7), most of which occur in two ~30-m-diameter clusters, and 2946 is a linear scatter of 57 pieces of debitage associated with five bifaces/biface fragments and (two each) stemmed points, slug scrapers, and edge-modified flake tools. The lithic artifacts largely occur in a low area between topographically inverted Lavender stream deposits, and their current distributions may be the result of post-depositional transport by sheetwashing and wind. Site 3236 is a linear scatter of 25 flakes associated with three biface fragments, two simple flake tools, an obsidian

core, and a Stubby point. As at 2946, materials are in a linear scatter along the channel/channel margin, and their distribution may be, at least in part, a product of postdepositional movement.

Light Blue Channel Sites

The Light Blue channel (Figures 4.6 and 4.20) is a meandering, low-energy distributary in the eastern ORB delta. It consists of fine to medium sands with occasional pea gravels and represents one of the youngest distributaries in the delta (see Chapter 3). The channel runs along the eastern margin of the RLF–mudflat interface in and out from under the dune fringe before turning to

TABLE **4.8**. Selected Attributes and Contents of Sites Associated with the Light Blue Channel in the Proximal Old River Bed Delta.

Site No. (42To-)	Projectile Points[a]	Area (m²)	No. of Tool Classes[b]	No. of Tools	Pieces of Debitage	Tool:Debitage Ratio	Artifact Density[c]
1000	S, E, A	116,789	6	37	~300	.12	.47
1153	S, E	78,333	6	30	~300	.10	.42
1155	—	11	2	3	8	.38	100.00[d]
1157	S	2,480	4	5	~60	.08	2.62
1161	S, E, A	18,285	5	8	~50	.16	0.42
1162	—	37,812	1	2	~500	<.01	1.33
1163	S, E, A	92,313	6	11	~250	.04	0.34
1164	—	3,487	1	1	~250	<.01	1.02
1165	—	226	2	2	15	.13	7.52
1166	S, A	748	2	3	7	.43	1.21
1169	—	474	3	4	~50	.08	11.40
1170	—	4,773	3	3	~50	.06	1.11
1171	—	2,615	1	2	~50	.04	1.41
1172	S, A	57,207	6	20	~100	.20	.21
1182	S, E, A	159,970	7	26	~700	.04	.45
1352	M, A	37,800	3	10	~250	.04	.69
1357	S	12,155	2	5	~100	.05	.86
1358	S, E, A	201,972	9	80	~750	.11	.42
1383	S, E	83,483	2	18	~125	.17	.17
1671	S	8,538	3	9	~150	.06	1.86
1674	S	11,238	2	2	8	.25	.09
1675	M	5,942	2	3	16	.19	.32
1677	S	16,922	6	16	25	.64	.24
1678	S, E	14,301	4	12	38	.32	.35
1679	S, E	9,571	2	9	26	.35	.37
2767	S	15,848	6	22	~50	.44	.45
Total (mean)		993,293	(3.8)	343	~4,228	(.17)	(5.18)

[a] S = Great Basin Stemmed and lanceolate; E = Early Holocene stemmed, Stubby, and Pinto; A = Archaic dart and/or arrow; M = Great Basin/Early Holocene stemmed.
[b] General functional classes/types (see Table 4.1). Some tallies include artifact classes that were identified in the field and not collected.
[c] Number of artifacts per 100 m².
[d] In omitting this cluster of 11 artifacts across 11 m², the mean density is 1.39 artifacts per square meter.

the northwest, where it splits into two primary distributaries that continue to the north. In some areas it has been deflated down to the level of the neighboring mudflats, and in other places along the dune margin where it has been protected and only recently exposed, the channel is topographically inverted 1–1.5 m above the surrounding terrain. Archaeological survey in and adjacent the Light Blue channel identified 26 sites, including 18 sites that appear to be wholly PA in age (Table 4.8). The remaining eight sites represent palimpsests where Paleoarchaic materials are covered with a veneer of Archaic/Fremont period artifacts and/or fire-cracked rock in dunes built atop and along the channel (see Madsen et al. 2000; Schmitt, Madsen, Hunt, Callister, Jensen, and Quist 2002). Together, PA sites in the Light Blue channel tend to consist of dispersed scatters containing few tools and tool classes with low T:D ratios (see Tables 4.3 and 4.8).

Site 1153 is a large artifact scatter containing six small artifact concentrations on the outer (southern) margin of a horseshoe meander; site 1161 lies to the immediate north atop the channel and along its inner bend, and the two sites may be related (Figure 4.20). Identified artifacts at 1153 consist of various bifaces, nine WST points, one Stubby, scrapers and gravers (including three combination scraper/graver tools), and a debitage assemblage (n = ~300) dominated by FGV core-reduction detritus. Although 1153 may have served as a campsite, its large size and overall low artifact density suggest that it may be the result of multiple occupations where different tasks were performed. In more northerly portions of the Light Blue channel, three sites (1677, 1678, and

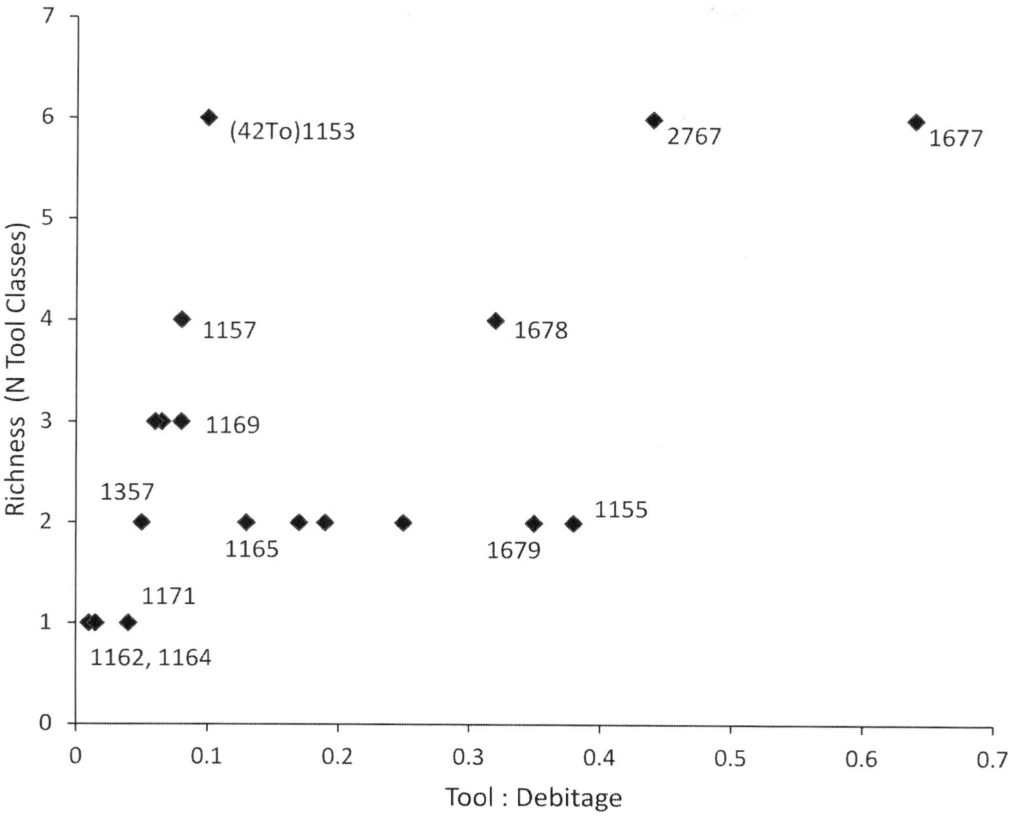

FIGURE 4.21. Scatterplot of the relationship between the number of tool classes and tool:debitage ratios for sites associated with the Light Blue channel. The labeled plots represent the primary sites discussed in the text.

2767) are large, sparse linear scatters on and adjacent slightly elevated distributary segments. Each contains small yet diverse tool assemblages and moderate to high T:D ratios (Table 4.8; Figure 4.21), and they likely represent materials lost or discarded while harvesting marsh resources. Site 1677 contains three projectile points, five bifaces, two scrapers, two drills, and a graver and a uniface, and the 1678 tool kit ($n = 12$) includes two Pinto points, one Silver Lake point, seven bifaces/biface fragments, one graver, and a slug scraper. Twenty-two tools representing seven classes were observed at 2767 in association with approximately 50 flakes. The tools include bifaces, WST points/point fragments, and a host of scrapers and edge-modified flakes, and the debitage consists of relatively equal numbers of core-reduction and biface-thinning debris. In addition, site 1679 also appears to manifest a procurement location, but seven of the nine observed tools are projectile points, and the primary procurement activity appears to have been the hunting of wetland game.

At least four sites are dominated by lithic debitage with small, limited tool kits and likely represent brief visits that centered on stone tool production. Site 1162 is a linear (~500-m) artifact scatter along the Light Blue

channel margin (Figure 4.20) containing two fragmentary bifaces and approximately 500 flakes dominated by FGV and obsidian core-reduction debris. A few widespread artifact concentrations occur, including a dense cluster of flakes at the northern end of the site (Madsen et al. 2000), and it likely resulted from multiple reduction events. Site 1164 is a smaller and denser cluster of some 250 flakes and a unifacially modified flake atop a topographically inverted sand and gravel "island." It contains abundant core-reduction detritus along with some thinning debris that reflect the initial production of bifacial tools. Sites 1171 and 1357 are along the eastern margin of the Light Blue channel and currently rest on deflated/eroded surfaces. Site 1171 contains two distal projectile point (blade) fragments and about 50 flakes dominated by FGV core-reduction debris, and 1357 consists of ~100 flakes (largely core-reduction detritus) associated with four WST projectile point fragments and a broken biface.

Sites 1155, 1165, and 1169, based on site sizes and artifact types and densities, appear to have largely served as processing locations. Site 1155 is a small concentration of eight flakes associated with a chert bifacially modified flake tool and two bifaces, and 1165 consists of a

TABLE 4.9. Selected Attributes and Contents of Sites Associated with Various Channels in the Proximal Old River Bed Delta.

Site No. (42To-)	Channel Association	Projectile Points[a]	Area (m²)	No. of Tool Classes[b]	No. of Tools	Pieces of Debitage	Tool:Debitage Ratio	Artifact Density[c]
3141	Gold	S, E	5,401	2	6	9	.67	.28
3142	Gold	S, E	28,431	6	38	~300	.13	1.19
1921	Limestone	S, E	2,587	3	12	27	.44	1.51
1922	Limestone	S, E	8,289	4	18	19	.95	.45
3140	Buff/Royal Blue	S	8,903	4	11	44	.25	.62
1689	Brown	S, E, A	16,520	3	7	~150	.05	.95
1683	Light Green	S, A	6,319	4	12	35	.34	.74
1353	Unnamed	S, E	1,414	4	5	19	.26	1.27
1354	Unnamed	S	2,089	4	13	27	.48	1.91

[a] S = Great Basin Stemmed and lanceolate; E = Early Holocene stemmed, Stubby, and Pinto; A = Archaic dart and/or arrow.
[b] General functional classes/types (see Table 4.1). Some tallies include artifact classes that were identified in the field and not collected.
[c] Number of artifacts per 100 m².

small cluster of 15 flakes and two combination (scraper/graver) tools atop partially exhumed stream deposits. Site 1169 contains a uniface, two side scrapers, a beak scraper, and approximately 50 pieces of debitage; the relative abundance of lithic detritus at 1169 suggests that some tool manufacture and maintenance occurred in concert with scraping tasks. Although 1155 and 1169 are currently on deflated surfaces along the channel margins, these areas may have been elevated patches of dry ground where local marsh resources were brought for initial processing.

Site 1383 is a massive (~83,500-m²), diffuse scatter of approximately 125 flakes, five bifaces, and 13 WST points (largely fragments) and may have served as a hunting locus that witnessed tool fabrication and maintenance. Similarly, 1671 is a smaller and more densely clustered array of ~150 flakes associated with five WST projectile point fragments and a blade core where activities appear to have also centered on hunting and lithic reduction. Finally, 1170 consists of some 50 pieces of debitage, one uniface, one scraper, and a hammerstone on and adjacent a severely eroded Light Blue stream segment. While the presence of debitage (especially core-reduction detritus) indicates that hammerstones were a common component of the forager tool kit throughout the ORB delta, the 1170 specimen represents the only hammer discovered in a wholly PA context; given the delta's impoverished lithic terrain, a stone hammer was doubtless a highly curated, indeed vital, piece of equipment that was not casually left behind.

Other Channel Sites

Nine PA sites were recorded during brief field investigations along portions of six additional channels (Table 4.9). Site 3140 (see Figure 4.12) is a sparse artifact scatter dispersed over a 182-×-87-m area. It occurs on the western edge of the Buff channel and sits on/adjacent the Royal Blue channel, making it unclear as to which channel it is associated with. Cultural materials include 44 pieces of obsidian, FGV, and cryptocrystalline silicate ($n = 7$) debitage and 11 tools (five stemmed points/point fragments, three scrapers, and two bifaces), and 3140 appears to represent a task site. Sites 1921 and 1922 (see Figure 4.4) were recorded along the Limestone channel (~10,500–10,000 ^{14}C BP; see Chapter 3) north of Black Channel A2. Artifacts at 1921 are associated with the broad northern end/lobe of a prominent, partially exhumed gravel-filled distributary, and at 1922 materials are on the inner bend of a channel meander; the channel surface has been deflated, but it once may have been elevated above the surrounding terrain (Schmitt et al. 2003). Both sites contain small artifact assemblages (Table 4.9) that extend out onto the mudflats and likely represent areas where foragers staged procurement activities in the neighboring marshlands.

The Gold channel flows on a westerly course in the central portion of the delta, and based on radiocarbon age estimates and its relationship to other distributaries, it represents one of the earliest deltaic channels in the ORB system (see Chapter 3). Two PA sites were recorded during brief field investigations along the Gold channel distributaries (Table 4.9). Site 3141 is a sparse scatter of nine flakes, five projectile points/point fragments, and one biface on the southwest margin of a distributary bend (Page et al. 2008). This diffuse scatter occupies an area of 5,401 m², but many of the artifacts may be in secondary contexts due to extensive postdepositional dispersal. Site 3142 (see Figure 4.8) is a large, complex deflated lag of FGV and obsidian flakes and tools situated within and alongside a high-energy

TABLE 4.10. Approximate Ages of Channel Formation and the Earliest Age Estimates for Associated Paleoarchaic Sites in the Proximal Old River Bed Delta.

Channel Association	No. of Sites (Site Nos. [42To-])	Channel Age (^{14}C BP)[a]	Relative Site Age (^{14}C BP)[b]
Gold	2 (all)	~11,300–10,500	<10,500
Black	36 (all)	~11,000–10,300	<10,300
Yellow	2 (3224, 3226)	~10,300–10,100	≤10,300–10,100
Yellow	11 (all but above)	~10,300–10,100	<10,100
Limestone	2 (all)	~10,500–10,000	<10,000
Green	7 (all)	~10,300–9800	<9800
Blue	2 (3233, 3234)	~10,000–9500	≤10,000–9500
Lime	1 (3520)	~9800–>9200	≤9800–9300
Blue	8 (all but above)	~10,000–9500	<9500
Light Blue	4 (1000, 1674, 1677, 1358)	~9800–8800	≤9800–8800
Lime	1 (3522)	~9800–>9200	>9200
Lavender	1 (2943)	~9100–9000	≤9100–9000
Lavender	10 (all but above)	~9100–9000	<9000
Light Blue	22 (all but above)	~9800–8800	<8800

[a] Age estimates of channel formation/flow based on ^{14}C results and relationships to other geomorphic features and/or dated channel deposits (see Chapter 3).
[b] Maximum (oldest) age estimates based on the age of channel formation and site contexts.

channel remnant associated with a small Gold channel gravel "island." It contains abundant debitage and tools (including stemmed points, gravers, 20 scrapers, and a crescent; Table 4.9) and appears to represent a camp. However, given the complex array of younger and older channels in this area, it is possible that the younger channels provided groundwater-fed marsh vegetation in the older channels, and materials on this high-ground island may represent a palimpsest of visits.

Site 1689 is a large, relatively dense palimpsest of PA and younger materials along the eastern margin of the Brown channel, on the deflated flats immediately north of the dunes that rim the RLF plain. A single site (1683) was identified along the Light Green channel in the central delta (see Figure 4.3), and two sites (1353 and 1354) were recorded along an unnamed channel in the eastern delta atop RLF deposits (see Figure 4.20). Sites 1353 and 1354 represent relatively small scatters of lithic detritus associated with four tool classes that include stemmed points, and 1683 contains 35 flakes and 12 tools representing four types, including a Humboldt point that marks a later visit.

SITE CHRONOLOGY

None of the sites in the proximal ORB delta yielded charcoal, subsistence remains, or any other residues that might provide radiocarbon age estimates. The overall context and the presence of stemmed points, crescents, and/or Cody knives clearly indicate that these sites are PA in age (e.g., Grayson 1993; Jones and Beck 1999), but more precise age estimates on the timing and duration of human occupations remain ambiguous. Based on the results of radiocarbon assay of organic materials from various channels, channel relationships, and the types and locations of PA sites, it is only possible to identify the earliest date at which foragers may have inhabited the various deltaic channels. For example, it is reasonable to assume that a campsite contained wholly within a channel represents an occupation of exhumed dry ground that postdates channel formation, while a large, diffuse lithic scatter dominated by edged tools in and adjacent to a low-energy distributary may represent procurement activities that occurred during channel flow. However, inferences on the limiting ages of occupations remain speculative as it is unclear exactly when channels/channel segments became topographically inverted. It is also unclear when, where, and for how long associated groundwater-fed marshes occurred in most areas.

Maximum age estimations for PA sites in the proximal delta are presented in Table 4.10. The majority of these sites ($n = 99$; 90 percent) are contained within channel margins (most of which are topographically inverted) and appear to represent occupations of elevated ground that postdate channel formation. All sites associated with the Black, Gold, Limestone, and Green channels are included, as are most sites along the Blue, Lavender, and Light Blue channels. Sites in the Black channel postdate its formation and must have been created after ~10,300 ^{14}C BP. Because they overlook portions of the Light Blue channel and other early Holocene distributaries, it is also possible that they mark forager activities that occurred more than 1,500 ^{14}C years later.

The remaining 11 sites, based on their context, content, and distribution, represent occupations that may have occurred during streamflow (Table 4.10). Six of these sites (3224 and 3225 [Yellow], 2555 [Blue], 2943 [Lavender], and 1674 and 1677 [Light Blue]) are large, linear scatters dominated by tools. Each consists of very diffuse scatters contained wholly within channels, and they appear to represent resource-acquisition locations. Although it is possible that these visits occurred following stream formation, the overall context and the types and frequencies of artifacts suggest that items were lost or discarded while pursuing resources in shallow wetlands, most likely during the waning years of channel flow. Site 1000 consists of linear scatters of obsidian and FGV artifacts that border the Light Blue channel, and it, too, appears to have been occupied during streamflow. Artifacts are most abundant on a neighboring gravel-filled, partially exhumed distributary (likely Black; see Figure 4.20) that runs parallel to the Light Blue channel, and it is possible that this area provided an elevated levee for staging resource-procurement and other activities while the Light Blue stream was active. A very similar pattern is evident at nearby site 1358. It is a massive scatter containing fire-cracked rock and later projectile point types and abraded obsidian and FGV artifacts, which are most abundant along a pronounced gravel-filled distributary that directly overlooks the Light Blue stream (Schmitt, Madsen, Hunt, Callister, Jensen, and Quist 2002:Figure 6).

Two neighboring sites (3223 and 3224) occur on Blue distributaries that predate the formation of the main terminal channel and may have been occupied during the flow of the main Blue stream (10,000–9500 ^{14}C BP; Table 4.10). Site 3233 is a linear artifact scatter along an early distributary immediately south and west of the main terminal channel, and 3234 is a small camp along an older meander. It may be associated with a natural levee formed when water was flowing in the terminal Blue channel, or it may have been deposited in association with the older stream itself after channel flow. Regardless, the context of these sites suggests that they may be older than sites wholly contained within the main Blue distributary, which flowed until about 9500 ^{14}C BP.

The context and content of site 3520 are compelling. This small camp is atop topographically inverted deposits along an inner oxbow bend of the Lime channel. Although some materials occur in an earlier channel meander, most are clustered atop elevated deposits that overlook the main terminal channel (see Figure 4.16). This location may have served as favorable dry ground subsequent to main Lime channel flow, but artifact distributions and setting suggest that 3520 may have been inhabited during the latter years of channel formation.

PALIMPSESTS OF PA AND YOUNGER MATERIALS

With few exceptions, the majority of the sites containing PA and later materials are clustered along the dunes that mark the RLF–mudflat transition. Of the 15 sites containing Archaic/Fremont projectile points, one is on the mudflats in a small cove ~150 m north of the dunes (1689 [Brown channel]) and 11 are along the eastern delta margin in association with the Black and Light Blue channels where they meander through and immediately adjacent the margin of the RLF plain. The Holocene dunes that developed along this interface offered one of the few areas in the valley bottom that would have provided attractive resources to later foragers (e.g., Indian ricegrass [*Achnatherum hymenoides*], globe mallow [*Sphaeralcea* sp.], and small game), and it is not surprising that later materials are largely restricted to these areas (see, for example, Figure 4.20). The three remaining multicomponent sites (1683 [Light Green] and 2947 and 2948 [Lavender]) are atop the mudflats far removed from the dunes, and it appears that later foragers ventured out onto the mudflats on some occasions; this especially is the case at 2948, where a complete Rocker Side-notched point and a small side-notched point were discovered.

DEBITAGE MATERIAL TYPES

Debitage frequencies at PA sites in the proximal ORB delta are variable due to differences in the types and duration of activities undertaken. Nonetheless, all of the sites contain lithic detritus from tool manufacture and/or maintenance and are similar to most hunter-gatherer assemblages in the region as, with few exceptions, flakes represent the most abundant artifact class. Although a small sample of flakes was collected for analysis (see Chapter 5), debitage from a larger sample of sites is presented here to further examine the types and abundances of raw materials that were brought into the ORB wetlands. It is reasonable to assume that flake debris was usually left at its place of production, and in most instances it represents waste generated from working cores, large flake blanks, and bifaces. Unlike formal tools that moved from place to place and may have been acquired through trade (e.g., Shott 1994; see also Bamforth 2009), waste flakes can offer specific glimpses of

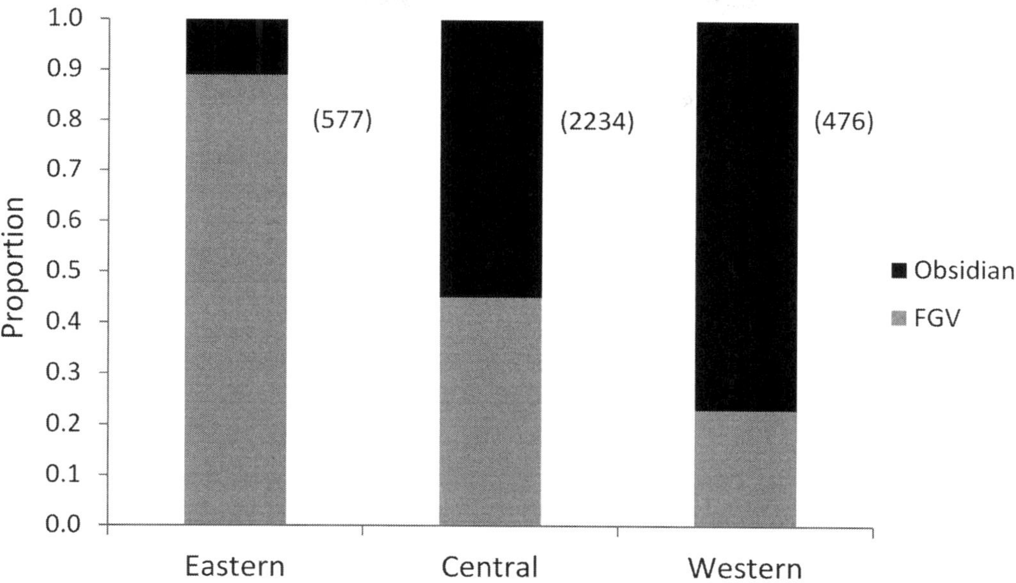

FIGURE 4.22. Relative proportions of fine-grained volcanic (FGV) and obsidian debitage from sites in the eastern, central, and western portions of the proximal Old River Bed delta (see Table 4.11). Parentheses bracket the total number of FGV and obsidian flakes in each aggregate.

the types of utilitarian materials that were brought into the delta for daily tasks.

Debitage tallies by material type from 42 sites across nine channels in the proximal delta are presented in Table 4.11. Like the PA tools, the waste flakes are dominated by FGV and obsidian, which together constitute 99 percent of the lithic materials and occur in nearly equal proportions when viewing the sample as a whole. However, when examining the proportions of these two materials in site aggregates in eastern, central, and western portions of the proximal delta there are some marked differences (Figure 4.22). Sites along the Black and Light Blue channels in the eastern delta are dominated by FGV detritus, where it constitutes almost 89 percent of the assemblage. Sites along channels in the central delta tend to contain a relatively even mixture of FGV and obsidian (especially Yellow and Blue channel sites), and the relative abundances of these materials are significantly different from those to the east ($\chi^2 = 351.82$; $df = 1$; $p < .001$). In contrast, site assemblages along the western reaches of the delta are dominated by obsidian and differ from site aggregates in the central delta ($\chi^2 = 71.08$; $df = 1$; $p < .001$) and especially the FGV-dominant sites to the east ($\chi^2 = 451.36$; $df = 1$; $p < .001$).

Since these data were largely collected from field observations and specimens were not subject to geochemical analysis, the specific sources remain unknown; however, based on the large sample of sourced artifacts (see Chapter 7), it is assumed that the vast majority of

the FGV is from the Flat Hills area and the obsidian was acquired from the Topaz Mountain source. While some of the variability in the proportions of FGV and obsidian may be due to chronological differences, it appears that proximity to toolstone played a key role in raw material acquisition. For foragers occupying the eastern delta the Flat Hills area represented the closest source of quality toolstone, while the nearest source area for people in the western reaches of the delta would have been Topaz Mountain. These source areas were approximately equidistant for foragers in channels in the central portions of the delta, and it appears that some occupants focused on Flat Hills FGV (e.g., 3521 [Yellow], 3142 [Gold]), some largely exploited Topaz Mountain glass (e.g., 2558 and 2559 [Green]), and others may have ventured to both (e.g., 3233 [Blue]). It is possible that some of the variability in toolstone reflects different routes into the delta (e.g., people moving in from the southwest loading up with Topaz Mountain obsidian), but because there are channels of the same age in different areas, it appears that toolstone forays were occasionally undertaken to areas outside the ORB marshlands, and these trips often considered the shortest and most unchallenging avenue.

DISCUSSION AND SUMMARY

Archaeological sites in the proximal ORB delta include many forms indicative of long-term occupations and short-term logistical use, and in this respect they are

TABLE 4.11. Number of Flakes by Material Type from a Sample of Sites in the Proximal Old River Bed Delta.

Area, Channel, and Site No. (42To-)	Fine-Grained Volcanics		Obsidian		Other[a]		Total
	n	%	*n*	%	*n*	%	
East							
Black Channel							
1368	16	80	4	20	—	—	20
1369[b]	30	100	—	—	—	—	30
1668	18	90	—	—	2	10	20
1669	6	86	—	—	1	14	7
1670	25	86	4	14	—	—	29
1673	18	90	—	—	2	10	20
1680	10	91	1	9	—	—	11
1682	5	83	1	17	—	—	6
1860	~100	100	—	—	—	—	~100
1876	7	100	—	—	—	—	7
1924	108	82	23	18	—	—	131
Subtotal	343	90	33	9	5	1	381
Light Blue Channel							
1161	29	63	14	30	3	6	46
1165	11	73	3	20	1	7	15
1671	128	85	16	11	6	4	150
Subtotal	168	80	33	16	10	5	211
Central							
Yellow Channel							
2951	24	44	30	56	—	—	54
2953	18	82	4	18	—	—	22
2954	7	70	3	30	—	—	10
3219	49	43	62	54	4	3	115
3221	34	52	31	48	—	—	65
3224	18	55	15	45	—	—	33
3521	43	78	9	16	3	5	55
Subtotal	193	55	154	44	7	2	354
Green Channel							
2551	47	39	72	59	3	2	122
2552	46	38	76	62	—	—	122
2553	6	46	7	54	—	—	13
2558	2	1	131	98	1	1	134
2559	125	30	296	70	2	<1	423
Subtotal	226	28	582	71	6	1	814
Blue Channel							
3228	9	37	15	63	—	—	24
3229[c]	64	40	97	60	1	<1	162
3231	26	44	33	56	—	—	59
3232	10	77	3	23	—	—	13
3233	51	46	61	54	—	—	112
3234	41	47	46	53	—	—	87
3235	128	47	142	53	—	—	270
Subtotal	329	45	397	55	1	<1	727
Buff/Royal Blue Channel							
3140	21	48	16	36	7	16	44
Gold Channel							
3142	194	69	82	29	5	2	281
West							
Lime Channel							
3520	16	23	55	77	—	—	71
3522	7	30	16	70	—	—	23
Subtotal	23	24	71	76	—	—	94
Lavender Channel							
2943	5	22	18	78	—	—	23
2947[c]	9	9	87	91	—	—	96
2948	50	26	138	73	1	1	189
2949	20	39	30	59	1	2	51
3236	7	28	18	72	—	—	25
Subtotal	91	24	291	76	2	<1	384
Total	1,588	48	1,659	51	43	1	3,290

[a] Includes quartzite, cryptocrystalline silicate, and/or limestone.

[b] Debitage from a 10-x-10-m² surface collection unit.

[c] Debitage counts by type are from sample transects.

TABLE 4.12. Number of General Paleoarchaic Site Types by Channel in the Proximal Old River Bed Delta.

Site Type	Black	Yellow	Green	Blue/Lime	Lavender	Light Blue	Other	Total
				Channel Association(s)				
Camp	5	1	3	4	3	1	1	18
Acquisition	3	5	1	4	2	3	2	20
Processing	3	—	—	—	1	3	—	7
Lithic reduction/fabrication	7	2	—	—	—	4	—	13
Hunting	1	—	1	—	—	3	—	5
General/multiple tasks	14	6	2	4	3	4	4	37
Total	33	14	7	12	9	18	7	100

similar to the hundreds of known Archaic, Fremont, and Late Prehistoric hunter-gatherer sites that rim the southern Great Salt Lake Desert. Although changes in Holocene climates, biotic communities, and demography clearly modified post-PA landscapes and forager land use (e.g., Grayson 2011; Madsen et al. 2001; Schmitt et al. 2004), most differences between regional PA and younger records are not necessarily reflections of changes in what people did but, rather, technological differences in how they did it. Specifically, open PA sites consist entirely of scattered flaked-stone artifacts that can be difficult to interpret, while many Archaic and later sites contain organic residues, grinding stones, ceramic vessels, cooking rocks, and/or other innovations (cf. Thomas 1988b:Table 73) that offer more explicit evidence regarding the behaviors that created them. However, analyses of PA artifact types, abundances, and distributions allow some inferences on the different kinds of activities undertaken in the ORB wetlands and in a few instances the relative duration of these events. These inferred activities include sites containing large numbers of artifacts and tool types that likely represent camps, purported lithic-reduction loci dominated by flakes, and a variety of often large and diffuse task sites that appear to have resulted from logistical resource-procurement and/or -processing activities.

Eighteen PA sites contain abundant and relatively dense concentrations of tools and flaked-stone detritus and appear to have served as base camps (Table 4.12). Some consist of large and diverse assemblages with discrete artifact concentrations and likely witnessed prolonged stays (e.g., 1686 [Black], 2559 [Green], and 3237 [Lavender]), while others are smaller in size with higher T:D ratios and probably represent short-term occupations (e.g., 3219 [Yellow], 3234 [Blue], and 3520 [Lime]). Regardless of the duration of occupation, a number of these camps were established on high ground immediately overlooking wetland patches (e.g., 1920 and 1371 near the end of Black Channel A2), where the occupants appear to have pursued adjoining resources.

It comes as no surprise that most of the sites reflect one or more tasks, and it is reasonable to assume that they were created by small mobile groups and/or by individual men and women while exploiting patches of food, construction materials, and other resources. At least 20 of these sites appear to have largely served as acquisition loci (Table 4.12). While variable in size, most consist of large, very diffuse scatters dominated by tools (e.g., 3224 [Yellow], 2554 [Blue], and 2943 [Lavender]); they tend to occur along the margins of partially exhumed stream deposits and the associated mudflats (e.g., 2949; Figure 4.19); and in a few instances they appear to mark productive patches subject to multiple, intensive use episodes that likely included overnight stays (e.g., Krispy Kreme Vista [Black A2]). Based on their size and content, seven sites appear to represent briefly occupied areas where resources were processed. Most of these sites occur in association with the Black and Light Blue channels (Table 4.12), including the small and typologically diverse tool clusters at sites 1877, 1923, and 1155, which provide the most persuasive evidence for processing activities.

Thirteen relatively small, often concentrated site assemblages are dominated by debitage and seem to represent tasks that centered on lithic reduction and fabrication. Like the processing locations, most of these sites are associated with elevated segments of the Black and Light Blue channels along the delta's eastern margin. These include a number of locations (e.g., 1385, 1170 [Black], 1164, and 1171 [Light Blue]) that contain small assemblages of cores, broken bifaces, and/or edge-modified flakes associated with abundant lithic detritus and probably mark men's activities. In addition, five sites appear to have served as hunting areas, which may have also been occupied by men. Each tool assemblage is dominated by projectile points/point fragments, and three of these—2558 (Green) and 1383 and 1671 (Light Blue)—contain abundant stone debitage that also reflects recurrent tool fabrication and maintenance.

The low-energy channel streams that make up the majority of the ORB delta created a vast, meandering wetland system with relatively few areas suitable for habitation or for activities not directly associated with foraging in the marsh itself. Most of these channels (e.g., Blue, Lavender, and Light Green) were created sometime between about 10,000 and 8800 ^{14}C BP (see Table 4.10), although wetlands in the low areas of entrenched stream deposits may have persisted until desertification took hold in the waning years of the early Holocene (~8500 ^{14}C BP [e.g., Grayson 2000a; Madsen et al. 2001; Schmitt and Lupo 2012]). Save for a few resource-acquisition loci, most of the archaeological materials associated with exhumed channels were deposited after stream flow and date to this latter part of this period. However, marsh/wetland habitats in the ORB delta were present prior to this time, and there may be older materials as well.

Although a number of other sites are apparently associated with low-energy channels, a definitive spatial (and chronological) relationship to the natural levees along the channel margins is less clear. Most are very large sites, and they have low-energy channels either running through them or running immediately adjacent to them, but there is no clear association of lithic debris with the channels themselves. Rather, artifacts are widely scattered, with no apparent concentrations evident in any one location; 1172 and 1383 (Light Blue) are good examples of this kind of site. There are a couple of possible explanations for these distributional patterns. These sites may be a product of the deflational/erosional forces that are exhuming the low-energy channels, with wind and wave action in ephemeral ponds scattering debris across a wide area, or they may be the result of actual foraging activities in the marsh/wetlands along the channels. While overall artifact densities and distributions tend to suggest that the latter is the case, whether this reflects tasks performed in association with stream-flow or occupations of subsequently exhumed deposits remains ambiguous.

In comparing the ORB delta sites with other Great Basin PA records there are predictable similarities and differences. For example, many assemblages have been recovered from large sites/localities that indicate long-term stays and/or multiple visits (e.g., Beck and Jones 2009; Simms and Lindsay 1989; Smith 2007; Willig 1988). These sites commonly contain low T:D ratios and an array of hunting, scraping, cutting, and/or perforating implements, and in this respect they mirror most campsites in the proximal delta. There are, however, significant differences in surface artifact densities and overall abundances that almost certainly reflect differential access to toolstone. The delta was a lithic-impoverished labyrinth of active streams and marshlands in a massive basin where stone-procurement forays were doubtless costly and difficult endeavors. Access in and out of the wetlands was largely restricted to meandering, mostly southern and eastern routes throughout much of its history, and, while patch-to-patch movement within the marsh was likely frequent, logistical forays to areas outside the delta probably occurred only after extended periods of stay. As such, people appear to have been frugal with lithic resources, since most ORB sites, including base camps, contain comparatively far fewer artifacts than other PA sites, with artifact densities rarely exceeding three items per 100 m^2 (see, for example, Tables 4.5 and 4.6). Conversely, and while distances were often great (Jones, Beck, Jones, and Hughes 2003), foragers occupying pluvial lake margins had significantly less arduous access to lithic sources (i.e., skirting lakeshores rather than trudging through vast brackish marshes and active streams), including locally available cherts (see also Goebel 2007), and artifact densities here are markedly higher than those in the delta. For example, surface collections conducted at the Sunshine Locality in 1993 (400 m^2) and 1997 (Operation 2; 300 m^2) recovered 1,116 pieces of debitage (Beck and Jones 2009), which equates to almost 160 items per 100 m^2, and arbitrary surface collection of a total of 20 m^2 in four Parman localities yielded 2,611 flakes, which reflects more than 13,000 specimens per 100 m^2 (Smith 2007).

Investigations of Great Basin PA sites have largely focused on artifact technology, transport, and typology, especially stemmed projectile points, and few provide site-oriented data regarding structure, density, and/or inferred function. This is not to say that these studies are deficient; quite the contrary is true. Because many of these assemblages consist of large artifact samples from sites that were subject to multiple and/or intensive habitations, such artifact-oriented analyses are warranted and have provided important information on regional chronology and human adaptations and technology (e.g., Beck and Jones 1990a, 2007, 2009; Beck et al. 2002; Fagan 1988; Smith and Kielhofer 2011). Sites in the proximal ORB delta provide similarly useful information on where people acquired toolstone and how they used it, but they also offer discrete and unique glimpses of a range of PA forager behaviors, including procurement and fabrication activities that in many instances appear to represent single events.

5

Lithic Analysis

Charlotte Beck and George T. Jones

This chapter reports on the lithic artifacts collected from the proximal Old River Bed (ORB) delta that comprises Dugway Proving Ground (referred to hereafter as simply the "ORB delta"). A total of 2,287 lithic artifacts were collected from 136 surface sites and 246 isolate localities between 1999 and 2008. The analysis of 1,357 artifacts collected up through 2003 was reported by Jones, Beck, and Kessler (2003), while that of the remaining 930 was reported by Beck and Jones (2010b). Because the analysis protocol was modified in the latter analysis, we refer here to these two assemblages as ORB1 and ORB2. For this chapter the revised protocol was applied to both assemblages, and results are reported for the ORB assemblage as a whole.

Apart from a small number of Archaic period diagnostics, the majority of specimens are attributed to the Paleoarchaic period, which broadly dates to 11,000–8500 ^{14}C BP (Beck and Jones 1997, 2010a). The collected sample (see Chapter 4 for a discussion of collection protocols) favors formal tools, those exhibiting retouch flaking; flakes and other manufacture by-products were not systematically collected, and thus their occurrences across individual site assemblages cannot be regarded as statistically representative.

General Analysis Protocol and Results

The present analysis was conducted in two phases. In the first phase, all artifacts were identified using a set of general technological categories (Table 5.1), and we organize our descriptions in large part by these categories. Additionally, each artifact was characterized as to actual weight, raw material, and degree of weathering. Raw material categories include chert, obsidian, fine-grained volcanics (FGVs; e.g., andesite, dacite, rhyolite, and basalt), quartzite, quartz, and limestone. The term *chert* is used here as a general term referring to sedimentary cryptocrystalline silicates, such as flint, agate, jasper, and chalcedony (see Luedtke 1978:414). The differentiation of obsidian and FGV artifacts by geochemical source is reported in Chapter 6.

Table 5.2 shows the representation of artifact specimens by technological category and raw material. As is evident from this table, the overwhelming majority of the ORB artifacts are formal tools; less than 14 percent of the assemblage is composed of debitage. Although field observations suggest that the ORB artifact record is dominated by flake debitage (as is the case for all of the Paleoarchaic assemblages we have analyzed from eastern and central Nevada), the collection strategy was designed to obtain only a small sample of these artifacts. Therefore, our focus in this chapter is primarily on formal tools, although debitage will be considered in the discussion of biface reduction.

In the second phase, formal tools and cores were subjected to more detailed analysis. The unifaces and bifaces have been identified with categories that have proved broadly useful in characterizing Paleoindian assemblages across the continent. As we found in our analysis of the artifact assemblage from the Sunshine Locality in eastern Nevada (Beck and Jones 2009), the ORB tool assemblage is similar to those from Paleoindian sites in the Great Plains as well as the Great Lakes/northeastern United States, where some of

TABLE 5.1. General Technological Categories Used to Identify the Old River Bed Artifacts.

Technological Category	Definition
Cortical flake/blade	A flake that exhibits cortex on its dorsal surface and no edge retouch. The distinction between primary and secondary cortical flakes is not made here given the ambiguity of this distinction.
Interior flake/blade	A flake that exhibits no cortex and no edge retouch.
Uniface	An artifact with retouch on only one face.
Biface	An artifact (not a projectile point) with retouch on two opposing faces.
Projectile point	A biface that possesses consistent manufacturing properties of "point," "haft," and "bilateral symmetry." These items were analyzed separately because of their chronological significance.
Core	An irregularly shaped mass bearing negative flake scars originating from one or more platforms.
Split pebble/cobble	A core or core fragment produced by the bipolar technique (Crabtree 1972), as distinct from other kinds of cores and core fragments.
Chunk	A thick, angular piece produced during the manufacturing process that exhibits no classic attributes of a flake or core.
Natural	A natural object that has been transported to a site by humans ("manuport") or utilized in its natural state.

TABLE 5.2. Technological Categories Recognized in the Old River Bed Assemblage.

Technological Category	Chert		Obsidian		Fine-Grained Volcanics		Quartzite		Limestone		Quartz		Total
	n	%	n	%	n	%	n	%	n	%	n	%	
Cortical flake	4	2.5	11	1.2	13	<.1							28
Interior flake	22	13.6	97	10.2	153	13.4	5	19.2			1	100.0	278
Interior blade flake			1	.1	3	<.1							4
Uniface	31	19.1	63	6.6	167	14.7	3	11.5	1	25.0			265
Biface	58	35.8	169	17.7	358	31.4	11	42.3					596
Projectile point	35	21.6	599	62.7	438	38.5	3	11.5	2	50.0			1,077
Core	10	6.2	7	.7	4	<.1	1	3.8					22
Chunk	2	1.2	5	.5	1	<.1			1	25.0			9
Split pebble/cobble			2	.2	1	<.1							3
Worked pebble			1	.1									1
Hammerstone					1	<.1	1	3.8					2
Ground stone							1	3.8					1
Manuport							1	3.8					1
Total	162	100.0	955	100.0	1,139	100.0	26	100.0	4	100.0	1	100.0	2,287

the most detailed analyses have been conducted (e.g., Deller and Ellis 1992; Grimes and Grimes 1985; Storck 1997; Witthoft 1952). Our presentation of the ORB assemblage follows the approach used in the Sunshine analysis. The categories are defined in Table 5.3, and the number of specimens in each category is shown by raw material in Table 5.4. Some of these categories crosscut the technological categories of "uniface" and "biface." For example, although most scrapers are unifacial, about a third of them are bifacial. Each of these categories will be discussed separately.

For most tool categories, several metric variables were recorded, including maximum length, maximum width, and maximum thickness. In addition, an ordinal variable, size category, was also recorded. Size categories are based on plan area (square centimeters); the area of the smallest category measures .296 cm^2 (coded as "1"), and each successive category is 1.5 times the area of the previous one (Figure 5.1). The largest category (Category 14) has an area of 57.67 cm^2; any artifact exceeding this size was designated "15." With the exception of production bifaces (see Table 5.3), only complete artifacts were given size designations. We had a specific purpose in mind for identifying the production bifaces with size

TABLE 5.3. Definitions of Formal Tool Categories Utilized in the Old River Bed Assemblage Analysis.

Tool Category	Definition
Amorphous uniface	An artifact with retouch on only one face. Many of the tools defined below are unifacial, including gravers and scrapers, but these tools can also be bifacial. Thus we have called all unifaces that cannot be identified as these tools amorphous unifaces.
Amorphous biface	A biface that has no particular or standardized form and thus cannot be identified with any specific biface category.
Core	A mass bearing negative flake scars originating from one or more platforms. Four types of cores are recognized: *Amorphous*: shape is irregular, and flake scars originate from various locations on the surface. *Pebble*: small cores made on pebbles, which often retain cortex. *Discoidal*: a core with a circular, relatively flat face bounded by a continuous striking platform (Gramly 1990:20). *Prismatic blade*: conical and wedge-shaped source of prismatic blades (see Collins 1999a; Collins and Lohse 2004).
Chisel	An artifact with a long, narrow projection ending in a rounded or blunt point.
Combination tool	An artifact exhibiting two or more different tools. A number of different combination tools are recognized, including various combinations of scrapers and gravers (see definitions below).
Crescent	An artifact of crescentic shape. The edges often exhibit grinding at the midpoint (Gramly 1990:18). Generally bifacial but also can be unifacial.
Drill	Implement with a projection that is used for perforating. Identified by edge beveling on alternating faces (Gramly 1990:20).
Graver	A flake or blade with a small, sharp projection (sometimes called a spur) created by unifacial or bifacial retouch. Often exhibits multiple spurs (Gramly 1990:26).
Ground stone	A slab or cobble that has been smoothed by grinding (mano or metate) (Gramly 1990:27).
Indeterminate biface	Recognizable as a biface but too fragmentary to categorize.
Indeterminate uniface	Recognizable as a uniface but too fragmentary to categorize.
Knife	An implement of variable form used with a slicing motion to cut substances. May be bifacial, unifacial, or unretouched (Gramly 1990:30).
Notch	A small, narrow concavity on the edge of a flake or shaped tool, usually unifacial (Gramly 1990:34).
Point blade	The tip and/or blade section of a projectile point. Four types are recognized: *Western Stemmed Tradition (WST) blade*: recognizable as part of a WST point, but not necessarily the particular type. *WST/Short Stem*: recognized only as from a non-Archaic point. *Archaic*: identifiable as part of an Archaic point, but not to a particular type. *Indeterminate*: recognizable only as a point blade.
Preform	A biface that was destined to have become a particular type of tool, especially a specific type of projectile point (Gramly 1990:42).
Prismatic blade	A removal from a core (prismatic blade core) that has a length at least twice its width. A pronounced single or double arris runs parallel to the main axis making the cross section triangular or quadrilateral (Gramly 1990:44).
Production biface	Biface identifiable with the production sequence for Western Stemmed projectile points.
Projectile point	A biface that possesses consistent manufacturing properties of "point," "haft," and "bilateral symmetry." These items were analyzed separately from other bifaces because of their chronological utility.
Scraper	A flaked-stone tool, generally made on a flake or blade, having bold unifacial or bifacial retouch on one or more margins (Gramly 1990:49).
Unique uniface	A uniface that cannot be identified with any particular class.

TABLE 5.4. Formal Tool Categories Represented in the Old River Bed Assemblage.

Tool Category	Chert		Obsidian		Fine-Grained Volcanics		Quartzite/Quartz		Limestone		Total
	n	%	*n*	%	*n*	%	*n*	%	*n*	%	
Projectile Points											
Classifiable	32	3.1	560	54.9	424	41.6	2	.2	2	.2	1,020
Unclassifiable	3	5.3	38	67.9	14	25.0	1	1.8			56
Unknown			1	100.0							1
Subtotal	35	3.2	599	55.6	438	40.7	3	.3	2	.2	1,077
Bifaces											
Production biface	7	5.8	9	7.4	103	85.1	2	1.7			121
Knife	2	20.0			7	70.0	1	10.0			10
Crescent	7	28.0	5	20.0	13	52.0					25
Scraper	7	10.1	19	27.5	42	60.9	1	1.4			69
Graver	3	13.6	8	36.4	11	50.0					22
Combination tool	2	33.3	1	16.7	3	50.0					6
Drill	3	15.8	6	31.6	10	52.6					19
Chisel			1	33.3	2	66.7					3
Amorphous biface	4	6.7	13	21.7	40	66.7	3	5.0			60
Indeterminate biface	23	9.0	107	41.8	122	47.7	4	1.6			256
Unique biface					5	100.0					5
Subtotal	58	9.7	169	28.4	358	60.1	11	1.8			596
Unifaces											
Crescent					2	100.0					2
Scraper	17	9.6	46	26.0	113	63.8			1	.6	177
Graver	4	28.6	3	21.4	7	50.0					14
Notch					1	100.0					1
Chisel					1	100.0					1
Combination tool	6	18.8	6	18.8	18	56.3	2	6.3			32
Amorphous uniface	4	13.8	3	10.3	21	72.4	1	3.4			29
Indeterminate uniface			4	44.4	5	55.6					9
Subtotal	31	11.7	62	23.4	168	63.4	3	1.1	1	.4	265
Other Categories											
Core	10	45.5	7	31.8	4	18.2	1	4.5			22
Debitage	26	8.4	109	35.2	169	54.5	6	1.9			310
Chunk	2	22.2	5	55.6	1	11.1			1	11.1	9
Hammerstone					1	50.0	1	50.0			2
Ground stone							1	100.0			1
Manuport							1	100.0			1
Split pebble/cobble			2	66.7	1	33.3					3
Worked pebble			1	100.0							1
Total	162	7.1	954	41.7	1,140	49.8	27	1.2	4	<.1	2,287

categories, although most are incomplete, and address this purpose in the section on these tools.

The projectile points have been identified with a typology originally presented by Jones, Beck, and Kessler (2003) and modified by Beck and Jones (2010b). This typology is discussed in detail in the section on projectile points.

CONDITION OF ASSEMBLAGE

One of the most distinctive aspects of the ORB assemblage is the degree of weathering damage evident on most of the artifacts. As the analysis of the ORB1 assemblage proceeded we realized how much such damage constrained the observations possible. We therefore derived a classification of weathering damage measured

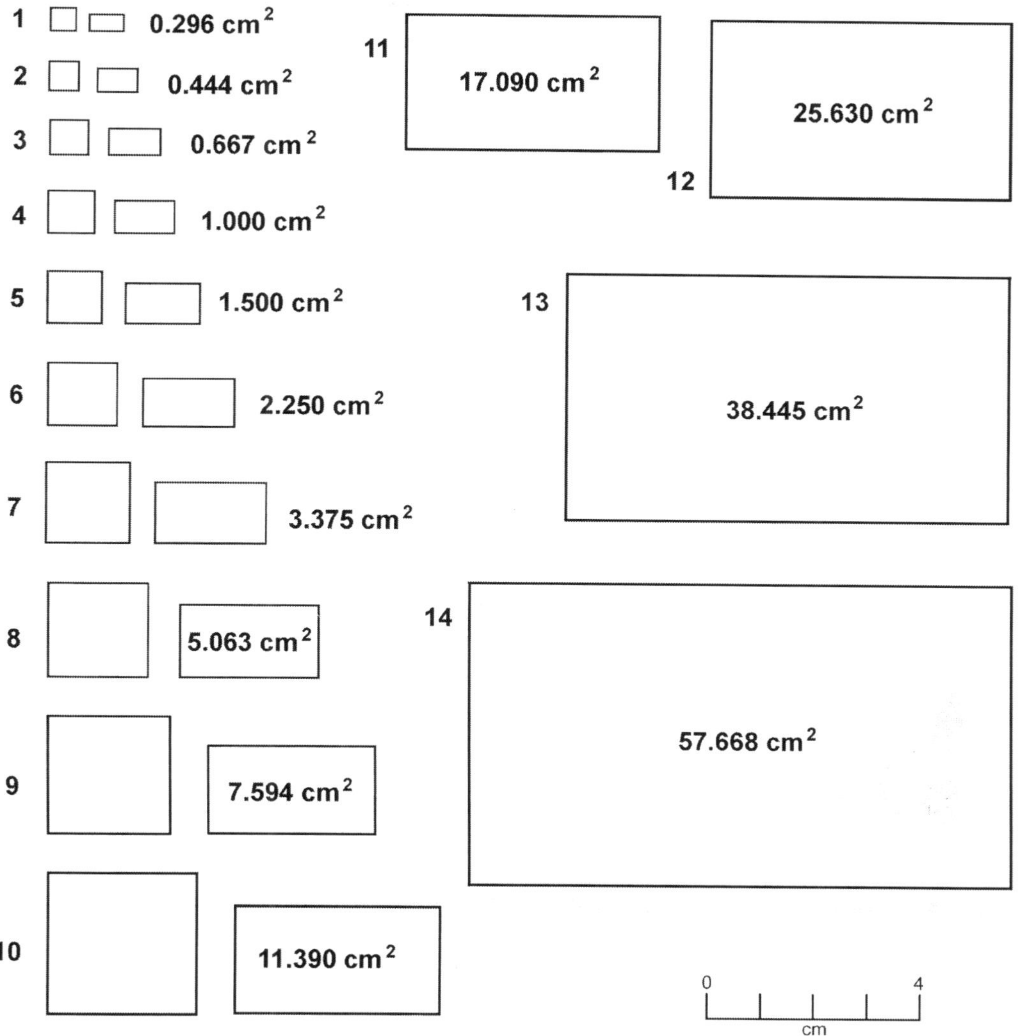

FIGURE 5.1. Size categories recorded in the Old River Bed artifact analysis.

on an ordinal scale of "none to minimal," "medium," "heavy," and "extreme" (Figure 5.2). This assessment is obviously subjective, but it gives a general picture of artifact condition. For instance, "none to minimal" weathering refers to those artifacts that have sharp edges and clear surface flake scars and on which use-wear, retouch, and resharpening are generally observable under 10–100× power (although these vary with material type). Those artifacts with "medium" weathering have dulled edges and visible flake scars with lightly rounded arises, and use-wear and resharpening are not observable under 10–100× power. Artifacts with "heavy" weathering have rounded edges and visible flake scars with heavily rounded arises. Artifacts with "extreme" weathering often appear almost like rounded pebbles; flake scars are not visible at all.

Table 5.5 shows the distribution of the ORB arti-

facts across the weathering classes, separated by raw material. Of the 2,048 specimens for which condition was evaluated, 1,812 (88.5 percent) exhibit at least medium weathering damage. More than half ($n = 1,265$, 61.8 percent) show heavy to extreme weathering damage. There are clear differences in weathering damage across raw materials. Quartzite and chert, for example, exhibit the least amount of damage, while obsidian exhibits the most. An especially high proportion of obsidian artifacts (41.0 percent) shows extreme weathering damage. A chi-square test reveals a significant association between weathering damage and raw material ($\chi^2 = 583.44$; $df = 15$; $p < .001$).

It is apparent that chert and quartzite (although the sample size of the latter material is very small) in the ORB delta weather differently than obsidian and FGV. If weathering was caused by wind abrasion, it

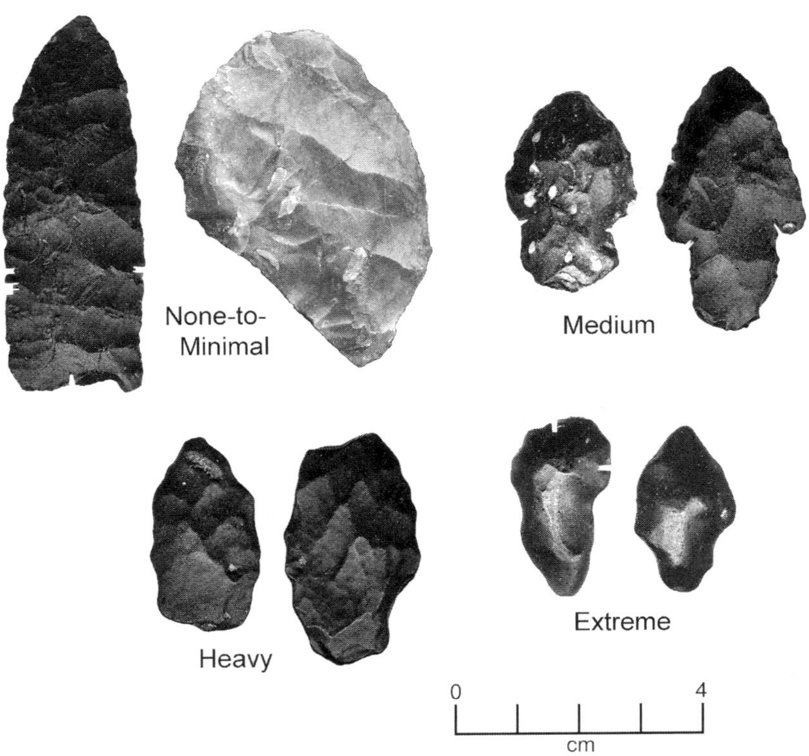

FIGURE 5.2. Examples of weathering damage on Old River Bed artifacts.

TABLE 5.5. Weathering Categories by Raw Material.

| | Abrasion Category | | | | | | | | | Not | |
| Raw Material | None to Minimal | | Medium | | Heavy | | Extreme | | Subtotal | Recorded | Total |
	n	%	n	%	n	%	n	%			
Chert	90	58.9	50	32.7	9	5.9	4	2.6	153	9	162
Obsidian	65	7.4	166	18.8	291	32.9	362	41.0	884	71	955
Fine-grained volcanics	66	6.7	321	32.6	396	40.2	201	20.4	984	155	1,139
Quartzite	13	56.5	9	37.1	1	4.3			23	3	26
Limestone	1	33.3	1	33.3	1	33.3			3	1	4
Quartz			1	100.0					1		1
Total	235	11.5	548	26.8	698	34.1	567	27.7	2,048	239	2,287

stands to reason that all materials should be affected to some degree. For example, some years ago we examined a number of artifacts from the vicinity of Pleistocene Lake Tonopah, and although the damage caused by weathering varied by raw material, all were damaged to the degree that use-wear was not visible. Because such large proportions of ORB chert and quartzite artifacts have none to minimal weathering damage, we suspect that the source of damage may be a chemical process in which strong bases may differentially affect raw material,

having the least effect on quartzite and chert and the greatest on obsidian.

As stated above, the severity of weathering damage limited the kinds of features that could be observed on the ORB artifacts, particularly small-scale edge modification resulting from tool use and edge preparation such as grinding. In a very few cases we were able to observe that an artifact had been used or an edge was ground but not the type of use-wear or extent of edge grinding.

General Observations on the ORB Assemblage

The ORB assemblage differs from other Paleoarchaic assemblages in three ways: raw material composition, the size of the artifacts, and the degree of expedient manufacture present. In all of the Paleoarchaic assemblages we have studied from eastern and central Nevada, FGV is the predominant toolstone, followed by chert and obsidian. For example, in assemblages from eastern Nevada, FGV constitutes between 45 and 55 percent; chert, between 30 and 37 percent; and obsidian, between 13 and 19 percent. These patterns actually reflect differences in material use for specific categories. For instance, in Paleoarchaic assemblages across the Great Basin, FGV toolstone was favored for the manufacture of the large stemmed projectile points of the Western Stemmed Tradition (WST) except in those areas where obsidian was prevalent, such as in the northern Great Basin (Amick 1993, 1995; Beck and Jones 1990b). These points are less often made from chert, generally less than 7 percent. As Table 5.4 shows, however, the ORB projectile point assemblage is dominated not by FGV (40.7 percent) but by obsidian (55.6 percent); chert is extremely rare (3.2 percent), which is likely due to the rarity of this raw material in the ORB delta (see Chapter 6).

Where projectile points in Paleoarchaic assemblages elsewhere in the province are more commonly made from FGV and obsidian, such tools as crescents, scrapers, and gravers are most often manufactured from chert, though obsidian was sometimes substituted where available. For example, of the 245 crescents from the Sunshine Locality (Beck and Jones 2009), 231 (94.3 percent) are chert. Similarly, some 90 percent of other categories such as scrapers, gravers, drills, chisels, and combination tools are chert. But as Table 5.4 shows, this is not the case in the ORB assemblage. Of the 27 unifacial and bifacial crescents, only seven (25.0 percent) are manufactured from chert; 15 (55.6 percent) are made of FGV, and five (18.5 percent) are made from obsidian. Scrapers exhibit a similar pattern in which most are manufactured from FGV (63.0 percent); only 9.8 percent are chert. The rarity of chert is evident in every tool class in the ORB assemblage.

Raw material use is, of course, in some part a function of the proximity of toolstone sources. For example, if there are no obsidian sources available locally, as is the case for our eastern Nevada study area (e.g., Beck and Jones 2009; Jones, Beck, Jones, and Hughes 2003), other materials such as FGV will be used more heavily. If, on the other hand, obsidian is available locally, as is

the case in most of the northern Great Basin, this material may be preferred for most tools and be heavily represented in artifact assemblages. Obsidian, although not locally available, occurs less than 50 km from the ORB delta (Figure 5.3). The Topaz Mountain source lies about 50 km south of the ORB delta and makes up the majority (~60 percent) of the WST and Early Holocene obsidian artifacts sourced thus far (see Chapter 6). Most of the cobbles from this source, however, are fairly small and certainly were not appropriate for the manufacture of some of the largest production bifaces, which are more than 100 mm in length. On the other hand, 116 WST projectile points sourced are made from this material (Chapter 6), and thus Paleoarchaic peoples did make use of Topaz Mountain obsidian.

Much larger cobbles can be found at the Browns Bench source, some 240 km to the northeast. This source accounts for about 35 percent of the WST projectile points analyzed for chemical composition (see Chapter 6). Large cobbles are available from good-quality FGV sources, several of which are more locally available (Figure 5.3). Nearly three-quarters of the FGV artifacts sourced are from Flat Hills, which lies less than 50 km to the east (Chapter 6). Sources of chert are rare, and, as Page and Duke note in Chapter 6, most of the chert artifacts in the ORB assemblage are from unknown sources.

A complicating factor in assessing source use in the ORB delta is the considerable amount of tool reuse and recycling. As we discuss below, some of the projectile points identify with established WST types, such as Parman, Lake Mohave, and Cougar Mountain. Others, however, do not. Some of the WST points are similar in size to those in Paleoarchaic assemblages elsewhere, but a considerable number are somewhat smaller. Further, the most numerous of the unknown forms, which have been named "Dugway Stubbies," are especially small, on average between 20 mm and 35 mm in length. We believe that many of these points were manufactured from production biface fragments and blade segments of expended stemmed points, which would explain their small size and irregular shape. These points are at times almost completely unifacially flaked. Most of the smaller points recycled from material left by earlier visitors bear the source profile of these earlier folk, complicating their interpretation vis-à-vis mobility and material exchange.

Some of the other tools, such as scrapers and gravers, were also likely manufactured from scavenged raw material, which would explain the large proportion of obsidian given the distance to the sources represented.

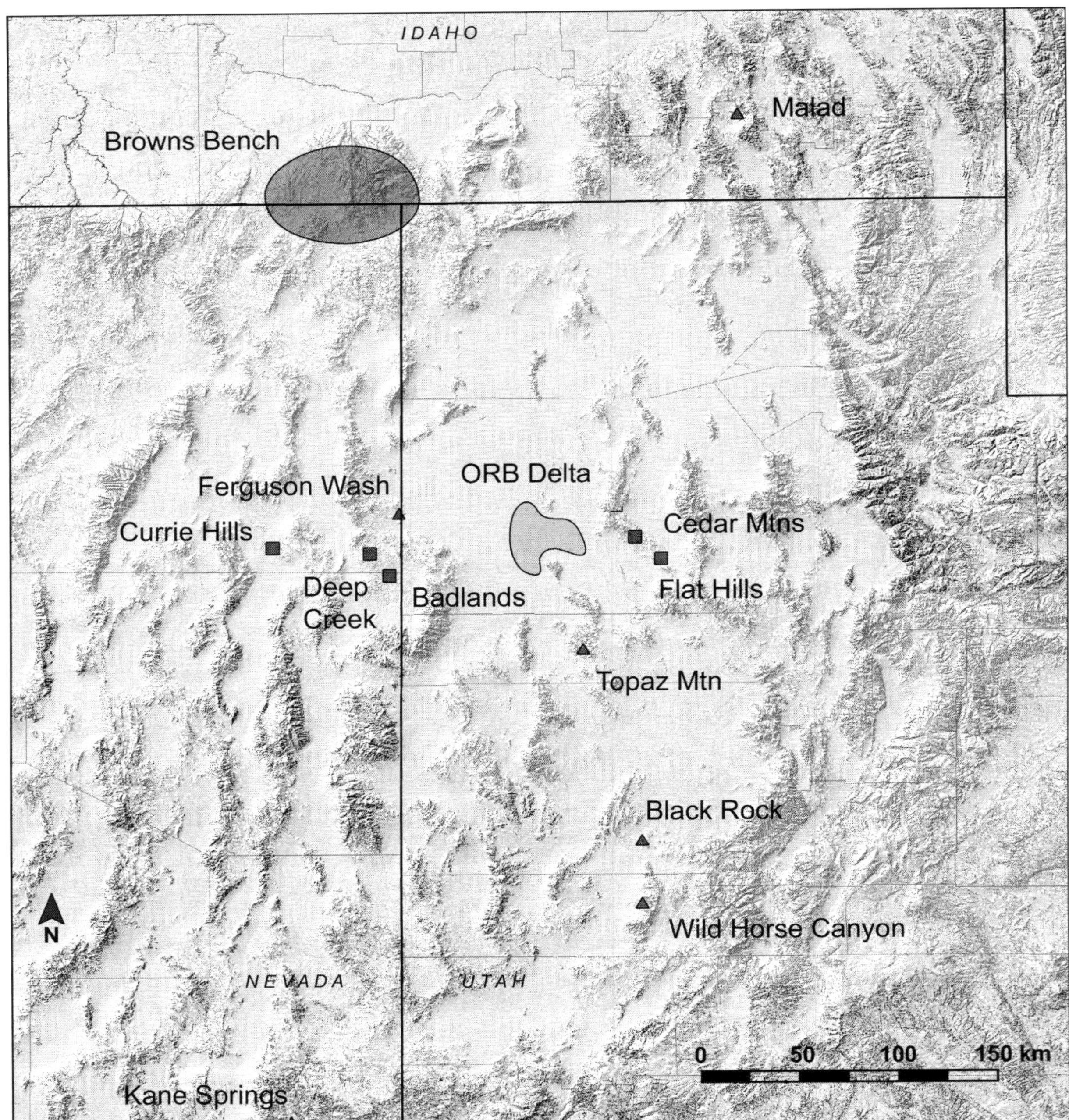

FIGURE 5.3. Locations of obsidian (*triangles and large circle*) and fine-grained volcanic (*squares*) sources in the vicinity of the Old River Bed (ORB) Delta.

The ORB consists of a number of raised streambeds that served as causeways into the marsh. Once people moved into the interior of the marsh, they would have been at some distance from toolstone sources and rather than leave to obtain more stone, may have scavenged biface and point fragments left by earlier occupants for the manufacture of new tools (see Duke 2011 for a similar argument).

PROJECTILE POINTS

Those stone tools that possess the manufactured properties of point, haft, and bilateral symmetry are identified as projectile points. The ORB projectile point sample, 1,077 specimens in all, contains many complete or nearly complete points along with fragments exhibiting enough diagnostic features to be identified by type. There are nine unfluted Great Basin Concave

Base (GBCB) points, while some 489 points represent Paleoarchaic types. Another 426 stemmed specimens do not possess diagnostic WST type morphologies but are believed to have Paleoarchaic (probably early Holocene) affiliation. We refer to these as Early Holocene (EH) types. One of these most closely resembles the Eden type, which is associated with the Great Plains Cody Complex. Twenty-four specimens are difficult to distinguish as WST or EH types and thus are categorized as WST/EH. Only 72 specimens identify with Archaic types. One of the remaining 57 points represents a form of unknown typological and temporal affiliation, while the other 56 are unclassifiable.

The Archaic projectile points conform to well-established types. In contrast, many Paleoarchaic specimens cannot be associated easily with recognized categories. As discussed below, this relates to the considerable level of opportunistic manufacture and material recycling at the ORB sites. Many of the projectile points in this sample exhibit both smaller size and less uniform morphology than we find in assemblages of comparable age elsewhere in the Great Basin, although this morphological irregularity seems to be true of the early point assemblage from Danger Cave as well (see Jennings 1957). This prompted Jones, Beck, and Kessler (2003) to devise a dimensional classification to organize this variability. This classification was used originally to distinguish all identifiable points in the ORB1 assemblage, including those of traditional WST types. The classification was revised by Beck and Jones (2010b), and this revised form is used here.

Analytical Protocol

The analytic protocol devised for the ORB projectile points draws on our previous studies of the ORB1 and ORB2 assemblages as well as our analysis of points from the Sunshine Locality in eastern Nevada (Beck and Jones 2009).

Metric Attributes

All points were measured on the set of attributes standardized by Thomas (1981; Figures 5.4 and 5.5; Table 5.6), with two exceptions. First, we differed from Thomas (1981) in the way in which proximal shoulder angle (PSA) and distal shoulder angle (DSA) were measured (Figure 5.5). In his examination of PSA and DSA for the Monitor Valley points Thomas selected the largest angle, between that on the left and that on the right, for measurement. Here, if both left and right angles were measureable, they were averaged for the specimen. If only one side was measureable, that side was considered

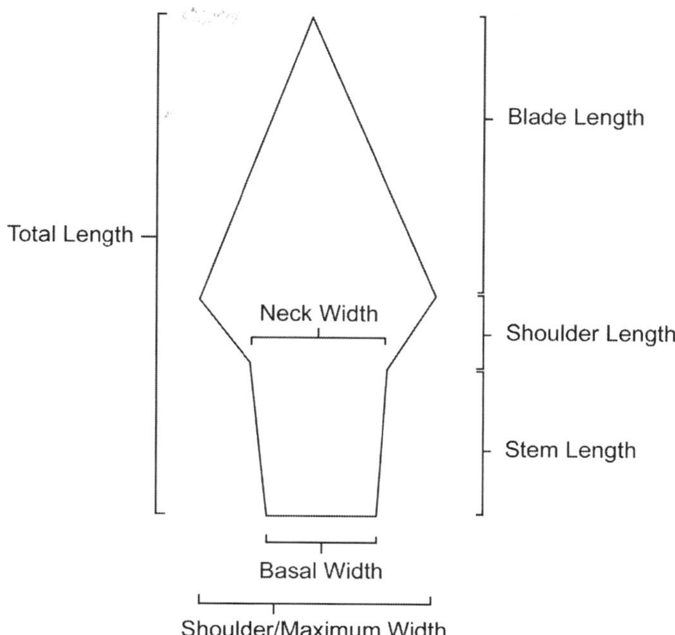

FIGURE 5.4. Metric attributes measured on the Old River Bed projectile points (from Beck and Jones 2009).

representative of the specimen. Second, as a substitute for L_M, which measures the distance from the widest point to the base, we took two separate measurements, L_B (blade length) and L_S (stem length; Figure 5.4). The points in the Sunshine assemblage that are attributable to four of the WST types represented in the ORB assemblage—Lake Mohave, Silver Lake, Cougar Mountain, and Parman—are shouldered and have stems that often are equal to the length of the blade. The DSA on most of these points is greater than 180°, creating a relatively large area between the point of maximum width and the neck, which we refer to as the shoulder length (SL). In these cases, SL = L_T (total length) – (L_B + L_S). Whereas DSA measures the *angle* of the shoulder, SL measures the *portion that the shoulder contributes to overall length*. We believed that it might be possible to differentiate points of some types on the basis of stem, shoulder, and blade length. Thus, for the WST points, L_B and L_S were measured as shown in Figure 5.4, and SL was calculated by the above formula. As was done for other artifact categories, actual thickness and actual weight were measured for all specimens.

Qualitative Attributes

In addition to raw material, which was recorded for all artifacts, and weathering damage, which was recorded for most, several qualitative attributes were measured

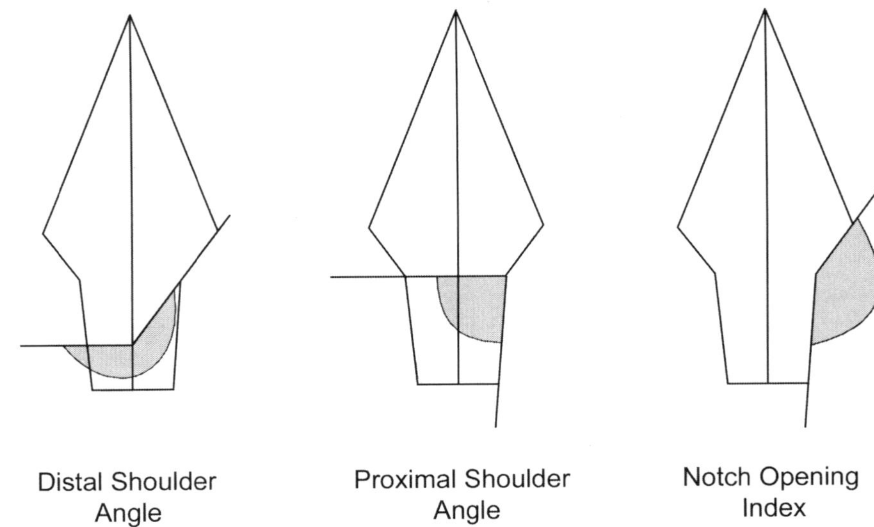

FIGURE 5.5. Angles measured on the Old River Bed projectile points (from Beck and Jones 2009).

TABLE 5.6. Quantitative Variables Recorded for the Old River Bed Projectile Points.

Variable	Comment/State
Total length (L_T)	As shown in Figure 5.4
Blade length (L_B)	As shown in Figure 5.4
Stem length (L_S)	As shown in Figure 5.4
Shoulder length (SL)	(Total length − [blade length + stem length])
Neck width	As shown in Figure 5.4
Maximum width	As shown in Figure 5.4 (same as blade width for Great Basin Stemmed Series points)
Basal width	As shown in Figure 5.4
Actual weight	
Proximal shoulder angle	As shown in Figure 5.5
Distal shoulder angle	As shown in Figure 5.5
Notch opening index	As shown in Figure 5.5

TABLE 5.7. Additional Qualitative Variables Recorded for the Old River Bed Projectile Points.

Variable	States
Raw material	Chert, obsidian, fine-grained volcanics, quartzite, limestone, quartz
Flaking pattern	As shown in Figure 5.6
Use-wear damage	Present, absent
Resharpening	Absent, tip/blade, barb, stem, combination, different tool
Lateral edge grinding	Present, absent
Beveling	Absent, blade, stem, blade and stem
Fragment	Complete, tip/blade, barb, stem/base, combination

on all points (Table 5.7). These include flaking pattern, presence of use-wear damage, location of resharpening retouch, presence of lateral edge grinding, presence and location of beveling (on either blade or stem), and fragment (the portion of the point represented). Because of the extensive weathering damage on most specimens, however, we were often unable to detect the presence or absence of these attributes, especially use-wear and lateral edge grinding.

Flaking Pattern

Some researchers have suggested that particular flaking patterns are associated with points of distinct morphologies. For example, in Paleoarchaic assemblages from eastern and central Nevada (e.g., Beck and Jones 1997, 2009), points of the WST types most often exhibit broad collateral flaking. Therefore, the flaking pattern (Figure 5.6) was recorded for all points.

Use-Wear, Resharpening, and Lateral Edge Grinding

Because of the considerable damage caused by weathering, we were able to record use-wear and lateral edge grinding only as present or absent. Even so, in most cases we were unable to evaluate these attributes. We were, however, able to evaluate resharpening on a number of specimens. On projectile points we recorded the portion of the point that has been resharpened. Figure 5.7A shows a Butte Valley Corner-notched point that has been resharpened on the tip, blade, barb, and stem. The resharpening is evident in the invasive flake scars that overlie previous manufacturing scars or breaks. The Expanding Stem point in Figure 5.7B has been resharpened along the blade and barb, but eventually the

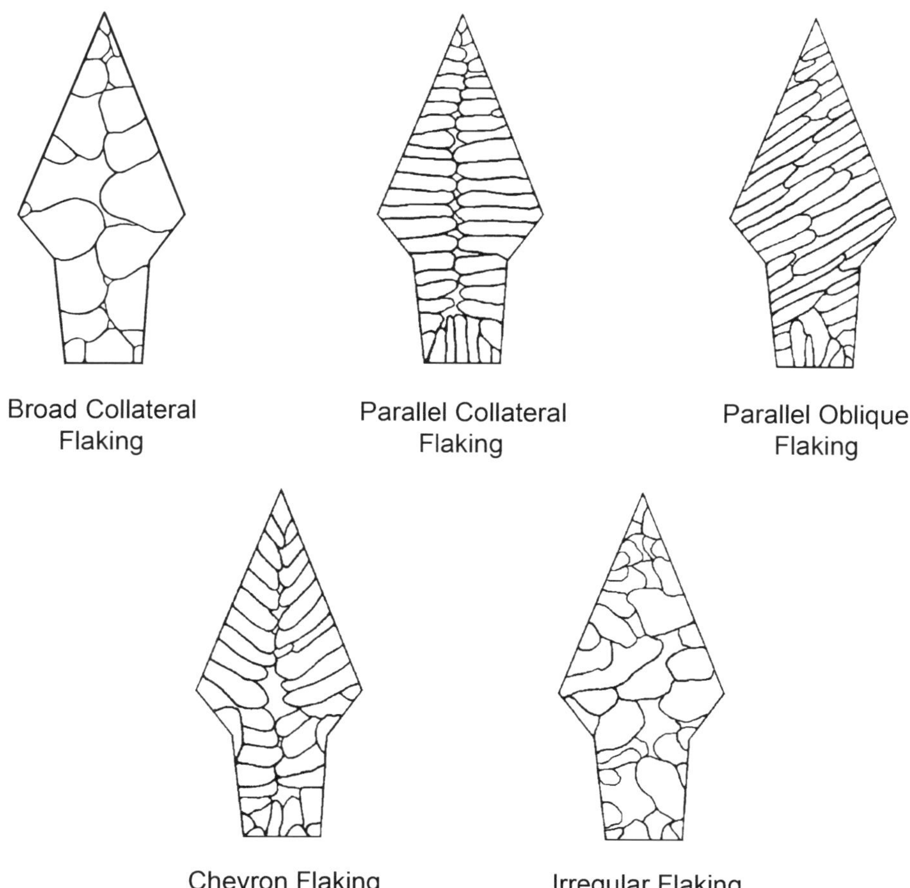

Broad Collateral
Flaking

Parallel Collateral
Flaking

Parallel Oblique
Flaking

Chevron Flaking

Irregular Flaking

FIGURE 5.6. Flaking patterns recorded for the Old River Bed projectile points (from Beck and Jones 2009).

tip was formed into a graver bit. The resharpening of projectile points into different tools is not common in the ORB assemblage, but a number of cases do occur. These are noted on the projectile point figures.

The Elko point in Figure 5.7C has been resharpened on the tip, blade, and stem; the barbs have both been broken, but no resharpening scars are evident along the breaks. Finally, one of the points of the crescent in Figure 5.7D was broken off, and resharpening has taken place along the lower edge of the remaining wing fragment, including a portion of the break.

Beveling

Beveled retouch has been reported to be a distinctive feature of some WST projectile points (e.g., Tuohy 1969). In our analysis of the Sunshine Locality projectile points, we identified a number of specimens with retouch on alternate margins ("different edges of opposite faces" [Storck 1997:47]), creating a beveled or diamond-shaped cross section. Although most com-

monly a feature of the blade, beveling appears occasionally on the stem. Beveling is believed by some to be the result of unifacial resharpening (e.g., Deller and Ellis 1992:49). Tuohy suggested that beveling was part of the process of the burination of point tips, in the creation of what he calls a "squared-off chisel bit" (1969:139). Our results from previous studies indicate that beveling occurs most often on Lake Mohave points, although it is observed occasionally on other WST points as well. The chisel tip, however, occurs almost exclusively on Lake Mohave points, and there is a significant statistical association between this feature and beveling on these points in the Sunshine assemblage. Even so, as it is unclear whether beveling was used to resharpen a tip or to enhance the function of a hafted biface tip, we recorded it separately from other evidence of resharpening on the ORB points. This permits us to examine (1) if beveling occurs on one type more frequently than it does on others and (2) if it might be an important aspect of point function (as Tuohy suggested).

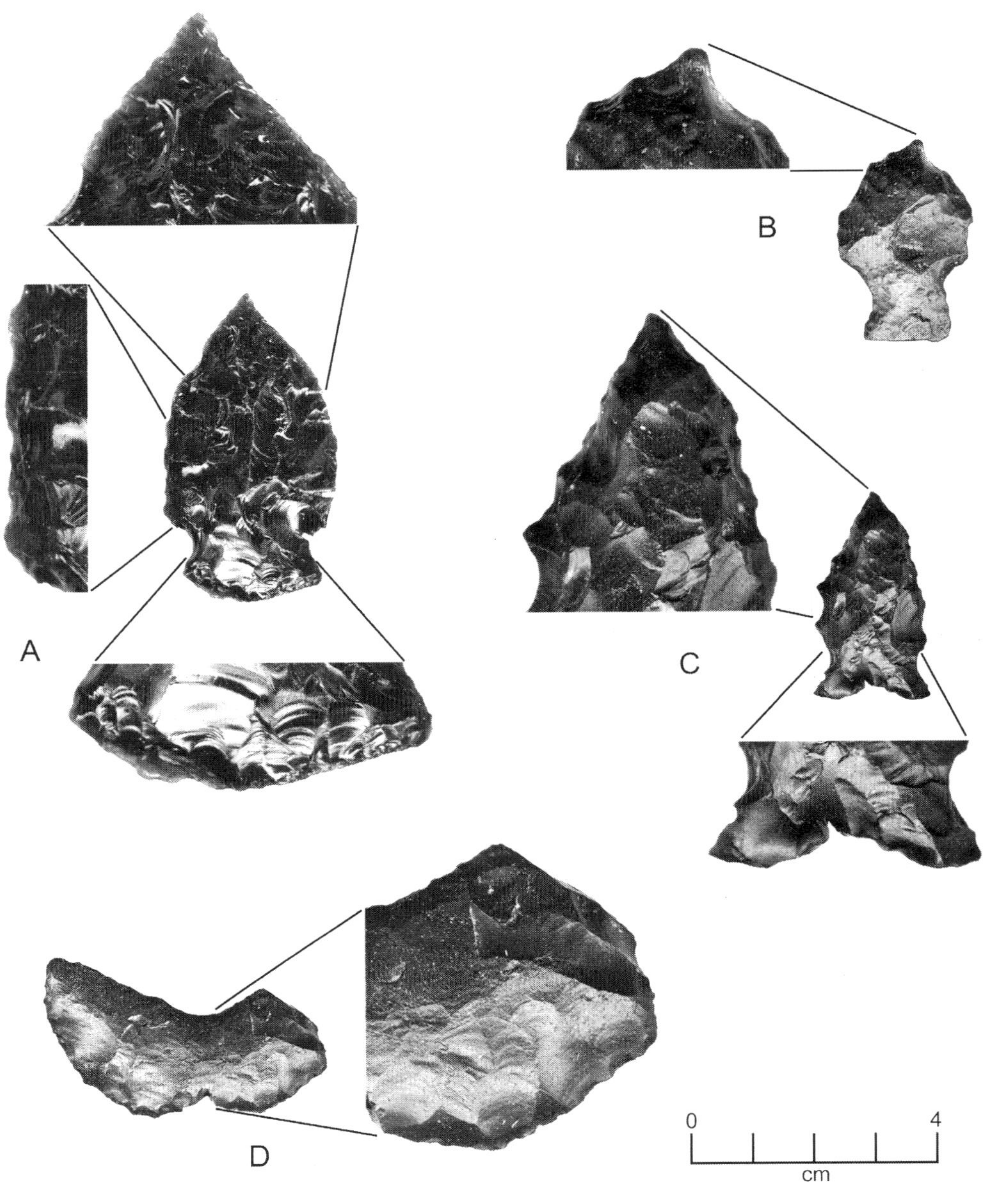

FIGURE 5.7. (*A–D*) Examples of resharpening on Old River Bed artifacts.

Projectile Point Classification

Jones, Beck, and Kessler (2003) devised a paradigmatic classification for the ORB1 assemblage, which was revised by Beck and Jones (2010b). The original classification was based on four dimensions (Table 5.8). Classes are formed by the intersection of these dimensions; that is, a class definition comprises four modes, one from each dimension. For example, a point with a square shoulder and a contracting and extremely long stem with a rounded base was assigned to a different class than one with a square shoulder and a parallel, square stem with a flat base. The classification, then, produces a total of 192 monothetic classes. The advantage of dimen-

sional classifications is that they produce classes that are exhaustive and completely comparable because each is based on the same set of criteria. In practice, however, they tend to overdivide a data field. Moreover, dimensional classes do not generally align well with conventional categories, which typically are polythetic units. Such is the case when this classification is applied to points of traditional WST types. For example, a Lake Mohave point might have a square or sloping shoulder with a sharp angle, its stem may be medium to extremely long, and its base may be rounded or flat. Thus, there are eight possible classes into which a Lake Mohave point might fit. Such overdivision of these traditional types

TABLE 5.8. Dimensional Classification Created by Jones, Beck, and Kessler (2003).

Dimension	Modes
1. Shape of the shoulder The shoulder is the area of transition between the hafting element (stem) and the cutting/piercing element (blade). In stemmed forms, the shoulder is isolated by two angles, one at the neck and the other at the corner of the blade. Lanceolate points exhibit the latter angle only. The distal shoulder angle and notch opening index (Thomas 1981) are applicable to descriptions of this dimension.	a. Square. The transition between stem and blade makes a sharp angle, with distal shoulder angle close to 180°. b. Sloping with sharp angle. c. Sloping with weak angle (shallow arc). d. None (applies to lanceolate points).
2. Shape of stem margins This variable describes the stem margins as they descend proximally from the neck toward the base.	a. Contracting. The width of the stem decreases; basal width < neck width. b. Parallel. The stem margins are parallel; basal width = neck width. c. Expanding. The width of the stem increases; basal width > neck width.
3. Length of stem	a. Extremely long. Stem length is more than twice that of the width of the stem at the neck. b. Long. Stem length is 1.25 to 2.0 times greater than (neck width + basal width)/2. c. Square. Stem length is equal to neck width. d. Wide. Stem length is less than (neck width + basal width)/2.
4. Stem base shape	a. Rounded. Corners of stem base are removed to form a convex arc. b. Flat. c. Concave. d. Indented or bifurcate.

added no information in the previous analysis (such as improving the precision of the chronology produced by Jones, Beck, and Kessler [2003]), and thus the dimensional classes for these points are not used here. Instead, we employ definitions derived from our analysis of the Sunshine WST points.

One of the goals in the Sunshine analysis was to more specifically define the WST types: Cougar Mountain, Lake Mohave, Parman, and Silver Lake (Haskett is not represented in the Sunshine assemblage). The points were first identified with types based on the original type definitions, and then points of each type were compared with one another. Although it would have been preferable to proceed as did Thomas (1981) in his creation and revision of the key for Great Basin Archaic points, statistical grouping of the points themselves proved unsuccessful because there is overlap in almost every variable. For example, Figure 5.8 shows scatterplots of measurements taken on a set of relatively complete points from eastern Nevada reported by Beck and Jones (2009). This assemblage, which we call the Eastern Nevada Comparative Collection, contains 101 points, 79 of which are relevant here: 18 Cougar Mountain, 30 Lake Mohave, 17 Silver Lake, and 14 Parman. As these scatterplots show, there is considerable overlap among points of the four types, with a few exceptions.

In Figure 5.8C, for example, Silver Lake points are fairly distinct with respect to stem length, and when this variable is plotted against total length, these points cluster at the lower left corner, while Lake Mohave points are also distinct, separating from Parman and Cougar Mountain. On the other hand, if all points had been plotted using the same symbol (as opposed to a distinct symbol for each type), the distribution would appear to be continuous.

When points of each type were compared statistically, however, a number of significant differences were revealed, which allowed us to define what we called "distinctive features" for several of the types. These features were combined with qualitative aspects to formulate definitions for the types. These definitions are used here in combination with general definitions from Jones, Beck, and Kessler (2003) to identify WST points in the ORB assemblage, although in some cases the metrics in the definitions can only be suggestive, given the diminutive size of some of the ORB points. For this analysis the revised classification was also applied to the ORB1 points. Table 5.9 shows both the general definitions given by Jones, Beck, and Kessler (2003) and the distinctive features derived by Beck and Jones (2009) for the Lake Mohave, Silver Lake, Parman, and Cougar Mountain types. Because Haskett

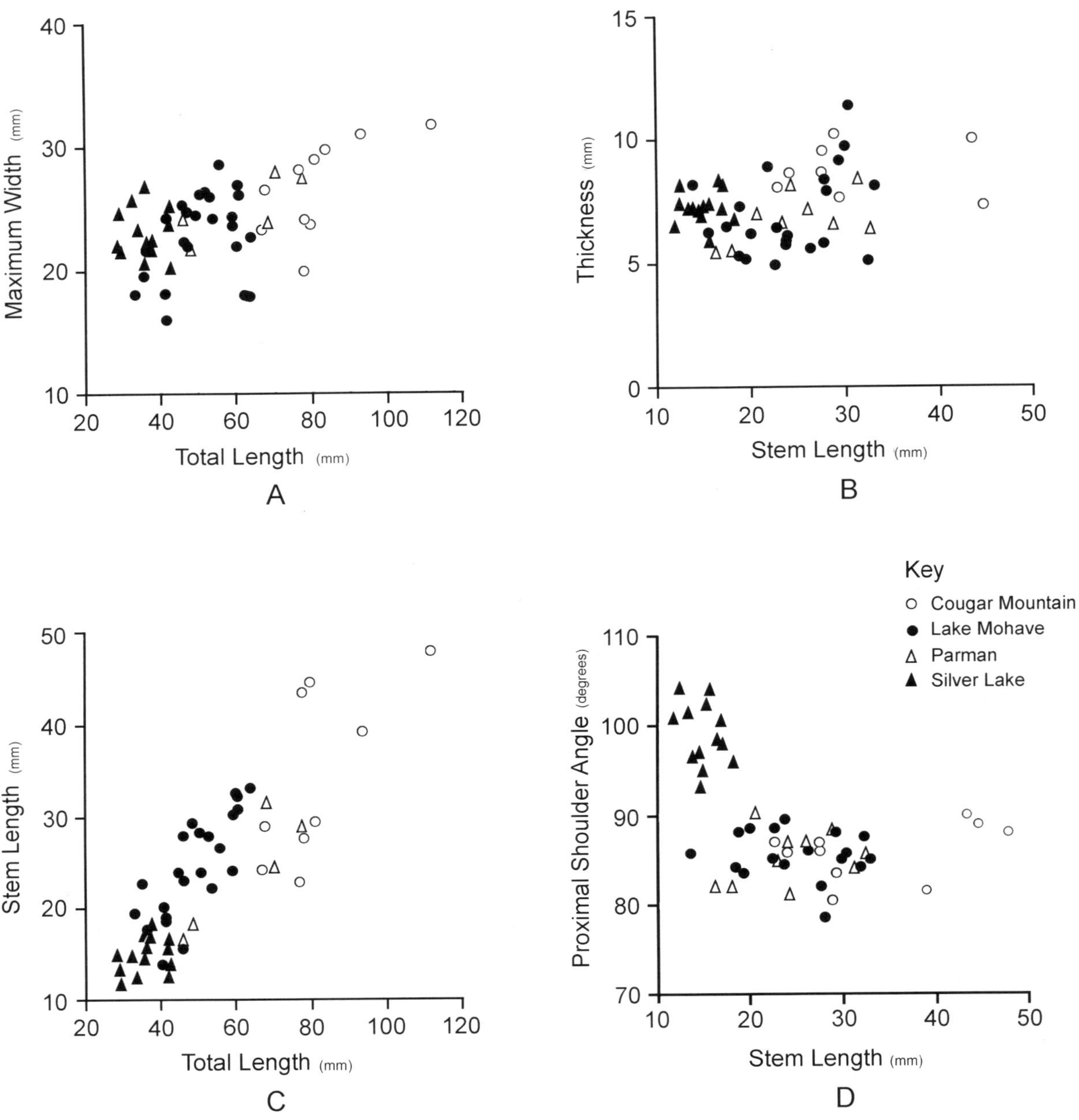

FIGURE 5.8. (*A–D*) Scatterplots of measurements taken on Great Basin Stemmed Series points in the Eastern Nevada Comparative Collection (from Beck and Jones 2009).

points were not included in our Sunshine analysis, Table 5.9 includes the definition presented by Jones, Beck, and Kessler (2003:15). We retain the classes produced by Jones, Beck, and Kessler for points that could not be identified with an existing type. These include Square Stem, Expanding Stem, Contracting Stem, and Dugway Stubby (Table 5.9). Finally, Table 5.9 includes definitions for other established types, including Eden, Pinto, and Butte Valley Corner-notched (see Beck and Jones 2009). Archaic points are identi-

fied with types using Thomas's (1981) key. Table 5.10 summarizes these types by raw material for the ORB assemblage.

Unfluted Great Basin Concave Base Points

While no fluted points have been recovered in the proximal ORB delta, nine lanceolates were collected (Figure 5.9; Table 5.10). These points are included in the Great Basin Concave Base Series (Pendleton 1979) and are referred to here as GBCB-UF. The quantitative data for

Table 5.9. Definitions of Projectile Point Types Recognized in the Old River Bed Assemblage.

Point Type	Definition
Terminal Pleistocene/Early Holocene Types	
Unfluted Great Basin Concave Base	Lanceolate point with parallel or slightly convex sides and concave bases. No stemming evident. Often included in the Great Basin Concave Base Series.
Lake Mohave	*General Definition*: Sharply angled sloping shoulders, long contracting stem, rounded or flat base (Jones, Beck, and Kessler 2003:15). *Distinctive Features*: Blades that are triangular in shape with beveling and/or chisel tip (Beck and Jones 2009:129).
Silver Lake	*General Definition*: Abrupt shoulder, parallel or expanding stem, rounded or flat base (Jones, Beck, and Kessler 2003:15). *Distinctive Features*: Stems that are ≤13.5 mm in length, with a proximal shoulder angle ≥91° (Beck and Jones 2009:129).
Parman	*General Definition*: Abrupt shoulder, parallel stem margin, rectangular or square stem, rounded or flat base (Jones, Beck, and Kessler 2003:15). *Distinctive Features*: Pronounced shoulders, lenticular cross section, and parallel to contracting stem (overlaps completely with Lake Mohave but can be used after Lake Mohave has been distinguished as above) (Beck and Jones 2009:129).
Cougar Mountain	*General Definition*: Weakly shouldered, parallel or gently contracting stem, stem very long in relation to width, flat or rounded base (Jones, Beck, and Kessler 2003:15). *Distinctive Features*: Stems that are ≥35 mm in length, blades that have shoulder values ≥15 mm, midsections that have distal shoulder angles ≥240°, and neck widths ≥20 mm (Beck and Jones 2009:129).
Haskett	*General Definition*: Shoulder region is a single angle formed by intersection of the stem and blade; contracting, very long stem; flat or rounded base (Jones, Beck, and Kessler 2003:15).
Western Stemmed Tradition (WST) blade	Point blade of a Great Basin Stemmed Series point unidentifiable to type.
WST stem	Point stem of a Great Basin Stemmed Series point unidentifiable to type.
WST/Short Stem	Some specimens, due to their heavy weathering and fragmentary nature, could not be distinguished between Great Basin Stemmed or Early Holocene types (see below) and thus are identified with this more inclusive category.
Early Holocene Types	
Eden	Relatively narrow and parallel-sided with slightly indented parallel stems (occasionally produced only by grinding), straight bases, and highly controlled serial pressure flaking that is perpendicular to the edges and meets evenly in the middle, producing a distinct diamond cross section. Associated with the Cody Complex (Gibbon and Ames 1998:169).
Square Stem (Jones, Beck, and Kessler 2003:26–27)	*Definition*: Shoulder is square or sloping with an abrupt or intermediate angle, square stem with parallel margins, flat or slightly convex base. *Subtypes*: Four subcategories are differentiated on the basis of the size of the stem and the shape of the shoulder. A fifth subtype is used for specimens too fragmentary to assign to one of the other four subtypes.
Expanding Stem (Jones, Beck, and Kessler 2003:27–28)	*Definition*: Sloping shoulders with abrupt to weak stem–shoulder angle, expanding stem, rounded or flat base. *Description*: Members of this class have expanding stem margins. Some specimens exhibit this stem shape as a result simply of possessing a pair of constrictions that face each other, which were flaked to isolate the stem. Other examples have fully flaked stems, and this morphology was clearly the intended product. Most of these points are expediently manufactured using biface fragments or flake blanks. Many exhibit considerable longitudinal curvature and cross-section asymmetry. Flaking is typically localized steep retouch; little fully facial flaking was performed in shaping these points. *Subtypes*: Two subtypes are recognized on the basis of differences in size only.

Point Type	Definition
Contracting Stem (Jones, Beck, and Kessler 2003:26)	*Definition*: Shoulder is sloping with a continuous rather than abrupt stem–shoulder angle, the stem contracts, square or wide stem, flat base. *Description*: The class includes well-flaked bifaces and specimens made expediently from (recycled) biface fragments. Examples of the latter often exhibit a stem that has been isolated from the blade segment by removal of a single, steep-faced lunate flake. This creates a contracting stem with gently arcing margins and sloping shoulders. On most specimens the base incorporates a transverse fracture facet. Whether the break was part of the original "preform" or occurred during or following manufacture is difficult to judge because of the weathered conditions of most of the specimens. The blade segment is comparatively wide; the blade plan may be triangular or spatulate. *Subtypes*: Two subtypes are recognized on the basis of size only.
Dugway Stubby (Jones, Beck, and Kessler 2003:23–24)	*Definition*: Square or sloping shoulder with an abrupt or intermediate shoulder angle, slightly contracting or parallel stem margins, square stem with round or flat base. *Description*: "Stubbies" are small with very abbreviated blade segments. In most instances, they are less than 1.5 cm in length; the stem and shoulder represent between one-half and two-thirds of the length of the point. The plan of the blade varies between an equilateral and wide isosceles triangle. The blade is often rounded rather than obviously pointed, a consequence of breakage and subsequent weathering and/or resharpening. A few examples exhibit unifacial retouch on the distal section of the blade, producing a convex plan. Rather than serving as projectile tips, these specimens may have completed their use lives as scrapers. Their weathered condition precludes complete evaluation of this possibility, however. *Subtypes*: Four subtypes were defined on the basis of flaking technique, blank type, facial flaking, relative blade size, and symmetry (see Jones, Beck, and Kessler 2003:23–24). A fifth subtype was created for those specimens that cannot be identified with one of the other four subtypes.
Butte Valley Corner-notched (Beck and Jones 2009:127)	*Definition*: Corner-removed, expanding stem, convex to straight base, very large neck width relative to length and width. *Description*: This type was defined by Beck and Jones (2009:Figure 6.35) based on specimens found in Butte and Long valleys in eastern Nevada. Neck widths of points of this type average from ca. 21 mm, large relative to their length, which averages about 43 mm.
Pinto	*Definition*: Stemmed points with a concave stem base and parallel stem margins (Jones, Beck, and Kessler 2003:29). Basgall and Hall (2000:261) define this type relative to Elko Series points using Thomas's (1981) criteria: basal width >10.0 mm; proximal shoulder angle ≤100° or notch opening angle <80°.
Short Stem	Points assigned to this category are believed to be representative of one of the above Early Holocene types, but specific type identification is not possible.
Archaic Types Elko, Northern Side-notched, Humboldt, Gatecliff, Rocker Side-notched, Rosegate, Small Side-notched	Points were assigned to these types using Thomas's (1981) key. One additional category, Archaic blade, refers to point blades that are obviously from an Archaic point but for which not enough information exists to assign to a type.
Unknown Type	Points that cannot be identified with any known type or any type defined by Jones, Beck, and Kessler (2003). These are generally unique, and thus no new types were formulated for them.
Unclassifiable	Points that are too weathered and/or fragmentary to assign to a specific type.

TABLE 5.10. Projectile Point Types by Raw Material Represented in the Old River Bed Assemblage.

Type	Chert n	Chert %	Obsidian n	Obsidian %	Fine-Grained Volcanics n	Fine-Grained Volcanics %	Quartzite n	Quartzite %	Limestone n	Limestone %	Total
Concave Base											
Unfluted Great Basin Concave Base	2	22.2	6	66.7	1	11.1					9
Subtotal	2	22.2	6	66.7	1	11.1					9
Western Stemmed Tradition (WST)											
Lake Mohave			14	48.3	15	51.7					29
Silver Lake	2	1.9	80	74.1	26	24.1					108
Parman	1	5.0	13	65.0	6	30.0					20
Cougar Mountain	1	5.6	5	27.8	12	66.7					18
Haskett			9	75.0	3	25.0					12
Cougar Mountain/ Haskett			4	30.8	9	69.2					13
WST stem/midsection	3	2.0	52	33.8	98	63.6			1	.7	154
WST blade	5	3.7	51	37.8	79	58.5					135
Subtotal	12	2.5	228	46.6	248	50.7			1	.2	489
WST/Early Holocene											
WST/Early Holocene	1	4.2	14	58.3	9	37.5					24
Subtotal	1	4.2	14	58.3	9	37.5					24
Early Holocene											
Butte Valley Corner-notched	1	6.3	7	43.8	8	50.0					16
Square Stem			15	29.4	33	64.7	2	3.9	1	2.0	51
Expanding Stem			21	48.8	22	51.2					43
Contracting Stem					6	100.0					6
Dugway Stubby			164	85.9	27	14.1					191
Eden					1	100.0					1
Pinto	2	5.4	21	56.0	14	37.8					37
Short Stem	1	1.2	42	51.9	38	46.9					81
Subtotal	4	.9	270	63.4	149	35.0	2	.5	1	.2	426
Archaic											
Elko	6	24.0	12	48.0	7	28.0					25
Northern Side-notched					1	100.0					1
Humboldt	1	7.7	7	53.8	5	38.5					13
Rocker Side-notched	1	14.3	3	42.9	3	42.9					7
Gatecliff					1	100.0					1
Rosegate			7	100.0							7
Small Side-notched			1	100.0							1
Archaic blade	5	29.4	12	73.3							17
Subtotal	13	19.4	42	56.9	17	23.6					72
Unknown Type			1	100.0							1
Unclassifiable	3	5.4	38	67.9	14	25.0	1	1.8			56
Total	35	3.2	599	55.6	438	40.7	3	.3	2	.2	1,077

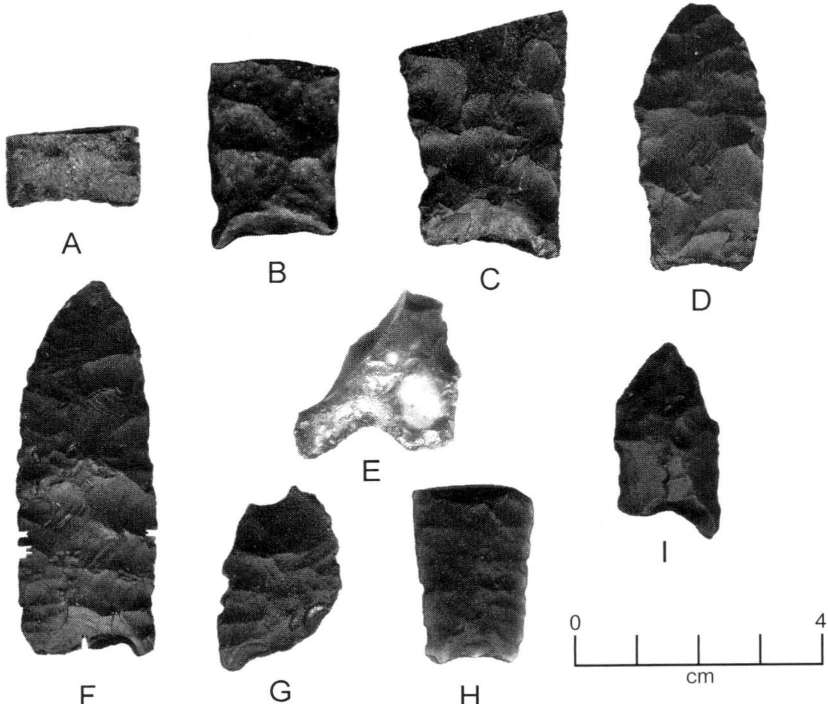

FIGURE 5.9. Unfluted Great Basin Concave Base points represented in the Old River Bed assemblage: (*A*) 42To1153, FS 17; (*B*) DPGIF 450; (*C*) DPGIF 520; (*D*) 42To2947, FS 28; (*E*) 42To1161, FS 5; (*F*) 42To2955, FS 1; (*G*) 42To3233, FS 2; (*H*) 42To3237, FS 76; (*I*) 42To2947, FS 3.

these points are in Table 5.11, while the qualitative data are presented in Table 5.12. Table 5.13 shows the site locations of the nine GBCB-UF points.

These points vary somewhat in size, although this variation is difficult to evaluate given the breakage and resharpening evident. For example, the specimens in Figure 5.9G and I show extensive resharpening of the tip and blade, considerably reducing overall length. Such extensive resharpening is commonly seen among fluted points, which can be resharpened repeatedly until virtually nothing is left but the hafting element and a tip. Although the base and lower blade of specimen 5.9G are broken, it is evident that the basal width of this point is consistent with most of the larger points, even though it is much shorter. This was likely the case as well for the point in 5.9I until one side of the base was resharpened, reducing its basal width.

The majority of these points are made from obsidian, although two are chert and one is FGV. This contrasts with patterns elsewhere, such as the Sunshine Locality, where chert is the predominant material for this point type. This material difference may relate to the scarcity of chert in the ORB region. In such instances obsidian serves as an alternative to chert for the manu-facture of both fluted and unfluted GBCB points, and thus the ORB sample conforms to that general pattern.

Little is known about the age of GBCB-UF points in the Great Basin, although because of their similarity to fluted points (both Clovis and non-Clovis), they are believed to be quite early. They are associated with radiocarbon dates from several sites such as Hogup Cave, but the dates from this site are not particularly helpful. This type of point (identified as Blackrock Concave Base by Aikens) occurs in strata 4, 6, 8, and 9, suggesting a possible age range between 8800 and 480 ^{14}C BP (Aikens 1970). We believe that at least some of these points follow fluted points in time (Beck and Jones 2009), but thus far there are no clear data to support this conjecture. The problem is that no standard set of criteria has been devised for an early unfluted concave-base form, and thus several different morphologies (although similar) that date to different times are likely being combined into one.

Western Stemmed Tradition Points

A total of 200 points have been identified with one of five WST types: Lake Mohave, Silver Lake, Parman, Cougar Mountain, and Haskett (Table 5.10). An additional 289

TABLE 5.11. Summary of Quantitative Data for Unfluted Great Basin Concave Base Points in the Old River Bed Assemblage.

Variable	Statistic	Value
Total length (mm)	n	3
	Range	32.7–62.2
	Mean	46.40
	SD	14.862
Medial length (mm)	n	3
	Range	27.7–60.5
	Mean	43.77
	SD	16.410
Basal width (mm)	n	7
	Range	15.7–26.0
	Mean	20.99
	SD	3.435
Basal indentation (mm)	n	9
	Range	1.0–6.3
	Mean	3.03
	SD	1.696
Maximum width (mm)	n	3
	Range	21.4–23.3
	Mean	22.05
	SD	1.083
Thickness (mm)	n	9
	Range	4.9–6.3
	Mean	5.46
	SD	.575

TABLE 5.13. Site Locations for Old River Bed Unfluted Great Basin Concave Base Points.

Site	No. of Points
42To1153	1
42To1161	1
42To2947	2
42To2955	1
42To3233	1
42To3237	1
Isolates	2
Total	9

TABLE 5.12. Distribution of Old River Bed Unfluted Great Basin Concave Base Points Across Eight Qualitative Variables.

Variable and State	Value	
	n	%
Raw Material		
Chert	2	22.2
Obsidian	6	66.7
Fine-grained volcanics	1	11.1
Total	9	100.0
Flaking Pattern		
Broad collateral	3	33.3
Parallel collateral	1	11.1
Parallel/oblique	2	22.2
Parallel/indeterminate	1	11.1
Indeterminate	2	22.2
Total	9	100.0
Beveling		
Absent	5	55.6
Indeterminate	4	44.4
Total	9	100.0
Lateral Edge Grinding		
Indeterminate	9	100.0
Total	9	100.0
Use-Wear		
Indeterminate	9	100.0
Total	9	100.0
Resharpening		
Absent	2	22.2
Tip/blade	2	22.2
Blade	1	11.1
Blade/base	2	22.2
Indeterminate	2	22.2
Total	9	100.0
Size Category		
8	1	11.1
9	1	11.1
10	1	11.1
11	1	11.1
Indeterminate	5	55.6
Total	9	100.0
Weathering Damage		
Minimal	3	33.3
Medium	2	22.2
Heavy	2	22.2
Extreme	2	22.2
Total	9	100.0

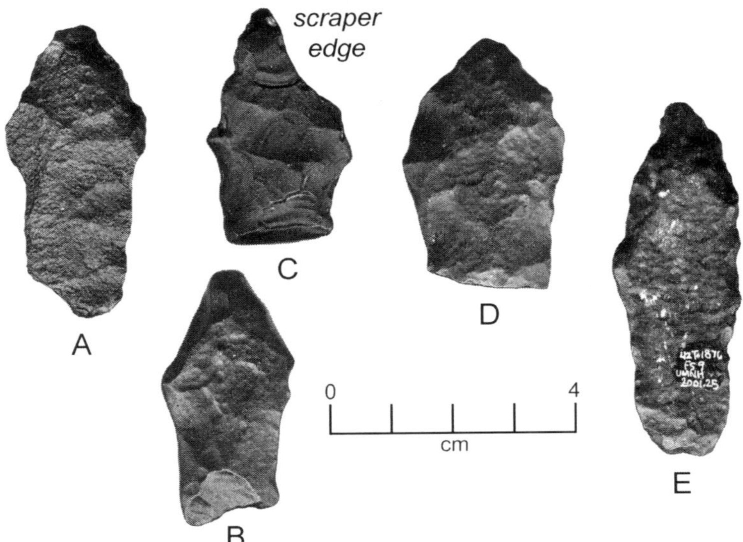

FIGURE 5.10. Examples of Lake Mohave points in the Old River Bed assemblage: (*A*) 42To0385, FS3; (*B*) 42To1686, FS 36; (*C*) DPGIF 735; (*D*) 42To1875, FS 24; (*E*) 42To1876, FS 9.

fragments cannot be identified with specific types but are identifiable as stems, midsections, or blades from these points (WST stems/midsections, WST blades; Table 5.10). Table 5.14 summarizes the quantitative data for all 489 WST points, while Table 5.15 summarizes the qualitative data. Table 5.16 shows the site locations of these points. These tables also include data for the WST/EH points, which could not be clearly identified with a particular stemmed type.

Lake Mohave Points

Twenty-nine specimens have been identified with the Lake Mohave type (Figures 5.10, S-5.1; Tables 5.10, 5.14–5.16). Fifteen of these points are made from FGV toolstone, which, although typical for WST points in other regions of the eastern and central Great Basin, is atypical for stemmed points at ORB. As Table 5.10 shows, the majority of all other WST points, except for Cougar Mountain and Cougar Mountain/Haskett stems, are manufactured from obsidian. Interestingly, however, the majority of both WST blades and WST stems/midsections are FGV. Flaking pattern was observable on 18 of the 29 Lake Mohave points, and the majority of these show broad collateral flaking, which is the typical pattern for WST points. Use-wear and edge grinding are largely indeterminate, but resharpening was observable on 13 specimens. Three were resharpened into scrapers after breakage (see, for example, Figure 5.10C).

A number of ORB Lake Mohave points are somewhat smaller overall than others of this type from east-

ern and central Nevada. For example, the mean length of the 10 Lake Mohave points from the Eastern Nevada Comparative Collection is 49.2 mm, while mean stem length is 24.3 mm, and neck width is 19.9 mm. The mean length for the ORB specimens is 38.7 mm, while mean stem length is 15.9 mm, and neck width is 15.1 mm (Table 5.14).

As noted above, beveling on alternate margins is prevalent on the Lake Mohave points from the Sunshine Locality, while this attribute is only occasionally observable on points of the other WST types in that assemblage. More than half (56.3 percent) of the Sunshine Lake Mohave points are beveled on the blade. When trying to answer the question of why beveling is so prevalent on these points and not on those of other types, we concluded that it might have something to do with the diamond shape of the blade. Some specimens were manufactured in this form to begin with, but others appear to have been resharpened into this form (Beck and Jones 2009:190). Regarding the latter, we found several Lake Mohave points that had been created through resharpening of a WST stem fragment (see Beck and Jones 2009:Figure 6.23). The focus of this resharpening was the creation of the diamond-shaped blade. Interestingly, there is one of these in the ORB assemblage as well (see Figure S-5.1bb), which lends support to our hypothesis. Twelve of the 20 ORB points for which beveling could be evaluated are beveled on the blade; two of these are also beveled on the stem (Table 5.15).

In addition to beveling, "chisel tips" are also prevalent on the Lake Mohave points from the Sunshine

TABLE 5.14. Summary of Quantitative Data for Western Stemmed Tradition (WST) Points in the Old River Bed Assemblage.

					Point Type					
Variable	Statistic	Lake Mohave	Silver Lake	Parman	Cougar Mountain	Haskett	Cougar Mountain/ Haskett Stem	WST Blade	WST Stem/ Midsection	WST/ Early Holocene
Total length (mm)	n	8	58	6	6	2	—	—	—	2
	Range	26.4–58.4	20.0–49.0	28.6–43.2	71.7–127.8	87.1–88.1				22.4–25.8
	Mean	38.68	29.46	34.22	100.71	87.60				24.08
	SD	10.552	6.085	4.967	19.109	.707				2.440
Blade length (mm)[a]	n	19	65	4	7	—	—	4	—	2
	Range	8.9–26.4	7.6–30.8	11.8–17.1	21.1–60.4			23.7–38.6		12.2–14.3
	Mean	16.92	16.85	14.43	46.90			29.11		13.25
	SD	4.773	4.872	2.202	12.983			6.525		1.485
Shoulder (mm)[a]	n	8	47	4	6	—	—	—	—	2
	Range	1.4–8.2	9.5–13.6	2.5–5.0	6.8–33.2					2.6–5.9
	Mean	5.49	3.85	3.88	19.38					4.20
	SD	2.415	3.128	1.315	10.439					2.333
Stem length (mm)[a]	n	10	71	19	9	—	10	—	—	3
	Range	5.8–23.9	5.0–16.7	11.7–25.7	21.4–63.9		41.0–64.9			5.7–7.6
	Mean	15.86	9.67	16.43	36.91		48.93			6.47
	SD	5.367	2.204	4.102	11.300		8.737			1.013
Maximum width (mm)	n	17	50	4	11	6	—	15	1	2
	Range	15.8–23.2	13.6–55.8	15.5–24.5	20.9–48.9	19.1–29.3		15.9–37.6		21.5–22.3
	Mean	19.85	21.18	18.95	31.2	25.48		22.47	25.1	21.85
	SD	2.272	6.084	3.877	7.616	4.188		5.449		.566
Thickness (mm)	n	28	104	20	18	12	13	130	153	21
	Range	3.0–8.0	3.5–9.4	4.9–11.1	5.0–29.7	5.1–11.1	5.6–10.7	3.3–14.9	3.6–10.7	3.5–9.4
	Mean	5.34	6.23	7.33	9.76	8.19	7.45	6.96	6.44	6.14
	SD	1.049	1.107	1.577	5.577	1.964	1.487	1.923	1.489	1.365
Basal width (mm)	n	9	48	10	8	5	11	—	69	3
	Range	7.6–16.5	11.5–20.4	8.2–17.0	10.2–18.8	6.8–19.9	9.1–17.5		6.7–29.9	12.0–15.7
	Mean	13.82	15.16	13.54	14.90	13.62	14.43		14.42	13.32
	SD	2.851	2.343	2.952	3.126	4.970	2.493		3.551	2.031
Neck width (mm)[a]	n	19	91	18	6	—	1	2	4	3
	Range	9.8–18.6	10.9–18.4	10.9–25.0	5.3–24.4			21.9–22.3	17.5–24.7	12.7–16.1
	Mean	15.09	14.55	16.42	18.21		21.3	22.05	20.36	13.94
	SD	1.943	1.815	3.864	6.777			.283	3.055	1.841
Proximal shoulder angle (°)	n	16	96	19	10	8	12	2	19	2
	Range	74.0–97.0	81.5–128.0	81.5–115.0	82.5–87.5	79.0–89.0	82.5–90.0	21.9–22.3	80.0–89.0	77.0–79.0
	Mean	84.78	103.80	99.40	84.25	84.88	85.54	22.05	84.95	78.00
	SD	5.825	10.147	9.124	1.687	3.204	2.407	.283	2.608	1.414
Distal shoulder angle (°)[a]	n	23	90	14	14	—	—	1	3	2
	Range	213.0–262.5	102.5–265.0	170.0–237.0	231.5–255.5				213.0–245.0	215.0–215.5
	Mean	240.63	226.60	219.43	246.14			245.0	228.33	215.25
	SD	13.790	21.0928	20.776	8.177				16.042	.354

[a] Not measured for Haskett points because there is no point of inflection separating the blade from the stem.

TABLE 5.15. Distribution of Old River Bed Western Stemmed Tradition (WST) Points Across Eight Qualitative Variables.

| | Point Type | | | | | | | | | | | | | | | |
| | Lake Mohave | | Silver Lake | | Parman | | Cougar Mountain | | Haskett | | Cougar Mountain/ Haskett Stem | | WST Blade | | WST Stem/ Midsection | | WST/ Early Holocene | |
Variable and State	n	%	n	%	n	%	n	%	n	%	n	%	n	%	n	%	n	%
Raw Material																		
Chert			2	1.9	1	5.0	1	5.6					5	3.7	3	1.9	1	4.2
Obsidian	14	48.3	80	74.1	13	65.0	5	27.8	9	75.0	4	30.8	51	37.8	52	33.8	14	37.5
Fine-grained volcanics	15	51.7	26	24.1	6	30.0	12	66.7	3	25.0	9	69.2	79	58.5	98	63.6	9	58.3
Limestone															1	.6		
Total	29	100.0	108	100.0	20	100.0	18	100.0	12	100.0	13	100.0	135	100.0	154	100.0	24	100.0
Flaking Pattern																		
Broad collateral	11	37.9	39	36.1	6	30.0	11	61.1	7	58.3	12	92.3	36	26.7	31	20.1	1	4.2
Oblique	6	20.7	2	1.9									1	.7				
Chevron					1	5.0												
Irregular			27	25.0	7	35.0	6	33.3	1	8.3	1	7.7	27	20.0	9	5.8	2	8.3
Collateral/parallel			1	.9	1	5.0			1	8.3					2	1.3	2	8.3
Collateral/oblique									2	16.7								
Collateral/unifacial (UF)			1	.9														
Collateral/irregular			3	2.8					1	8.3					2	1.3		
Parallel/oblique															1	.6		
Chevron/irregular			2	1.9														
Irregular/UF	1	3.5	3	2.8													1	4.2
None (UF)			2	1.9														
Indeterminate	11	37.9	28	25.9	5	25.0	1	5.6					71	52.6	109	70.8	18	75.00
Total	29	100.0	108	100.0	20	100.0	18	100.0	12	100.0	13	100.0	135	100.0	154	100.0	24	100.0
Beveling																		
Absent	8	27.6	71	65.7	8	40.0	7	38.9	5	41.7			70	51.9			14	58.3
Blade	10	34.5	6	5.6			1	5.6					4	3.0			1	4.2
Stem							1	5.6										
Blade/stem	2	6.9																
Indeterminate	9	31.0	31	28.7	12	60.0	9	50.0	7	58.3	13	100.0	61	45.2	154	100.0	9	37.5
Total	29	100.0	108	100.0	20	100.0	18	100.0	12	100.0	13	100.0	135	100.0	154	100.0	24	100.0
Lateral Edge Grinding																		
Present	1	3.4	3	2.8	2	10.0	3	16.7	5	41.7	3	23.1			9	5.8		
Indeterminate	28	96.6	105	97.2	18	90.0	15	83.3	7	58.3	10	76.9	135	100.0	145	94.2	24	100.0
Total	29	100.0	108	100.0	20	100.0	18	100.0	12	100.0	13	100.0	135	100.0	154	100.0	24	100.0

	n	%	n	%	n	%	n	%	n	%	n	%	n	%	n	%	n	%
Use-Wear																		
Present					1	5.0			1	8.3			4	3.0				
Indeterminate	29	100.0	108	100.0	19	95.0	18	100.0	11	91.7	13	100.0	131	97.0	154	100.0	24	100.0
Total	29	100.0	108	100.0	20	100.0	18	100.0	12	100.0	13	100.0	135	100.0	154	100.0	24	100.0
Resharpening																		
Absent	4	13.8	8	7.5			3	16.7	7	58.3	3	23.1	6	4.4	3	1.9	2	8.3
Tip/blade (TB)	4	13.8	7	6.5	1	5.0	3	16.7					14	10.4			2	8.3
Barb			2	1.9														
Stem			3	2.8	1	5.0			3	25.0	2	15.4			4	2.6		
TB/barb	4	13.8	11	10.2	3	15.0							1	.7				
TB/stem			4	3.7			2	11.1									1	4.2
Barb/stem			1	.9														
TB/different tool (DT)													2	1.5				
TB/barb/stem	2	6.9	8	7.4	2	10.0												
TB/barb/DT			4	3.7														
Stem/DT			1	.9														
TB/barb/stem/DT			1	.9														
DT	3	10.3	1	.9			1	5.6					1	.7				
Indeterminate	12	41.4	57	52.8	13	65.0	9	50.0	2	16.7	8	61.5	111	82.2	147	95.5	19	79.2
Total	29	100.0	108	100.0	20	100.0	18	100.0	12	100.0	13	100.0	135	100.0	154	100.0	24	100.0
Size Category																		
6			1	.9														
7	2	6.9	12	11.1														
8	2	6.9	30	27.8	4	20.0												
9	2	6.9	16	14.8	3	15.0												
10	2	6.9	1	.9					1	8.3								
12							2	11.1	2	16.7								
13							3	16.7										
14							1	5.6										
Indeterminate	21	72.4	48	44.4	13	65.0	12	66.7	9	75.0	13	100.0	135	100.0	154	100.0	24	100.0
Total	29	100.0	108	100.0	20	100.0	18	100.0	12	100.0	13	100.0	135	100.0	154	100.0	24	100.0
Weathering Damage																		
Minimal	4	13.8	11	10.2	3	15.0	2	11.1	3	25.0			11	8.2	5	3.2	2	8.3
Medium	4	13.8	8	7.4	2	10.0	9	50.0	3	25.0	5	38.5	20	14.8	29	18.8	7	29.2
Heavy	7	24.1	37	34.9	10	50.0	5	27.8	3	25.0	6	46.2	66	48.9	70	45.5	6	25.0
Extreme	14	48.3	48	44.4	5	25.0	2	11.1	3	25.0	2	15.4	30	22.2	47	30.5	6	25.0
Not recorded			4	3.7									8	5.9	3	1.9	3	12.5
Total	29	100.0	108	100.0	20	100.0	18	100.0	12	100.0	13	100.0	135	100.0	154	100.0	24	100.0

TABLE 5.16. Site Locations for Old River Bed Western Stemmed Tradition (WST) Points.

Site	Lake Mohave	Silver Lake	Parman	Cougar Mountain	Haskett	Cougar Mountain/Haskett	WST Blade	WST Stem/Midsection	WST/Short Stem
04DM02				1					
04DM03				1					
08DM30				1					
42To0385	1								
42To1000		2					5	1	2
42To1152			1						
42To1153				1	1		3	4	
42To1157				1					
42To1161	1								
42To1163								1	
42To1166		1							
42To1171									2
42To1172							1		
42To1173									1
42To1177							1		
42To1182	1						1	6	
42To1352									2
42To1353								1	
42To1354				1		1	1	2	1
42To1356								1	
42To1357								2	
42To1358	1						2	5	1
42To1368							1		
42To1369		1					1	4	2
42To1370				1		1		2	
42To1371		3	1				2		
42To1383							2	9	
42To1666							1		1
42To1668								1	
42To1669							1	1	
42To1671					1		2	1	
42To1672							1	3	
42To1673							1	1	
42To1674		1							
42To1677								1	
42To1678		1							
42To1679		1	1				1		
42To1681							1		
42To1682							2		
42To1683		2		1				1	
42To1684							4	4	
42To1685		4	1			1	3	2	
42To1686	5	4					4	1	
42To1687	1	3					1		
42To1688		4	1						
42To1689									1
42To1859	1								
42To1860							1		
42To1861		1					2	3	
42To1862							1		
42To1872	2	3	1		1		9	10	2
42To1873	1	1	2				3	2	
42To1874								1	
42To1875	1					1	3	2	

Table 5.16. (cont'd.) Site Locations for Old River Bed Western Stemmed Tradition (WST) Points.

Site	Lake Mohave	Silver Lake	Parman	Cougar Mountain	Haskett	Cougar Mountain/Haskett	WST Blade	WST Stem/Midsection	WST/Short Stem
42To1876	1								
42To1878		1		1		1		2	
42To1920							2	1	
42To1921		1					3		
42To1922		1							
42To1923							1		
42To1924	1	2					2		
42To2551		1		1			3	2	
42To2552	1	1					1	1	
42To2553						1	1	1	
42To2554		3					1	1	
42To2555		1					2		
42To2556		3	1				2	2	
42To2557		1							
42To2558	1	3						2	
42To2559		5	1			1	4	2	1
42To2767	1	1						2	
42To2943								1	
42To2944		1							
42To2945		3							
42To2946		1							
42To2948	1	1							
42To2949	1	2							
42To2951		2						2	
42To2952	1						1	1	
42To2953								1	
42To2954								2	
42To2955				1					
42To2957								1	
42To3140					1	1		3	
42To3141							1	1	
42To3142					1	1			
42To3219		1	3				2	2	1
42To3220							1	2	
42To3221							3		
42To3222				1			1	1	
42To3223						1	1		1
42To3224							2	1	
42To3225							1	2	
42To3226		1				1	3	3	
42To3228		2	1						1
42To3229		1						2	
42To3230		7	1	1			3	7	
42To3231		1						1	
42To3233		3						4	
42To3234		2		1	1		1		
42To3235		3					5	7	
42To3237		8							
42To3238		3						1	
42To3520					3		6	3	1
42To3522		1			1				
Isolates	6	9	5	4	2	2	21	18	4
Total	29	108	20	18	12	13	135	154	24

TABLE 5.17. Beveling and Chisel Tips on Lake Mohave Points in the Old River Bed Assemblage.

Variable	No. of Points
Beveling on the Blade	
Absent	8
Present	12
Indeterminate	9
Total	29
Chisel Tip	
Absent	10
Present	12
Indeterminate	7
Total	29
Beveling and Chisel Tip	
Both present	4
Both absent	4
One absent, one present	6
One present, one indeterminate	5
One absent, one indeterminate	6
Both indeterminate	4
Total	29

Locality; 10 of the 13 points where the tip is present have chisel tips. In fact, most of the chisel tips occur on those points that have beveled blades, an association that proved to be significant ($\chi^2 = 13.268$; $df = 6.0$; $p = .039$ [Beck and Jones 2009:191]). Twelve of the 22 ORB Lake Mohave points with tip sections exhibit chisel tips (Table 5.17). However, both attributes are present on only four specimens. Unfortunately there are 15 points on which either or both beveling and chisel tip were not observable. A chi-square test reveals no significant association between these two variables for the ORB Lake Mohave points.

Silver Lake Points

Silver Lake points ($n = 108$) are the most numerous of the WST points in the ORB assemblage, representing 54.0 percent of the 200 that could be assigned to a specific type (Figures 5.11, S-5.2; Tables 5.10, 5.14–5.16). This is contrary to most Paleoarchaic assemblages, except perhaps in the Mojave section of the Great Basin. Nearly three-quarters of these points are manufactured from obsidian (74.1 percent), and correspondingly more than 80 percent have either heavy or extreme weathering damage. Even so, 43 of the 51 that could be evaluated (84.3 percent) show evidence of resharpening, and seven of these were eventually transformed into a different tool (see, for example, Figure 5.11E). More than half

(55.0 percent) of the 80 points that could be evaluated exhibit broad collateral flaking on at least one face, and another 43.0 percent are irregularly flaked on at least one face. Several are actually unifacial. A small number ($n = 6$) are beveled on the blade. Many of the ORB Silver Lake points are smaller, on the average, than their counterparts in other Paleoarchaic assemblages. For example, although mean total length and basal width are similar to those in the Sunshine assemblage, neck width and thickness are smaller.

Parman Points

Twenty points were assigned to the Parman type (Figures 5.12, S-5.3; Tables 5.10, 5.14–5.16). Like Silver Lake points, the majority (65.0 percent) are manufactured from obsidian. In our studies at the Sunshine Locality, points of the Parman type show the greatest within-type variation, which is also the case for the ORB specimens. This is likely due to the problematic definitions given for this type. The most distinctive features are a pronounced shoulder and a roughly parallel stem with a rounded base. As Figures 5.12 and S-5.3 show, although these features are present on all specimens, there is still a large range of variation in both size and form. Given this variation, the extreme weathering evident, and the fragmentary nature of most of the specimens, there is little that can be said regarding these points.

Cougar Mountain and Haskett Points

Points of the Cougar Mountain and Haskett types are considered together because they tend to grade into one another in form. They are quite similar in both size and form, with the primary difference being that there is an inflection point at the intersection between the blade and stem on Cougar Mountain points, creating a shoulder, while this is not the case for Haskett points, which have no shoulder. Thus, a long stem fragment can be identified specifically as one or the other only if the midsection is present. Eighteen points have been identified as Cougar Mountain (Figures 5.13, S-5.4); 12, as Haskett (Figures 5.14, S-5.5); and 13, as Cougar Mountain/Haskett stems (Figures 5.15, S-5.6). Quantitative data are presented in Table 5.14; and qualitative data, in Table 5.15. Site locations are shown in Table 5.16.

Interestingly, Cougar Mountain and Haskett points are opposite with respect to the raw material from which they are made: Cougar Mountain points are predominantly FGV, while Haskett points are predominantly obsidian (Table 5.15). Cougar Mountain/Haskett stems mirror the raw material patterns of Cougar Mountain rather than Haskett. There are other differences as well.

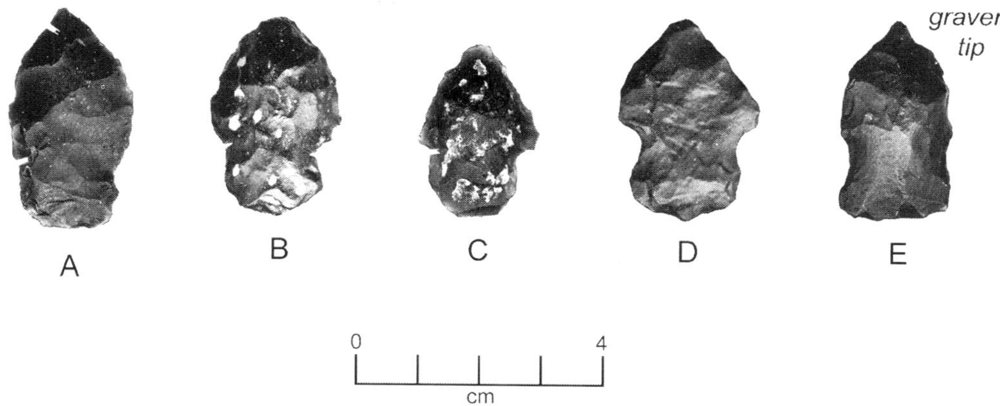

FIGURE 5.11. Examples of Silver Lake points in the Old River Bed assemblage: (*A*) DPGIF 733; (*B*) 42To1686, FS 43; (*C*) 42To2559, FS 59; (*D*) 42To3238, FS 27; (*E*) 42To2949, FS 26.

FIGURE 5.12. Examples of Parman points in the Old River Bed assemblage: (*A*) 42To1685, FS 11; (*B*) 42To1872, FS 19; (*C*) DPGIF 208; (*D*) DPGIF 2415, FS 1.

Cougar Mountain points exhibit predominantly broad collateral flaking, although about a third have irregular flaking; more than 90 percent of the Haskett points, however, are collaterally flaked on at least one face. Here, Cougar Mountain/Haskett stems mirror Haskett. In addition five of the eight Cougar Mountain points that could be evaluated are resharpened, while this is the case for only three of 10 Haskett points. These differences in flaking and resharpening could, of course, be related to the difference in raw material, but chi-square tests reveal no association between raw material type and either flaking pattern or resharpening.

One thing that is notable for both Cougar Mountain and Haskett points is their size. In fact, in total length the ORB Cougar Mountain points are considerably *longer* than those in the Eastern Nevada Comparative Collection, with a mean length of 100.71 mm for the former (Table 5.14), as compared with 81.69 mm for the latter. On the other hand, the mean neck width for the ORB specimens is *smaller* than that in the Eastern Nevada Comparative Collection, 18.21 mm as compared with 31.27 mm.

WST Blades, WST Stems/Midsections, and WST/EH Points

Finally, 289 points are believed to represent the WST but are too fragmentary or weathered to be assigned to a specific type. These have been designated as WST blades ($n = 135$) and WST stems/midsections ($n = 154$). An additional 24 points were assigned to the category WST/EH, as they could be distinguished from Archaic points. Quantitative and qualitative data for these points are shown in Tables 5.14 and 5.15, respectively. Site locations can be found in Table 5.16.

FIGURE 5.13. Examples of Cougar Mountain points in the Old River Bed assemblage: (*A*) 42To2551, FS 30; (*B*) DPGIF 2529; (*C*) DPGIF 2577.

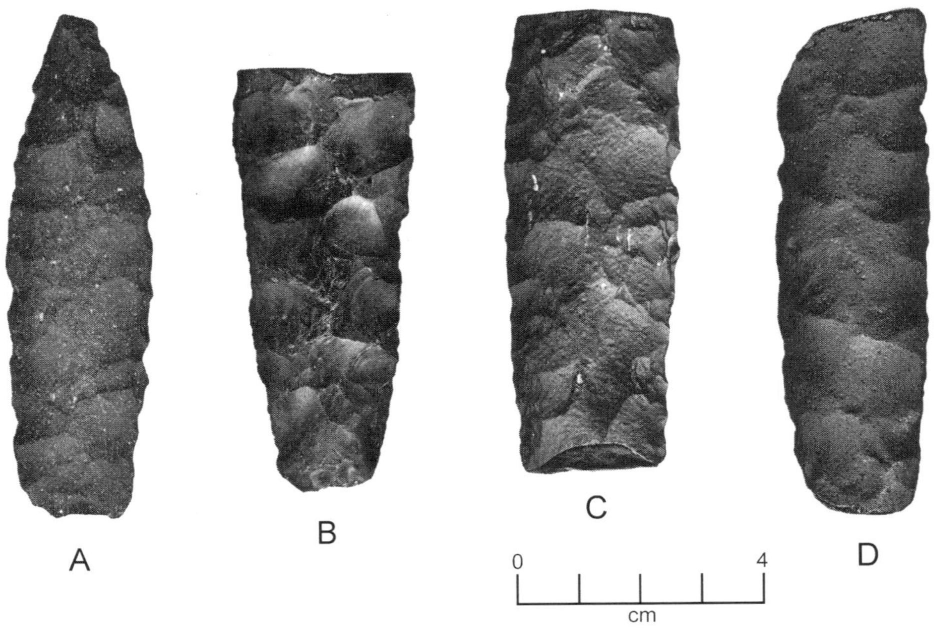

FIGURE 5.14. Examples of Haskett points in the Old River Bed assemblage: (*A*) DPGIF 2453; (*B*) 42To1671, FS 6; (*C*) 42To3140, FS 11; (*D*) 42To3234, FS 21.

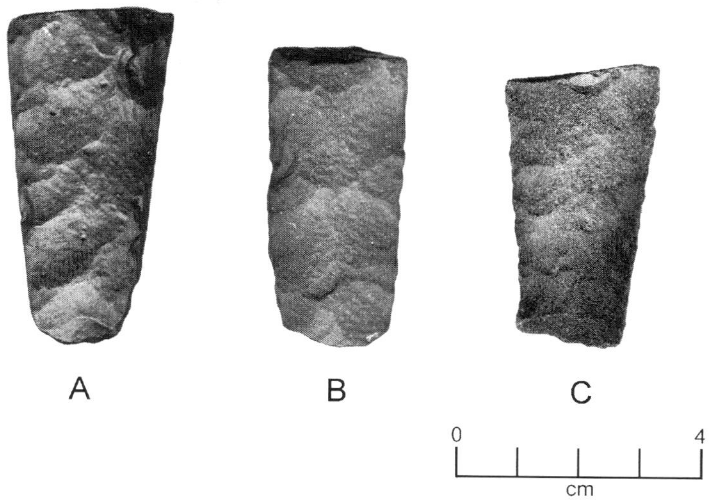

FIGURE 5.15. Examples of Cougar Mountain/Haskett stems in the Old River Bed assemblage: (*A*) 42To2553, FS 3; (*B*) 42To3142, FS 31; (*C*) 42To1370, FS 3.

Early Holocene Points

Seven types are considered here to be "Early Holocene" types: Square Stem, Expanding Stem, Contracting Stem, Dugway Stubby, Butte Valley Corner-notched, Pinto, and Eden, of which Pinto and Eden are the only traditional types. Square Stem, Expanding Stem, Contracting Stem, and Stubby are all types defined by the classification created by Jones, Beck, and Kessler (2003). The Butte Valley Corner-notched type was defined by Beck and Jones (2009), based on points originally found in Butte Valley. Finally, 81 fragments were identifiable only as EH points and are characterized as "Short Stem."

Square Stem Points

Jones, Beck, and Kessler (2003) define Square Stem points as having a shoulder that is square or sloping with an abrupt or intermediate angle. Four subtypes were differentiated within the Square Stem type on the basis of stem size and degree of shouldering (Jones, Beck, and Kessler 2003:26).

Square Stem 1. Square Stem 1 (SS1) specimens have a wide stem, in absolute terms, with parallel margins. Their bases are flat, but the angle of the base in these specimens lies slightly off the transverse plane. They have a short sloping shoulder, and their blades are narrow in relation to the width of the stem. Thirteen specimens have been identified with this subtype (Figures 5.16A–B, S-5.7).

Square Stem 2. Square Stem 2 (SS2) specimens have narrower stems than SS1. The stem is parallel-sided or

slightly contracting; the base is flat and often incorporates a transverse fracture facet. In most of these, biface shaping was subsequent to fracture. Most examples have strongly developed shoulders, with an abrupt stem angle (square shoulder) and a strong proximal blade angle. These points exhibit careful bifacial flaking and have symmetrical cross sections, but a number of them are nevertheless quite thick. This implies either a good deal of resharpening of the specimens or that they were made from thick blanks distinct from SS1. Most of these specimens appear to be made from "scavenged" biface fragments. Their narrow shoulder reflects the size of the original blank or resharpening. Nine specimens have been characterized as SS2 (Figures 5.16C–D, S-5.8).

Square Stem 3. Square Stem 3 (SS3) specimens differ from SS2, primarily, in having a sloping shoulder formed by a shallow DSA and rounded proximal blade angle. As in the case of SS2, most of these specimens appear to be made from "scavenged" biface fragments. Their narrow shoulder reflects the size of the original blank or resharpening. Fourteen points have been identified with this subtype (Figures 5.16E–F, S-5.9).

Square Stem 4 and 5. The six Square Stem 4 (SS4) specimens possess smaller stems (Figures 5.16G–H, S-5.10). A fifth subcategory is used here for the nine specimens too fragmentary to assign to one of the above four subtypes (Figures 5.16I–J, S-5.11).

Summary. These subtypes are retained in the present analysis as well as in the data tables so that their validity

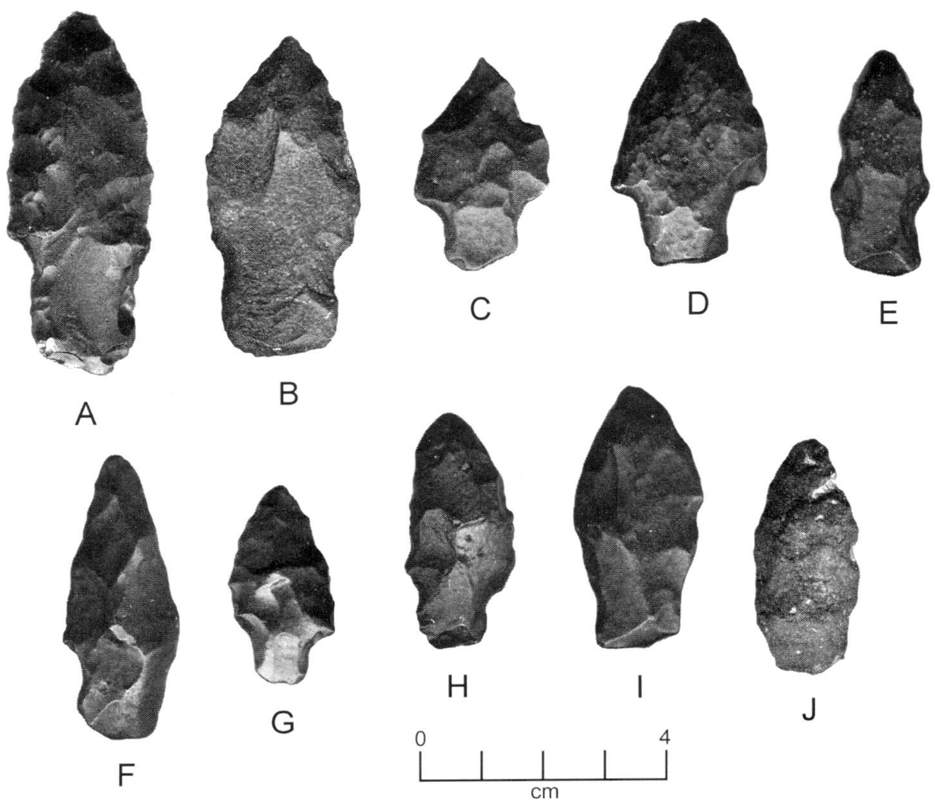

FIGURE 5.16. Examples of Square Stem points in the Old River Bed assemblage. Square Stem 1: (*A*) 42To1922, FS 1; (*B*) 42To3229, FS 11. Square Stem 2: (*C*) DPGIF 706; (*D*) DPGIF 829. Square Stem 3: (*E*) 42To1371, FS 102; (*F*) DPGIF 846. Square Stem 4: (*G*) 42To1681, FS 4; (*H*) 42To2948, FS 20. Square Stem 5: (*I*) 42To1687, FS 9; (*J*) 42To3230, FS 107.

can be evaluated using the entire ORB assemblage, which consists of 51 points (Table 5.10).

Although Jones, Beck, and Kessler (2003) do not specify metric criteria for these subtypes, given the definitions, significant differences among them would be expected in neck width, DSA, maximum width, and, possibly, stem length. Figure 5.17 shows histograms of all four of these variables. The distributions of DSA and stem length are fairly continuous, but those of maximum width and, to some extent, neck width are bimodal. A comparison of these graphs with the value ranges for neck width and maximum width shown in Table 5.18 suggests that the divisions within both of these variables are between subtypes SS3 and SS4 on the lower end and subtypes SS1 and SS2 on the upper.

Statistical comparisons were made among SS1–SS4 on all quantitative variables using analysis of variance. Table 5.19 shows the results for those variables where at least one significant difference exists. The difference between SS1 and subtypes SS3 and SS4 suggested by the histogram is confirmed for neck width and maximum width (although the difference in the latter is significant only at the .06 level). That between SS2 and SS4 is also confirmed for neck width. Significant differences exist as well between SS2 and SS3 in DSA and between SS1 and SS4 in basal width. These results, then, suggest that only two subtypes exist: SS1 and SS2, on the one hand, and SS3 and SS4, on the other. However, because Tables 5.18 and 5.20 present data by original subtype, we continue our discussion with respect to these four subtypes.

SS1–SS4 are remarkably similar with respect to the qualitative attributes (Table 5.20). For example, FGV toolstone predominates over obsidian, except for SS4, where these two toolstone types are equally represented; this pattern is contrary to the overall raw material pattern for the ORB projectile points, in which obsidian predominates. The large majority of these points (across subtypes) exhibit heavy to extreme weathering damage. Flaking, primarily irregular, is also

TABLE 5.18. Summary of Quantitative Data for Square Stem Points in the Old River Bed Assemblage.

Variable	Statistic	Square Stem 1	Square Stem 2	Square Stem 3	Square Stem 4	Square Stem 5
				Point Subtype		
Total length (mm)	n	9	5	6	4	3
	Range	30.8–60.4	35.5–43.8	33.1–47.3	32.3–39.1	30.0–54.0
	Mean	45.52	40.01	38.92	35.39	41.07
	SD	8.843	3.667	4.917	3.171	12.108
Blade length (mm)	n	10	4	7	3	6
	Range	19.0–37.4	22.7–30.5	18.3–30.8	21.1–27.3	17.9–42.0
	Mean	29.04	28.16	25.44	23.68	30.15
	SD	5.792	3.703	4.546	3.239	9.155
Stem length (mm)	n	9	7	11	4	5
	Range	4.5–17.4	6.4–17.8	7.2–18.4	5.7–11.4	9.2–16.4
	Mean	11.23	10.96	10.85	8.50	12.52
	SD	3.833	3.958	3.193	2.331	3.049
Maximum width (mm)	n	6	2	7	2	3
	Range	18.0–24.1	16.2–25.9	16.2–19.8	14.3–17.7	18.8–21.6
	Mean	22.46	21.03	18.06	15.98	20.48
	SD	2.218	6.894	1.416	2.369	1.475
Thickness (mm)	n	13	9	14	6	9
	Range	4.7–9.0	3.4–9.6	4.4–8.0	4.2–7.5	4.5–10.0
	Mean	6.74	6.42	5.93	6.08	7.34
	SD	1.454	1.779	.965	1.422	1.663
Basal width (mm)	n	9	4	10	5	2
	Range	6.1–21.7	12.8–15.3	9.1–16.7	9.0–13.2	14.0–21.1
	Mean	15.95	13.69	12.96	11.42	17.55
	SD	4.230	1.103	1.934	1.540	5.020
Neck width (mm)	n	11	8	14	6	5
	Range	13.5–18.3	12.4–19.0	11.9–15.9	8.9–13.9	13.3–19.8
	Mean	16.21	14.76	13.76	11.71	15.67
	SD	1.676	2.430	1.209	1.989	2.448
Proximal shoulder angle (°)	n	12	8	13	4	7
	Range	85.0–110.0	77.5–105.5	81.5–111.0	88.0–103.0	86.0–96.0
	Mean	94.50	93.75	91.31	95.50	90.43
	SD	6.234	8.485	7.729	7.594	3.322
Distal shoulder angle (°)	n	10	7	11	4	4
	Range	191.5–242.0	193.5–233.5	217.0–255.0	210.0–238.0	237.0–250.5
	Mean	225.80	215.29	237.64	223.50	242.88
	SD	18.332	15.676	10.870	12.021	5.662

fairly consistent across the subtypes. Table 5.21 gives the locations for Square Stemmed points.

Although the points identified as SS2, SS3, and SS4 do not necessarily resemble points of any known type, SS1 points are similar to Windust points in the Great Basin and the Columbia Plateau as well as Cody/Scottsbluff points on the Plains. Although Windust points were prevalent during the early Holocene, several terminal Pleistocene dates are associated with them (Table 5.22; but see Goebel and Keene 2014). Consequently,

they are commonly included in the WST. The Cody Complex dates between ~9500 and ~8000 ^{14}C BP on the Plains (Holliday 2000), suggesting that this form occurs earliest in the Intermountain West (see discussion in Chapter 1).

Expanding Stem Points

Jones, Beck, and Kessler (2003:28) describe the Expanding Stem points as having sloping shoulders with an abrupt-to-weak stem-to-shoulder angle. Stems have

TABLE 5.19. Results of Analysis of Variance Test (*p*) Among Four Square Stem Subtypes for Four Variables.

Variable and Subtype	Point Subtype			
	Square Stem 1	Square Stem 2	Square Stem 3	Square Stem 4
Neck Width				
Square Stem 1	1.000			
Square Stem 2	.449	1.000		
Square Stem 3	**.008**	1.000	1.000	
Square Stem 4	**.000**	**.017**	.135	1.000
Maximum Width				
Square Stem 1	1.000			
Square Stem 2	1.000	1.000		
Square Stem 3	**.060**[a]	1.000	1.000	
Square Stem 4	**.059**[a]	.461	1.000	1.000
Distal Shoulder Angle				
Square Stem 1	1.000			
Square Stem 2	.959	1.000		
Square Stem 3	.465	**.024**	1.000	
Square Stem 4	1.000	1.000	.675	1.000
Basal Width				
Square Stem 1	1.000			
Square Stem 2	1.000	1.000		
Square Stem 3	.178	1.000	1.000	
Square Stem 4	**.049**	1.000	1.000	1.000

Note: Significant values in bold.
[a] These results are considered significant even though they are slightly above .05.

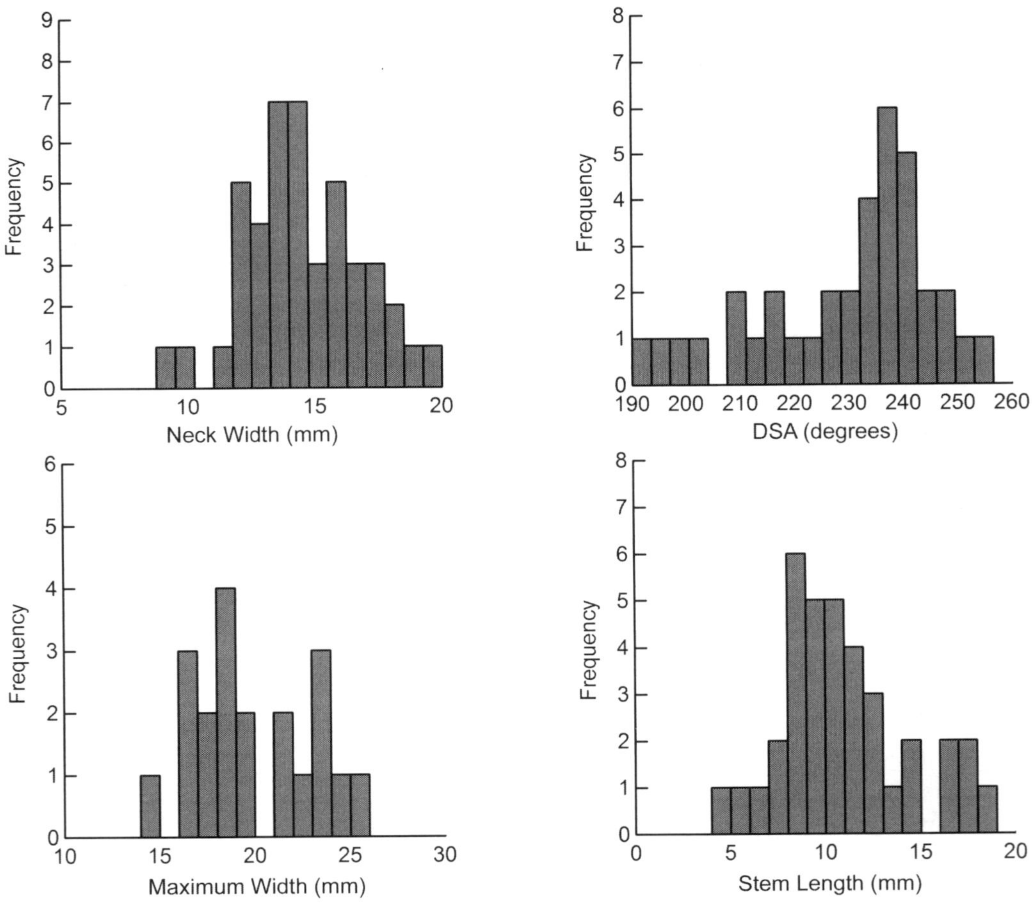

FIGURE 5.17. Histograms of four metric variables for Square Stem points in the Old River Bed assemblage. DSA = distal shoulder angle.

TABLE 5.20. Distribution of Old River Bed Square Stem Points Across Eight Qualitative Variables.

| | Point Subtype | | | | | | | | | |
| | Square Stem 1 | | Square Stem 2 | | Square Stem 3 | | Square Stem 4 | | Square Stem 5 | |
Variable and State	*n*	%	*n*	%	*n*	%	*n*	%	*n*	%
Raw Material										
Obsidian	3	23.1	1	11.1	5	35.7	3	50.0	3	33.3
Fine-grained volcanics	9	69.2	8	88.9	8	57.1	3	50.0	5	55.6
Quartzite	1	7.7			1	7.1				
Limestone									1	11.1
Total	13	100.0	9	100.0	14	100.0	6	100.0	9	100.0
Flaking Pattern										
Broad collateral	1	7.7			2	14.3	2	33.3	1	11.1
Irregular	11	84.6	7	77.8	6	42.9			1	11.1
Collateral/irregular			1	11.1						
Irregular/unifacial					1	7.1				
None (unifacial)									1	11.1
Indeterminate	1	7.7	1	11.1	5	35.7	4	66.7	6	66.7
Total	13	100.0	9	100.0	14	100.0	6	100.0	9	100.0
Beveling										
Absent	11	84.6	5	55.6	9	64.3	6	100.0	5	55.6
Blade					1	7.1				
Stem	1	7.7	4	44.4	4	28.6			4	44.4
Indeterminate	1	8.3								
Total	13	100.0	9	100.0	14	100.0	6	100.0	9	100.0
Lateral Edge Grinding										
Present	1	7.7								
Indeterminate	12	92.3	9	100.0	14	100.0	6	100.0	9	100.0
Total	13	100.0	9	100.0	14	100.0	6	100.0	9	100.0
Use-Wear										
Indeterminate	13	100.0	9	100.0	14	100.0	6	100.0	9	100.0
Total	13	100.0	9	100.0	14	100.0	6	100.0	9	100.0
Resharpening										
Absent	1	7.7			1	7.1				
Tip/blade (TB)	2	15.4	2	22.2	2	14.3			1	11.1
Barb	1	7.7								
Stem	1	7.7								
TB/barb	2	15.4			2	14.3	2	33.3		
TB/stem	1	7.7	1	11.1						
TB/barb/stem	1	7.7							2	22.2
Indeterminate	4	30.8	6	66.7	9	64.3	4	66.7	6	66.7
Total	13	100.0	9	100.0	14	100.0	6	100.0	9	100.0
Size Category										
7							1	16.7		
8	3	23.1	2	22.2	4	28.6	3	50.0	1	11.1
9	5	38.5	3	33.3	3	21.4	1	16.7	1	11.1
10	3	23.1	1	11.1					1	11.1
11	1	7.7								
Indeterminate	1	7.7	3	33.3	7	50.0	1	16.7	6	66.7
Total	13	100.0	9	100.0	14	100.0	6	100.0	9	100.0
Weathering Damage										
Minimal	3	23.1								
Medium	2	15.4	1	11.1			1	16.7	4	44.4
Heavy	5	38.5	5	55.6	6	42.9	2	33.3	3	33.3
Extreme	3	23.1	3	33.3	8	57.1	3	50.0	2	22.2
Total	13	100.0	9	100.0	14	100.0	6	100.0	9	100.0

TABLE 5.21. Site Locations for Old River Bed Early Holocene Projectile Points.

Site	Short Stem	Square Stem	Expanding Stem	Contracting Stem	Stubby	Butte Valley Corner-Notched	Eden
42To0394		1					
42To0962		1					
42To1000						1	
42To1153	1				1		
42To1161					1	1	
42To1163						1	
42To1182					2	1	
42To1353		1					
42To1357	2						
42To1358	2				2	1	
42To1368	1						
42To1369					1		
42To1371	1	2		1	6		
42To1383					1		
42To1669					1		
42To1671	1						
42To1672				1			
42To1675	2						
42To1676					1		
42To1679	1				2		
42To1681	1	1					
42To1684	1				3		
42To1685	1				3		
42To1686	2		1	2	11		
42To1687		1			2		
42To1688				1	11		
42To1858		1					
42To1859	1				1		
42To1861					3		
42To1862		2			8		
42To1872	2	1			2		
42To1873	1	1			6		
42To1874	1				1		
42To1875					2		
42To1876	1				1		
42To1877	1						
42To1878					1		
42To1920	1		1				
42To1921			1		1		
42To1922	1	1					
42To1924	1	1			9		
42To2551	1	3	1		5	2	
42To2553	1				1		
42To2554	1	1	3		4		
42To2555	2	1	2		4		
42To2556	1		3		5		
42To2557					1		
42To2558				1			
42To2559	1			2	14		
42To2943				1			
42To2945	1		1		4		
42To2946			1		2		

TABLE 5.21. (cont'd.) Site Locations for Old River Bed Early Holocene Projectile Points.

Site	Short Stem	Square Stem	Expanding Stem	Contracting Stem	Stubby	Butte Valley Corner-Notched	Eden
42To2947					2		
42To2948		1			1		
42To2949	1	1	2				
42To2951	1	1	1		1		
42To2952			1		1		
42To2953		1					
42To3141					1		
42To3142		1	1			1	
42To3219	3	2	1		3	1	
42To3221		1					
42To3222			1				
42To3225					1	1	
42To3226			2		4		
42To3228					2	1	
42To3229		3	1			1	
42To3230	10	7	3		6		
42To3231						1	
42To3233	2				1		
42To3234	5	1	1		6		
42To3235	4	2			5	2	
42To3236					1		
42To3237	3	2	2		5		
42To3238					1		
42To3520	2		1		3		
42To3522	1	1					
Isolates	15	7	9	1	25	1	1
Total	81	50	40	10	192	16	1

Note: Locations of Pinto points are shown in Table 5.33.

TABLE 5.22. Early Radiocarbon Dates Associated with Windust Points in the Intermountain West.

Region / Site	Date	Notes	Reference(s)
Columbia Plateau			
Marmes, WA	10,810 ± 275		Rice 1972
	10,750 ± 90		
Hatwai, ID	(pooled mean)		Ames et al. 1981
	10,796 ± 138	based on 5500-yr half-life	Sanders 1982
	11,120 ± 138	based on 5730-yr half-life	
Paulina Lake, OR	7930 ± 80	from component 2	Connolly and Jenkins 1999
	8210 ± 60		
	8460 ± 110		
	8540 ± 90		
	8670 ± 110		
	8680 ± 70		
	8880 ± 110		
	8980 ± 190		
	9060 ± 80		
	9820 ± 470	From component 1	
Great Basin			
Wildcat Canyon, OR	10,600 ± 200		Dumond and Minor 1983

TABLE 5.23. Summary of Quantitative Data for Expanding Stem Points in the Old River Bed Assemblage.

Variable	Statistic	Point Subtype	
		Expanding Stem 1	Expanding Stem 2
Total length (mm)	n	6	19
	Range	38.0–55.9	29.6–63.8
	Mean	43.55	40.93
	SD	6.500	9.937
Blade length (mm)	n	7	14
	Range	14.9–38.5	18.2–50.2
	Mean	28.47	29.45
	SD	7.596	10.180
Stem length (mm)	n	9	24
	Range	6.1–18.9	5.0–13.0
	Mean	10.66	8.27
	SD	4.147	1.865
Maximum width (mm)	n	3	6
	Range	20.7–24.9	15.7–22.4
	Mean	23.13	18.69
	SD	2.178	2.662
Thickness (mm)	n	14	29
	Range	5.1–8.0	3.4–8.8
	Mean	6.52	6.19
	SD	.879	1.604
Basal width (mm)	n	6	13
	Range	15.2–18.2	10.5–16.8
	Mean	16.78	12.71
	SD	1.207	1.995
Neck width (mm)	n	12	27
	Range	13.5–16.6	9.0–13.2
	Mean	14.75	11.26
	SD	.886	1.179
Proximal shoulder angle (°)	n	12	29
	Range	93.0–121.0	93.0–126.0
	Mean	106.99	104.67
	SD	9.171	7.604
Distal shoulder angle (°)	n	11	26
	Range	187.0–243.0	121.0–243.0
	Mean	216.91	218.98
	SD	14.775	23.021

expanding margins with a rounded to flat base. Forty-three points from the ORB assemblage have been assigned to this type.

Jones, Beck, and Kessler (2003) distinguish two subtypes based on size, although they do not report specific measurements that differentiate these two subcategories. Thus, for the analysis of the ORB2 points (Beck and Jones 2010b), histograms were produced for all quantitative measures and assessed for signs of visible clustering. Total length and neck width appeared to discriminate the best (Figure 5.18A). These graphs

TABLE 5.24. Results of Pooled-Variance *t*-Tests Between Expanding Stem 1 and Expanding Stem 2 Points for Five Variables Measuring Size.

Variable	t	df	p
Total length	.601	23	.554
Neck width	9.157	37	**<.001**
Stem length	2.309	31	**.028**
Basal width	4.584	17	**<.001**
Maximum width	2.480	7	**.042**

Note: Significant values in bold.

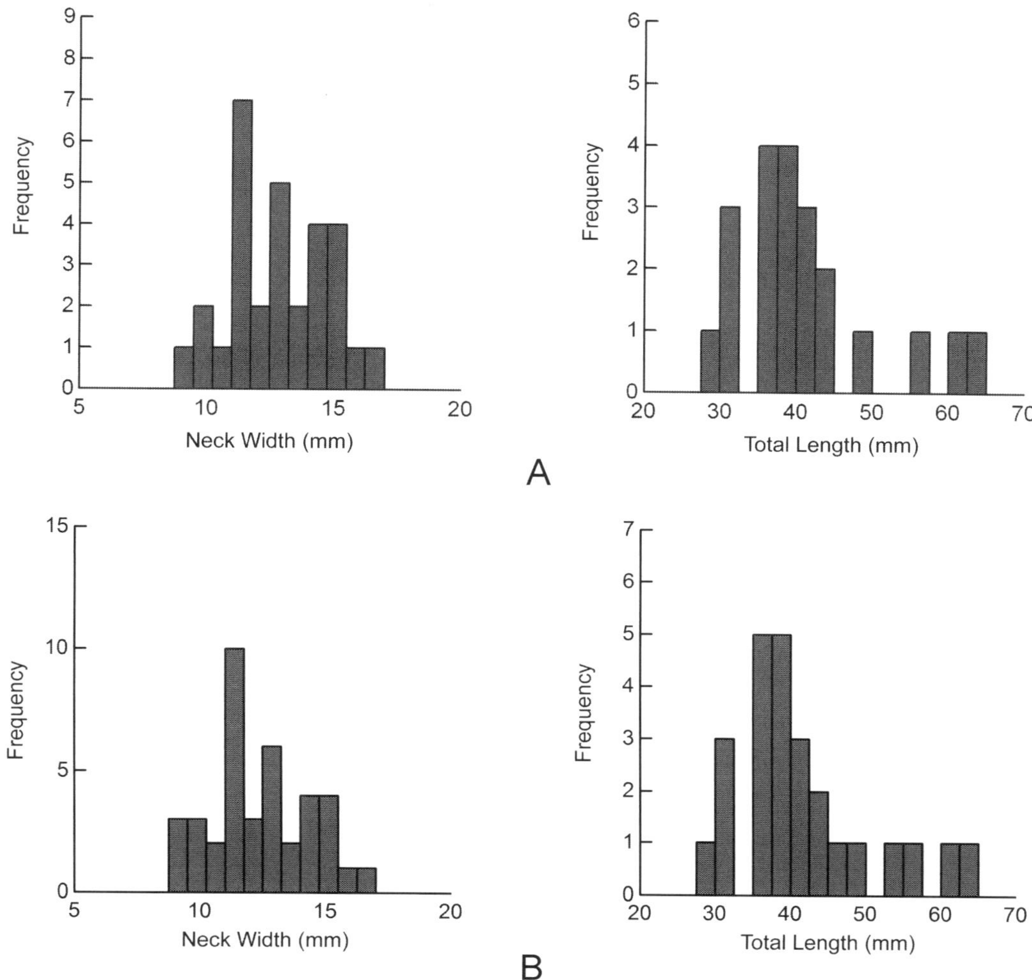

FIGURE 5.18. Histograms of Expanding Stem neck width and total length for (*A*) the ORB2 assemblage and (*B*) the combined ORB1 and ORB2 assemblages.

suggested a break for length at 45 mm and of neck width at 10.7 mm and 13.5 mm. Therefore, the subtypes were defined for the ORB2 points according to these breaks:

Expanding Stem 1 (ES1): total length < 45 mm and neck width < 13.5 mm
Expanding Stem 2 (ES2): total length ≥ 45 mm and neck width ≥ 13.5 mm

Histograms of neck width and total length for the combined ORB1 and ORB2 assemblages show the same modality (Figure 5.18B), and thus, on the basis of these two variables, there appears to be validity in the division of the Expanding Stem type into two subtypes. The above definitions, however, are not completely applicable to all of the Expanding Stem points. A number of specimens are broken, and thus total length is not measureable. In these cases type assignment was made using neck width alone. In several cases points had neck widths of less than 13.5 mm but total lengths greater than 45.0 mm. In these cases the points were assigned to ES2 based on neck width.

Fourteen points have been assigned to the ES1 subtype, and 29, to the ES2 subtype (Figures 5.19, S-5.12–S-5.13). As Table 5.23 shows, points of these two types appear to differ on several other quantitative measures of size, such as stem length and basal width. In fact, pooled-variance *t*-tests indicate that significant differences exist in neck width, stem length, basal width, and maximum width (Table 5.24). There is no significant difference in total length, however, because of the inclusion of several points with small neck widths but long blades in ES2.

There are qualitative differences between the two subtypes as well (Table 5.25). ES1 points are more often made from FGV toolstone (57.1 percent), while ES2 points are slightly more often made from obsidian (51.7 percent). In addition, about half of both subtypes

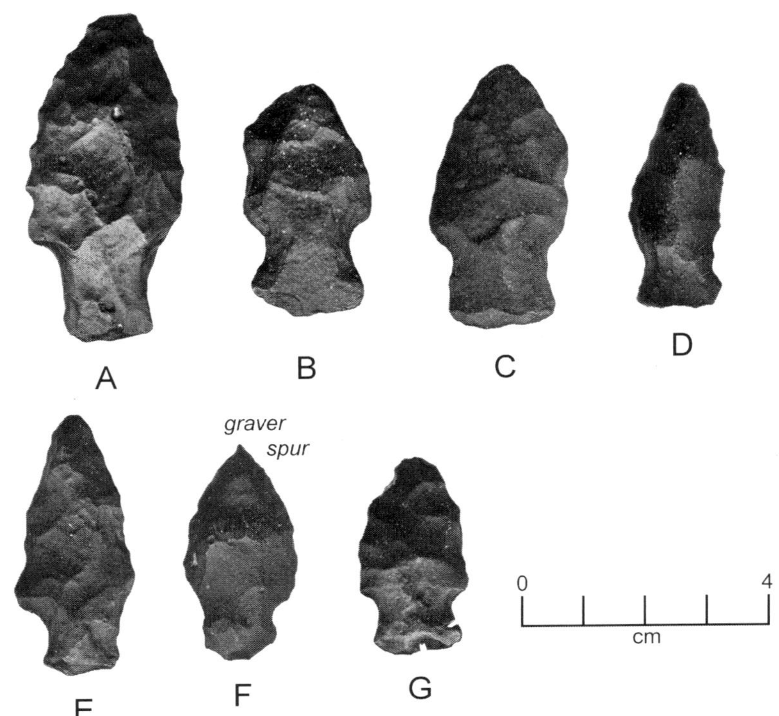

FIGURE 5.19. Examples of Expanding Stem points in the Old River Bed assemblage: (*A*) 42To2554, FS 19; (*B*) 42To3226, FS 41; (*C*) 42To3230, FS 91; (*D*) 42To3237, FS 70; (*E*) 42To3226, FS 13; (*F*) 42To3237, FS 74; (*G*) DPGIF 1559.

exhibit irregular flaking patterns, but all but one of the remaining ES2 points for which flaking was observable are collaterally flaked, while the remaining ES1 points show a variety of flaking patterns. About the same proportions of each subtype are resharpened. Two ES2 points were resharpened as gravers (see, for example, Figure 5.19F). Site locations of these points are shown in Table 5.20.

Contracting Stem Points

Six specimens were identified as Contracting Stem points (Table 5.10). Jones, Beck, and Kessler (2003:26) describe these points as well-flaked bifaces and specimens made expediently from (recycled) biface fragments. Examples of the latter often exhibit a stem that has been isolated from the blade segment by the removal of a single, steep-faced lunate flake. This creates a contracting stem with gently arcing margins and sloping shoulders. On most specimens the base incorporates a transverse fracture facet. Whether the break was part of the original "preform" or occurred during or following manufacture is difficult to judge because of the weathered condition of most specimens. The blade segment is comparatively wide, and its plan may be triangular or spatulate. Jones, Beck, and Kessler (2003) define two subtypes based on size.

Contracting Stem 1. The Contracting Stem 1 (CS1) group includes examples made from well-flaked bifaces and fragmentary blanks. Among the former, retouch was applied to the stem and shoulder regions to form a more or less continuous arc.

Contracting Stem 2. Members of the Contracting Stem 2 (CS2) subcategory are larger than those of CS1.

Summary. As no precise size dimensions were specified by Jones, Beck, and Kessler (2003), histograms were produced of five size measures that might distinguish these two subtypes (Figure 5.20). There are clear divisions in all but neck width, where the division is less clear. Based on these graphs the two subtypes should be defined as follows:

CS1: neck width ≤ 17.0 mm, basal width ≤ 12.5 mm, stem length ≤ 12.0 mm, maximum width ≤ 23.0, and thickness ≤ 8.0
CS2: neck width > 17.0 mm, basal width > 12.5 mm, stem length > 12.0 mm, maximum width > 23.0, and thickness > 8.0

Figure 5.21 shows the six Contracting Stem points. All but one specimen fit into one of these two categories

TABLE 5.25. Distribution of Old River Bed Expanding Stem Points Across Eight Qualitative Variables.

Variable and State	Point Subtype			
	Expanding Stem 1		Expanding Stem 2	
	n	%	*n*	%
Raw Material				
Obsidian	6	42.9	15	51.7
Fine-grained volcanics	8	57.1	14	48.3
Total	14	100.0	29	100.0
Flaking Pattern				
Broad collateral	1	7.1	7	24.1
Parallel collateral	2	14.3		
Irregular	6	42.9	11	37.9
Collateral/irregular			1	3.4
Parallel/oblique	1	7.7		
Irregular/unifacial	1	7.7		
None (unifacial)	1	7.7	1	3.4
Indeterminate	2	14.3	9	31.0
Total	14	100.0	29	100.0
Beveling				
Absent	10	71.4	17	58.6
Blade	1	7.1	3	10.3
Indeterminate	3	21.4	9	31.0
Total	14	100.0	29	100.0
Lateral Edge Grinding				
Present	1	7.1		
Indeterminate	13	92.9	29	100.0
Total	14	100.0	29	100.0
Use-Wear				
Indeterminate	14	100.0	29	100.0
Total	14	100.0	29	100.0
Resharpening				
Absent	1	7.1	1	3.4
Tip/blade (TB)			3	10.3
Barb	1	7.1	2	6.9
Stem	1	7.1	1	3.4
TB/barb	6	42.9	1	3.4
TB/stem			1	3.4
TB/barb/different tool			2	6.9
Indeterminate	5	35.7	18	62.1
Total	14	100.0	29	100.0
Size Category				
7			2	6.9
8	2	14.3	8	27.6
9	3	29.4	7	24.1
10	2	14.3	2	6.9
Indeterminate	7	50.0	10	34.5
Total	14	100.0	29	100.0
Weathering Damage				
Medium	2	14.3	8	27.6
Heavy	9	64.3	11	37.9
Extreme	3	21.4	10	34.5
Total	14	100.0	29	100.0

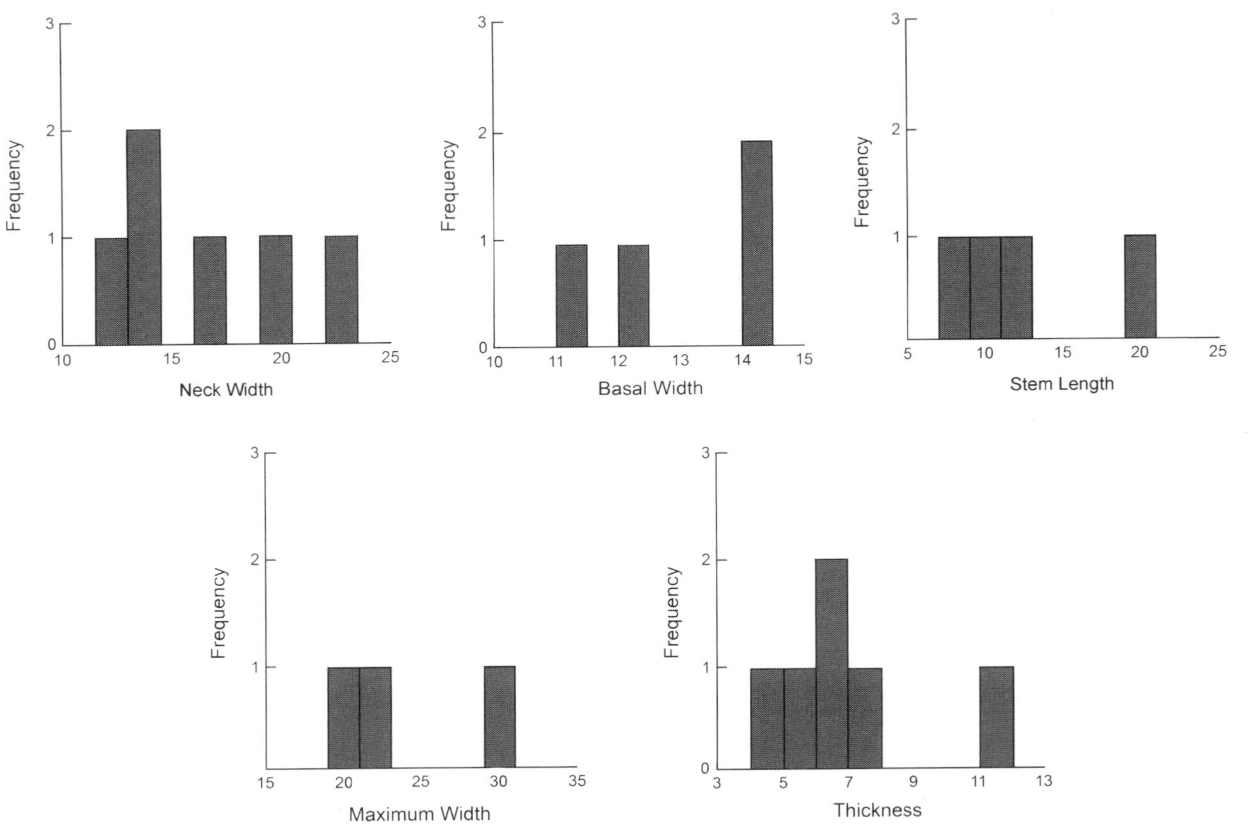

FIGURE 5.20. Histograms of five metric variables for Contracting Stem points in the Old River Bed assemblage.

FIGURE 5.21. Contracting Stem points in the Old River Bed assemblage. Contracting Stem 1: (*A*) 42To1371, FS 71; (*B*) 42To1686, FS 13; (*C*) 42To1688, FS 58; (*D*) 42To1672, FS 8. Contracting Stem 2: (*E*) DPGIF 879; (*F*) 42To1686, FS 67.

TABLE 5.26. Summary of Quantitative Data for Contracting Stem Points in the Old River Bed Assemblage.

| Variable | Statistic | Point Subtype | |
		Contracting Stem 1	Contracting Stem 2
Total length (mm)	n	2	—
	Range	32.3–32.9	
	Mean	32.55	
	SD	.424	
Blade length (mm)	n	3	—
	Range	19.5–29.8	
	Mean	24.28	
	SD	5.209	
Stem length (mm)	n	2	2
	Range	8.1–9.3	11.6–20.4
	Mean	8.65	16.00
	SD	.849	6.223
Maximum width (mm)	n	2	1
	Range	19.5–23.0	
	Mean	21.23	29.6
	SD	2.440	
Thickness (mm)	n	4	2
	Range	4.6–6.5	7.9–11.2
	Mean	5.65	9.55
	SD	.916	2.333
Basal width (mm)	n	2	2
	Range	11.1–12.3	14.0–14.4
	Mean	11.68	14.20
	SD	.813	.283
Neck width (mm)	n	4	2
	Range	12.1–16.4	19.1–22.8
	Mean	13.98	20.95
	SD	1.800	2.616
Proximal shoulder angle (°)	n	4	2
	Range	77.0–81.0	76.0–79.0
	Mean	79.0	77.50
	SD	1.826	2.121
Distal shoulder angle (°)	n	4	2
	Range	209.0–227.0	219.0–230.0
	Mean	218.5	224.5
	SD	7.550	7.778

TABLE 5.27. Distribution of Old River Bed Contracting Stem Points Across Eight Qualitative Variables.

| Variable and State | Point Subtype | | | |
| | Contracting Stem 1 | | Contracting Stem 2 | |
	n	%	n	%
Raw Material				
Fine-grained volcanics	4	100.0	2	100.0
Total	4	100.0	2	100.0
Flaking Pattern				
Irregular	1	25.0		
Indeterminate	3	75.0	2	100.0
Total	4	100.0	2	100.0
Beveling				
Absent	2	50.0		
Indeterminate	2	50.0	2	100.0
Total	4	100.0	2	100.0
Lateral Edge Grinding				
Indeterminate	4	100.0	2	100.0
Total	4	100.0	2	100.0
Use-Wear				
Indeterminate	4	100.0	2	100.0
Total	4	100.0	2	100.0
Resharpening				
Different tool			1	50.0
Indeterminate	4	100.0	1	50.0
Total	4	100.0	2	100.0
Size Category				
8	2	50.0		
Indeterminate	2	50.0	2	100.0
Total	4	100.0	2	100.0
Weathering Damage				
Heavy	4	100.0	2	100.0
Total	4	100.0	2	100.0

based on the measurements that could be taken. Specimen F in Figure 5.21 is consistent with three of the metric criteria for CS2 (neck width, basal width, and maximum width), and thus we include it in this category.

Quantitative data are given in Table 5.26, while qualitative data for Contracting Stem points are presented in Table 5.27. Site locations are in Table 5.20. Little can be said concerning the qualitative data. All six points are made from FGV toolstone, and all exhibit heavy weathering damage. Very little else is observable. Resharpening could be evaluated for one specimen, only because the edge of the blade had been transformed into another tool—a cutting edge of some sort (see Figure 5.21E).

Dugway Stubby Points

The Stubby designation was first applied to points that were found at Dugway Proving Ground and were off-handedly referred to as "Stubbies" because of their small size, often poor flaking, and asymmetrical shape. Jones, Beck, and Kessler provide the following description of the ORB1 Stubbies:

"Stubbies" take their name from their small size and very abbreviated blade segment. In most instances, they are less than 30 mm in length [Jones, Beck, and Kessler (2003) originally stated "less than 1.5 cm in length," which is in error]; the stem and shoulder represents between ½ and ⅔ of the length of the point. With the exception of S4, mean lengths are less than 30 mm. The plan of the blade varies between an equilateral and wide isosceles triangle. The blade is often rounded rather than obviously pointed, probably a consequence of breakage and subsequent weathering or resharpening (and weathering). A few examples exhibit unifacial retouch on the distal section of the blade, producing a convex plan. Rather than serving as projectile tips, these specimens may have completed their use-lives as scrapers. Their weathered condition precludes complete evaluation of this possibility, however [2003:23].

A total of 191 points of this type have been identified in the ORB assemblage (Table 5.10). Whereas other categories have been subdivided on the basis of form and size, four subcategories were created for Stubbies on the basis of flaking technique and blank type. Jones, Beck, and Kessler note:

> Only the first group (S1) are made by conventional bifacial flaking. The other groups appear to be manufactured more expediently. Despite their less standardized morphology, resulting from use of different kinds of blanks and application of different manufacture techniques, they appear to be designed to the same plan as the first group. The latter appear to be made from scavenged fragments [2003:23].

A fifth category (Stubby 5) was created for those points that could not be identified with any of the other four. The following subtype definitions are taken from Jones, Beck, and Kessler (2003:23–24).

Stubby 1. Examples in the Stubby 1 (S1) category (*n* = 43; Figures 5.22A–D, S-5.14) are made from facially flaked preforms. Flake scars are invasive, meeting at or passing the medial line, and cover the entire artifact. Both sides of the artifact display this flaking, giving specimens a biconvex cross section. In comparison with other Stubby groups, these specimens exhibit greater bilateral symmetry.

Stubby 2. The Stubby 2 (S2) specimens (*n* = 48; Figures 5.22E–H, S-5.15) exhibit less extensive facial flaking and generally are less symmetrical than the first group. They are typically made from biface fragments. These specimens often retain fracture facets, ridges, and flake arrises that relate to the blank's previous morphology, which apparently inhibited treatment with pressure flaking. Instead, shaping was accomplished by minimal retouch to achieve a point and to isolate a stem section.

Stubby 3. Members of the Stubby 3 (S3) group (*n* = 20; Figures 5.22I–K, S-5.16) are made from flake blanks and thus exhibit a curved longitudinal section. Like S2 examples, these S3 specimens are shaped mainly by discontinuous retouch, only some of which is invasive, and exhibit less bilateral symmetry than members of S1.

Stubby 4. In contrast to the other groups, members of the Stubby 4 (S4) group (*n* = 27; Figures 5.22L–O, S-5.17) exhibit large blade segments, equaling twice or more of the length of the stem. These large blades are comparatively wide as well. Stem sections, however, fall within the morphological range of the other groups. Examples are manufactured from either relatively thin bifacial preforms or thick, irregular biface fragments.

Stubby 5. The Stubby 5 specimens (*n* = 53; Figures 5.22P–R, S-5.18) are fragmentary or exhibit significantly less symmetry than other groups. Their metrical variation, however, falls within the range of the other Stubby variants.

Summary. Table 5.28 shows the quantitative data for all five subtypes, while Table 5.29 shows the qualitative data. The locations of Stubby points are shown in Table 5.20. Points within subtypes 1 through 4 are remarkably similar with respect to measures of size, with two exceptions. First, S4 points differ significantly from points of all three other subtypes in total length (Table 5.30). In addition, S1 and S2 points differ from S3 points in PSA.

As Table 5.29 shows, the points of all five subtypes are quite similar with respect to qualitative measures as well. The large majority are manufactured from obsidian, and thus nearly all exhibit medium to extreme weathering damage. Nearly all observable flaking patterns are collateral or irregular, and only three specimens are beveled. Although it is likely that a majority of these points have been resharpened, the extensive weathering damage precluded this determination on more than a few. Three S2 points were resharpened into gravers (see, for example, Figure 5.22G), while one

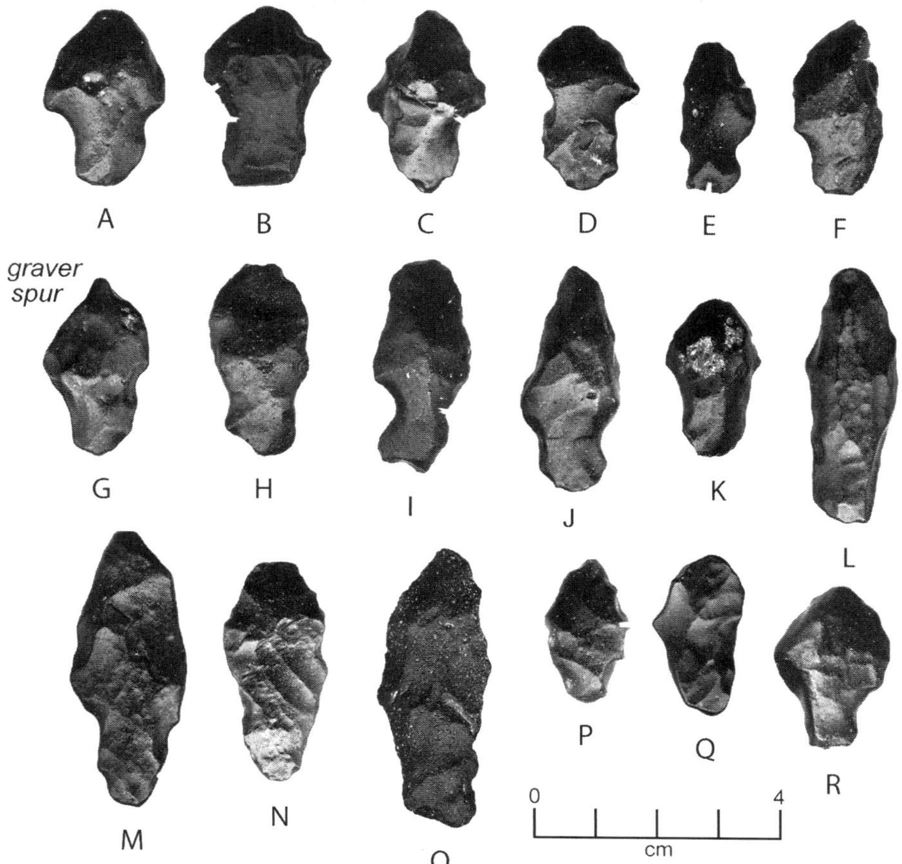

FIGURE 5.22. Examples of Stubby points in the Old River Bed assemblage. Stubby 1: (A) 42To1686, FS 39; (B) 42To1688, FS 60; (C) DPGIF 325; (D) 42To2551, FS 40. Stubby 2: (E) 42To1686, FS 25; (F) 42To1688, FS 31; (G) 42To1862, FS 2; (H) 42To2551, FS 60. Stubby 3: (I) 42To1688, FS 1; (J) 42To3520, FS 24; (K) 42To2559, FS 39. Stubby 4: (L) 42To1371, FS 97; (M) 42To1685, FS 8; (N) 42To3219, FS 19; (O) DPGIF 1397. Stubby 5: (P) 42To1686, FS 63; (Q) 42To1859, FS 6; (R) DPGIF 465.

S4 specimen was reworked into a beaked scraper (see Figure S-5.17u).

Butte Valley Corner-Notched Points

We formally defined the Butte Valley Corner-notched (BVCN) type in our report on the Sunshine Locality (Beck and Jones 2009:204–205), although we have referred to it previously (e.g., Beck and Jones 1990a). We found the first of these points on a Paleoarchaic surface site in Butte Valley, eastern Nevada, in 1986. Over the next several field seasons more were collected from Long and Jakes valleys as well as Butte Valley. More recently we collected several points of this type at sites in Coal Valley to the south.

Although named Butte Valley Corner-*notched*, these specimens might be better described as corner-*removed*. They have large triangular blades, very large neck widths, and expanding stems with convex bases (Beck and Jones 2009:204). Sixteen BVCN points have been identified in the ORB assemblage (Figures 5.23, S-5.19; Tables 5.10, 5.31–5.32). They appear to vary considerably with respect to stem shape, but the stems of some specimens have been broken (Figure 5.23A, D; S-5.19d, f, and h) and/or heavily resharpened (Figure S-5.19d, h, j–k, and n). Site locations for these points are shown in Table 5.20.

The most distinctive feature of BVCN points is their large neck width (mean = 19.29 mm), which is the largest of any WST or EH type; only Cougar Mountain is close, with a mean neck width of 18.21 mm. Neck widths, as well as other size measures, are fairly consistent between the ORB specimens and the eastern Nevada specimens

TABLE 5.28. Summary of Quantitative Data for Dugway Stubby Points in the Old River Bed Assemblage.

Variable	Statistic	Stubby 1	Stubby 2	Stubby 3	Stubby 4	Stubby 5
				Point Subtype		
Total length (mm)	n	24	25	12	16	7
	Range	20.9–34.7	21.8–33.9	24.0–37.1	26.3–46.0	23.9–33.8
	Mean	26.48	26.88	29.89	37.12	28.68
	SD	3.812	2.997	4.005	4.925	3.381
Blade length (mm)	n	24	27	12	13	10
	Range	6.6–18.1	6.9–18.7	11.2–26.1	9.9–25.5	11.9–19.8
	Mean	12.10	11.96	17.13	19.36	16.02
	SD	3.854	3.142	5.330	4.798	2.650
Stem length (mm)	n	29	38	16	19	21
	Range	5.1–13.8	5.7–24.7	5.3–14.1	6.6–18.9	4.0–21.6
	Mean	10.22	9.70	9.50	11.49	9.30
	SD	2.700	3.229	2.777	2.579	3.623
Maximum width (mm)	n	16	27	11	15	8
	Range	14.8–22.4	10.6–20.7	13.4–19.5	8.7–21.5	13.2–17.2
	Mean	17.58	16.03	16.40	16.19	15.00
	SD	2.512	2.591	1.641	3.229	1.398
Thickness (mm)	n	41	48	20	27	52
	Range	3.1–8.7	3.5–9.0	3.9–8.2	4.4–8.7	3.7–9.4
	Mean	6.03	5.80	5.86	6.60	6.24
	SD	1.176	1.255	1.010	1.180	1.303
Basal width (mm)	n	16	29	7	16	15
	Range	5.7–12.6	4.9–12.3	8.6–13.3	5.6–11.2	5.3–12.6
	Mean	10.38	9.64	10.29	9.00	9.29
	SD	1.788	1.812	1.909	1.345	2.113
Neck width (mm)	n	37	41	16	21	25
	Range	8.9–14.0	7.2–13.8	8.1–14.5	8.3–15.7	8.1–13.6
	Mean	11.62	10.91	10.97	11.43	11.00
	SD	1.355	1.632	1.936	1.817	1.752
Proximal shoulder angle (°)	n	39	42	18	23	24
	Range	80.5–109.0	69.5–105.5	78.0–122.5	73.0–108.0	73.0–122.0
	Mean	91.59	91.32	98.47	87.70	91.15
	SD	7.024	7.784	13.588	9.319	12.723
Distal shoulder angle (°)	n	38	42	19	21	26
	Range	133.0–257.0	187.5–250.0	208.5–241.0	222.5–250.5	195.0–250.5
	Mean	224.50	228.41	226.11	235.41	228.18
	SD	20.027	14.306	10.617	8.720	12.333

(Beck and Jones 2009:Table 6.28). Half (50.0 percent) of the ORB points are manufactured from FGV, and 43.7 percent are manufactured from obsidian. All 10 of those that could be evaluated are resharpened, some heavily so.

BVCN points tend to occur with WST, Pinto, and Windust points, and thus we have always maintained that they are early Holocene in age. Recently, Janetski et al. (2012) reported four points from North Creek Shelter in eastern Utah, which they named North Creek Stemmed, that are very similar to BVCN points. In fact, they fit all of the criteria for BVCN points. Since BVCN points have been found primarily in the eastern Great Basin, we believe that the two types represent the same entity. One North Creek Stemmed point was found in stratum IIg; a radiocarbon date of ~9690 ± 60 ^{14}C BP was obtained for stratum IIIa, just above IIg (Janetski et al. 2012:136). Most, however, were recovered from stratum IIIe, sandwiched between the date of ~9690 ± 60 ^{14}C BP from stratum IIIa and a date of ~9510 ± 80 ^{14}C BP from stratum IVa (Janetski et al. 2012:140). These are the first dates associated with this type, and they confirm that BVCN points are of early Holocene age.

TABLE 5.29. Distribution of Old River Bed Dugway Stubby Points Across Eight Qualitative Variables.

| | Stubby 1 | | Stubby 2 | | Stubby 3 | | Stubby 4 | | Stubby 5 | |
Variable and State	n	%	n	%	n	%	n	%	n	%
Raw Material										
Obsidian	41	95.3	39	81.2	18	90.0	19	70.4	47	88.7
Fine-grained volcanics	2	4.7	9	18.8	2	10.0	8	29.6	6	11.3
Total	43	100.0	48	100.0	20	100.0	27	100.0	53	100.0
Flaking Pattern										
Broad collateral	12	27.9	8	16.7	3	15.0	6	22.2	7	13.2
Parallel collateral	1	2.3					1	3.7	3	5.7
Oblique									1	1.9
Irregular	16	37.2	22	45.8	6	30.0	11	40.7	14	26.4
Collateral/irregular					2	10.0			3	5.7
Collateral/unifacial					2	10.0				
Parallel/chevron							1	3.7		
Oblique/irregular									1	1.9
Irregular/unifacial					3	15.0			1	1.9
None (unifacial)			1	2.1						
Indeterminate	14	32.6	17	35.4	4	20.0	8	29.6	23	43.4
Total	43	100.0	48	100.0	20	100.0	27	100.0	53	100.0
Beveling										
Absent	31	72.1	33	68.7	13	65.0	15	59.3	22	41.5
Blade	1	2.3					1	3.7		
Stem							1	3.7		
Indeterminate	11	25.6	15	31.3	7	35.0	10	37.0	31	58.5
Total	43	100.0	48	100.0	20	100.0	27	100.0	53	100.0
Lateral Edge Grinding										
Indeterminate	43	100.0	48	100.0	20	100.0	27	100.0	53	100.0
Total	43	100.0	48	100.0	20	100.0	27	100.0	53	100.0
Use-Wear										
Present	1	2.3								
Indeterminate	42	97.7	48	100.0	20	100.0	27	100.0	53	100.0
Total	43	100.0	48	100.0	20	100.0	27	100.0	53	100.0
Resharpening										
Absent	4	9.3					2	7.4		
Tip/blade (TB)	6	14.0	2	9.5	2	10.0			6	11.3
Stem	1	2.3								
TB/barb	2	9.7	5	10.4	1	5.0	3	11.1	2	3.8
TB/stem					2	10.0				
TB/barb/stem							1	3.7		
TB/barb/different tool							1	3.7		
Different tool			3	6.2						
Indeterminate	30	69.8	38	79.2	15	75.00	20	74.1	45	84.9
Total	43	100.0	48	100.0	20	100.0	27	100.0	53	100.0
Size Category										
4			1	2.1					1	1.9
6	2	4.7	3	6.2					1	1.9
7	18	41.9	19	39.6	5	25.0	2	7.4	6	11.3
8	11	27.9	14	29.2	8	40.0	7	25.9	2	3.8
9					1	5.0	6	22.3		
Indeterminate	12	27.9	11	22.9	6	30.0	12	44.4	43	81.1
Total	43	100.0	48	100.0	20	100.0	27	100.0	53	100.0
Weathering Damage										
Minimal	1	2.3								
Medium	1	2.3					4	14.8	5	9.4
Heavy	6	14.0	7	14.6	6	30.0	7	25.9	16	30.2
Extreme	35	81.4	41	85.4	14	70.0	15	59.3	31	58.5
Not recorded							1	3.7	1	1.9
Total	43	100.0	48	100.0	20	100.0	27	100.0	53	100.0

TABLE 5.30. Results of Analysis of Variance Test (*p*) Among Four Stubby Subtypes for Total Length and Proximal Shoulder Angle.

Variable and Point Subtype	Point Subtype			
	Stubby 1	Stubby 2	Stubby 3	Stubby 4
Total Length				
Stubby 1	1.000			
Stubby 2	1.000	1.000		
Stubby 3	.090	.178	1.000	
Stubby 4	**<.001**	**<.001**	**<.001**	1.000
Proximal Shoulder Angle				
Stubby 1	1.000			
Stubby 2	1.000	1.000		
Stubby 3	**.047**	**.032**	1.000	
Stubby 4	.599	.721	**.001**	1.000

Note: Significant values in bold.

Eden Point

One point from the ORB assemblage, found as an isolate (DPGIF 2523), has been identified with the Eden type (Figure 5.24; Tables 5.31–5.32). This type is most commonly seen in Plains Paleoindian assemblages as part of the Cody Complex, which dates between ~9500 and ~8800 ^{14}C BP (Frison 1998; Holliday 2000). The ORB point is made from FGV toolstone and is not as finely flaked as Plains examples (see, for example, Frison 1991:Figure 9.21). Instead of fine parallel collateral flaking, the flaking on the ORB specimen is a combination of broad and parallel collateral flaking, with many flake scars carrying past the center line. While Plains Eden points sometimes are quite long, upwards of 125–130 mm, they are more often shorter. The eight relatively complete examples from the Horner site, for example, range in length from 35.0 mm to 122.0 mm (Bradley and Frison 1987:Table 6.2), to which the ORB specimen, with a length of 93.5 mm (Table 5.31), is comparable. Some of the Horner specimens have been resharpened, but three that measure 65.6 mm, 66.0 mm, and 66.0 mm are not (Bradley and Frison 1987:Table 6.2). As is the case for the majority of the ORB points, resharpening could not be determined for the Eden point. This point is not the only Plains-like artifact to be found at the ORB; several Cody knives have been collected as well (see below). Also, as noted above, the Square Stem subtype 1 resembles Cody/Scottsbluff (as well as Windust) points.

Pinto and Elko Points

The final Early Holocene point type to be discussed is Pinto, which has a problematic definitional history. One of the primary problems has been the difficulty of distinguishing these points from those of the Little Lake, Gatecliff, and Elko series (see discussion of this issue in Beck and Jones 2009:206–208). We are concerned here with distinguishing Pinto from Elko points, and thus both will be described together, even though Elko is considered an Archaic type.

A fundamental difference between Pinto and Elko is that the former is stemmed and the latter is notched. Thus the similarity of Pinto to Elko as well as those of the other two series is superficial, based completely on morphology rather than technology. Several attempts have been made to distinguish Pinto from Elko points quantitatively, the most recent being by Basgall and Hall (2000). Based on statistical comparisons of 11 metric variables, Basgall and Hall (2000:261) present metric

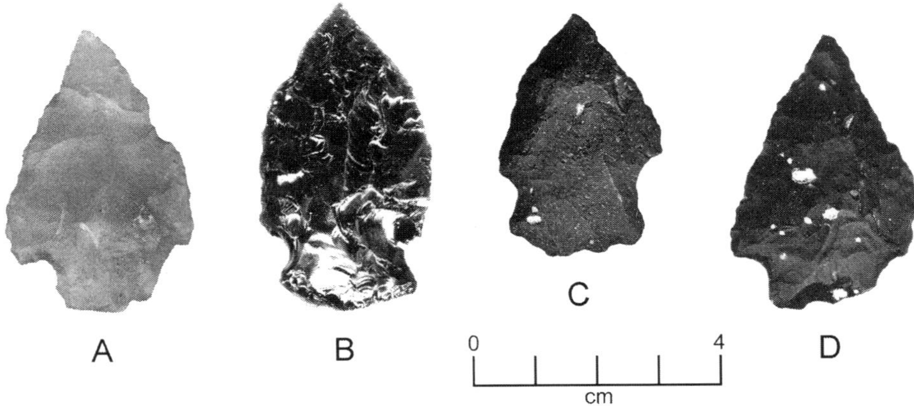

FIGURE 5.23. Examples of Butte Valley Corner-notched points in the Old River Bed assemblage: (*A*) 42To1161, FS 8; (*B*) 42To1182, FS 1; (*C*) 42To3229, FS 20; (*D*) 42To3228, FS 15.

TABLE 5.31. Summary of Quantitative Data for Butte Valley Corner-Notched, Eden, Pinto, Short Stem, and Elko Points in the Old River Bed Assemblage.

		Point Type				
Variable	Statistic	Butte Valley Corner-Notched	Eden	Pinto	Short Stem	Elko
Total length (mm)	n	6	1	16	13	9
	Range	30.2–52.6		21.3–59.4	18.0–36.9	28.7–50.2
	Mean	41.77	93.5	30.82	30.02	38.59
	SD	8.908		9.235	6.228	8.111
Blade length (mm)	n	8	—	12	9	6
	Range	17.8–38.1		11.0–30.9	7.2–27.9	17.9–38.4
	Mean	28.83		19.48	20.53	27.32
	SD	7.700		6.587	6.264	7.002
Stem length (mm)	n	9	—	27	15	18
	Range	6.6–21.6		5.5–14.0	3.3–32.5	5.2–15.7
	Mean	11.19		8.05	11.52	8.46
	SD	4.307		2.039	6.898	2.516
Maximum width (mm)	n	5	1	7	18	5
	Range	25.1–33.8		14.8–19.5	7.8–25.9	14.5–22.5
	Mean	28.74	22.9	17.10	17.34	18.34
	SD	3.669		1.596	3.914	2.990
Thickness (mm)	n	14	1	37	65	25
	Range	5.5–8.7		3.9–8.1	3.7–9.1	3.4–7.6
	Mean	7.21	5.8	5.53	6.46	5.50
	SD	.938		1.062	1.170	1.229
Basal width (mm)	n	5	1	20	12	9
	Range	11.5–22.9		10.8–27.6	4.7–18.6	12.3–19.2
	Mean	19.02	22.6	16.56	11.99	15.85
	SD	4.673		5.012	4.288	2.548
Neck width (mm)	n	9	—	25	13	21
	Range	17.4–21.6		10.3–26.4	8.2–18.3	11.0–17.8
	Mean	19.29		14.00	13.35	13.91
	SD	1.270		3.572	3.371	1.891
Proximal shoulder angle (°)	n	11	—	36	20	23
	Range	90.0–126.5		94.0–118.0	71.5–123.5	96.0–143.5
	Mean	105.09		103.04	91.80	119.63
	SD	11.018		6.306	15.502	10.620
Distal shoulder angle (°)	n	8	—	22	18	5
	Range	172.0–234.0		213.0–263.0	212.0–251.5	154.0–212.0
	Mean	206.31		232.41	232.06	180.0
	SD	20.858		11.819	9.769	27.019

definitions for both Pinto and Elko points, which are based on Thomas's (1981) definitions of Gatecliff and Elko points:

Pinto Series: BW > 10.0 mm, PSA ≤ 100° *or* NOA ≥ 80°
Elko Series: BW > 10.0 mm, 110° ≤ PSA ≤ 150° *or* NOA < 80°

where BW = basal width, PSA = proximal shoulder angle, and NOA = notch opening angle (the same as Thomas's 1981 notch opening index).

Since points of both types have basal widths greater than 10 mm, type assignment is actually based on the values of proximal shoulder angle and notch opening angle. These values for the ORB points assigned to the Pinto (n = 37; Figures 5.25, S-5.20) and Elko (n = 25;

TABLE 5.32. Distribution of Old River Bed Butte Valley Corner-Notched, Eden, Pinto, Short Stem, and Elko Points Across Eight Qualitative Variables.

Variable and State	Butte Valley Corner-Notched n	%	Eden n	%	Pinto n	%	Short Stem n	%	Elko n	%
Raw Material										
Chert	1	6.2			2	5.4	1	1.2	6	24.0
Obsidian	7	43.8			21	56.0	42	51.9	7	28.0
Fine-grained volcanics	8	50.0	1	100.0	14	37.8	38	46.9	12	48.0
Total	16	100.0	1	100.0	37	100.0	81	100.0	25	100.0
Flaking Pattern										
Broad collateral	2	12.5			4	10.8	12	14.8		
Parallel collateral	1	6.2			2	5.4	1	1.2	3	12.0
Oblique					1	2.7			1	4.0
Irregular	8	50.0			12	32.3	16	19.8	13	52.0
Collateral/irregular					1	2.7	1	1.2		
Parallel/oblique					1	2.7				
Parallel/irregular			1	100.0						
Irregular/unifacial	2	12.5					1	1.2		
Indeterminate	3	18.8			16	43.2	50	61.7	8	32.0
Total	16	100.0	1	100.0	37	100.0	81	100.0	25	100.0
Beveling										
Absent	10	62.5	1	100.0	21	56.8	37	45.7	11	44.0
Blade							1	1.2		
Indeterminate	6	37.5			16	43.2	43	53.1	14	56.0
Total	16	100.0	1	100.0	37	100.0	81	100.0	25	100.0
Lateral Edge Grinding										
Present	2	12.5					1	1.2		
Indeterminate	14	87.5	1	100.0	37	100.0	80	98.8	25	100.0
Total	16	100.0	1	100.0	37	100.0	81	100.0	25	100.0
Use-Wear										
Present	2	12.5			1	2.7	1	1.2	1	4.0
Indeterminate	14	87.5	1	100.0	36	97.3	80	98.8	24	96.0
Total	16	100.0	1	100.0	37	100.0	81	100.0	25	100.0
Resharpening										
Absent					1	2.7	5	6.2	1	4.0
Tip/blade (TB)	2	12.5			2	5.4	4	4.9		
Barb							1	1.2		
Stem	1	6.2					1	1.2	1	4.0
TB/barb	6	37.6			9	24.3			11	44.0
TB/barb/stem	2	12.5			4	10.8	2	2.5	3	12.0
Barb/stem							1	1.2		
Different tool					1	2.7	1	1.2		
Indeterminate	5	31.3	1	100.0	20	54.1	66	81.5	9	36.0
Total	16	100.0	1	100.0	37	100.0	81	100.0	25	100.0
Size Category										
7					6	16.2				
8					9	24.3	2	2.5	4	16.0
9	3	18.8			1	2.7	3	3.7	1	4.0
10	3	18.8							3	12.0
11					1	2.7				
Indeterminate	10	62.5	1	100.0	20	54.1	76	93.8	17	68.0
Total	16	100.0	1	100.0	37	100.0	81	100.0	25	100.0
Weathering Damage										
Minimal	5	31.3			4	10.8			15	60.0
Medium	6	37.5	1	100.0	1	2.7	7	8.6	4	16.0
Heavy	4	25.0			11	29.7	35	43.2	3	12.0
Extreme	1	6.2			21	56.8	23	28.4	3	12.0
Indeterminate							16	19.8		
Total	16	100.0	1	100.0	37	100.0	81	100.0	25	100.0

FIGURE 5.24. Example of an Eden type in the Old River Bed assemblage.

FIGURE 5.25. Examples of Pinto points in the Old River Bed assemblage: (A) 42To1668, FS 9; (B) 42To1689, FS 3; (C) DPGIF 526; (D) 42To3522, Fs 7; (E) DPGIF 2447.

Figures 5.26, S-5.21) types are shown in Table 5.33. As is evident from the values in this table, several points identified with the Pinto type for which NOA could not be calculated have PSA values slightly more than 100°, but not so much that they could be identified as Elko points. Similarly, three points with PSA values slightly less than 110° were identified with the Elko type. Additionally points for which neither PSA nor DSA could be measured (and thus notch opening index could not be calculated) were identified as Elko (Figure 5.26B, S-5.21f, i), based on overall morphology. The quantitative data for points of both types are shown in Table 5.31; the qualitative data are shown in Table 5.32.

Short Stem Points

Eighty-one points were too fragmentary or heavily weathered to assign to a specific EH type; these are designated as Short Stem points. Quantitative and qualitative data for these points are presented in Tables 5.31 and 5.32, respectively. Site locations can be found in Table 5.20.

Archaic, "Unknown," and Unclassifiable Points

Here we consider the Rocker Side-notched, Humboldt, Northern Side-notched, Gatecliff, Rosegate, and Small Side-notched types. Thirty points have been identified with Archaic types other than Elko, while another

FIGURE 5.26. Examples of Elko points in the Old River Bed assemblage: (A) 42To1173, FS 1; (B) 42To1358, FS 65; (C) 42To1358, FS 86; (D) 42To1666, FS 9; (E) DPGIF 193.

17 fragments have been identified as Archaic blades (Tables 5.10, 5.34–5.35).

Seven points were identified as Rocker Side-notched (Figure 5.27), a type defined by Holmer (1980:76) in his analysis of the projectile points from Sudden Shelter. Holmer describes these points as having "wide lance-olate blades…with moderately high horizontal side

TABLE 5.33. Proximal Shoulder Angle and Notch Opening Angle for Pinto and Elko Points in the Old River Bed Assemblage.

Point Type and Location	Field Specimen No.	Proximal Shoulder Angle (°)	Notch Opening Angle (°)	Point Type and Location	Field Specimen No.	Proximal Shoulder Angle (°)	Notch Opening Angle (°)
Pinto				DPGIF 811	—	101.0	—
42To1152	2	100.5	—	DPGIF 867	—	95.5	—
42To1177	4	100.0	140.0	DPGIF 876	—	96.0	140.0
42To1178	4	101.0	131.0	DPGIF 1932	—	95.0	—
42To1371	32	100.0	149.0	DPGIF 2447	—	98.0	130.0
42To1668	9	106.0	140.0	**Elko**			
42To1677	13	102.5	126.5	42To1172	19	121.0	—
42To1678	9	96.0	141.0	42To1173	1	122.5	—
42To1678	11	104.0	153.0	42To1173	4	110.0	—
42To1679	7	103.0	—	42To1182	2	132.0	—
42To1681	3	94.5	—	42To1352	3	128.5	—
42To1686	72	98.5	—	42To1352	5	112.0	—
42To1689	3	99.0	—	42To1358	10	—	—
42To1859	3	97.5	—	42To1358	33	122.0	—
42To2552	4	101.0	—	42To1358	45	120.0	—
42To2554	12	95.5	121.5	42To1358	65	—	—
42To2945	17	101.0	—	42To1358	86	143.5	—
42To2945	36	98.0	—	42To1367	2	122.0	—
42To2947	6	109.5	103.5	42To1666	9	115.0	51.0
42To3141	6	113.5	115.5	42To1672	4	105.0	—
42To3219	50	100.0	138.0	42To1689	8	109.5	—
42To3219	59	113.5	121.5	42To1875	23	120.0	40.0
42To3223	18	108.5	133.5	42To2945	32	127.0	—
42To3230	56	84.0	—	42To3230	52	108.5	—
42To3230	71	99.0	—	42To3230	80	112.5	—
42To3230	94	118.0	145.0	42To3230	95	120.0	—
42To3234	28	106.5	128.5	DPGIF 193	—	119.0	—
42To3237	32	94.0	135.0	DPGIF 209	—	114.0	—
42To3237	56	101.0	129.0	DPGIF 376	—	139.0	67.0
42To3237	58	99.0	116.0	DPGIF 869	—	112.0	—
42To3520	16	116.0	107.0	DPGIF 2413	—	127.0	85.0
42To3522	7	104.0	113.0				
DPGIF 526	—	102.0	128.0				

notches forming a stem that approaches semicircular in shape. The edges both above and below the notches form a continuous, smooth curve, broken only by the notches" (1980:76). One ORB specimen (Figure 5.27A) is a classic example of this type. The other six, however, either are broken or have been resharpened, and thus they deviate somewhat from this description.

Thirteen points were identified with the Humboldt type (Figures 5.28, S-5.22), one each with the Northern Side-notched and Gatecliff types (Figure 5.29), seven with the Rosegate type (Figure 5.30), and one simply as a Small Side-notched point (Figure 5.30). The latter does not resemble Desert Side-notched points but is considerably smaller than the Northern Side-notched and Rocker Side-notched specimens. Finally, the 17 fragments identified as "Archaic blades" are shown in Figures 5.31A–C and S-5.23. The locations of these points are shown in Table 5.36.

One point (Figure 5.31D), from 42To1000, has been designated as an "unknown type." It is very small and unique in its form. Because there is only one of these in the assemblage we are hesitant to name it as a new type. The remaining 56 points could not be identified with a type because they were too fragmentary or too heavily damaged by weathering and thus are designated as unclassifiable (Table 5.10). The locations of Archaic

TABLE 5.34. Summary of Quantitative Data for Remaining Archaic Points in the Old River Bed Assemblage.

Variable	Statistic	Rocker Side-Notched	Northern Side-Notched	Humboldt	Gatecliff	Rosegate	Small Side-Notched	Archaic Blade
Total length (mm)	n	3	—	5	—	3	1	—
	Range	26.5–61.4		27.5–36.6		17.0–28.8		
	Mean	40.17		31.59		24.35	22.3	
	SD	18.640		3.475		6.412		
Blade length (mm)[a]	n	3	—	—	—	3	—	1
	Range	15.4–47.3				13.3–23.7		
	Mean	28.33				19.07		17.1
	SD	16.784				5.292		
Stem length (mm)[a]	n	6	1	—	1	5	—	—
	Range	7.6–12.9				3.1–5.5		
	Mean	9.60	6.8		6.9	4.28		
	SD	2.025				1.105		
Maximum width (mm)	n	2	1	6	1	3	—	—
	Range	20.8–25.7		14.6–21.4		12.2–21.2		
	Mean	23.25	20.6	16.22	22.0	16.83		
	SD	3.465		2.599		4.506		
Thickness (mm)	n	7	1	13	1	7	1	17
	Range	4.8–7.6		3.4–14.8		.4–5.7		2.9–7.5
	Mean	6.13	2.9	5.47	4.7	3.65	4.3	4.67
	SD	.907		2.970		1.605		1.136
Basal width (mm)	n	5	1	9	1	2	—	—
	Range	18.4–20.0		9.2–15.4		5.1–10.6		
	Mean	19.26	15.0	12.74	9.9	7.85		
	SD	.720		2.194		3.889		
Neck width (mm)[a]	n	6	1	—	1	5	1	—
	Range	13.6–16.9				4.0–9.1		
	Mean	15.36	8.8		10.7	7.22	8.0	
	SD	1.354				2.469		
Proximal shoulder angle (°)[a]	n	7	1	—	1	5	1	—
	Range	126.0–159.5				101.5–125.0		
	Mean	142.14	171.0		79.5	112.50	141.0	
	SD	10.641				8.544		
Distal shoulder angle (°)[a]	n	4	—	—	1	3	—	—
	Range	178.5–224.5				127.5–172.0		
	Mean	205.75			211.5	156.67		
	SD	19.564				25.270		

[a] Not measured for Humboldt.

(except Elko), unknown, and unclassifiable points can be found in Table 5.36.

Discussion

There are several important features of the ORB projectile point assemblage that deserve further discussion. First, the primary occupation of the ORB delta ended in about 8500 ^{14}C BP as the region became nearly uninhabitable after this point, and thus, for the most part, the presence of Archaic, especially arrow, points in the region can be considered intrusive, deposited by later (likely mid- to late Holocene) people who might have wandered briefly through the area. The only series that could conceivably be legitimate components of these assemblages are Elko and Northern Side-notched, both of which date to ~8500 ^{14}C BP at Danger Cave. Notching may appear even earlier at North Creek Shelter in south-central Utah, where it could

TABLE 5.35. Distribution of Remaining Old River Bed Archaic Points Across Eight Qualitative Variables.

Variable and State	Rocker Side-Notched n	%	Northern Side-Notched n	%	Humboldt n	%	Gatecliff n	%	Rosegate n	%	Small Side-Notched n	%	Archaic Blade n	%
Raw Material														
Chert	1	14.3			1	7.7							5	29.4
Obsidian	3	42.9			7	53.8			7	100.0	1	100.0	12	70.6
Fine-grained volcanics	3	42.9	1	100.0	5	38.5	1	100.0						
Total	7	100.0	1	100.0	13	100.0	1	100.0	7	100.0	1	100.0	17	100.0
Flaking Pattern														
Broad collateral					1	7.7								
Parallel collateral					1	7.7							2	11.8
Oblique					1	7.7							2	11.8
Chevron	1	14.3											1	5.9
Irregular	3	42.9			1	7.7	1	100.0	5	71.4	1	100.0	9	52.9
Parallel/oblique			1	100.0	2	15.4								
Oblique/irregular													1	5.9
Indeterminate	3	42.9			7	53.8			2	28.6			2	11.8
Total	7	100.0	1	100.0	13	100.0	1	100.0	7	100.0	1	100.0	17	100.0
Beveling														
Absent	3	42.9			6	46.2	1	100.0	5	71.4	1	100.0	13	76.5
Indeterminate	4	57.1	1	100.0	7	53.8			2	28.6			4	23.5
Total	7	100.0	1	100.0	13	100.0	1	100.0	7	100.0	1	100.0	17	100.0
Lateral Edge Grinding														
Indeterminate	7	100.0	1	100.0	13	100.0	1	100.0	7	100.0	1	100.0	17	100.0
Total	7	100.0	1	100.0	13	100.0	1	100.0	7	100.0	1	100.0	17	100.0
Use-Wear														
Indeterminate	7	100.0	1	100.0	13	100.0	1	100.0	7	100.0	1	100.0	17	100.0
Total	7	100.0	1	100.0	13	100.0	1	100.0	7	100.0	1	100.0	17	100.0
Resharpening														
Absent	1	14.3			4	30.8			1	14.3			2	11.8
Tip/blade (TB)					2	15.4			1	14.3			2	11.8
Stem	1	14.3												
TB/barb							1	100.0					1	5.9
TB/barb/stem	2	28.6							1	14.3				
Different tool													1	5.9
Indeterminate	3	42.9	1	100.0	7	53.8			4	57.1	1	100.0	11	64.7
Total	7	100.0	1	100.0	13	100.0	1	100.0	7	100.0	1	100.0	17	100.0
Size Category														
5									2	28.6				
7					3	23.1			2	28.6	1	100.0		
8	1	14.3			2	15.4								
9	1	14.3			1	7.7	1	100.0						
11	1	14.3												
Indeterminate	4	57.1	1	100.0	7	53.8			3	42.9			17	100.0
Total	7	100.0	1	100.0	13	100.0	1	100.0	7	100.0	1	100.0	17	100.0
Weathering Damage														
Minimal	4	57.1			4	30.8			4	57.1			7	41.2
Medium	2	28.6	1	100.0	1	7.7	1	100.0					5	29.4
Heavy	1	14.3			7	53.8			2	28.6			5	29.4
Extreme					1	7.7			1	14.3	1	100.0		
Total	7	100.0	1	100.0	13	100.0	1	100.0	7	100.0	1	100.0	17	100.0

FIGURE 5.27. Rocker Side-notched points in the Old River Bed assemblage: (*A*) 42To2948, FS 16; (*B*) 42To1173, FS 6; (*C*) 42To1358, FS 56; (*D*) 42To1358, FS 70; (*E*) 42To1166, FS 2; (*F*) DPGIF 319; (*G*) DPGIF 363.

FIGURE 5.28. Examples of Humboldt points in the Old River Bed assemblage: (*A*) 42To1182, FS 4; (*B*) 42To1359, FS 6; (*C*) 42To1367, FS 2.

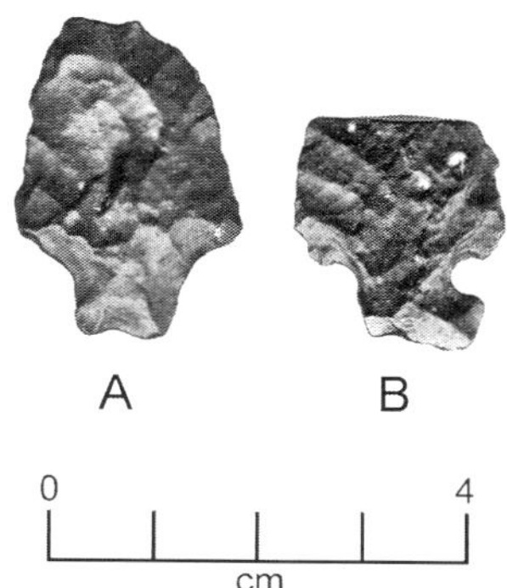

FIGURE 5.29. Gatecliff (*A*) and Northern Side-notched (*B*) points in the Old River Bed assemblage: (*A*) 42To1685, FS 39; (*B*) DPGIF 478.

FIGURE 5.30. Rosegate (*A–F*) and Small Side-notched (*G*) points in the Old River Bed assemblage: (*A*) 42To1358, FS 14; (*B*) 42To1358, FS 25; (*C*) 42To1372, FS 1; (*D*) DPGIF 224; (*E*) 42To1000, FS 22; (*F*) DPGIF 33; (*G*) 42To2948, FS 53.

FIGURE 5.31. Archaic blades (*A–C*) and point of unknown type (*D*) in the Old River Bed assemblage: (*A*) 42To1182, FS 6; (*B*) 42To1352, FS 4; (*C*) 42To1358, FS 48; (*D*) 42To1000, FS 1.

TABLE 5.36. Site Locations for Remaining Old River Bed Archaic Points.

Site	Rocker Side-Notched	Northern Side-Notched	Humboldt	Gatecliff	Rosegate	Small Side-Notched	Archaic Blade	Unknown Type
42To1000					2		1	1
42To1163			1				1	
42To1166	1							
42To1172							2	
42To1173	1							
42To1178			1					
42To1182			2				2	
42To1352							1	
42To1358	2				2		2	
42To1359			1				3	
42To1367			1					
42To1372					1			
42To1384			1					
42To1666							1	
42To1671							1	
42To1676								
42To1683			1				1	
42To1685			1	1				
42To2947								
42To2948	1					1		
42To3230			2					
42To3235							1	
Isolates	2	1	2		2		1	
Total	7	1	13	1	7	1	17	1

date to 9020 ± 70 ^{14}C BP (Janetski et al. 2012:140). It is more likely, however, that the ORB Elko and Northern Side-notched points date to the mid-Holocene or later. Therefore, the Archaic points will not be considered further in this chapter.

The second feature concerns the degree of breakage and resharpening, which in many cases is extreme. This is especially the case for EH points, with the exception of Stubbies, most of which are so damaged by weathering that breakage and resharpening only rarely could be evaluated. The only WST points that exhibit extensive resharpening are those of the Silver Lake type, which is true of the Sunshine Silver Lake points as well. The heavy breakage and resharpening on EH points by itself is not necessarily meaningful; many of the Archaic points exhibit heavy breakage and resharpening as well. But in concert with what we believe to be extensive recycling, this is suggestive of raw material conservation as a consequence of the relatively long distance to even the closest FGV sources. This is the third feature we discuss.

Many of the projectile points, especially Stubbies, appear to be manufactured from discarded artifacts left behind by earlier people who visited the area. As Jones, Beck, and Kessler (2003) state, two of the distinctive features of the ORB point assemblage are the small size of some of the artifacts and the degree of expedient manufacture present. Although some of the points that identify with established WST types are comparable in size to those found elsewhere across the Great Basin, others are smaller. This is true, for example, of many of the Silver Lake specimens. The most numerous of the newly defined types, Dugway Stubbies, are especially small, most less than 30 mm in length. We believe that many (if not most) of these points were manufactured from the blades of production biface fragments and expended points, which would explain their small size and irregular shape. Further, other tools, such as scrapers and gravers, were also likely manufactured from scavenged artifacts, which might explain the predominance of FGV and obsidian for tools that are more commonly made from chert. That chert was available to some extent, however, is evidenced by the 10 chert cores discussed below, although most of these cores are small.

Duke (2011:128), who has worked in the distal ORB to the north, argues that scavenging would not be expected to be a significant aspect of the technology of the region, even when combined with caching. He argues that the farther into the marsh people moved, the greater the risk of running out of raw material, especially as they would need to be able to find cached material. This would certainly be true of highly mobile people who were moving through the marsh rather quickly, but possibly not for later folks, who, Duke argues (and we agree), were less mobile, remaining in the marsh for longer periods of time. If people remained in the marsh for extended lengths of time, however, they would learn the landscape and likely be more able to relocate caches made either days or seasons before. It may be that there is not as much evidence of recycling in the distal ORB as we have observed in the proximal ORB assemblage. As noted above, Dugway Stubbies are the primary category we believe to be largely the result of recycling, making up ~21 percent of the projectile point assemblage. But these points represent less than 9 percent of the distal ORB assemblage (Duke 2011:193, Table 14). In contrast, however, Duke's "Bonneville" category (which appears to subsume our Square, Expanding, and Contracting Stem categories) and Pinto points constitute nearly 50 percent of his assemblage, whereas they represent less than 25 percent of the proximal ORB assemblage. Pinto is especially rare (4 percent). These differences may, in part at least, be explained by the use of different criteria between Duke and ourselves in assigning points to types, but probably not completely. Since we have considered Stubby to be an Early Holocene type along with Pinto, BVCN, and our short stemmed types, these differences may indicate, on the one hand, that Stubbies are different in age from the other EH types or, on the other hand, that recycling was just not as prevalent in the distal ORB. In any event, we believe that scavenging and caching *did* make up an important component of the technology during at least some period of time in the ORB marshes.

Seriation Analysis

Two seriation orders have previously been produced of ORB sites, the first by Jones, Beck, and Kessler (2003) for ORB1 and the second by Beck and Jones (2010b) for ORB2. Both are based on projectile point types. Here we present a final seriation order for the combined ORB1 and ORB2 assemblages.

Seriation is commonly used to develop chronological orders of archaeological assemblages. An order produced by seriation is one in which units (such as sites, strata, etc.) are ordered such that each unit is more similar to its neighbors than it is to any other unit (Dunnell 1970; Marquardt 1978). The similarity of units is judged on the basis of a set of temporally sensitive classes of artifacts, most often pottery types. The method is based on three assumptions: (1) The distribution of an artifact class is continuous through time; (2) the distribution of class frequencies forms a unimodal curve through time; and (3) assemblages that are most similar to one another in terms of the classes represented and their relative frequencies are the closest in time (Dunnell 1970).

An order produced by seriation, however, is not inherently chronological. For the order to be considered a chronological one, three conditions must be met (Dunnell 1970). First, the units being ordered must be of the same temporal duration. If, for example, one site represents 100 years of repeated occupation while the others represent just a few years, that site will cover multiple time periods represented individually by the other sites. As a consequence a greater number of types will be represented in its assemblage, and thus it will not fit anywhere in the seriation order.

Second, the units must come from the same geographical area. This means that the set of types used must be relevant to all of the assemblages to be ordered. That is, each type must be present in at least two assemblages. Finally, the artifact classes must be "historically significant" (Krieger 1944), meaning that they are based on attributes that vary more in time than in space. Historically significant classes are generally based on stylistic rather than functional attributes because style behaves in a predictable manner through time (i.e., unimodally) while this is not necessarily the case for function (Dunnell 1978). Some functional attributes, however, *are* temporally sensitive. For example, it has been demonstrated repeatedly that Great Basin Archaic projectile point types, which are based primarily on functional attributes (Beck 1998), behave historically (e.g., Thomas 1981) and can thus be successfully used in a seriation analysis.

A shortcoming of projectile points for use in seriation analysis is that they rarely occur in large numbers in the way that pottery sherds do. Thus, smaller samples must be used, which can create problems if the number of classes is large. In very small samples the presence of a single point of a particular type has much more impact than it does in a much larger sample; this can cause deviations from the seriation model (i.e., the unimodality of the type frequency distributions). Also, a site with a larger sample of points is likely to have more types represented than one with a small sample, simply because of sample size (see Jones et al. 1983), even if it represents the same duration of occupation as the other sites. Even so, successful seriation orders have been produced using projectile points (e.g., Beck 1984, 1998, 1999; Oetting 1994).

The first successful seriation using projectile point types in the Great Basin was presented by Beck (1984), in which 18 surface sites from the Steens Mountain region of southeastern Oregon were ordered using 16 projectile point types. Although more than 100 sites were collected as part of the Steens Mountain Prehistory Project (Beck 1984), most had very few or no projectile points. Because so many projectile point types were used to create the order, the sites ordered were limited to those with 10 or more identifiable projectile points in the assemblage. For the final order, projectile point types were combined into series (for example, Elko Corner-notched and Elko Eared were combined into the Elko Series; Gatecliff Split-Stem and Gatecliff Contracting Stem, into the Gatecliff Series; etc.) to improve the seriation solution. The resulting order was extremely good, with all but the Humboldt Series conforming to the seriation model (see Beck 1999:177, Figure 15.2).

Creating a seriation of ORB sites posed more of a challenge. The seriation produced by Beck (1984) was based on Archaic types, which span the period from ~8500 ^{14}C BP to historic times. This is important because there is time for types to reach their maximum occurrence and then be slowly replaced by others. For example, in the Steens seriation the maximum occurrence of the Large Side-notched Series occurs near the lower end, after which it decreases in frequency. At the same time, Gatecliff frequency is increasing to a maximum and then begins to decline. As it declines, another series, Large Stemmed, which is short-lived, peaks, at which time Large Side-notched has all but disappeared. The important point here is that these series, although they overlap in their distributions through time, have different temporal ranges.

The ORB seriation, however, is based exclusively on *Paleoarchaic* types, spanning no more than 3,000 years and, as we discuss below, probably less. The ORB seriation is based primarily on only two groups of types: WST types, most of which span about 3,000 years, and Early Holocene types, which are present for about 1,500 years and overlap completely with WST types during this period. Thus the same pattern of successive type replacement through time is not possible in this order.

A second difficulty with the ORB seriation is sample size. If we set the minimum sample number of projectile points per site at 10, as Beck (1984) did, we would severely limit the number of sites that could potentially be included in the seriation. Therefore, we lowered the

minimum to five, which increased the number of potential sites but also likely caused some deviations from the seriation model in the eventual order. For example, the largest sample size is 28 points, and thus each point represents 3.5 percent of the site assemblage, where each point in a sample of five represents 20 percent.

The seriation order of 28 sites based on 11 projectile point types/series is shown in Figure 5.32. Given all of the problems outlined above, this order is quite good. First, the two predominant classes (Silver Lake and Stubby) are distributed unimodally with few deviations. Second, the maximum occurrence of Silver Lake, as well as that of GBCB, is near the bottom, and the maximum for Stubby is midway. The distribution of other types, however, is hampered by small sample size. For example, there are too few occurrences of Cougar Mountain and Haskett points to be meaningful. Because their distributions were similar, we combined these two types as well as Cougar Mountain/Haskett stems, but the sample size is still very small. This is true as well for the Lake Mohave, Parman, BVCN, Pinto, and Contracting Stem types. The distributions of Square Stem and Expanding Stem, however, even though there are gaps, are fairly good.

To provide some sense of chronological placement for the ORB1 order, Jones, Beck, and Kessler (2003) make comparisons with projectile points from Danger Cave. The assemblage from this site has no classic WST points, but it does contain specimens that are similar to Dugway Stubbies in DII, reflecting an early Holocene age. Points similar to the ORB Contracting Stem points occur in DII and DIII, which date from the early to middle Holocene (Jennings 1957; Madsen and Rhode 1990). Pinto points also are represented in the DII levels at Danger Cave, which suggests that this type was in use prior to ~9,000 ^{14}C years ago (Jennings 1957; Madsen and Rhode 1990). In addition, as noted earlier, BVCN points (North Creek Stemmed) have been dated to between 9690 ± 60 ^{14}C BP and 9510 ± 80 ^{14}C BP at North Creek Shelter in eastern Utah (Janetski et al. 2012:140).

It would appear from these comparisons that a conservative age estimate of the recent end of the seriation is ~9,500 to ~8,500 ^{14}C years ago. It is difficult to evaluate the length of time represented by the seriation order, but the fact that many of the types are distributed throughout the order suggests that it represents a relatively short period of time. For example, both Pinto and Dugway Stubby points are scattered throughout the seriation order, and Contracting Stem points occur midway and toward the top. Square Stem 1 points are also scattered throughout the order but are more prevalent in the upper third. We noted earlier that points of this

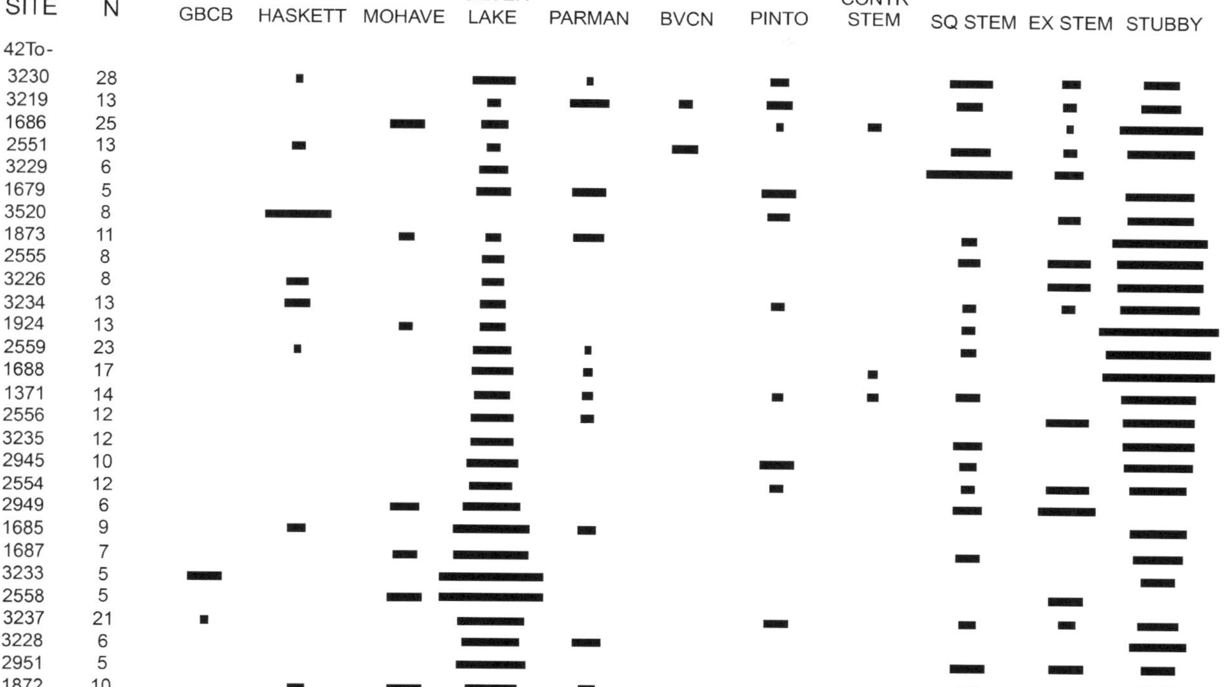

FIGURE 5.32. Seriation order of 28 Old River Bed site assemblages. GBCB = Great Basin Concave Base; BVCN = Butte Valley Corner-notched; CONTR = Contracting; SQ = Square; EX = Expanding.

sub-type are quite similar to those of the Windust type, which, as the dates in Table 5.12 suggest, could be terminal Pleistocene in age. Therefore, the oldest end of the order could be anywhere between ca. 10,800 and 10,000 ^{14}C BP, and thus the maximum time span is probably between 2,300 and 1,500 years.

CORRELATION OF ARCHAEOLOGICAL SITES WITH THE CHANNELS IN THE ORB DELTA

In Chapter 3 Madsen et al. discuss the formation and dating of various channels in the ORB delta. Channels are color-coded for discussion. Archaeological sites are associated with 11 of these channels (see Chapter 4). In an attempt to discern a temporal pattern of channel use by foraging populations, projectile points were summed by three type sets (Great Basin Concave Base, Western Stemmed, Early Holocene) for each color (Table 5.37). We discuss each channel separately.

Gold Channel

The Mango, Mocha, and Gold channels are the earliest in the channel sequence, but archaeological sites were found associated only with Gold. Based on dates on black mats within the channel, Madsen et al. (Chap-

ter 3) estimate the channel's age as between ~11,300 and ~10,500 ^{14}C BP, providing a limiting date range before which people were probably not present. A later date of ~9250 ^{14}C BP was obtained on a black mat that formed above the channel fill and spread out into the mudflats, suggesting an extensive marsh. Foragers may have used pockets of sand "islands" to make their way out into this marsh, although only two sites are associated with this channel. Table 5.37 shows the projectile point type sets represented at these sites. EH points are somewhat more prevalent than WST points, but the sample size is quite small and thus limits the interpretation of these results.

Black Channels

The Black channels are also believed to have formed relatively early, between ~11,000 and ~10,300 ^{14}C BP. At least nine inverted channels have been recognized in the ORB delta and probably provided causeways into the marshes that formed in the delta early in the Holocene. The largest number of archaeological sites ($n = 36$) are associated with these channels, although only 31 of these assemblages contain identifiable projectile points (Table 5.37). Since these channels are among the oldest

TABLE 5.37. Projectile Point Type Groups Represented in Site Assemblages by Channel Color.

Channel Color and Site	Point Type Group						Total
	Great Basin Concave Base		Western Stemmed Tradition		Early Holocene		
	n	%	n	%	n	%	
Gold							
42To3141			2	50	2	50	4
42To3142			2	40	3	60	5
Total			4	44	5	56	9
Black							
42To1356			1	100			1
42To1368			1	50	1	50	2
42To1369			6	86	1	14	7
42To1370			4	100			4
42To1371			6	35	11	65	17
42To1666			1	100			1
42To1668			1	50	1	50	2
42To1669			2	67	1	33	3
42To1672			4	80	1	20	5
42To1673			2	100			2
42To1681			1	25	3	75	4
42To1682			2	100			2
42To1684			8	67	4	33	12
42To1685			11	73	4	27	15
42To1686			14	44	18	56	32
42To1687			5	63	3	27	8
42To1688			5	29	12	71	17
42To1858					1	100	1
42To1859			1	25	3	75	4
42To1860			1	100			1
42To1861			6	67	3	33	9
42To1862			1	9	10	91	11
42To1872			26	84	5	16	31
42To1873			9	53	8	47	17
42To1874			1	33	2	67	3
42To1875			7	70	2	20	9
42To1876			1	33	2	67	3
42To1878			5	83	1	17	6
42To1920			3	60	2	40	5
42To1923			1	100			1
42To1924			5	31	11	69	16
Total			141	56	110	44	251
Yellow							
42To2951			4	50	4	50	8
42To2952			3	60	2	40	5
42To2953			1	50	1	50	2
42To2954			2	100			2
42To3219			8	40	12	60	20
42To3220			3	100			3
42To3221			3	75	1	25	4
42To3222			3	75	1	25	4
42To3223			2	67	1	33	3
42To3224			3	100			3
42To3225			3	60	2	40	5
42To3226			8	57	6	43	14
Total			43	59	30	41	73
Limestone							
42To1921			4	67	2	33	6
42To1922			1	33	2	67	3
Total			5	56	4	44	9
Green							
42To2551			7	37	12	63	19
42To2552			4	80	1	20	5
42To2553			3	60	2	40	5
42To2556			8	47	9	53	17
42To2557			1	50	1	50	2
42To2558			6	86	1	14	7
42To2559			13	43	17	57	30
Total			42	49	43	51	85

TABLE 5.37. (cont'd.) Projectile Point Type Groups Represented in Site Assemblages by Channel Color.

Channel Color and Site	Great Basin Concave Base n	Great Basin Concave Base %	Western Stemmed Tradition n	Western Stemmed Tradition %	Early Holocene n	Early Holocene %	Total
Red							
4DM02			1	100			1
4DM03			1	100			1
Total			2	100			2
Blue							
42To2554			5	33	10	67	15
42To2555			3	25	9	75	12
42To3228			3	50	3	50	6
42To3229			3	38	5	62	8
42To3231			2	67	1	33	3
42To3233	1	9	7	64	3	27	11
42To3234			5	26	14	74	19
42To3235			15	54	13	46	28
Total	1	1	43	42	58	57	102
Lime							
42To3520			12	63	7	37	19
42To3522			2	40	3	60	5
Total			14	58	10	42	24
Lavender							
42To2943			1	50	1	50	2
42To2944			1	100			1
42To2945			3	27	8	73	11
42To2946			1	50	1	50	2
42To2947	2	40			3	60	5
42To2948			2	50	2	50	4
42To2949			3	43	4	57	7
42To3236					1	100	1
42To3237	1	4	8	33	15	63	24
42To3238			4	80	1	20	5
Total	3	5	23	37	36	58	62
Buff							
42To3140			5	100			5
Total			5	100			5
Brown							
42To1689					1	100	1
Total					1	100	1
Light Blue							
42To1000			8	89	1	11	9
42To1153	1	8	9	75	2	17	12
42To1157			1	100			1
42To1161	1	25	1	25	2	50	4
42To1163			1	50	1	50	2
42To1166			1	100			1
42To1172			1	100			1
42To1182			8	73	3	27	11
42To1357			2	50	2	50	4
42To1358			8	62	5	38	13
42To1383			11	92	1	8	12
42To1671			4	80	1	20	5
42To1674			1	100			1
42To1675					2	100	2
42To1677			1	50	1	50	2
42To1678			1	33	2	67	3
42To1679			3	43	4	57	7
42To2767			4	100			4
Total	2	2	65	69	27	29	94

Note: Channels are arranged from oldest (Gold) to youngest (Light Blue). Percentages are rounded to the nearest whole number.

and largest, it is not surprising that there are large numbers of WST and EH points in assemblages associated with them, with WST being the most abundant.

Yellow Channel

One date of ~10,130 ^{14}C BP was obtained from a sample in the upper fill of this channel. Thirteen sites are associated with the Yellow channel, 12 of which have identifiable projectile points (Table 5.37). WST points are predominant here, representing nearly 59 percent of the point sample.

Limestone Channel

The Limestone channel, which could only be traced a short distance, is estimated to have formed between ~10,500 and ~10,000 ^{14}C BP, based on a date obtained on *Sphaerium* sp. shells. Only two sites are associated with this channel, and both have identifiable projectile points (Table 5.37). As in the case of the Gold channel, the sample size is small, and WST points predominate over EH points.

Green Channel

The Green channel is believed to have formed in ~10,290 ^{14}C BP. This date and a later date of ~10,000 ^{14}C BP are from samples taken from black mats in the channel fill, and Madsen et al. (Chapter 3) suggest that these mats may represent groundwater-fed marsh growth within an abandoned channel. If this is the case, people could have utilized the resources within this marsh by at least ~10,000 ^{14}C BP. Seven sites are associated with the Green channel (Table 5.37). Here we see a shift to roughly equal proportions of WST and EH points on the average, but several sites have great numbers of WST points, while for others the opposite is the case.

Red Channel

Although geomorphically uninvestigated, four archaeological locations are associated with the Red channel, two of which have one WST point each (Table 5.37).

Blue and Lime Channels

A black mat in the fill of the Blue channel ranges between ~9750 and ~9450 ^{14}C BP. Madsen et al. (Chapter 3) state that although it has yielded no radiocarbon dates, the Lime channel may be the same age. Eight sites associated with identifiable projectile points were located along the Blue channel, and two, along the Lime channel (Table 5.37). Most of these points are of either WST or EH types, although one assemblage contains a GBCB point. As is evident from Table 5.37, for three Blue channel sites and one Lime channel site WST

points are in the majority, while for five other sites the opposite is the case. Overall, however, EH points are slightly more abundant for Blue channel sites, and WST points, for Lime channel sites. If the sites from these two channels are combined, EH points prevail (54 percent).

Lavender Channel

Only one date, ~9010 ^{14}C BP, is associated with the Lavender channel, and Madsen et al. (Chapter 3) state that it is unclear whether it dates channel formation or groundwater flow. Twelve sites, 11 with identifiable projectile points, are associated with this channel (Table 5.37). Interestingly, three of the eight GBCB points in the ORB assemblage were found at sites along this channel. EH points increase to 60 percent for sites associated with this channel.

Buff Channel

The Buff channel is believed to have formed about ~9200 ^{14}C BP and is associated with only one site (Table 5.37). This assemblage contains five WST points.

Brown Channel

The Brown channel is younger than the Buff channel, since the latter is crosscut by the former. No radiocarbon dates are given, however. One site is associated with this channel, and it contains one EH point (Table 5.37).

Light Blue Channel

There are six dates for the Light Blue channel, ranging from ~9850 to ~8800 ^{14}C BP, "suggesting a prolonged period of channel filling by postflow groundwater-fed marsh systems" (Madsen et al., Chapter 3, this volume). Although 26 sites are associated with this channel, only 19 of the assemblages contain identifiable projectile points (Table 5.37). Interestingly, there is a sharp shift here in the proportion of WST vs. EH points, with the former accounting for 69 percent of the total, the largest representation of WST points among all of the channels. The preponderance of WST forms and the paucity of EH points are odd, especially given that this channel was the latest to form.

Discussion

Theoretically, assuming that all channels were equally productive and accessible from their inception until ~8500 ^{14}C BP (which was probably not the case), and also that the maximum occurrence of WST points was during the latest Pleistocene and the very early Holocene, these points would be expected to constitute greater proportions of the earliest channels, and EH points should make up the greatest proportion of the

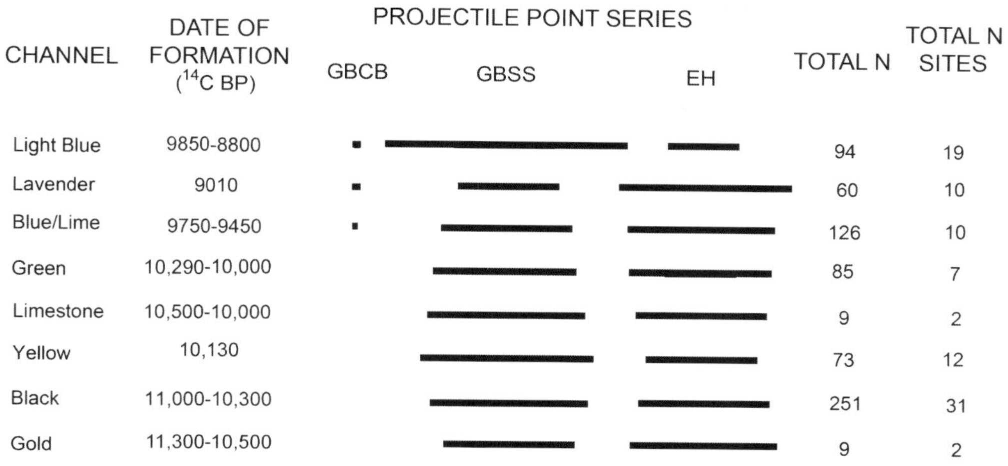

CHANNEL	DATE OF FORMATION (^{14}C BP)	PROJECTILE POINT SERIES			TOTAL N	TOTAL N SITES
		GBCB	GBSS	EH		
Light Blue	9850-8800				94	19
Lavender	9010				60	10
Blue/Lime	9750-9450				126	10
Green	10,290-10,000				85	7
Limestone	10,500-10,000				9	2
Yellow	10,130				73	12
Black	11,000-10,300				251	31
Gold	11,300-10,500				9	2

FIGURE 5.33. Summary schematic of channel dates and associated projectile points. GBCB = Great Basin Concave Base; GBSS = Great Basin Stemmed Series; EH = Early Holocene.

later channels as the use of WST forms waned. Figure 5.33 presents a summary schematic of the dated channel information and associated projectile points. Leaving aside those channels that have very small samples of points (Brown, Buff, and Red), there does appear to be a progression from a greater use of WST points early on to a greater use of EH points later, except, of course, for the Blue channel. In spite of this trend, however, there are still complicating factors.

First, these channels are not isolated from one another but crisscross in a rather complex pattern. Thus, an individual harvesting resources along one channel might have encountered a junction with a second and proceeded along that channel. This would complicate the temporal pattern.

Second, ^{14}C dating indicates that many channels formed over long periods of time, as open channels with flowing water, abandoned channels with open quiet water, shallow water marshes, and wetland meadows. Thus, sites along these channels may have formed as palimpsests of occupations associated with differential use of the varying stages of channel development.

Third, WST and EH points do not occur during separate periods, such as is the case for Archaic types. The use of EH points is completely contained within the period of later use of WST points for the period under consideration. This, of course, is one of the factors that is problematic for the seriation order.

Finally there is the issue of reuse and recycling, which also has implications for the seriation order. We have argued that considerable reuse and resharpening went on during the human occupation of the ORB delta, evidenced by (1) the extensive resharpening of projectile points; (2) the transformation of some projectile points into other tools, such as scrapers and gravers; and (3) the manufacture of other point forms (such as Stubbies) from artifacts that had been discarded by previous visitors. Such reuse and recycling implies that discarded points and bifaces that were large enough to reuse or to refashion into other forms would have been retrieved by new visitors upon discovery, especially as marshes became more extensive and foragers remained in them for longer periods of time. Therefore, an assemblage may contain both WST points and Stubbies, but this does not necessarily mean that the people responsible for the assemblage actually *manufactured* the WST points but, instead, may have *scavenged* them for future use. Depending upon how prevalent this practice was, the implication for the seriation is that it may not be telling us what we think it is.

BIFACES

Exclusive of projectile points, 596 bifaces were identified in the ORB assemblage (Tables 5.2, 5.4, and 5.38). In this section production bifaces, knives, drills, unique bifaces, amorphous bifaces, and indeterminate bifaces will be discussed. Crescents, scrapers, gravers, and chisels will be discussed in subsequent sections.

Production Bifaces and Biface Reduction

The most characteristic artifacts identifiable with Paleoarchaic assemblages are large contracting-stem points, such as Cougar Mountain, which are the end products of a production sequence that, in assemblages we have examined, is so standardized that it is identifiable even in those that contain no finished points (Beck and Jones

TABLE 5.38. Bifaces in the Old River Bed Assemblage by Raw Material.

| | Chert | | Obsidian | | Fine-Grained Volcanics | | Quartzite | | Total | |
Biface Category	n	%	n	%	n	%	n	%	n	%
Amorphous biface	4	6.7	13	21.7	40	66.7	3	5.0	60	100.0
Bifacial scraper	7	10.1	19	27.5	42	60.9	1	1.4	69	100.0
Bifacial graver	3	13.6	8	36.4	11	50.0			22	100.0
Bifacial scraper/graver	2	40.0	1	20.0	2	40.0			5	100.0
Bifacial chisel			1	33.3	2	66.7			3	100.0
Bifacial chisel/scraper					1	100.0			1	100.0
Cody knife	1	12.5			6	87.5	1	12.5	8	100.0
Crescent[a]	7	28.0	5	20.0	13	52.0			25	100.0
Drill	3	15.8	6	31.6	10	52.6			19	100.0
Knife	1	50.0			1	50.0			2	100.0
Production biface	7	5.8	9	7.4	103	85.1	2	1.7	121	100.0
Unique biface					5	100.0			5	100.0
Indeterminate biface	23	9.0	107	41.8	122	47.7	4	1.6	256	100.0
Total	58	9.7	169	28.4	358	60.1	11	1.8	596	100.0

[a] Two crescents are unifacial and thus not included here.

TABLE 5.39. Summary of Quantitative Data for Production Bifaces, Knives, Drills, Unique Bifaces, and Amorphous Bifaces in the Old River Bed Assemblage.

| | | Class | | | | | |
Variable	Statistic	Production Biface	Cody Knife	Knife	Drill	Unique Biface	Amorphous Biface
Total length (mm)	n	10	5	—	6	2	10
	Range	36.3–127.5	63.4–100.7		21.6–74.3	40.5–53.8	29.0–74.9
	Mean	74.37	77.30		40.65	47.15	47.81
	SD	31.962	14.450		18.909	9.405	15.413
Maximum width (mm)	n	15	6	1	8	2	14
	Range	19.1–54.9	20.6–45.7		12.8–35.4	20.9–33.0	14.9–52.8
	Mean	32.06	33.58	52.0	20.64	26.95	29.80
	SD	10.352	8.967		9.014	8.556	10.138
Maximum thickness (mm)	n	121	8	2	18	5	60
	Range	4.8–18.3	6.8–10.6	8.4–9.8	3.1–10.1	6.9–13.6	3.6–19.9
	Mean	9.24	8.84	9.10	6.82	8.82	10.08
	SD	2.383	1.309	.990	1.903	2.864	3.987
Actual weight (g)	n	121	8	2	18	5	60
	Range	3.1–68.5	8.2–54.1	19.2–37.9	.7–19.1	6.8–24.8	2.3–60.8
	Mean	13.90	23.60	28.51	5.27	14.57	16.69
	SD	11.284	14.531	13.237	4.792	7.934	14.905

2009; Beck et al. 2002; Jones, Beck, and Kessler 2003). These bifaces, and thus stemmed points, are made predominantly from FGV toolstone but also of obsidian where it is available; they are much less often made from chert. In this respect the ORB production biface assemblage does not differ from other Paleoarchaic assemblages across the Great Basin (Table 5.38).

There are 121 production bifaces in the ORB as-semblage. The descriptive statistics for these artifacts are presented in Table 5.39, and their site locations are given in Table 5.40. As Table 5.38 shows, 103 (85.1 percent) are manufactured from FGV. The majority of the 200 ORB stemmed points that could be identified with types, however, are made not from FGV (35.5 percent) but of obsidian (62.5 percent; see Tables 5.10 and 5.15). On the other hand, the opposite is the case for the 289

TABLE 5.40. Site Locations for Old River Bed Amorphous Bifaces, Knives, Drills, Production Bifaces, and Unique Bifaces.

Site	Class					
	Amorphous Biface	Cody Knife	Hafted Knife	Drill	Production Biface	Unique Biface
42To1000	1			2		
42To1157				1		
42To1168						1
42To1172					1	
42To1173	2					
42To1178				1		
42To1352	1					
42To1354					1	
42To1358	1			1		
42To1368	1				1	
42To1369					3	
42To1370					2	
42To1371	3				6	
42To1383					3	
42To1385					1	
42To1666	2				1	
42To1668					2	
42To1676						
42To1677				2	2	
42To1678	1				2	
42To1679					1	
42To1680	2				1	
42To1682					3	
42To1683					3	
42To1685			1	1	4	
42To1686	4				3	
42To1687						
42To1688	1				4	
42To1689					1	
42To1859					1	
42To1861	2					
42To1872	2	1			2	
42To1873	1				1	
42To1874					1	
42To1875					3	
42To1876	1				3	
42To1877	1					
42To1878					2	
42To1920	5			1	3	1
42To1921	1				1	
42To1922					2	
42To1924	4			2	5	
04DMO3					1	1
07DM01	1					
42To2475			1			
42To2551					1	
42To2554					2	
42To2556					2	
42To2559	2					
42To2766					1	
42To3140					1	
42To3142	1				6	
42To3219	2			1		
42To3220					1	
42To3222	1			1		
42To3223						1
42To3224	1				1	
42To3226	1				3	
42To3228	2					
42To3233					1	
42To3234					2	
42To3235		1			2	
42To3237				1		
42To3238					1	
42To3520					1	
Isolates	13	10		5	26	1
Total	60	12	2	19	121	5

Note: Locations for scrapers, gravers, chisels, and combination tools are given in sections specific to these categories.

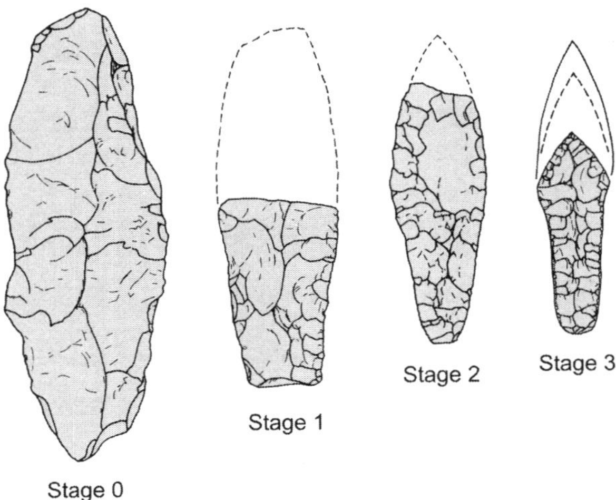

FIGURE 5.34. Biface reduction sequence used to produce Great Basin Stemmed Series projectile points (after Beck and Jones 2009:90, Figure 5.3).

TABLE 5.41. Biface Reduction Stage Definitions.

Stage	Definition
0	*Early-stage reduction.* Overall plan (shape) can be highly irregular and amorphous, flake scars are few and are widely spaced, and offset (sinuosity of edge) is extremely wide or nonexistent.
1	*Middle-stage reduction.* Overall plan is somewhat irregular to semisymmetrical, flake scars are widely and/or variably shaped, and offset is wide.
2	*Late-stage reduction.* Overall plan is semisymmetrical to symmetrical, flake scars are somewhat closely and/or semiregularly spaced (pressure flaking may appear but is not common), and offset is moderate.
3	*Finished point.* Overall plan is regular and symmetrical, flake scars are closely and/or quite regularly spaced, and offset is close; lateral edge grinding is evident on the stem.

Source: After Callahan 1979.

WST stems and blades, 177 (61.2 percent) of which are made from FGV toolstone. These patterns could indicate that finished obsidian points were brought into the area and replaced in FGV toolstone, although obsidian is available from the Topaz Mountain source less than 100 km to the south (Figure 5.3).

Although most Paleoarchaic assemblages contain bifaces from the stemmed point production sequence, they differ with respect to the degree to which these bifaces were reduced. For example, at Limestone Peak Locality 1, a site in Jakes Valley of eastern Nevada (Beck and Jones 1990a, 2009; Beck et al. 2002), the bifaces represent primarily middle- and late-stage reduction, whereas at the Knudtsen site in Grass Valley, central Nevada, bifaces are more indicative of early- to middle-stage reduction (Beck et al. 2002). These bifaces also vary considerably in size from site to site, depending primarily on the original flake blank but also on subsequent breakage and resharpening.

Previously we have used two measures to examine the degree of reduction represented in an assemblage: a traditional stage sequence modified from that originally designed by Callahan (1979:10–11) and a thinning index devised by Johnson (1981). In theory, a biface is processed through several stages of reduction before it reaches the form of the final product—in this case, a stemmed point. But as Muto (1971) pointed out many years ago, biface reduction takes place along a continuum, and thus stage assignment can be somewhat arbitrary. The advantage of stage classifications, however, is that they are generally based on multiple criteria, typically taking into account variation in plan and cross-section symmetry, edge linearity, distribution and scale of flake scars, and so forth, all features that change directionally in biface production. Thus, the assignment of an artifact to a stage is not based on a single variable.

The Johnson Thinning Index (JTI), on the other hand, has the advantage of measuring reduction along a continuum but is based on only two variables: weight and plan area. This index, which is the ratio between weight and plan surface area, is based on the assumption that the knapper will maximize biface surface area while minimizing thickness, and thus early-stage bifaces will be thicker, and thus heavier, than those of later stages. As Johnson (1981:13) points out, however, thickness depends on initial blank size, and thus a small early-stage biface may have the same thickness as a large late-stage biface. The ratio of weight to plan surface area eliminates the effects of initial blank size and therefore provides a more useful way to measure thinning during the reduction process. As reduction proceeds, then, the value of the index decreases. This index has an advantage over single measures of size, such as thickness, length, and width, since it can be applied to the fragmentary specimens that make up the majority of archaeological assemblages. These two independent indices, stage assignments and JTI, complement one another and act to limit bias that may be introduced by simply using one or the other.

The stage classification (Figure 5.34; Table 5.41) creates four stages of reduction. The stage definitions, first presented by Beck et al. (2002), are based on form, number and shape of flake scars, edge sinuosity, and thickness. These definitions, however, are qualitative

TABLE 5.42. Old River Bed Production Bifaces by Stage and Raw Material.

Extent of Reduction	Biface Stage	Raw Material				
		Chert	Obsidian	Fine-Grained Volcanics	Quartzite	Total
Early stage	.0		1			1
Early stage	.5			1		1
Middle stage	1.0		2	7		9
Middle stage	1.5	3		17	1	21
Late stage	2.0	2	1	23		26
Late stage	2.5		1	1		2
Subtotal		5	5	49		60
Indeterminate	Indeterminate	2	4	54	1	61
Total		7	9	103	2	121

rather than quantitative, and thus the assignment of a biface to a stage category remains subjective. That is, one analyst may emphasize edge sinuosity over form, while another may do just the opposite. In the analysis reported by Beck et al. (2002), for example, each biface was examined by two analysts, who occasionally disagreed on the stage assignment, although assignments were never more than one stage apart. In these cases the two assignments were averaged; if the two assignments were stages 1 and 2, for instance, the biface was designated as stage 1.5. The averaged category, however, indicates only that the biface has been reduced to a greater degree than stage 1 but to a lesser degree than stage 2. We use this expanded set of stage categories here, assigning bifaces to .5, 1.5, and 2.5 when their degree of reduction appeared to be in between the standard stage categories (e.g., 0, 1, and 2). In general, then, we consider stages 0 and .5 to represent early-stage reduction; 1 and 1.5, middle-stage reduction; and 2 and 2.5, late-stage reduction.

The subjectivity of biface stage analysis can also lead the analyst to make assignments differently over time. Since we do not conduct this type of analysis on a regular basis, to ensure that our assignments for the ORB bifaces were consistent with those we have made in the past, we examined photographs of the 330 bifaces from Little Smoky Quarry and their stage assignments that were reported by Beck et al. (2002).

Because of the fragmentary nature of the ORB production bifaces as well as the considerable weathering damage they exhibit, only 60 of the 121 production bifaces could confidently be identified with stage of manufacture (Figure 5.35; Table 5.42). As Table 5.42 shows, sample sizes of chert, obsidian, and quartzite bifaces are very small, and stage assignments do not differ markedly across raw material types. Therefore, bifaces of different materials were combined for subsequent analyses. The majority (78.3 percent) of the 60 bifaces are in stages 1.5 and 2.0, suggesting middle- to late-stage reduction.

As noted above, the Johnson Thinning Index is the ratio of weight to plan surface area. Instead of calculating the actual plan area of each biface, a time-consuming exercise, size category, which is based on plan area, was used as a proxy measure in the calculation of a modified thinning index (MTI) (see Beck and Jones 2010b; Jones, Beck, and Kessler 2003). Each successive category is 1.5 times that of the previous one (Figure 5.1). For the computation of the MTI, size categories were converted to their corresponding plan area (termed size plan area; Table 5.43), and a ratio of weight to size plan area was then computed. Although not quite the same as the JTI but more like "average" JTI values, the results of the MTI are, in general, analogous to those of the JTI.

MTI was calculated for all 121 ORB production bifaces. For comparison, MTI was also computed for the 105 of the 135 WST point blades (blades of WST points that could not be identified as to type; see Tables 5.10, 5.14–5.15). By definition, these point blades represent the final stage of reduction (stage 3). Figure 5.36 shows the distribution of MTI for production bifaces (A) and WST blades (B). As would be expected, the mean for production bifaces (1.083) is somewhat larger than that for WST blades (.807), a difference that is significant (pooled-variance $t = 7.204$; $df = 225$; $p < .001$), but there is considerable overlap between the two distributions. This suggests that the ORB production bifaces represent primarily mid- to late-stage reduction, supporting the results of the stage classification.

The few production bifaces that are complete range between 36.5 and 127.5 mm in length (Table 5.39), which is comparable to bifaces in other Paleoarchaic assemblages. However, given that the largest complete Cougar Mountain point is 127.8 mm in length (see Table 5.14), there had to have been much larger bifaces in

FIGURE 5.35. Examples of Stage 1 (*A*), Stage 1.5 (*B–C*), Stage 2 (*D–E*), and Stage 2.5 (*F*) production bifaces in the Old River Bed assemblage: (*A*) 42To2554, FS 30; (*B*) 42To3140, FS 7; (*C*) DPGIF 2411; (*D*) 42To1682, FS 1 and 2; (*E*) 42To3142, FS 34; (*F*) DPGIF 1632.

TABLE 5.43. Corresponding Plan Area for Size Categories Recorded in the Old River Bed Artifact Analysis.

Size Category	Plan Area (cm²)
1	.296
2	.444
3	.667
4	1.000
5	1.500
6	2.250
7	3.375
8	5.063
9	7.594
10	11.390
11	17.090
12	25.630
13	38.445
14	57.668
15	>57.668

the proximal ORB at some time. A few large early- to middle-stage production bifaces from this region are known to be in the hands of private collectors. One such specimen, which is approximately 23 cm long (Figure 5.37), was manufactured on a large flake blank with most of the flaking done on a single side. This biface and one quite similar to it in the same collection further attest to early-stage biface reduction in the proximal ORB.

Another large early- to middle-stage production biface, 189 mm long, was recovered from the distal ORB (Figure 5.38A; see Duke 2011). Although it is relatively thin, this biface is similar in form and size to early- to middle-stage bifaces in other Paleoarchaic assemblages, such as those shown in Figure 5.38B–C, from Little Smoky Quarry in eastern Nevada and Cowboy Rest Creek Quarry in central Nevada, respectively. Side 1 of the ORB biface exhibits a single facet that initiates from the right edge and intercepts the left edge, superficially resembling an overshot scar, but small step fracture scars at what would be the negative bulb make it impossible to discern this morphology. We believe, however, that this large "scar" actually represents the ventral side of the original side-struck flake blank from which the biface was manufactured. Large side-struck flakes detached from boulders were commonly used as blanks for the manufacture of production bifaces (see Bennett and Jones 2012). Figure 5.39A shows such a side-struck flake that was detached from a boulder at the Cowboy Rest Creek Quarry. Flake scars on an FGV boulder at the Cedar Mountain B quarry (Figure 5.39B) immediately east of the ORB delta may be the result of such initial flake removal.

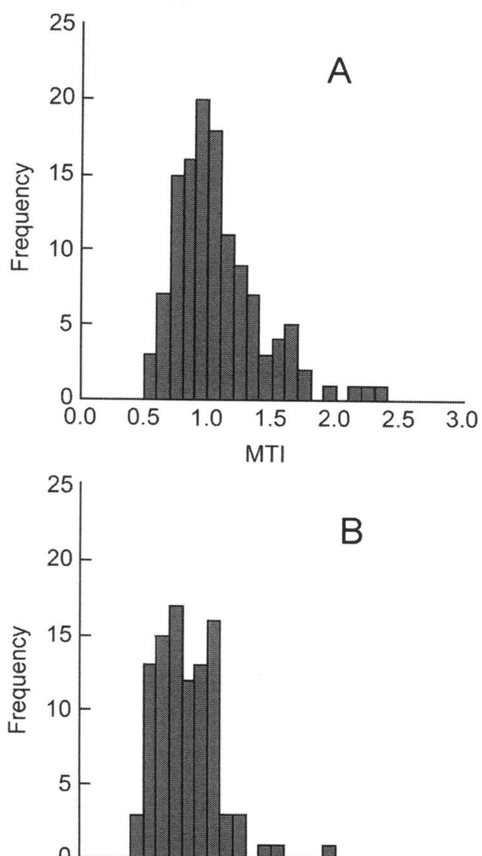

FIGURE 5.36. Distribution of modified thinning index (MTI) values for Old River Bed production bifaces and Western Stemmed Tradition (WST) blades.

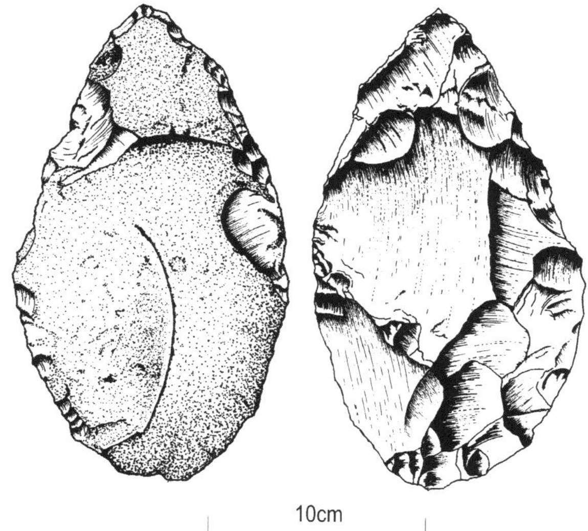

FIGURE 5.37. Early- to middle-stage biface collected from the distal Old River Bed delta (in a private collection). Drawing by Sergio Ayala.

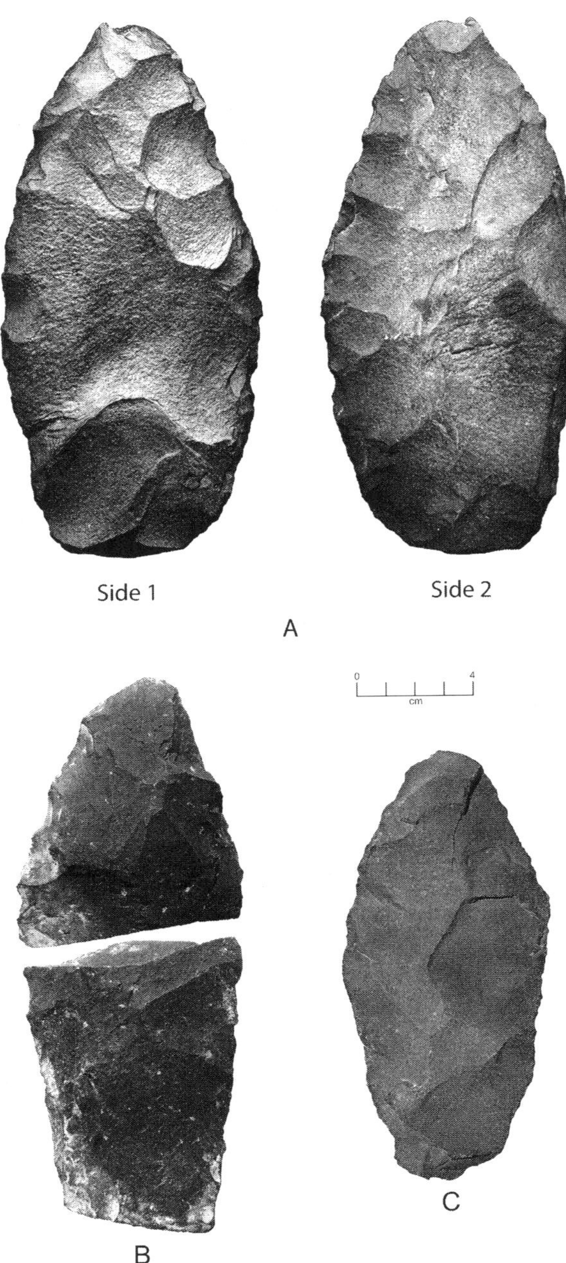

Figure 5.38. Early- to middle-stage bifaces from the distal Old River Bed delta (*A*), Little Smoky Quarry in eastern Nevada (*B*), and Cowboy Rest Creek Quarry in central Nevada (*C*).

Even though there are few early-stage bifaces in the ORB assemblage, the presence of flakes with cortex, several of which are quite large, suggests that some primary reduction took place in the delta. However, a flake sample of only 310 specimens limits what can be discerned concerning stages of reduction. Debitage at each site was generally counted and recorded but not collected (see Chapter 4). For the most part there was no effort to describe technological attributes in the field, which limits our ability to describe how Paleoarchaic foragers in the ORB delta used the stone materials available to them. With this proviso, however, we present limited analysis results of the small sample of flakes available to us.

Of the 310 flakes in the assemblage, 28 have cortex on their dorsal surface (we note that the amount of cortex varies; the dorsal surface is not necessarily completely covered with cortex). Most of the flakes are incomplete, and thus size category was recorded for only 129, 12 of which are cortex flakes (Table 5.44). For the most part, flake sizes are consistent with mid- to late-stage reduction, especially in light of the size of some of the late-stage bifaces (see Figure 5.35). There are 17, however, that are quite large (greater than size 10), most of which are FGV, and five of these are cortex flakes; two are size 11 (one chert, one FGV), and three are size 13 (all FGV). Although these flakes could have resulted from the early-stage reduction of production bifaces, they could just as easily be from the reduction of generalized cores for the production of other tools, such as scrapers. In most Paleoarchaic assemblages these tools are generally made from chert, but they are mostly manufactured from FGV in the ORB (see Tables 5.2 and 5.4). We examine both possibilities below.

In the analysis of flakes, several attributes were recorded that we believed might yield information concerning the type of reduction, biface or general core: type of platform (cortex, single-facet, multifacet), platform lipping (present, absent), and platform angle (90°, <90°). Biface-reduction flakes (BRFs) are generally defined as having multifaceted platforms, platform lipping, and acute platform angles. In our analysis of the 1,847 flakes with platforms from the Sunshine Locality, we examined the co-occurrence of all pairs of these attributes for cortex, single-facet, and multifacet platforms using chi-square. The combination of acute platform angle and the presence of platform lipping was significant for both single- and multiple-facet platforms across all material types. Consequently we expanded the definition of BRF to include single-facet platforms with platform lipping and acute platform angles (see discussion in Beck and Jones 2009:91–95). Although the ORB flake assemblage is far from a statistically representative sample of debitage, we use it here to make some preliminary observations regarding the type and stage of core reduction.

Only 64 of the 310 ORB flakes have platforms. Of these platforms, five are cortex, 10 are single-facet, and 43 are multifacet. A total of 38 flakes with single-

A

B

FIGURE 5.39. Fine-grained volcanic boulders at the Cowboy Rest Creek Quarry in central Nevada (*A*) and the Cedar Mountain B quarry immediately east of the Old River Bed delta (*B*).

or multiple-facet platforms have both acute platform angles and lipping and thus can be characterized as BRFs (Table 5.45). Eleven flakes with single- or multiple-facet platforms have platform angles near 90° and no lipping; these were characterized as general core flakes (GCFs).

Six additional flakes were designated as possible BRFs, two with multiple-facet platforms and acute platform angles but indeterminate lipping, one with a partial platform exhibiting both an acute platform angle and lipping, and three without platforms but exhibiting

TABLE 5.44. Size Categories of Old River Bed Complete/Nearly Complete Flakes.

Flake Type and Size Category	Chert	Obsidian	Fine-Grained Volcanics	Quartzite	Total
Cortex Flakes					
6			2		2
8	1	3			4
9	1				1
10					
11	1		1		2
13			3		3
Total	3	3	6		12
Interior Flakes					
2			1		1
4	1		2		3
5	1	1			2
6		9	5		14
7	2	6	17		25
8	4	10	14		28
9	2	7	13		22
10	2		8		10
11			5		5
12			5		5
13				2	2
Total	12	33	70	2	117

TABLE 5.45. Biface-Reduction and General Core Flakes Represented in the Old River Bed Debitage.

Platform Type	Cortex	Biface Reduction	Possible Biface Reduction	General Core	Possible General Core	Indeterminate	Total
Cortex	5						5
Single facet		3		7			10
Multifacet		35	2	4		2	43
Partial			1			1	2
Missing			3		4	243	250
Total	5	38	6	11	4	246	310

both acute platform angles and lipping. Four flakes without platforms but with angles near 90° and no lipping were designated as possible GCFs. Finally, three flakes with complete or partial platforms could not be characterized.

In sum, 44 (68.9 percent) of 64 flakes characterized are probably the result of biface reduction, while 15 (23.4 percent) are likely general core-reduction flakes. The percentage of BRFs varies by material: 25 (78.1 percent) of 32 FGV flakes, 15 (75.0 percent) of 20 obsidian flakes, and four (57.1 percent) of seven chert flakes are the result of biface reduction. Three of the large flakes previously discussed are BRFs (one size 11, two size 12), and one, size 13, is a GCF (Table 5.46). All of these are FGV. Although the sample of flakes is not representative, these results suggest that both early-stage biface and general core reduction likely took place in the ORB.

A cumulative graph of BRF and GCF size shows that, overall, obsidian flakes are smaller than those of chert and FGV (Figure 5.40). This is consistent with the hypothesis that finished obsidian points were brought into the ORB, refurbished, reused, and eventually discarded, although flakes of a much smaller size would be expected if this were the case. However, it is possible

TABLE 5.46. Size Categories and Raw Material for Biface-Reduction and General Core-Reduction Flakes in the Old River Bed Assemblage.

Flake Type and Size Category	Raw Material			
	Chert	Obsidian	Fine-Grained Volcanics	Total
Biface Reduction[a]				
5		1		1
6		3		3
7		1	2	3
8	2	1	3	6
9	1	2	3	6
10			2	2
11			1	1
12			2	2
Total	3	8	13	24
General Core Reduction[b]				
6		1	1	2
7			1	1
8			1	1
9	1	1	1	3
13			1	1
Total	1	2	5	8

Note: Size category could be measured for only 32 of the 59 biface-reduction and general core flakes.
[a] Includes possible biface-reduction flakes.
[b] Includes possible general core flakes.

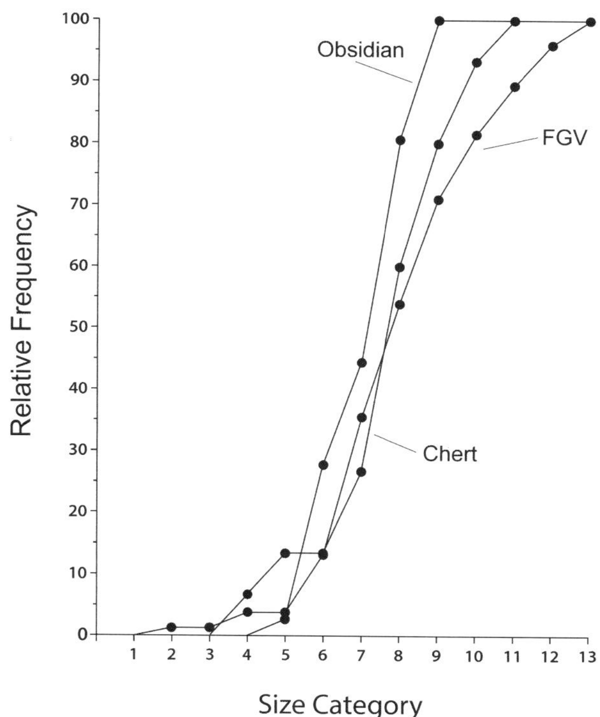

FIGURE 5.40. Cumulative frequency graphs of FGV, obsidian, and chert biface-reduction flake and general core flake size categories.

that smaller flakes were no longer there to be collected. Experiments conducted by Young (2008) indicate that while larger artifacts remain anchored in place on the ORB mudflats, small flakes (<.5 cm wide) may have been rapidly removed from surface sites once they were exposed to wind deflation. This suggests that the size categories of flakes in recorded surface sites have been biased by the likely removal of the smallest flakes. This is undoubtedly true to some extent, but other work by Carter et al. (2004) suggests that even on sites exposed relatively recently in dune blowouts, roughly half of all flakes fall in the 2- to 5-cm size range.

Although the meager flake evidence indicates that at least some early-stage biface reduction took place in the ORB delta, the majority of flakes are more consistent with middle- to late-stage reduction. This would be expected if foragers utilizing resources in the ORB marshes followed a typical Paleoarchaic pattern of performing much of the early-stage reduction at the toolstone source. As Beck et al. (2002) found, however, the degree to which bifaces were reduced depends on travel costs and thus the distance to be traveled upon departure from a quarry. In that study Beck et al. found

that reduction at Little Smoky Quarry in eastern Nevada proceeded further than that at Cowboy Rest Creek Quarry in central Nevada. Nearly two-thirds of the bifaces at Cowboy Rest Creek Quarry were in stages 0–1.0, while only a third of those from Little Smoky Quarry were in these stages. Using a central place foraging model they argue that this difference was due to the disparity in anticipated travel distance, ~9 km in the case of the former and ~60 km in the case of the latter.

As Page and Duke show in Chapter 6, FGV artifacts in assemblages from the eastern portion of the delta are predominantly of material from quarries east of the delta, while the opposite is the case for assemblages in the western portion. Average travel distances between sites and quarries, then, are somewhere in between those stated above, and thus the expectation is that foragers brought primarily middle-stage bifaces into the delta. But given that residence time in the marshes is believed to have increased over time, people would periodically need to leave and travel to toolstone sources for raw material. The question, then, arises as to whether at some point the pattern of biface manufacture and use changed to one in which production bifaces were manufactured with the intent of using them as cores to produce flake blanks for other tools. That is, rather than

TABLE 5.47. Retouch and Use-Wear Observable on Old River Bed Biface-Reduction and General Core-Reduction Flakes by Raw Material.

Flake Type and Raw Material	Retouch			Use-Wear		
	Present	Absent	Indeterminate	Present	Absent	Indeterminate
Biface Reduction[a]						
Chert	1	3		1	1	2
Obsidian	3	11	1		1	14
Fine-grained volcanics	7	17	1		7	18
Total	11	31	2	1	9	34
General Core[b]						
Chert		3			1	2
Obsidian	1	4				5
Fine-grained volcanics	3	4		1	1	5
Total	4	11		1	2	12

[a] Includes possible biface-reduction flakes.
[b] Includes possible general core flakes.

reducing bifaces optimally for the travel distance ahead, these artifacts were left in their earliest stages for transport into the marshes.

The use of bifaces as cores is believed to have been the case for Clovis, which Bradley et al. argue is an integrated complex of technologies that they refer to as the "Clovis techno-complex" (2010:7). Bradley et al. state that bifacial production was central to this complex as it was used to produce the majority of primary flake blanks. However, this does not appear to have been the case in any of the Paleoarchaic assemblages we have studied. The WST technology represented in these assemblages is not an integrated system but one in which biface reduction in FGV toolstone is directed primarily toward the end product: the long-stemmed projectile point. Very little use was made of FGV flakes resulting from the manufacture of production bifaces in these assemblages. Most other tools were produced through the reduction of chert cores.

However, chert is extremely rare in the distal ORB, constituting only 7.1 percent of the entire assemblage. Therefore, by necessity obsidian and FGV flakes were used, both expediently and to make formal tools such as scrapers, gravers, and even crescents. Table 5.47 shows the proportion of BRFs and GCFs exhibiting retouch and use-wear by raw material. BRFs do not differ appreciably with respect to retouch across materials, but GCFs do, with 42.9 percent of FGV flakes showing retouch whereas 20 percent of obsidian flakes and no chert GCFs are retouched. Little can be said concerning use-wear as this variable was observable on so few flakes.

The original flake blank is not evident for most of the formal tools, but of the 35 scrapers for which the blank platform is still visible, most of which are FGV, 24 (68.6 percent) are made on BRFs, and 11 (45.8 percent), on GCFs. The number of gravers for which the original blank could be evaluated is extremely small, but half are on GCFs. These data indicate the use of FGV BRFs as expedient and formal tool blanks, but they also point to the use of flakes from generalized cores.

Analyses of production bifaces and a limited flake sample suggest that the lithic-procurement tactics of ORB inhabitants were like those of Paleoarchaic foragers elsewhere in the province, particularly in those areas where FGV toolstone sources dominated the lithic terrane. That is, a significant portion of biface reduction occurred at the quarry. Some early- to middle-stage production bifaces were carried into the ORB delta along with minimally modified flake blanks. These no doubt occasionally served as cores from which flake tools were made, and they may also have served as heavy-duty tools before ultimately being transformed into stemmed points. Still, had this biface-as-core production tactic dominated, we would expect a far greater number of biface fragments in the ORB assemblages in light of the relatively high failure rate associated with FGV biface manufacture (see Bennett and Jones 2012). Even with high levels of scavenging of biface fragments, core-like chunks should be more common than they appear in these assemblages. One way to put this is that the Paleoarchaic approach to lithic procurement/production is un-Clovis-like. Certainly some early-stage bifaces were transported, but probably most were well shaped at the quarry, reduced to a blank/preform stage at which point they would be less vulnerable to breakage.

Figure 5.41. Cody knives (*A–F*) and possible Cody knives (*G–H*) in the Old River Bed assemblage: (*A*) 42To1872, FS 1; (*B*) DPGIF 216; (*C*) DPGIF 2525; (*D*) DPGIF 1400; (*E*) DPGIF 205; (*F*) DPGIF 2451; (*G*) 42To3235, FS 76; (*H*) DPGIF 2455.

Knives

Ten artifacts have been identified as knives (Tables 5.38–5.39). Of these, eight are Cody knives (*n* = 6) or possible Cody knives (*n* = 2; Figure 5.41). Six are made from FGV toolstone, one is made from quartzite, and one is made from chert. The latter (Figure 5.41B), which is made from high-quality butterscotch-colored chert, is a remarkable specimen in that it is complete and shows minimal, if any, weathering damage. It is quite large, more than 100 mm in length, and exhibits collateral flaking with pressure retouch around the edges. Slight abrasive use-wear damage is observable along the

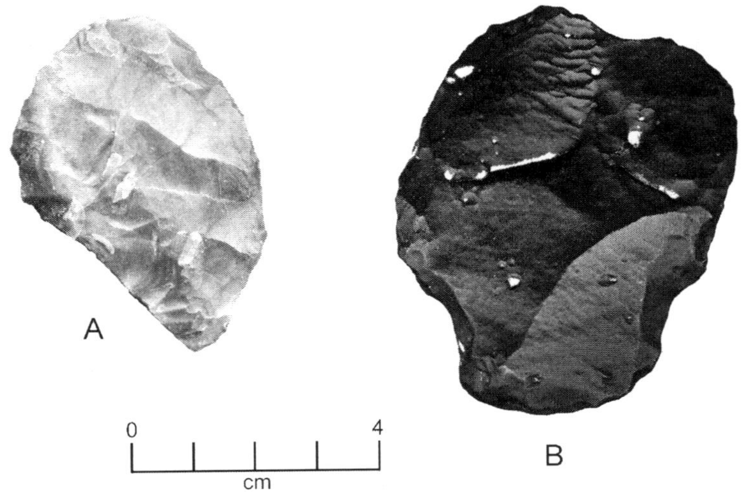

FIGURE 5.42. Possible knives in the Old River Bed assemblage: (*A*) 42To1685, FS 6; (*B*) DPGIF 2475.

straight edge from the tip down about two-thirds the length of the blade as well as below the curvature at the tip down the opposite edge of the blade. Light lateral edge grinding is observable on the stem.

The six FGV specimens exhibit medium to heavy weathering damage, while the quartzite knife is only minimally weathered. The specimen in Figure 5.41A represents a tool (perhaps a stemmed point) that was broken and subsequently resharpened into a Cody knife. The artifact in Figure 5.41H was fashioned on the stem and midsection of a broken tool; unifacial flaking occurs along the lateral edges, and the point has been snapped. Use-wear was not detectable on any of these artifacts or the specimen made from quartzite.

Two specimens have been categorized as possible knives (Figure 5.42). These are also large (Tables 5.38–5.39). One, which has a haft segment, is manufactured from FGV and is irregularly flaked. It is heavily weathered, and thus neither use-wear nor resharpening could be recorded. The second, represented only by the blade, is made of high-quality chert and exhibits broad, oblique flaking. There is minimal weathering damage, and crushing is evident on the high points all along the edge. Site locations for the 10 knives are shown in Table 5.40.

Drills

There are 19 drills in the ORB assemblage (Figure 5.43; Tables 5.38–5.40). As Figure 5.43 shows, these artifacts vary considerably in shape and size. Following Gramly (1990:20), however, drills are recognized less on the basis of morphology than on edge beveling and/or the pattern of use-wear; beveling and use-wear occur on al-ternate margins of the drill bit, indicating a rotary motion. Although beveling was recognizable on all of the ORB specimens, use-wear was observable only on one. About half of the drills are manufactured from FGV toolstone, six are obsidian, and three are chert. Most specimens on which the base is present show evidence of hafting.

Unique Bifaces

Five bifaces were identified as unique (Figure 5.44; Tables 5.38–5.40). All are made from FGV toolstone. All but one specimen (Figure 5.44A) are heavily abraded, and thus the presence of use-wear was impossible to determine. The biface in Figure 5.44A is ground in both concavities, similar to the grinding seen on crescents (see below). In addition, there are burin breaks on two corners, also similar to crescents, which often have this type of break on the points.

Specimen C in Figure 5.44 is similar in its haft shape and its size to two Pinto points (see, for example, Figure 5.25E), although its neck width is somewhat larger. After breakage this specimen was resharpened along its distal end, possibly into a scraper. However, we chose to categorize it as a unique biface rather than as a Pinto point or a scraper since it cannot be confidently identified with either category.

Amorphous Bifaces

The amorphous bifaces category includes those bifaces that are irregularly shaped and cannot be identified with any other biface category. There are 60 of these tools in the ORB assemblage (Figure 5.45; Tables 5.38–5.40). The majority are made from FGV toolstone, although 13

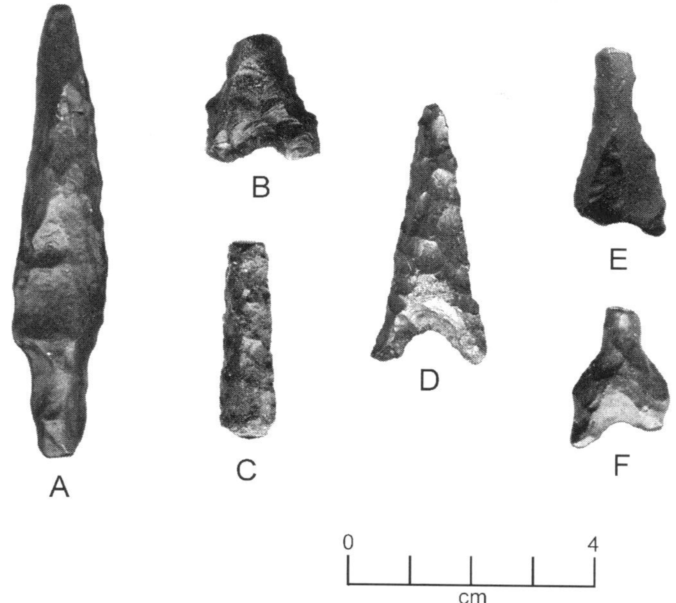

FIGURE 5.43. Examples of drills in the Old River Bed assemblage: (*A*) 42To1685, FS 1; (*B*) 42To1178, FS 6; (*C*) 42To1920, FS 18; (*D*) 42To1358, FS 67; (*E*) 42To3237, FS 72; (*F*) DPGIF 838.

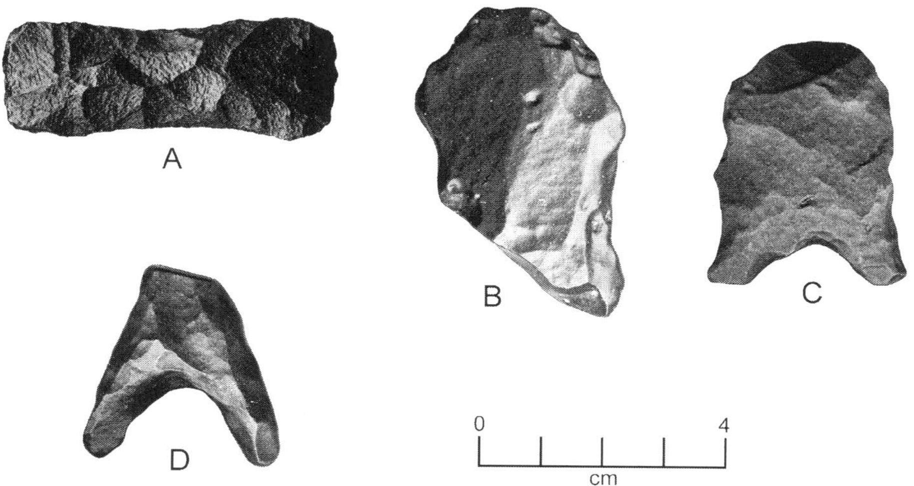

FIGURE 5.44. Unique bifaces in the Old River Bed assemblage: (*A*) 04DM03, FS 3; (*B*) 42To1920, S 22; (*C*) 42To3223, FS 9; (*D*) DPGIF 611. (Not shown: 42To1168, FS 1.)

are obsidian and four are chert. One specimen (Figure 5.45B) is made from a very thin piece of tabular chert that is almost completely covered with cortex on both faces. Use-wear was indeterminate for all but one of these artifacts. All but two show medium to extreme weathering damage.

Indeterminate Bifaces

The remaining 256 bifaces could not be assigned to a category because they are either too fragmentary or too heavily weathered. Most are made from either FGV (*n* =

122, 47.7 percent) or obsidian (*n* = 107, 41.8 percent), but 23 (9.0 percent) are made from chert, and four (1.6 percent) are quartzite.

CRESCENTS

Besides stemmed points, the most diagnostic tools in Paleoarchaic assemblages are crescents. These tools are most often bifacially flaked, but they also occur in the form of unifacially retouched flakes. They range widely in size and in Paleoarchaic assemblages across the Great Basin are almost always made from chert (Amick 1995,

FIGURE 5.45. Examples of amorphous bifaces in the Old River Bed assemblage: (*A*) 42To2559, FS 99; (*B*) 07DM01; (*C*) ISO2, FS2; (*D*) 42To3142, FS 12; (*E*) 42To3520, FS 10.

1999; Beck and Jones 2009). In addition, they are generally ground along both concave and convex edges (i.e., edges A and B; see Figure 5.46B).

In our discussion of the crescents from the Sunshine Locality we state that these tools are found almost exclusively in the Great Basin and occur only occasionally outside that province. This statement is no longer valid since increasing numbers of crescents have been found in California, especially in the Channel Islands (e.g., Erlandson et al. 2008). Most recently Erlandson et al. (2011) reported on three Channel Island sites where crescents were collected. One of these sites dates to almost 11,000 ^{14}C BP, making it the oldest site found on the California coast.

Still, the large majority of crescents have been found in the Great Basin. Even so, they do not occur in all Paleoarchaic assemblages, and there are only a few locations where they occur in large numbers. The largest sample that has been analyzed to date comprises 563 crescents from the Black Rock Desert that are held by private collectors. This assemblage was examined by Amick (1999). In our analysis of the 245 crescents from the Sunshine Locality (the largest sample from a single locality), Amick's analysis provided a background for comparison. Here we use Amick's results as well as those from the Sunshine analysis as background for comparison with the ORB crescents.

Crescents tend to be present in Paleoarchaic assemblages deposited near lakes and marshes (Davis 1978; Grayson 2011), and thus speculation as to their function has focused on activities in these contexts. Suggestions range from cutting and scraping implements to hafted weapons. Our analysis protocol devised for the Sunshine crescents, which we also use here, was designed to focus on attributes that possibly could help to determine which, if any, of the proposed functions have merit.

Several metric attributes were measured, including length, width at the midsection (see Figure 5.46A), thickness, and actual weight. Type of raw material and size category were recorded as previously described for the general artifact analysis. Cortex was recorded as present or absent. The presence of edge grinding was recorded for both concave and convex edges (Figure

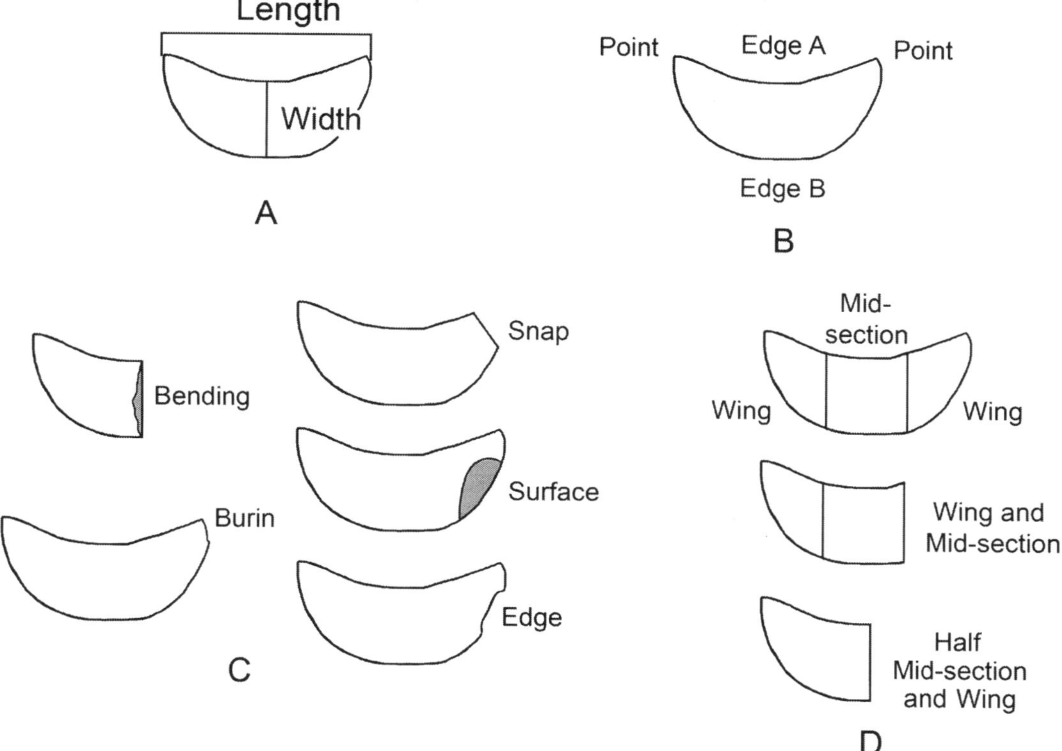

FIGURE 5.46. Protocol for Old River Bed crescent analysis: (*A*) length and width measurements; (*B*) location at which edge grinding, breaking, and/or use-wear may occur; (*C*) breakage types; (*D*) portion represented.

5.46B). In the Sunshine analysis we were able to distinguish the extent of this grinding, that is, whether it is light or heavy, but because of the weathering damage on most of the ORB specimens, this was not possible; instead we simply recorded whether grinding was present or absent. Use-wear and resharpening were recorded as present or absent. Again, because of weathering damage we were rarely able to evaluate use-wear. Breakage types are shown in Figure 5.46C. Locations of use-wear, resharpening, and breakage were recorded as indicated in Table 5.48 (see Figure 5.46B). The fragment represented (that is, whether it is complete or partial) was recorded as indicated in Figure 5.46D and Table 5.48. Flaking pattern was determined as shown in Figure 5.47.

A total of 18 crescents (Figure 5.48) and nine possible crescents (Figure 5.49) are represented in the ORB assemblage. This is not a particularly large sample given the overall assemblage size. On the other hand, among the nearly 13,000 Paleoarchaic artifacts collected from sites in Butte Valley, eastern Nevada, there are only three artifacts that could even possibly be crescents. Thus, the presence of at least 18, if not 27, crescents in the ORB delta is significant. Although we have distinguished between those artifacts that can definitely be identified as crescents and those that can only be said to be possible

crescents, we present analysis results of all 27 together. We distinguish only between those that are bifacial (*n* = 25) and those that are unifacial (*n* = 2). The quantitative data for these artifacts can be found in Table 5.49. The distribution of the crescents across seven qualitative variables is shown in Table 5.50.

As is the case for most crescents, nearly all (>90 percent) of those from the Black Rock Desert and Sunshine Locality are manufactured from chert (Amick 1999; Beck and Jones 2009). As might be expected given the previous discussions, this is not the case for the ORB crescents; only seven (25.9 percent) are made from chert (Table 5.50), while the majority (55.6 percent) are made of FGV. As is evident from Figures 5.48 and 5.49, the best-made crescents as well as those that exhibit the least weathering damage are almost all chert.

All but two of the ORB crescents are bifacially flaked (i.e., exhibit complete bifacial shaping). The remaining two (Figure 5.48M and Figure 5.49I) are made on what Amick (1999) refers to as edged flakes or flake blanks. The latter is a flake with retouched edges, while the former is well shaped and collaterally flaked on one face.

Interestingly, given that many of the tools in the ORB assemblage are smaller than their counterparts in other Paleoarchaic assemblages, most of the crescents

TABLE 5.48. Analytic Protocol for the Old River Bed Crescents.

Variable	Alternative States
Fragment (Figure 5.48D)	1. Complete/nearly complete 2. Midsection 3. Wing 4. Midsection + wing 5. Half midsection + wing
Burin fracture (Figure 5.48C)	1. Present, one wing 2. Present, both wings 3. Present, edge A adjacent to point 4. Present, edge B adjacent to point 5. Absent 6. Indeterminate
Bending fracture (Figure 5.48C)	1. Present, midsection 2. Present, one wing 3. Present, both wings 4. Present, midsection and one wing 5. Present, midsection and both wings 6. Absent 7. Indeterminate
Other breaks (Figure 5.48C)	1. Snap on tip of wing 2. Surface 3. Edge 4. Combination of 1, 2, and/or 3 5. None 6. Indeterminate
Location of use-wear (Figure 5.48B)	1. None 2. One point 3. Both points 4. Edge A 5. Edge B 6. Both edges 7. One point + edge A 8. One point + edge B 9. One point + both edges 10. Both points + edge A 11. Both points + edge B 12. Both points + both edges 13. Into another tool 14. Indeterminate
Location of resharpening (Figure 5.48B)	1. None 2. One point 3. Both points 4. Edge A 5. Edge B 6. Both edges 7. One point + edge A 8. One point + edge B 9. One point + both edges 10. Both points + edge A 11. Both points + edge B 12. Both points + both edges 13. Into another tool 14. Indeterminate

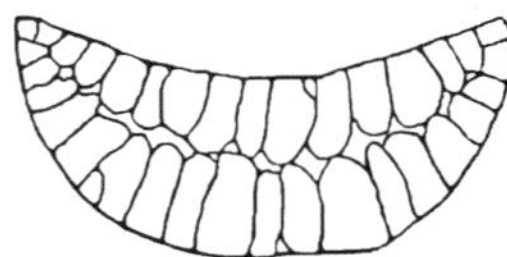

FIGURE 5.47. Flaking patterns recorded for Old River Bed crescents.

TABLE 5.49. Summary of Quantitative Data for Crescents in the Old River Bed Assemblage.

		Class	
Variable	Statistic	Bifacial Crescent	Unifacial Crescent
Total length (mm)	n	8	1
	Range	27.0–63.6	
	Mean	51.84	39.5
	SD	11.479	
Maximum width (mm)	n	15	2
	Range	11.2–26.1	11.2–24.0
	Mean	17.40	17.60
	SD	3.879	9.051
Maximum thickness (mm)	n	23	2
	Range	4.4–14.0	5.3–10.7
	Mean	7.31	8.00
	SD	2.150	3.818
Actual weight (g)	n	25	2
	Range	2.3–13.7	3.4–14.0
	Mean	6.65	8.69
	SD	3.431	7.545

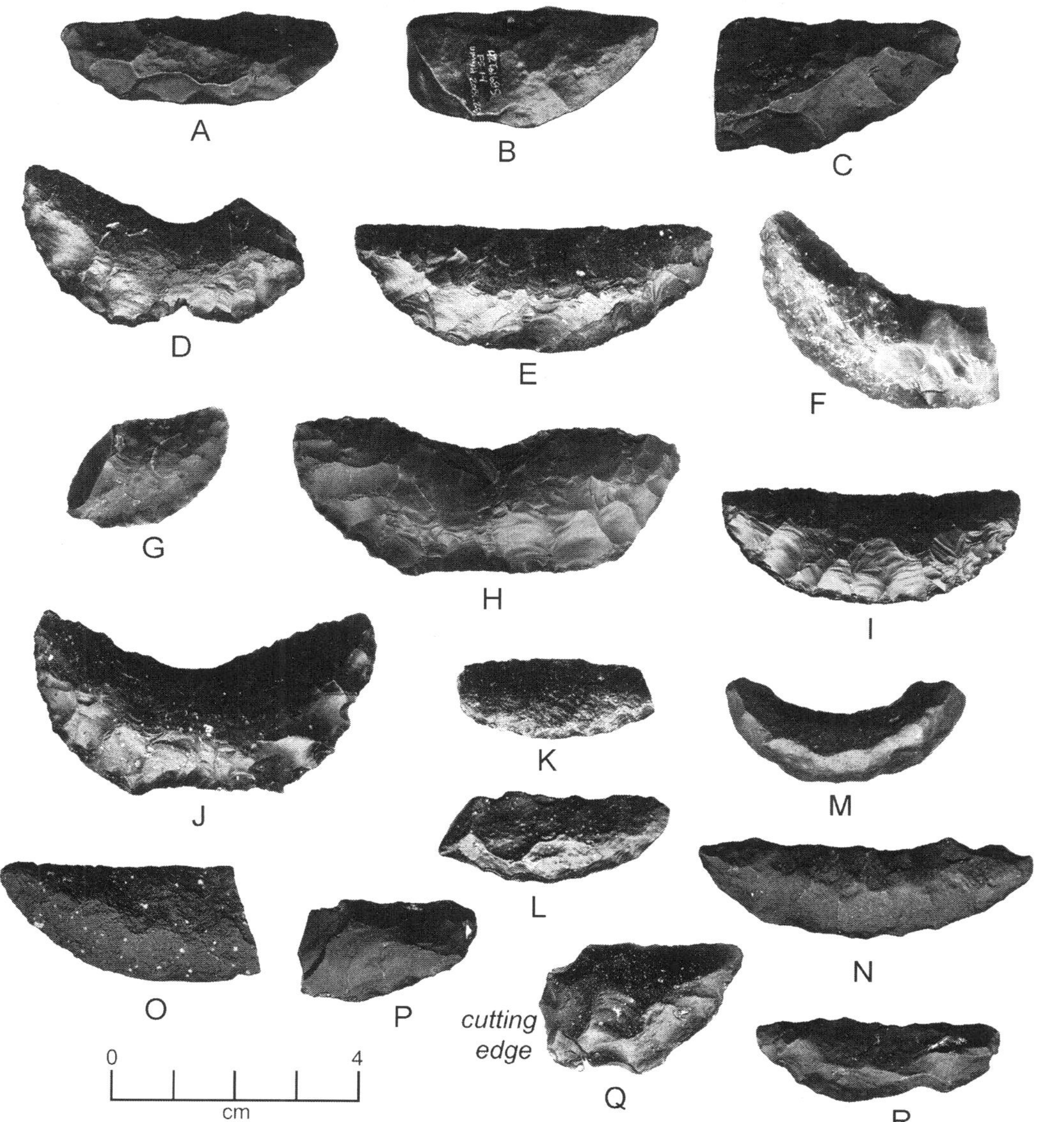

FIGURE 5.48. Crescents represented in the Old River Bed assemblage: (*A*) 42To1672, FS 7; (*B*) 42To1685, FS 14; (*C*) 42To1688, FS 57; (*D*) 42To1924, FS 52; (*E*) DPGIF 328; (*F*) DPGIF 329; (*G*) DPGIF 362; (*H*) DPGIF 607; (*I*) DPGIF 643; (*J*) DPGIF 901; (*K*) 42To3219, FS 18; (*L*) 42To3522, FS 1; (*M*) DPGIF 2528; (*O*) DPGIF 2526; (*P*) 42To2551, FS 8; (*Q*) 42To3520, FS 29; (*R*) DPGIF 2530.

are quite large. For example, among the 78 crescents in the Sunshine assemblage for which size category was recorded, nearly half are of size 9; only about a quarter are larger, in categories 10 and 11. In contrast, about half (53.3 percent) of the 15 ORB specimens for which size was recorded (estimated for most fragments) are size 10 or 11 (Table 5.50). As stated above, many uses have been suggested for crescents, including surgical instruments (Amsden 1937:51; Wardle 1913:656), cutting implements (Cressman et al. 1942; Daugherty 1956; Davis 1978:61; Gifford and Schenck 1926), ornaments or amulets (Davis 1978:61; Warren 1967), and "composite

tools used by women for scraping, peeling, trimming, incising, gouging, and sawing" (Davis 1978:61–62). Probably the most prevalent suggestion is that crescents were hafted transversely (as shown in Figure 5.50B) and used as weapons for waterside hunting of large birds (Clewlow 1968; Davis 1973, 1978; Tadlock 1966). Amick (1999:164) suggests that some of the use-wear, resharpening, and breakage patterns he has observed are consistent with a transversely hafted projectile (but more likely as depicted in Figure 5.50C because the burin breaks extend from the point down edge B; see below). For example, he states that "use-damage on

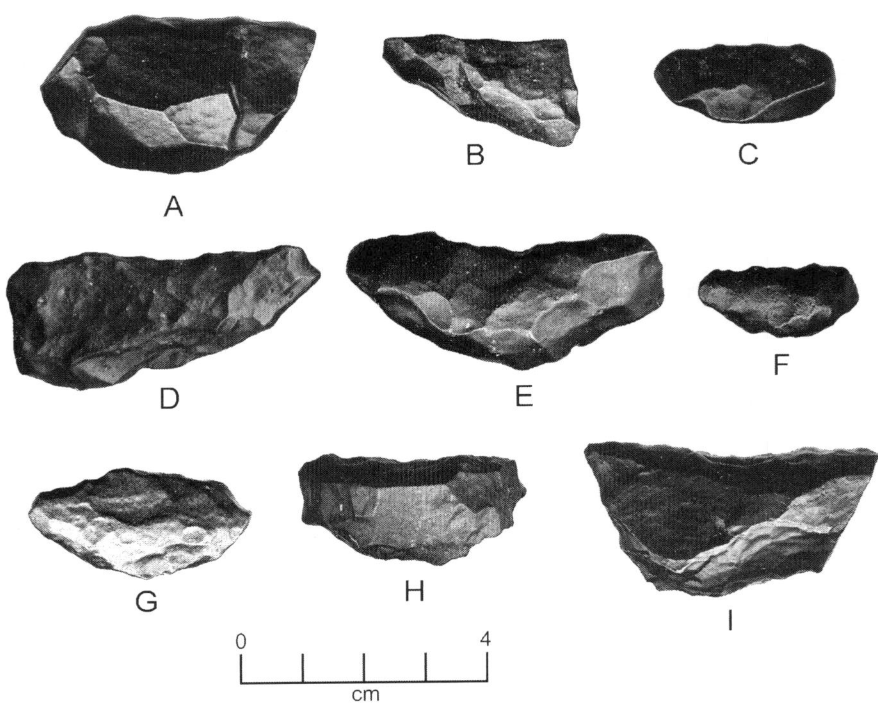

FIGURE 5.49. Possible crescents represented in the Old River Bed assemblage: (*A*) 42To1685, FS 12; (*B*) DPGIF 379; (*C*) DPGIF 476; (*D*) DPGIF 727; (*E*) DPGIF 897; (*F*) 42To3142, FS 8; (*G*) 42To3224, FS 4; (*H*) 42To3226, FS 11; (*I*) 42To3239, FS 1.

finished crescents commonly involves bending breaks or burination of the tips," which is "consistent with the wear and damage expected from transverse hafting and projectile impact, as well as cutting and slotting activities" (1999:164). On the other hand, he also observes that some of the same manufacturing, resharpening, and breakage patterns, such as burin fractures on the tips, resharpening of the tips into sharp projections, and bending fractures, are consistent with handheld or possibly longitudinally hafted tools (Figure 5.50A) used for cutting, slotting, and wedging of soft (e.g., animal hides and flesh and plant fiber) as well as hard material. Our results from the Sunshine analysis are similarly ambiguous.

The most convincing evidence of hafting is the fact that the inner and outer margins of crescents (edges A and B, Figure 5.46B) most often exhibit edge grinding. Among the Sunshine crescents for which edge grinding could be evaluated, about three-quarters were ground on at least edge A or edge B, with most ground on both. A quarter of these were not ground. Most of the latter were broken, possibly during manufacture, which would explain the lack of edge grinding. However, a few of these were complete except for burin fractures on the points or surface breaks, suggesting that they had

been used. Thus, our results support the hypothesis that the majority of the Sunshine crescents were hafted, although a few may have been handheld.

Weathering damage and, in a few cases, breakage precluded the recording of edge grinding on all but 12 of the ORB specimens, but as Table 5.50 shows, evidence of grinding on both edges is present on nine, and an additional three have grinding on at least one edge. These results are suggestive of hafting.

Weathering damage also prohibited the evaluation of use-wear on most specimens. Two crescents (Figure 5.48E, H) exhibit abrasion on both points, extending onto edge A. Another specimen (Figure 5.48J) also exhibits abrasion on both points, but in this case it extends onto edge B. We had more success in evaluating resharpening, which we were able to evaluate for 11 of the 27 crescents. Resharpening on three specimens is evident on both edges A and B of one wing (Figures 5.48H and 5.49F–G). Three others are resharpened on edge B (Figure 5.48D–E, Q), with that on one extending along the bending break on the wing. A sixth has been resharpened on one point and the adjacent edge B (Figure 5.48Q). In addition, a different tool has been created along the bending break of this specimen. Finally, four show no detectable resharpening.

TABLE 5.50. Distribution of Old River Bed Crescents Across Seven Qualitative Variables.

| | Class | | | | |
| Variable and State | Bifacial Crescent | | Unifacial Crescent | | Total |
	n	%	*n*	%	
Raw Material					
Chert	7	28.0			7
Obsidian	5	20.0			5
Fine-grained volcanics	13	52.0	2	100.0	15
Total	25	100.0	2	100.0	27
Flaking					
Collateral	12	48.0	1	50.0	13
Parallel/oblique	2	8.0			2
Irregular	3	12.0	1	50.0	4
Edge retouch only					
Indeterminate	8	32.0			8
Total	25	100.0	2	100.0	27
Edge Grinding (A/B)					
Present/present	9	36.0			9
Present/indeterminate	1	4.0			1
Indeterminate/present	2	8.0			2
Indeterminate/indeterminate	13	52.0	2	100.0	15
Total	25	100.0	2	100.0	27
Resharpening					
One wing + edges A and B	3	12.0			3
Edge B	3	12.0			3
Point + edge B, different tool	1	4.0			1
None	4	16.0			4
Indeterminate	14	56.0	2	100.0	16
Total	25	100.0	2	100.0	27
Weathering Damage					
Minimal	8	32.0			8
Medium	4	16.0			4
Heavy	10	40.0	1	50.0	11
Extreme	3	12.0	1	50.0	4
Total	25	100.0	2	100.0	27
Fragment					
Complete	4	16.0	1	50.0	5
Complete minus both points	1	4.0			1
Complete minus one point	3	12.0			3
Midsection + wing	9	36.0	1	50.0	10
Half midsection + wing	3	12.0			3
Wing only	4	16.0			4
Midsection only	1	4.0			1
Total	25	100.0	2	100.0	27
Size Category					
7	1	4.0			1
8	2	8.0	1	50.0	3
9	4	16.0			4
10	3	12.0			3
11	5	20.0			5
Indeterminate	10	40.0	1	50.0	11
Total	25	100.0	2	100.0	27

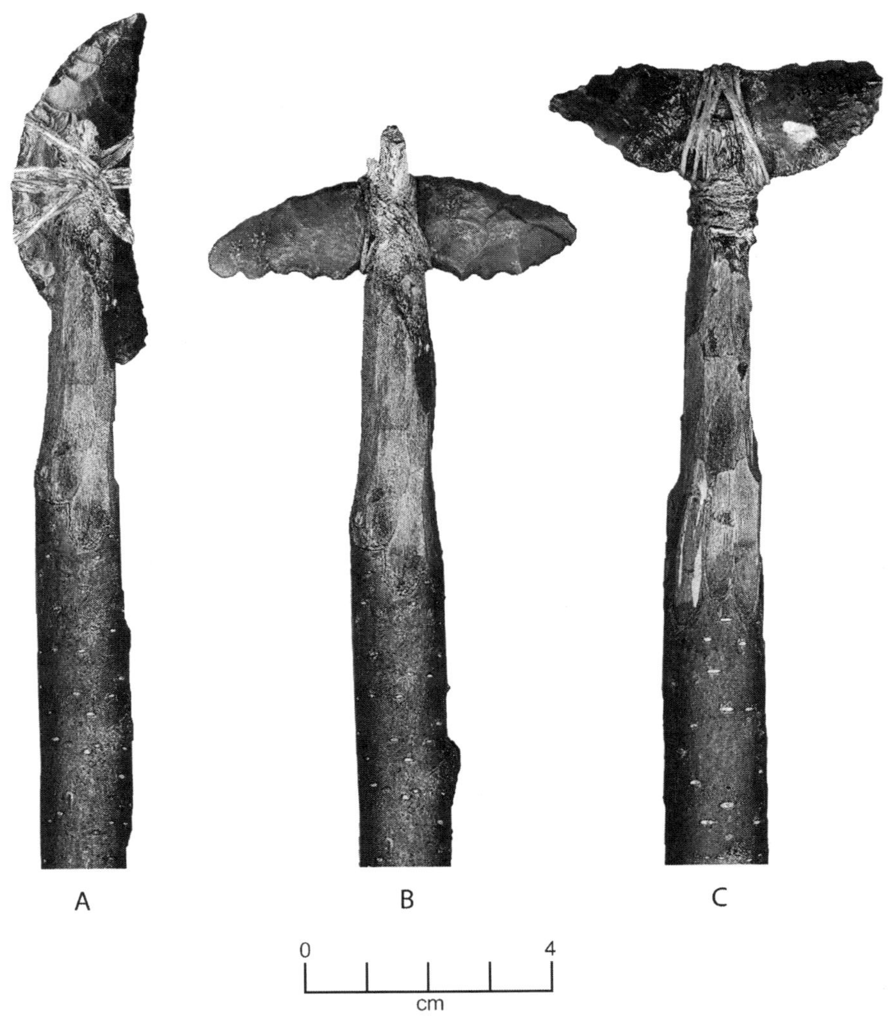

FIGURE 5.50. (*A–C*) Possible hafting modes of crescents (from Beck and Jones 2009:107, Figure 5.6).

Although these data indicate that the crescent points were used and resharpened, whether these tools were handheld or hafted is unclear, although edge grinding favors the latter. Breakage patterns also support hafting. Table 5.50 shows the fragments represented in the crescent assemblage (see Figure 5.46D). As this table shows, breakage is more often between the wing and midsection than at the center of the midsection. That is, the fragments of single wing, midsection, and midsection plus wing all show this breakage pattern. Only three (11.1 percent) were broken in half. This pattern supports hafting in which bending forces are concentrated at the edge of the haft rather than inside it.

Nearly all of these fragments exhibit bending breaks (Table 5.51). In addition, four of the 19 (21.1 percent) for which the presence of burin breaks could be evalu-

ated exhibit these on at least one point, extending down edge B. Finally, 21 (91.3 percent) of the 23 specimens that could be evaluated have other types of breaks, such as snapped tips and/or surface or edge impact fractures. These breakage patterns support the hypothesis that the ORB crescents were hafted.

In summary, the ORB crescents are manufactured primarily from FGV toolstone, followed by obsidian and finally by chert, which is contrary to the pattern that has been observed elsewhere. In fact, FGV crescents are quite rare in other assemblages across the Great Basin. Further, use-wear, resharpening, and breakage patterns all suggest that the points of these artifacts constituted the tool. Edge grinding and breakage patterns also suggest that they were hafted. These patterns overall, however, are more indicative of the hafting exhibited in

TABLE 5.51. Site Locations and Breakage for Old River Bed Crescents.

Site	Field Specimen No.	Break Type and Location		
		Burin	Bending	Other
42To1672	7	Indeterminate	Absent	Snap (point)
42To1685	12	Indeterminate	Midwing	Snap (point), surface
42To1685	14	Point + edge B	Midsection	Indeterminate
42To1688	57	Absent	Midwing	2 surface (B), edge (A)
42To1924	52	Absent	Midwing	Surface (B)
42To2551	1	Absent	Midwing	2 edge (A, B), surface (B)
42To3142	8	Absent	Absent	Snap (point)
42To3219	18	Absent	Both points	Snap (both points)
42To3224	4	Absent	Absent	Snap (both points), 2 edge (B)
42To3226	11	Indeterminate	Absent	Snap (both points), edge (B), surface (B)
42To3239	1	Indeterminate	Midwing	Indeterminate
42To3520	29	Absent	Midwing	Snap (point)
42To3522	1	Point + edge B	Midsection	Snap (B)
DPGIF328	—	Point + edge B	Absent	Surface (point), edge (B)
DPGIF329	—	Absent	Midsection	Snap (point)
DPGIF362	—	Absent	Midwing	Snap (point)
DPGIF379	—	Indeterminate	Midwing	Indeterminate
DPGIF476	—	Indeterminate	Midwing	Surface (B)
DPGIF607	—	Absent	Absent	Snap (point), edge (A)
DPGIF643	—	Absent	Absent	Absent
DPGIF727	—	Indeterminate	Indeterminate	Indeterminate
DPGIF897	—	Absent	Point–midsection	Surface (B)
DPGIF901	—	Absent	Absent	2 edge (B, A near point)
DPGIF1938	—	Absent	Absent	Snap (both points)
DPGIF2526	—	Absent	Midsection	Absent
DPGIF2528	—	Point + edge B	Absent	Snap (point)
DPGIF2530	—	Indeterminate	Absent	Snap (point), edge (B)

Figure 5.50A or 5.50C rather than Figure 5.50B. If these tools were used to stun birds (hafted as in Figure 5.50B), as several researchers have suggested, it is hard to imagine how the points would have been damaged in the way they are, especially by burin fractures that appear to originate at the point and extend down edge B. The ORB data are consistent with the crescent data from the Sunshine Locality, which suggest that these points were used in cutting activities.

It is notable that approximately half ($n = 14$) of the ORB crescents were found as isolates rather than on sites (Table 5.51). This is not the case for any other tool type. During the early Holocene the ORB delta was essentially one giant marsh, interspersed with causeways and dry patches that allowed foragers access to its resources. The crescents found as isolates likely represent individuals working inside the marsh at special-purpose locations where they left little other artifactual evidence.

SCRAPERS

A scraper is defined by Gramly as a flaked-stone tool that is "often made on a flake or blade, having bold unifacial retouch on one or more margins" (1990:49). Scrapers, however, can also be bifacial and often occur in combination on the same artifact with other tools, such as gravers or chisels. In addition to the general measurements taken on all formal tools, several specific variables were measured for scrapers. First, overall shape was recorded as shown in Figure 5.51A. As scrapers can be hafted, several variables that have been shown to relate to hafting were also recorded, including the presence or absence of (1) proximal thinning, (2) haft preparation (see Clark 1958:149), and (3) a transverse break (see Grimes and Grimes 1985:41; Figure 5.51C). Haft wear would also be useful here (see Beyries 1988:219–220; Hayden 1985), but because of the heavily weathered condition of the artifacts, such

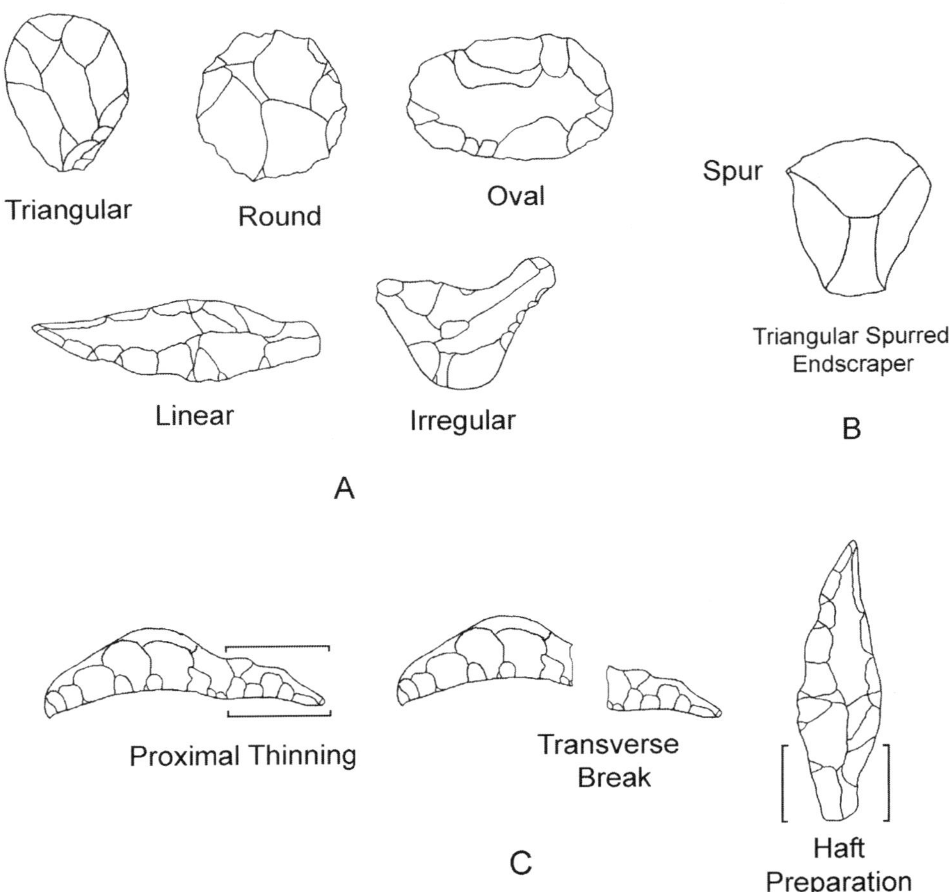

FIGURE 5.51. Qualitative attributes recorded for the Old River Bed scrapers: (*A*) shape; (*B*) presence of spur; (*C*) evidence of hafting.

wear is rarely visible. Finally, it has been noted that Paleoindian end scrapers often have a triangular shape with a spur at the intersection of the working and lateral edges, as shown in Figure 5.51B (e.g., Frison 1991; Morrow 1997; O'Brien 1984; Rogers 1986; Shott 1995). Therefore, the presence or absence of a spur was recorded for end scrapers.

Gramly describes a number of different types of scrapers found in Paleoindian assemblages, six of which have been identified in the ORB assemblage; one additional category not described by Gramly, beaked scrapers (Deller and Ellis 1992), was recognized as well (Table 5.52). We also use the concept of "employable unit" or EU, which was first used by Knudson (1973:17), to refer to the proportion of a continuous edge or projection deemed appropriate for use in performing a specific task (see also Knudson 1979). The EU concept has more recently been used by Beck and Jones (2009, 2010b), Frison (1991), Jones et al. (Jones, Beck, and Kessler 2003), and Shott (1996) and considers the uti-

lized edge or surface rather than the object as the tool (see also Dunnell 1978). For each EU, edge shape, edge angle, and use-wear, when visible, were recorded. We then used the EUs in combination with the scraper definitions for the identification of combination tools (e.g., end scraper/graver; Figure 5.52) to assign objects to a technological category.

A total of 284 scrapers have been identified in the ORB assemblage (Table 5.53). Each of the seven scraper categories is discussed separately in the following pages. Regarding combination tools, only the scraper portion is discussed. For example, Figure 5.52C shows a combination concave/convex scraper/graver. In the discussion of concave/convex scrapers, only that portion of the combination tool is discussed; the graver portion is addressed in the section on gravers. Most of the scrapers, as well as the other tools, occur in both unifacial and bifacial form. However, these two forms do not differ in the variables measured, such as raw material, shape, and so on, and thus they are combined for discussion.

Table 5.52. Scraper Categories Used in the Old River Bed Artifact Analysis.

Scraper Category	Definition
End scraper	A unifacial tool, usually made on a flake or blade, with a broad working edge opposite the bulbar scar (Gramly 1990:21).
Side scraper	A unifacial tool, usually made on a flake, with removals (scraper retouch) on one or both lateral edges (Gramly 1990:50).
Combination end/side scraper	A unifacial tool, usually made on a flake or blade, with a broad working edge opposite the bulbar scar as well as scraper retouch on one or both lateral edges (Gramly 1990:16).
Concave scraper	Usually unifacial, this artifact exhibits one or more broad concavities on its edge that have been created by scraper retouch (Gramly 1990:16). Also called spokeshaves.
Combination concave/convex scraper	A tool with no set form, exhibiting concave and/or straight to convex scraper edges. Cannot be identified as an end scraper or side scraper.
Circular scraper	A uniface, usually based on a flake, that has a circular shape and scraper retouch completely around its periphery. Also termed a discoidal scraper (Gramly 1990:14).
Slug scraper	A narrow, slug-shaped, unifacial, flaked-stone tool having steep edge angles and a high dorsal surface that is often rounded. Also commonly referred to as a limace or flake shaver; less commonly referred to as a groover, bar, narrow side scraper, unifacially flaked drill, or awl (Gramly 1990:32).
Beaked scraper	Artifact type varying greatly in form, having a cutting or scraping tip that is plano-convex in cross section, usually steeply retouched, and frequently quite thick (Storck 1997:65).

Note: Although many of these definitions specify unifacial tools, scrapers are sometimes bifacial, which is the case in the Old River Bed assemblage.

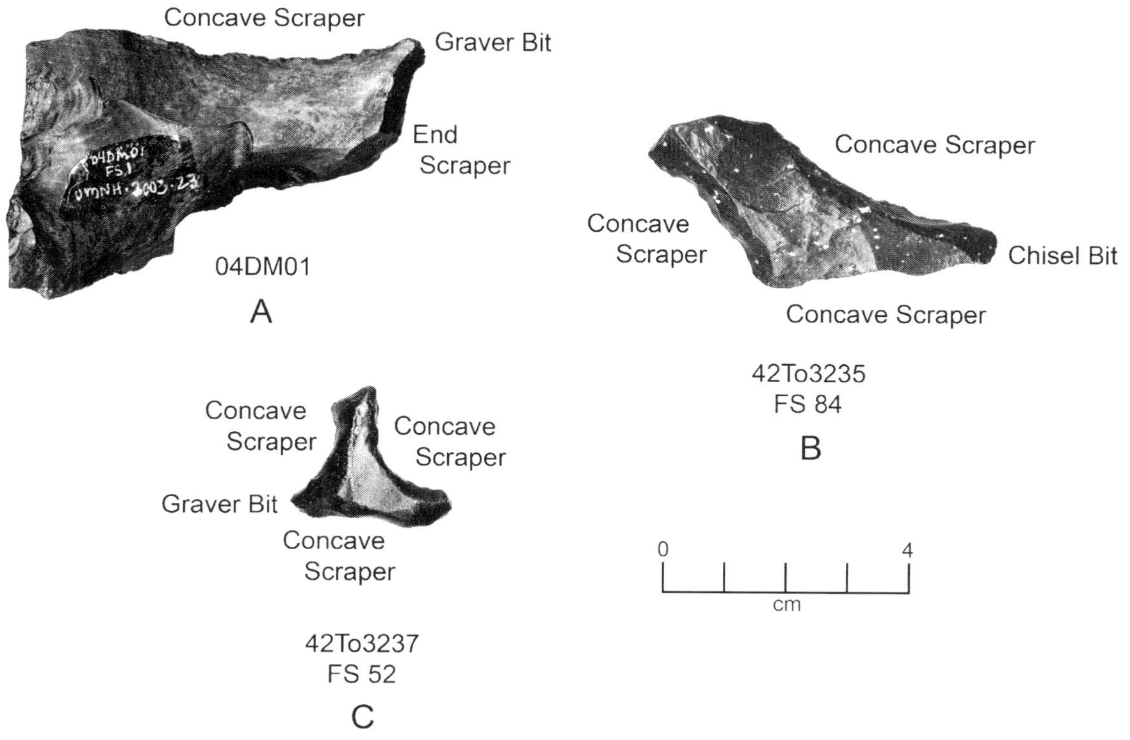

Figure 5.52. Examples of combination tools with multiple employable units: (*A*) concave/end scraper/graver; (*B*) concave scraper/chisel; (*C*) concave scraper/graver.

TABLE 5.53. Scraper Categories Represented in the Old River Bed Assemblage by Raw Material.

Tool Type and Scraper Category	Chert		Obsidian		Fine-Grained Volcanics		Quartzite		Limestone		Total
	n	%	n	%	n	%	n	%	n	%	
Single Tools											
End scraper	10	16.4	7	11.5	44	72.1					61
Side scraper	2	8.3	9	38.5	13	54.2					24
End/side scraper[a]	6	11.1	8	14.8	38	70.4	1	1.9	1	1.9	54
Concave/convex scraper	2	14.3	6	42.9	6	42.9					14
Circular scraper	2	25.0	2	25.0	4	50.0					8
Slug scraper			19	44.2	24	55.8					43
Beak scraper	1	2.4	14	33.3	27	64.3					42
Combination Tools											
End scraper/graver			1	33.3	2	66.7					3
Side scraper/graver	6	50.0	2	16.7	3	25.0	1	8.3			12
End/side scraper/graver	1	16.7	2	33.3	3	50.0					6
Concave/convex scraper/graver	1	16.7	2	33.3	2	33.3	1	16.7			6
End/concave/convex scraper					1	100.0					1
End/concave/convex scraper/graver			1	100.0							1
Side/concave/convex scraper/graver					1	100.0					1
Side scraper/chisel					4	100.0					4
Concave/convex scraper/chisel					1	100.0					1
Slug scraper/graver					3	100.0					3
Total	31	10.9	73	25.7	176	62.0	3	1.0	1	.4	284

[a] End/side scrapers are treated as single rather than combination tools.

End Scrapers and Side Scrapers

End scrapers (Figure 5.53) and side scrapers (Figure 5.54) occur alone as single tools ($n = 61$ and $n = 24$, respectively), as well as in combination with one another ($n = 54$; Figure 5.55; Table 5.53). In addition, both occur in combination with gravers ($n = 21$) and concave/convex scraper/gravers ($n = 2$; Table 5.53). Finally, there is one end/concave/convex scraper as well as four side scraper/chisels, bringing the total to 167. Quantitative data for these artifacts are presented in Table 5.54, and qualitative data are shown in Table 5.55.

Interestingly, end and side scrapers differ with respect to raw material. Single-tool end scrapers are overwhelmingly made from FGV toolstone ($n = 44$, 72.1 percent), and while about half of the single-tool side scrapers are made of FGV, a considerable percentage are made of obsidian (Table 5.53). End/side scrapers are more similar to end scrapers in that 70.4 percent are made from FGV. End and end/side scrapers (including combination tools) also account for more than 80 percent of scrapers manufactured from chert.

The majority of both end and side scrapers are large (Tables 5.54–5.55). For the 114 that were measureable

(including the combination tools), 78 (67.8 percent) are of size 10 or larger (Table 5.55).

EUs were identified primarily by edge shape and edge preparation rather than use-wear since the presence or absence of the latter was observable on only 17 specimens. The number of EUs, in large part, corresponds with the formal tool; that is, all but one of the 59 single-tool end scrapers for which EUs could be determined have only one, the scraping edge on the end. Thirty-four of the 53 single-tool end/side scrapers have only two, one corresponding to the end scraper and the other corresponding to the side scraper. This is not true for single-tool side scrapers, however, half of which have two or more EUs.

By definition, end scrapers are manufactured to obtain a steep angle on the working edge. This holds for most of the ORB specimens, where the majority ($n = 93$ [includes end/side scrapers and combination tools], 81.2 percent) have angles greater than 60°; 64 (43.0 percent) have angles between 81 and 90° (Table 5.55).

Table 5.56 shows the presence or absence of haft preparation, thinning of the proximal end, and transverse breaks on end and side scrapers. Including com-

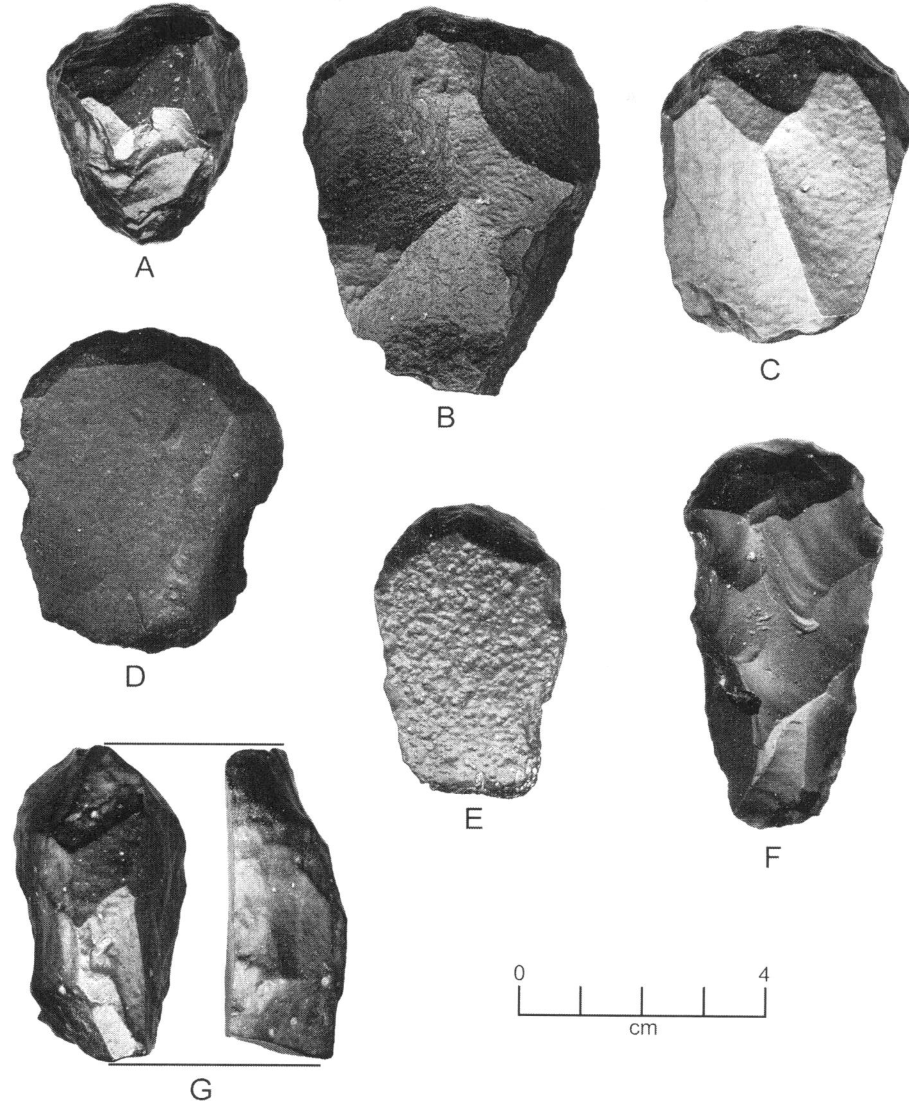

FIGURE 5.53. Examples of end scrapers in the Old River Bed assemblage: (*A*) 42To2554, FS 20; (*B*) 42To3142, FS 33; (*C*) 42To3142, FS 28; (*D*) 42To3142, FS 39; (*E*) 42To3142, FS 1; (*F*) 42To3520, FS 113; (*G*) DPGIF 1399.

binination tools, 15 end, 13 side, and 19 end/side scrapers and eight combination tools show no evidence of any of these three variables and thus were likely not hafted. One end scraper, three end/side scrapers, and one side scraper/graver show the presence of all three variables and thus were very likely hafted. Two of the criteria are present on a number of the other specimens, which is still fairly good evidence of hafting. There are several, however, on which only one is present (the other two are either absent or indeterminate); we have designated these as "possibly hafted." Thus, overall, just over half (*n* = 65) of the 121 end and side scrapers for which haft-

ing evidence could be evaluated (54.2 percent) appear to have been hafted or possibly hafted.

As noted above, it has been argued that triangular spurred end scrapers are typical of Paleoindian assemblages (see, for example, Figure 5.56). Table 5.57 shows the distribution of the 124 end scrapers across shape categories as well as the presence or absence of a spur. Only 22 (22.4 percent) of the 103 end scrapers for which both shape and presence of spur could be evaluated exhibit spurs; 16 of these are triangular, two are irregular, and four are fragmentary. The latter may have been triangular in shape before they were broken. This pattern is

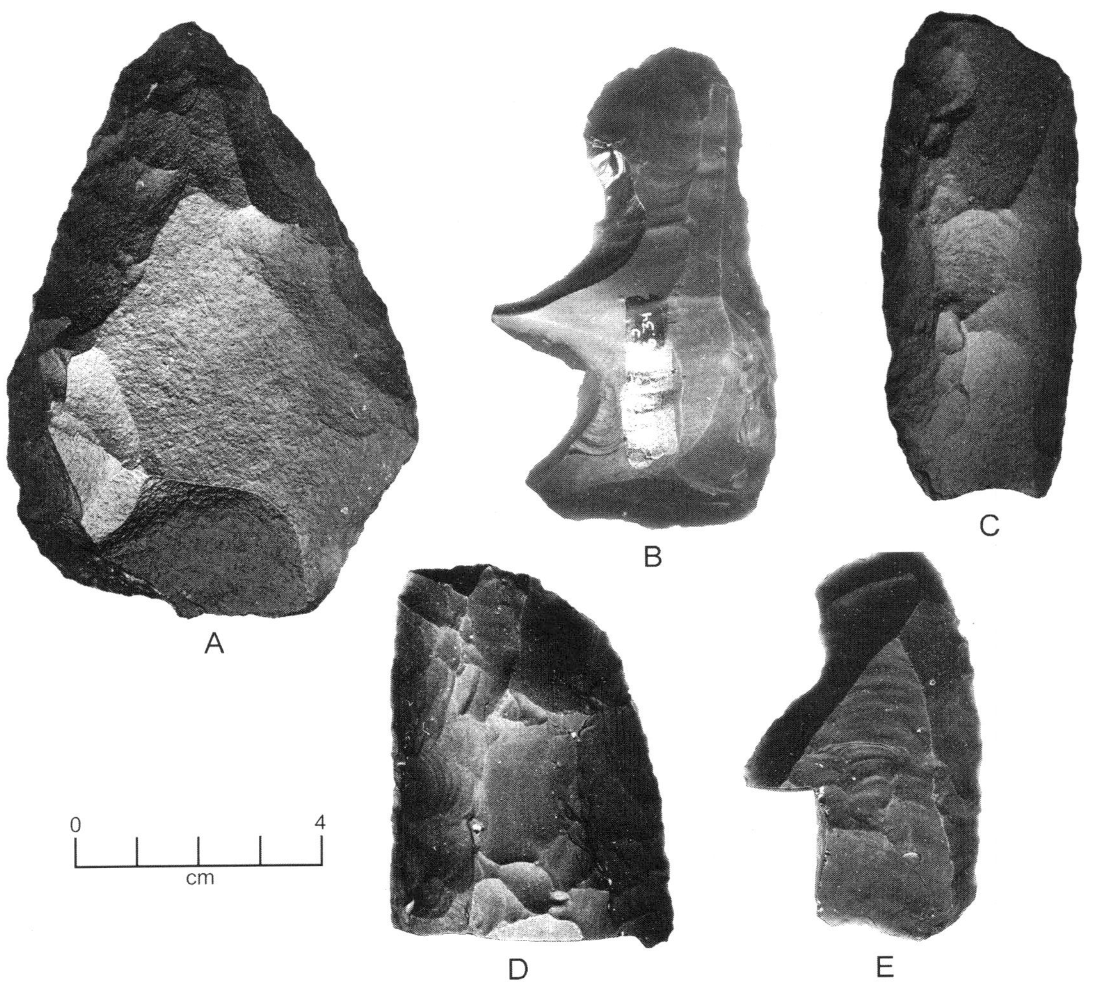

Figure 5.54. Examples of side scrapers in the Old River Bed assemblage: (*A*) 42To2554, FS 51; (*B*) 42To2955, FS 3; (*C*) 42To3140, FS 5; (*D*) 42To3520, FS 92; (*E*) 42To3520, FS 88.

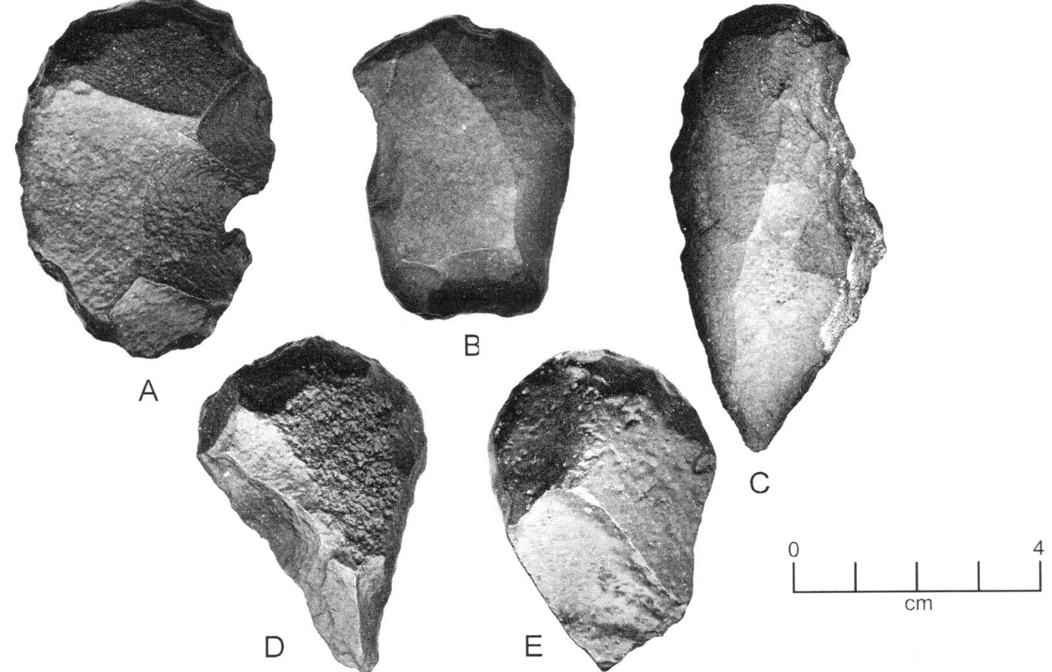

Figure 5.55. Examples of end/side scrapers in the Old River Bed assemblage: (*A*) 42To2554, FS 29; (*B*) 42To3142, FS 10; (*C*) 42To3520, FS 101; (*D*) 42To3142, FS 26; (*E*) 42To3142, FS 5.

TABLE 5.54. Summary of Quantitative Data for End and Side Scrapers in the Old River Bed Assemblage.

Scraper Category	Statistic	Variable		
		Length (mm)	Width (mm)	Thickness (mm)
End Scraper	N	32	38	60
	Range	19.0–69.5	12.8–62.8	3.3–30.0
	Mean	46.85	36.70	12.40
	SD	12.761	10.476	5.248
Side Scraper	N	11	12	24
	Range	24.0–97.8	13.9–69.6	5.6–17.2
	Mean	59.38	29.95	10.31
	SD	21.329	16.715	3.403
End/Side Scraper	N	28	36	54
	Range	26.4–76.6	3.2–61.6	5.7–23.4
	Mean	51.40	33.02	11.22
	SD	12.816	12.182	3.810
End/Concave/Convex Scraper	N	1	1	1
	Range	47.1	37.5	8.3
	Mean	—	—	—
	SD	—	—	—
End Scraper/Graver	N	1	2	3
	Range	30.8	18.8–41.3	9.8–18.8
	Mean	—	30.05	14.37
	SD	—	15.910	4.501
End/Side Scraper/Graver	N	5	3	6
	Range	27.1–62.8	24.6–38.4	4.1–12.6
	Mean	42.62	33.37	9.48
	SD	13.765	7.620	3.375
End/Concave/Convex Scraper/Graver	N	—	—	1
	Range	—	—	11.3
	Mean	—	—	—
	SD	—	—	—
Side Scraper/Graver	N	6	8	12
	Range	25.2–49.4	9.2–32.5	3.8–14.7
	Mean	37.87	22.60	8.10
	SD	9.234	7.855	3.356
Side/Concave/Convex Scraper/Graver	N	—	1	1
	Range	—	27.1	11.2
	Mean	—	—	—
	SD	—	—	—
Side Scraper/Chisel	N	1	3	4
	Range	49.1	17.9–21.8	8.2–17.7
	Mean	—	20.00	12.13
	SD	—	1.967	4.034

TABLE 5.55. Distribution of Old River Bed End and Side Scrapers Across Seven Qualitative Variables.

Variable and State	Scraper Category																			
	End Scraper		Side Scraper		End/Side Scraper		End Scraper/ Graver		Side Scraper/ Graver		End/Side Scraper/ Graver		End/ Concave/ Convex Scraper		End/ Concave/ Convex Scraper/ Graver		Side/ Concave/ Convex Scraper/ Graver		Side Scraper/ Chisel	
	n	%	n	%	n	%	n	%	n	%	n	%	n	%	n	%	n	%	n	%
Shape																				
Triangular	30	49.2	2	8.3	13	24.1	1	33.3	2	16.7	3	50.0								
Linear	1	1.6	4	16.7	6	11.1			1	8.3	1	16.7							1	25.0
Oval	7	11.5	4	16.7	15	27.8													1	25.0
Irregular	1	1.6	3	12.5	5	9.3			3	25.0	2	33.3	1	100.0	1	100.0	1	100.0	1	25.0
Fragment	17	27.9	11	45.8	15	27.8	2	66.7	6	50.0									1	25.0
Circular	5	8.2																		
Total	61	100.0	24	100.0	54	100.0	3	100.0	12	100.0	6	100.0	1	100.0	1	100.0	1	100.0	4	100.0
Size Category																				
5	1	1.6																		
6											1	8.3								
7	2	3.3	3	12.5	1	1.9			1	8.3	1	16.7								
8	3	4.9	1	4.2			1	33.3	2	16.7										
9	7	11.5	3	12.5	5	9.3			2	16.7	1	16.7							1	25.0
10	6	9.8			7	13.0			2	16.7	2	33.3							1	25.0
11	19	31.1	3	12.5	12	22.2	1	33.3			1	16.7	1	100.0						
12	8	13.1	3	12.5	5	9.3					1	16.7			1	100.0				
13	1	1.6			2	3.7														
14			1	4.2																
15			1	4.2																
Indeterminate	14	23.0	9	37.5	22	40.7	1	33.3	4	33.3							1	100.0	2	50.0
Total	61	100.0	24	100.0	54	100.0	3	100.0	12	100.0	6	100.0	1	100.0	1	100.0	1	100.0	4	100.0
Cortex																				
Absent	51	83.6	23	95.8	48	88.9	3	100.0	10	83.3	6	100.0	1	100.0	1	100.0	1	100.0	4	100.0
<50%	7	11.5	1	4.2	3	5.6			1	8.3										
>50%	2	3.3			3	5.6														
Indeterminate	1	1.6							1	8.3										
Total	61	100.0	24	100.0	54	100.0	3	100.0	12	100.0	6	100.0	1	100.0	1	100.0	1	100.0	4	100.0

	n	%	n	%	n	%	n	%	n	%	n	%	n	%	n	%	n	%	n	%
Use-Wear																				
Absent	2	3.3			1	1.9			1	8.3										
Present	5	8.2	5	20.8	1	1.9			1	8.3					1	100.0				
Indeterminate	54	88.5	19	79.2	52	96.3	3	100.0	10	83.3	6	100.0	1	100.0			1	100.0	4	100.0
Total	61	100.0	24	100.0	54	100.0	3	100.0	12	100.0	6	100.0	1	100.0	1	100.0	1	100.0	4	100.0
Maximum Angle (°)																				
31–40			2	8.3	1	1.9														
41–50	3	4.9	3	12.5	2	3.7	1	33.3	1	8.3										
51–60	4	6.6	2	8.3	5	9.3			1	8.3	1	16.7	1	100.0	1	100.0				
61–70	13	21.3	5	20.8	9	16.7			1	8.3	1	16.7							1	25.0
71–80	10	16.4	4	16.7	10	18.5	2	66.7	1	8.3										
81–90	23	37.7	6	25.0	23	42.6			5	41.7	3	50.0					1	100.0	3	75.0
Indeterminate	1	1.6			2	3.7														
Not recorded	7	11.5	2	8.3	2	3.7			3	25.0	1	16.7								
Total	61	100.0	24	100.0	54	100.0	3	100.0	12	100.0	6	100.0	1	100.0	1	100.0	1	100.0	4	100.0
Employable Units																				
1	58	95.1	12	50.0															2	50.0
2	1	1.6	10	41.7	34	63.0	2	66.7	7	58.3	1	16.7							2	50.0
3			1	4.2	17	31.5	1	33.3	3	25.0	3	50.0	1	100.0	1	100.0				
4			1	4.2	2	3.7					2	33.3					1	100.0		
5									2	16.7										
Indeterminate	2	3.3			1	1.9														
Total	61	100.0	24	100.0	54	100.0	3	100.0	12	100.0	6	100.0	1	100.0	1	100.0	1	100.0	4	100.0
Weathering Damage																				
Minimal	11	18.0	5	20.8	7	13.0			6	50.0										
Medium	21	34.4	10	41.7	19	35.2			2	16.7	4	66.7					1	100.0	2	50.0
Heavy	16	26.2	4	16.7	17	31.5	2	66.7	2	16.7	1	16.7	1	100.0					2	50.0
Extreme	6	9.8	4	16.7	5	9.3	1	33.3	1	8.3										
Not recorded	7	11.5	1	4.2	6	11.1			1	8.3	1	16.7			1	100.0				
Total	61	100.0	24	100.0	54	100.0	3	100.0	12	100.0	6	100.0	1	100.0	1	100.0	1	100.0	4	100.0

TABLE 5.56. Presence of Haft Preparation, Proximal Thinning, and Transverse Breaks on Old River Bed End and Side Scrapers.

Haft Preparation/ Proximal Thinning/ Transverse Break	Scraper Category										Total	Hafting
	End Scraper	End/Side Scraper	End Scraper/Graver	End/Side Scraper/ Graver	End/Concave/ Convex Scraper	End/Concave/ Convex Scraper/Graver	Side Scraper	Side Scraper/ Graver	Side/ Concave/Convex Scraper/Graver	Side Scraper/ Chisel		
A/A/A	15	19		3	1		13	4			55	NH
P/P/P	1	3						1			5	H
P/P/A	7	4	1	1			1	1		1	16	H
P/A/P	2	1									3	H
P/A/A	2	2		1							5	PH
I/A/A	6	3								1	10	I
I/P/A	3	3									6	PH
I/P/P		1									1	H
P/I/P	2										2	H
P/I/A	1										1	PH
I/I/P	9	9	1				1	1	1	2	24	PH
I/I/A	9	3					1	1			14	I
I/I/I	3	5	1				6	4			19	I
NR/NR/A				1			2				3	I
NR/NR/P	1	1									2	PH
NA/NA/NA						1					1	NH
Total hafted	28	24	2	1	—	—	2	3	1	3	64	
Total not hafted	15	19	—	3	1	1	13	4	—	—	56	
Total indeterminate	18	11	1	2	—	—	9	5	—	1	47	
Overall total	61	54	3	6	1	1	24	12	1	4	167	

Note: A = absent; P = present; I = indeterminate; NR = not recorded; NA = not applicable; NH = not hafted; H = hafted; PH = possibly hafted.

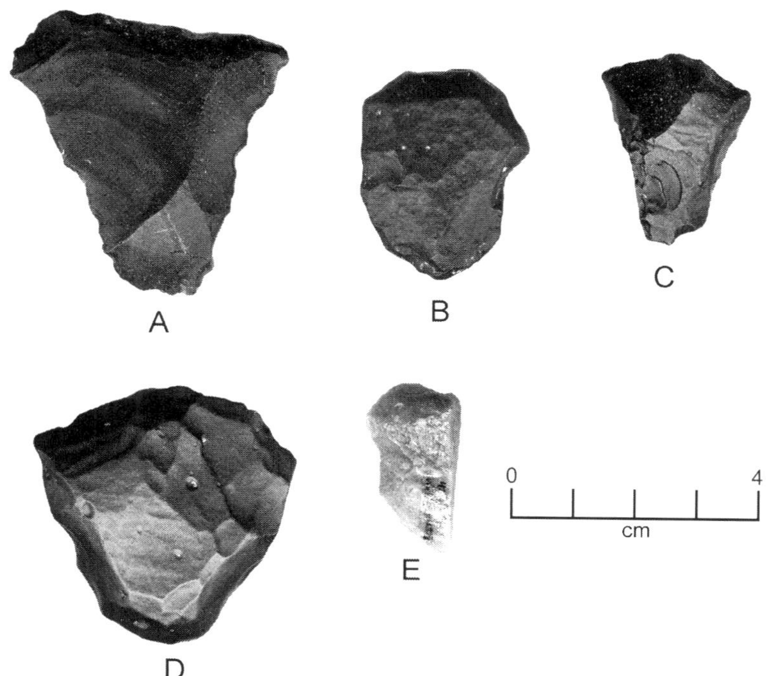

FIGURE 5.56. Examples of triangular spurred end scrapers in the Old River Bed assemblage: (*A*) 42To3570, FS 74; (*B*) 42To1924, FS 90; (*C*) 42To3239, FS 10; (*D*) 42To3142, FS 7; (*E*) 42To1686, FS 61.

TABLE 5.57. Shape and Presence of Spurs on Old River Bed End Scrapers.

| | Spur | | | | |
Shape	Present	Absent	Subtotal	Indeterminate	Total
Triangular	16	26	42	5	47
Linear		5	5	3	8
Oval		21	21	1	22
Irregular	2	5	7	1	8
Fragment	4	19	23	11	34
Circular		5	5		5
Total	22 (22.4%)	81 (78.6%)	103	21	124

Note: Table includes end/side scrapers, end scraper/gravers, and end/side scraper/gravers.

true of the Sunshine assemblage as well, suggesting that triangular spurred end scrapers are not predominant in Great Basin Paleoarchaic assemblages. The locations of end and side scrapers are shown in Table 5.58.

Concave/Convex Scrapers

Concave scrapers, also commonly called spokeshaves, and combination concave/convex scrapers are defined primarily on the basis of edge shape and the fact that they do not identify with either the end or side scraper categories. Most of the concave scrapers occur in combination with convex scrapers, and thus we use the category name of concave/convex scrapers for all of these specimens. There are 24 of these artifacts, 10 of which are combination tools (Figure 5.57; Tables 5.53, 5.59–5.60). Nearly half are made from FGV, but three are made from chert. As is evident from an examination of Figure 5.57, these tools vary considerably in size and shape (Table 5.60). Size categories, for example, range from 7 to 14. Maximum angle varies as well, from between 21 and 30° to greater than 80°.

Concave/convex scrapers are generally irregular in shape and thus were likely not hafted. For this reason the three variables related to hafting (haft preparation, proximal thinning, and transverse break) were not recorded for these tools. The majority of the ORB

Table 5.58. Site Locations for Old River Bed Scrapers.

Site	End Scraper	End/Side Scraper	Side Scraper	Concave/ Convex Scraper	Circular Scraper	Slug Scraper	Beak Scraper	Scraper/ Graver	Scraper/ Chisel	End/Concave/ Convex Scraper	Total
42To1000	1	1	1		2		1				6
42To1152								1			1
42To1153	2	1				1	2	2			8
42To1161						1					1
42To1163	1		1					1			3
42To1165								2			2
42To1166							1				1
42To1169		1	1				1				3
42To1172		1	2					1			4
42To1182	3										3
42To1354	1						1				2
42To1358	6	3	1	2				2			14
42To1368	1										1
42To1369								1			1
42To1371		1				1	2				4
42To1384			1								1
42To1668		1									1
42To1670							1				1
42To1671		2									2
42To1675							1				1
42To1677						1	1				2
42To1678						1					1
42To1680	1										1
42To1682		1									1
42To1683						1					1
42To1684						1	1	2			4
42To1685		3					1	1	1		6
42To1686	3	1	2	1		2	4	2	1		16
42To1688		1	1					2			4
42To1858							1				1
42To1860	1						1				2
42To1861			1				1	1			3
42To1862	3				1			1			5
42To1872	2			1	1		1		1		6
42To1873			1								1
42To1874						2					2
42To1875							1				1
42To1876								1			1
42To1877			1								1
42To1920	2						1	1	2		6
42To1922	1										1
42To1923		1					1				2
42To1924	2	3	1	1	1						8
04DMO1								1			1
42To2551		1									1
42To2552				1							1
42To2554	1	1	1								3
42To2556	1	2				1					4

TABLE 5.58. (cont'd.) Site Locations for Old River Bed Scrapers.

Site	End Scraper	End/Side Scraper	Side Scraper	Concave/ Convex Scraper	Circular Scraper	Slug Scraper	Beak Scraper	Scraper/ Graver	Scraper/ Chisel	End/Concave/ Convex Scraper	Total
42To2559						1					1
42To2767	2	1		1			1				5
42To2945						1					1
42To2946						2					2
42To2947						1					1
42To2949							2				2
42To2955			1	1							2
42To3140			2				1				3
42To3142	16	4									20
42To3219		1			1	3		2			7
42To3222						1	1				2
42To3223									1		1
42To3224	1					2					3
42To3225		1									1
42To3226	2										2
42To3228		1					1				2
42To3229		1	1								2
42To3230							1				1
42To3231				1							1
42To3233		2		2			1				5
42To3234		2				1	1				4
42To3235				1		1	2	1			5
42To3237		1				5		2			8
42To3238				2		1	1				4
42To3239	1							1			2
42To3520	1	5	4				1	1			12
42To3522						2		1			3
Isolates	6	10	1		2	10	5		1	1	36
Total	61	54	24	14	8	43	42	32	5	1	284

TABLE 5.59. Summary of Quantitative Data for Concave/Convex Scrapers in the Old River Bed Assemblage.

Variable	Statistic	Concave/ Convex Scraper	Concave/ Convex Scraper/ Graver	Concave/ Convex Scraper/Chisel	End/Concave/ Convex Scraper	End/Concave/ Convex Scraper/Graver	Side/Concave/ Convex Scraper/ Graver
Length (mm)	n	3	—	1	1	—	—
	Range	36.1–119.7					
	Mean	72.63		57.8	47.1		
	SD	42.784					
Width (mm)	n	3	—	1	1	—	1
	Range	14.5–25.1					
	Mean	20.13		21.3	37.5		27.1
	SD	5.331					
Thickness (mm)	n	13	5	1	1	1	1
	Range	4.7–13.9	5.0–9.8				
	Mean	10.11	7.72	9.4	8.3	11.3	11.2
	SD	3.166	2.217				

TABLE 5.60. Distribution of Old River Bed Concave/Convex Scrapers Across Seven Qualitative Variables.

Variable and State	Concave/Convex Scraper		Concave/Convex Scraper/Graver		Concave/Convex Scraper/Chisel		End/Concave/Convex Scraper		End/Concave/Convex Scraper/Graver		Side/Concave/Convex Scraper/Graver	
	n	%	n	%	n	%	n	%	n	%	n	%
Shape												
Linear	2	14.3										
Crescentic			1	16.7								
Irregular	7	50.0	5	83.3	1	100.0	1	100.0	1	100.0	1	100.0
Fragment	5	35.7										
Total	14	100.0	6	100.0	1	100.0	1	100.0	1	100.0	1	100.0
Size Category												
7	1	7.1	2	33.3								
8	1	7.1										
9	1	7.1										
10	1	7.1	1	16.7	1	100.0						
11			1	16.7			1	100.0				
12	1	7.1	1	16.7					1	100.0		
14	1	7.1										
Indeterminate	8	57.1	1	16.7							1	100.0
Total	14	100.0	6	100.0	1	100.0	1	100.0	1	100.0	1	100.0
Cortex												
Absent	14	100.0	5	83.3	1	100.0	1	100.0	1	100.0	1	100.0
>50% dorsal			1	16.7								
Total	14	100.0	6	100.0	1	100.0	1	100.0	1	100.0	1	100.0
Use-Wear												
Present	2	14.3							1	100.0		
Indeterminate	12	85.7	6	100.0	1	100.0	1	100.0			1	100.0
Total	14	100.0	6	100.0	1	100.0	1	100.0	1	100.0	1	100.0
Employable Units												
1	4	28.6										
2	6	42.9	2	33.3								
3	3	21.4	3	50.0			1	100.0	1	100.0		
4	1	7.1			1	100.0					1	100.0
5			1	16.7								
Total	14	100.0	6	100.0	1	100.0	1	100.0	1	100.0	1	100.0
Weathering Damage												
Minimal	3	21.4	1	16.7								
Medium	4	28.6	2	33.3							1	100.0
Heavy	3	21.4	2	33.3			1	100.0				
Extreme	1	7.1										
Not recorded	3	21.4	1	16.7	1	100.0			1	100.0		
Total	14	100.0	6	100.0	1	100.0	1	100.0	1	100.0	1	100.0
Maximum Angle (°)												
21–30	1	7.1										
31–40	1	7.1	1	16.7								
41–50			1	16.7								
51–60	3	21.4					1	100.0	1	100.0		
61–70	2	14.3										
71–80	2	14.3	2	33.3								
81–90	5	25.7	2	33.3	1	100.0					1	100.0
Total	14	100.0	6	100.0	1	100.0	1	100.0	1	100.0	1	100.0

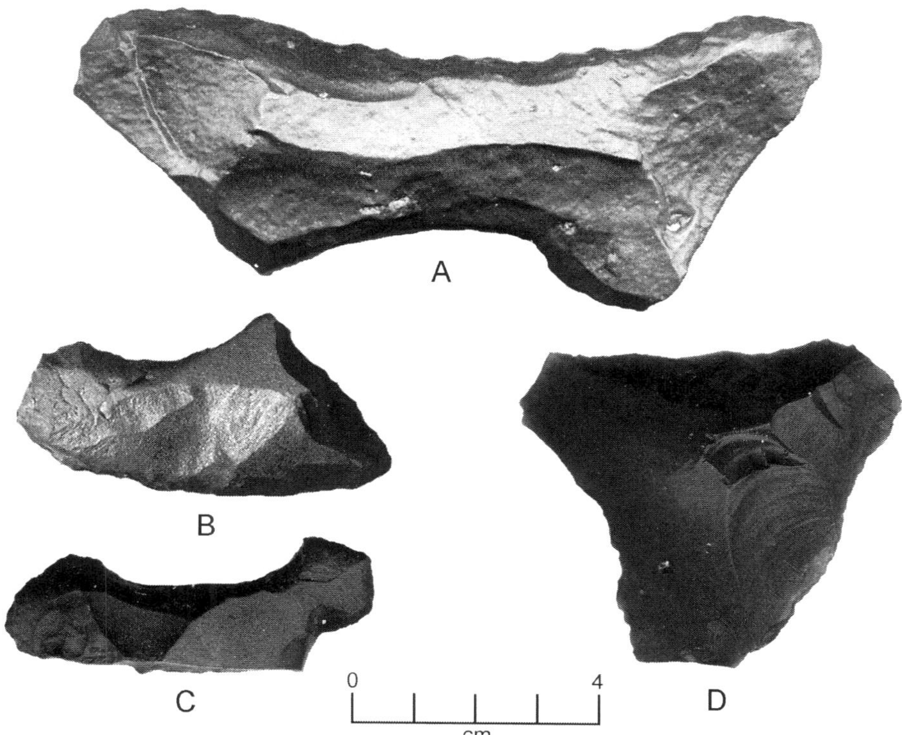

FIGURE 5.57. Examples of concave/convex scrapers in the Old River Bed assemblage: (*A*) 42To2552, FS 2; (*B*) 42To2767, FS 12; (*C*) 42To3233, FS 7; (*D*) 42To2955, FS 6.

concave/convex scrapers that are not fragments (13 of 16, 81.3 percent) are irregularly shaped. The locations of concave/convex scrapers are shown in Table 5.58.

Circular Scrapers

Circular scrapers are generally circular or oval in shape with retouch around all or most of the periphery. In contrast to most other scraper forms, circular scrapers do not often occur in combination with other tools. There are eight circular scrapers in the ORB assemblage (Figure 5.58; Table 5.53), none of which occur in combination with other tools. Half are manufactured from FGV; two, from obsidian; and two, from chert (Table 5.53). Quantitative data for these tools are shown in Table 5.61, and qualitative data are shown in Table 5.62. Locations are shown in Table 5.58.

Gramly suggests that circular scrapers are, in most instances, "resharpened and reworked endscrapers with their bulbs of percussion flaked away" (1990:14). This certainly could be the case for at least some of the ORB specimens, such as those in Figure 5.58C–D. Others, however, cannot be evaluated given their heavy weathering damage. On the average, both size and maximum angle are smaller for circular scrapers than for end scrapers, but the smaller sizes and angles fall within the

ORB ranges for end scrapers (see Table 5.55). The circular scrapers all have a single EU (Table 5.62), which in each case extends continuously around the perimeter.

Slug Scrapers

A total of 46 slug scrapers have been identified in the ORB assemblage (Figure 5.59; Table 5.53). These tools have also been referred to as "limaces" (Gramly 1982:37–38, 1990:32) and "flake shavers" (Grimes and Grimes 1985:40). The term *slug scraper* refers to their morphology—long, linear (generally unifacial) tools with steep edge angles; in cross section they are either flat-based or slightly curved. Slug scrapers sometimes occur in combination with gravers, three of which are represented in the ORB assemblage.

The majority (*n* = 27, 58.7 percent) of slugs are made from FGV (Table 5.53). It comes as no surprise that all 44 complete specimens are linear in shape, and the majority have steep angles, but they range considerably in size; the range in all three metric size variables (Table 5.61) as well as in size categories (Table 5.62) is quite large. Two forms have been recognized: a bipointed form and one on which the distal (working) end is blunted (Gramly 1990; Grimes and Grimes 1985). None of the ORB specimens, however, have a blunted end.

FIGURE 5.58. Examples of circular scrapers in the Old River Bed assemblage: (*A*) 42To1862, FS 10; (*B*) 42To1872, FS 9; (*C*) DPGIF 804; (*D*) 42To3219, FS 20; (*E*) DPGIF 685.

TABLE 5.61. Summary of Quantitative Data for Circular, Slug, and Beaked Scrapers in the Old River Bed Assemblage.

Variable	Statistic	Scraper Category					
		Circular Scraper	Slug Scraper	Slug Scraper/Graver	Beaked Scraper Type A	Beaked Scraper Type B	Beaked Scraper (Not Typed)
Length (mm)	n	7	22	2	7	12	1
	Range	23.9–43.2	23.3–75.7	46.0–98.8	28.1–44.5	29.0–75.8	
	Mean	32.91	52.95	72.40	36.07	45.65	52.8
	SD	7.567	14.587	37.335	6.251	15.428	
Width (mm)	n	7	34	3	13	11	2
	Range	24.2–39.6	10.2–25.6	16.5–25.0	8.9–30.4	17.4–41.4	19.4–25.7
	Mean	29.90	17.27	21.00	19.44	24.75	22.55
	SD	5.213	4.776	4.272	5.531	8.938	4.455
Thickness (mm)	n	8	43	3	17	14	10
	Range	6.0–11.3	5.3–21.0	9.0–19.6	6.1–14.5	4.8–13.7	6.1–11.2
	Mean	9.05	10.50	14.50	8.76	8.44	8.14
	SD	2.060	4.197	5.311	2.400	2.588	1.786

Table 5.63 shows the presence of haft preparation, proximal thinning, and transverse breaks on the ORB slug scrapers. Based on the same criteria used for end and side scrapers, the large majority ($n = 31$) of the 37 (83.8 percent) specimens that could be evaluated appear to have been hafted or possibly hafted, a considerably greater proportion than is the case for end and side scrapers (see Table 5.56). Based on the specimens from the Bull Brook site in Massachusetts, which are primarily those with blunted ends, Grimes and Grimes (1985:40) suggested that these tools were used to shave or gouge hard materials and referred to them as flake shavers. Bipointed specimens are present at Bull Brook as well and are included by Grimes and Grimes in the

same "flake shaver" category; it is possible that pointed specimens were also used to gouge or groove hard materials. Unfortunately, use-wear could not be recorded for any of the 46 ORB specimens.

Beaked Scrapers

The 42 tools categorized as "beaked scrapers" in this analysis have been referred to variously as "narrow end scrapers" (Deller and Ellis 1992:60–63), "groovers" (Deller 1980:67), "chisels" (Mason 1981), and "beaks" or "beaked scrapers" (Ellis and Deller 1982; Storck 1997). This difference in terminology is likely due to the variation in the form and size of these objects. Recognizing this wide variation, we use the term *beaked scrapers*

TABLE 5.62. Distribution of Old River Bed Circular, Slug, and Beaked Scrapers Across Seven Qualitative Variables.

Variable and State	Scraper Category											
	Circular Scraper		Slug Scraper		Slug Scraper/Graver		Beaked Scraper Type A		Beaked Scraper Type B		Beaked Scraper (Not Typed)	
	n	%	*n*	%	*n*	%	*n*	%	*n*	%	*n*	%
Shape												
Triangular							1	5.9	4	28.6	1	9.1
Linear			41	95.3	3	100.0	7	41.2	6	42.9	1	9.1
Circular	5	62.5										
Oval	3	37.5										
Irregular							3	17.6	3	21.4	1	9.1
Fragment			2	4.7			6	35.3	1	7.1	8	72.7
Total	8	100.0	43	100.0	3	100.0	17	100.0	14	100.0	11	100.0
Size Category												
5			1	2.3								
6			2	4.7								
7			4	9.3			2	11.8			1	9.1
8	2	25.0	3	7.0			6	35.3	5	35.7		
9	2	25.0	6	14.0	1	33.3			3	21.4	1	9.1
10	2	25.0	7	16.3			1	5.9	1	7.1		
11	1	12.5	4	9.3					1	7.1		
12					1	33.3			2	14.3		
Indeterminate	1	12.5	16	37.2	1	33.3	8	47.1	2	14.3	9	81.8
Total	8	100.0	43	100.0	3	100.0	17	100.0	14	100.0	11	100.0
Use-Wear												
Absent	1	12.5										
Present									1	7.1	1	9.1
Indeterminate	7	87.5	43	100.0	3	100.0	17	100.0	13	92.9	10	90.9
Total	8	100.0	43	100.0	3	100.0	17	100.0	14	100.0	11	100.0
Employable Units												
1	8	100.0	12	27.9			13	76.5	5	35.7	8	72.7
2			15	34.9	1	33.3			5	35.7	1	9.1
3			12	27.9	1	33.3	3	17.6	4	28.6	2	18.2
4					1	33.3						
Indeterminate			4	9.3			1	5.9				
Total	8	100.0	43	100.0	3	100.0	17	100.0	14	100.0	11	100.0
Maximum Angle (°)												
11–20							1	5.9				
31–40	1	12.5	1	2.3			2	11.8	1	7.1	1	9.1
41–50			1	2.3								
51–60	3	37.5	4	9.3					2	14.3	1	9.1
61–70			4	9.3			1	5.9	2	14.3		
71–80	3	37.5	2	4.7			3	17.6	2	14.3		
81–90	1	12.5	22	51.2	3	100.0	3	17.6	5	35.7	4	36.4
Indeterminate			6	14.0			2	11.8			3	27.3
Not recorded			3	7.0			5	29.4	2	14.3	2	18.2
Total	8	100.0	43	100.0	3	100.0	17	100.0	14	100.0	11	100.0
Cortex												
Absent	8	100.0	41	95.3	3	100.0	16	94.1	13	92.9	11	100.0
<50% dorsal							1	5.9	1	7.1		
Indeterminate			2	4.7								
Total	8	100.0	43	100.0	3	100.0	17	100.0	14	100.0	11	100.0
Weathering Damage												
Minimum			1	2.3					1	7.1	1	9.1
Medium	3	37.5	5	11.6	2	66.7	9	52.9	4	28.6	4	36.4
Heavy	2	25.0	19	44.2	1	33.3	4	23.5	7	50.0	4	36.4
Extreme	3	37.5	10	23.3			3	17.6	2	14.3	2	18.2
Not recorded			8	18.6			1	5.9				
Total	8	100.0	43	100.0	3	100.0	17	100.0	14	100.0	11	100.0

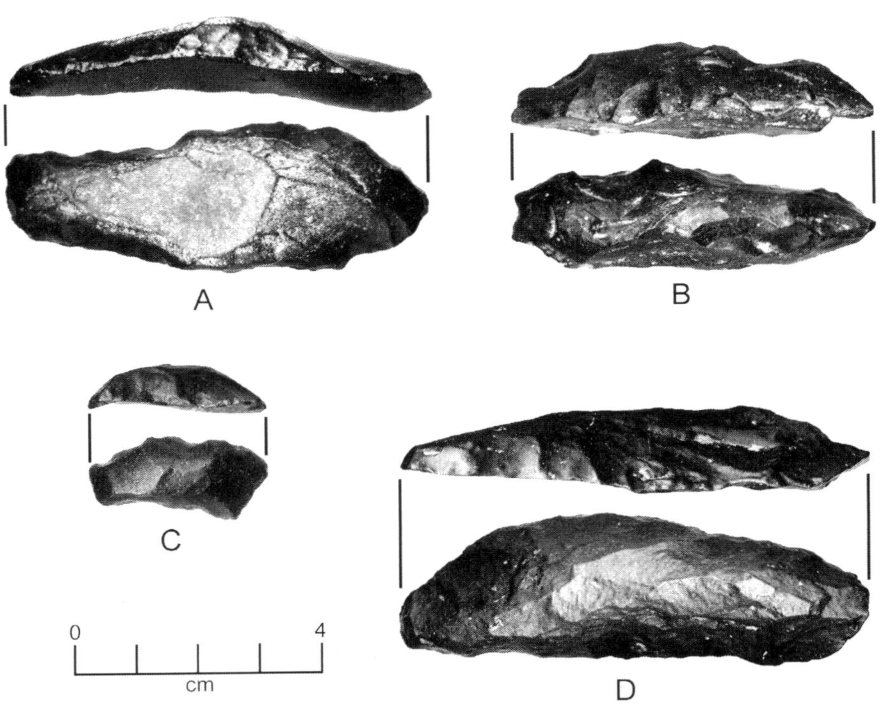

Figure 5.59. Examples of slug scrapers in the Old River Bed assemblage: (*A*) DPGIF 54; (*B*) ISO1, FS 1; (*C*) 42To3237, FS 13; (*D*) 42To3522, FS 5.

Table 5.63. Presence of Haft Preparation, Proximal Thinning, and Transverse Breaks on Old River Bed Slug Scrapers.

Haft Preparation/Proximal Thinning/Transverse Break	Slug Scraper	Slug Scraper/ Graver	Total	Hafting
A/A/A	5		5	NH
P/P/P	4		4	H
P/P/A	13	2	15	H
P/A/A		1	1	PH
A/A/P	1		1	PH
A/P/A	1		1	PH
I/A/A	2		2	I
I/A/P	1		1	PH
I/P/A	4		4	PH
I/I/A	3		3	I
I/I/P	4		4	PH
I/I/I	3		3	I
NR/NR/A	1		1	I
NA/NA/NA	1		1	NH
Total hafted	28	3	31	
Total not hafted	6		6	
Total indeterminate	9		9	
Overall total	43	3	46	

Note: A = absent; P = present; I = indeterminate; NR = not recorded; NA = not applicable; NH = not hafted; H = hafted; PH = possibly hafted.

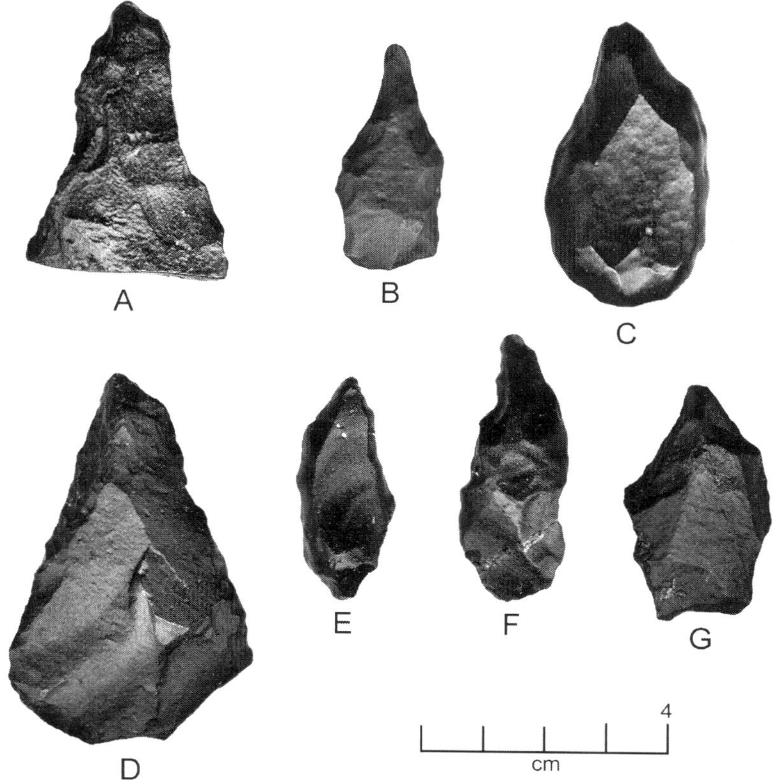

FIGURE 5.60. Examples of beaked scrapers in the Old River Bed assemblage. Type A: (*A*) 42To3235, FS 114; (*B*) 4To3520, FS 96. Type B: (*C*) DPGIF 618; (*D*) 42To3140, FS 13; (*E*) 42To3238, FS 13; (*F*) 42To3235, FS 26. Cannot be distinguished: (*G*) 42To3233, FS 32.

loosely and distinguish these from tools we refer to as "chisels." As defined by Storck, beaked scrapers have a "cutting or scraping tip which is plano-convex in cross-section, usually steeply retouched, and frequently quite thick, hence the term 'beaked'" (1997:65). Two forms of beaked scrapers are recognized in the ORB assemblage: (1) those for which the focus of use appears to have been limited to the tip or "beak," which we refer to as type A (*n* = 17; Figure 5.60A–B), and (2) those on which retouch occurs around the circumference to create scraper edges (*n* = 14; Figure 5.60C–F), which we refer to as type B. Eleven specimens were not identifiable to either A or B (Figure 5.60G).

Twenty-seven beaks are made from FGV; 14, from obsidian; and one, from chert (Table 5.53). As is evident from Figure 5.60 as well as Tables 5.61–5.62, the beaked scrapers range considerably in size, but type B beaks are, on the average, larger than type A beaks. More than half (60.7 percent) of the 28 for which edge angle was evaluated have angles greater than 70°. The majority have only a single EU (the beak), but nearly 40 percent have two or more.

Gravers and Notch

Sixty-six gravers have been identified in the ORB assemblage; 30 occur in combination with scrapers (Figure 5.61; Tables 5.64–5.65). A large portion of ORB gravers are made from FGV toolstone (45.5 percent), followed by obsidian (28.8 percent). A relatively large number (*n* = 15, 22.7 percent) are also made from chert. Single-tool gravers are, on average, smaller than the combination scraper/gravers, likely because the primary tool in the latter case is the scraper rather than the graver.

Most of the ORB gravers are single-spur gravers (Table 5.65), but a number have two or more spurs. The maximum angle of the spur ranges from 31 to near 90°. On average, the number of EUs is greater for scraper/gravers than for single-tool gravers, but of course this relates, at least in part, to the fact that the former are combination tools. Use-wear is indeterminate on all but one graver, which is worn.

One single-tool notch is present in the ORB assemblage (Figure 5.61G; Tables 5.64–5.65). It is made from obsidian, has a single EU, and is relatively small (size 7).

FIGURE 5.61. Examples of gravers (*A–F*) and the single notch (*G*) in the Old River Bed assemblage: (*A*) 42To3520, FS 56; (*B*) 42To2559, FS 95; (*C*) 42To2559, FS 83; (*D*) 42To1872, FS 52; (*E*) 42To3520, FS 17; (*F*) 42To3142, FS 13; (*G*) 42To3520, FS 35.

Use-wear is indeterminate. Site locations for the gravers and notch are shown in Table 5.66.

Chisels

Nine artifacts have been identified as possible chisels (Figure 5.62; Tables 5.67–5.68). These tools are identified by a long, narrow projection that ends in a rounded or blunt point. Five of these tools occur in combination with scrapers (Figure 5.62E–H), while four are single-tool artifacts. Eight of the ORB chisels are made from FGV; the ninth is obsidian. Edge angles are high; all but one are above 70°, but the large majority are above 80°. Use-wear was not recordable on any of these artifacts. Chisel locations are shown in Table 5.66.

Amorphous Unifaces

Amorphous unifaces are those unifaces that cannot be identified with another uniface category. There are 29 of these tools in the ORB assemblage (Table 5.68), primarily made of FGV (Figure 5.63). Use-wear is present on one and indeterminate for the other 28. Site locations for these tools are presented in Table 5.66.

Cores and Blades

There are 22 cores in the ORB assemblage (Figures 5.64 and 5.65; Table 5.69). They range widely in size but overall are relatively small given the size of many of the ORB tools. It is notable that nearly half (*n* = 10) are made from chert, a material that, as noted repeatedly, is

TABLE 5.64. Summary of Quantitative Data for Gravers and the Notch in the Old River Bed Assemblage.

		Tool Type		
Variable	Statistic	Graver	Scraper/ Graver	Notch
Length (mm)	*n*	13	14	—
	Range	25.5–51.7	25.2–98.8	
	Mean	37.05	43.99	
	SD	9.228	18.812	
Width (mm)	*n*	17	16	—
	Range	6.4–35.0	9.2–41.3	
	Mean	21.37	25.60	
	SD	7.476	8.866	
Thickness (mm)	*n*	29	28	1
	Range	1.1–11.1	3.8–19.6	
	Mean	7.50	9.37	2.7
	SD	2.238	3.647	

quite rare in the overall assemblage. Most of these are among the smallest of the cores and could easily represent the use of pebbles and small cobbles gathered from a streambed. We note, however, that the largest core, although still not that large (Figure 5.64F), is also made from chert.

Ten of the cores are referred to here as amorphous cores (Figure 5.64D–E), which are also commonly called generalized reduction cores. These were likely used for the detachment of flakes for expedient use.

TABLE 5.65. Distribution of Old River Bed Gravers and Notch Across Five Qualitative Variables.

Variable and State	Tool Type						Total	
	Graver		Scraper/Graver		Notch			
	n	%	n	%	n	%	n	%
Raw Material								
Chert	7	19.4	8	25.0			15	22.4
Obsidian	11	30.6	8	25.0	1	100.0	20	29.9
Fine-grained volcanics	18	50.0	14	43.8			30	44.8
Quartzite			2	6.3			2	3.0
Total	36	100.0	32	100.0	1	100.0	67	100.0
Size Category								
6			1	3.1			1	1.5
7	5	13.9	4	12.5	1	100.0	10	14.9
8	11	30.6	3	9.4			14	20.9
9	6	16.7	4	12.5			10	14.9
10	5	13.9	5	15.6			10	14.9
11	1	2.8	3	9.4			4	6.0
12			4	12.5			4	6.0
Indeterminate	8	22.2	8	25.0			14	20.9
Total	36	100.0	32	100.0	1	100.0	67	100.0
Employable Units								
1	19	52.8			1	100.0	20	29.9
2	7	19.4	13	40.6			17	25.4
3	4	11.1	12	37.5			17	25.4
4	4	11.1	4	12.5			8	11.9
5	1	2.8	3	9.4			4	6.0
Indeterminate								
Not recorded	1	2.8					1	1.5
Total	36	100.0	32	100.0	1	100.0	67	100.0
Spurs								
1	22	61.1	24	75.0			46	68.7
2	8	22.2	5	15.6			13	19.4
3		25.6	1	3.1			3	4.5
4	2	5.6					2	3.0
5	1	2.8					1	1.5
Not recorded	1	2.8	2	6.3			1	1.5
Not applicable					1	100.0	1	1.5
Total	34	100.0	32	100.0	1	100.0	65	100.0
Weathering Damage								
Minimal	2	5.6	7	21.9			9	13.4
Medium	7	19.4	11	34.4	1	100.0	18	26.9
Heavy	19	52.8	8	25.0			26	38.8
Extreme	5	13.9	2	6.3			7	10.5
Not recorded	3	8.3	4	9.4			7	10.5
Total	36	100.0	32	100.0	1	100.0	67	100.0

Site	Category					
	Graver	Scraper/ Graver	Notch	Chisel	Chisel/ Scraper	Amorphous Uniface
42To1152		1				
42To1153	3	2				
42To1157						1
42To1161	2					
42To1163	1	1				
42To1165		2				
42To1169						1
42To1172		1				2
42To1173						1
42To1182	1					1
42To1358		2		1		2
42To1369		1				
42To1371	1					3
42To1668	1					
42To1674						1
42To1677	1					1
42To1678	1					
42To1683						2
42To1684		2				
42To1685		1			1	
42To1686	3	2				1
42To1688	3	2		1		
42To1689	1					
42To1860	1					
42To1861		1				1
42To1862		1				
42To1872	1			1		
42To1873	1					
42To1876		1				1
42To1877					1	
42To1878	1					
42To1920					1	
42To1924				1		3
04DMO1		1				
42To2556	1					
42To2559	2					
42To2947	1					
42To3140						1
42To3142	2					
42To3219	1	2				
42To3223	1	1				
42To3224						1
42To3229						
42To3235		1			1	1
42To3237	1	2				
42To3239		1				
42To3520	3	1	1			1
42To3522		1				
Isolates	2				1	4
Total	36	30	1	4	5	29

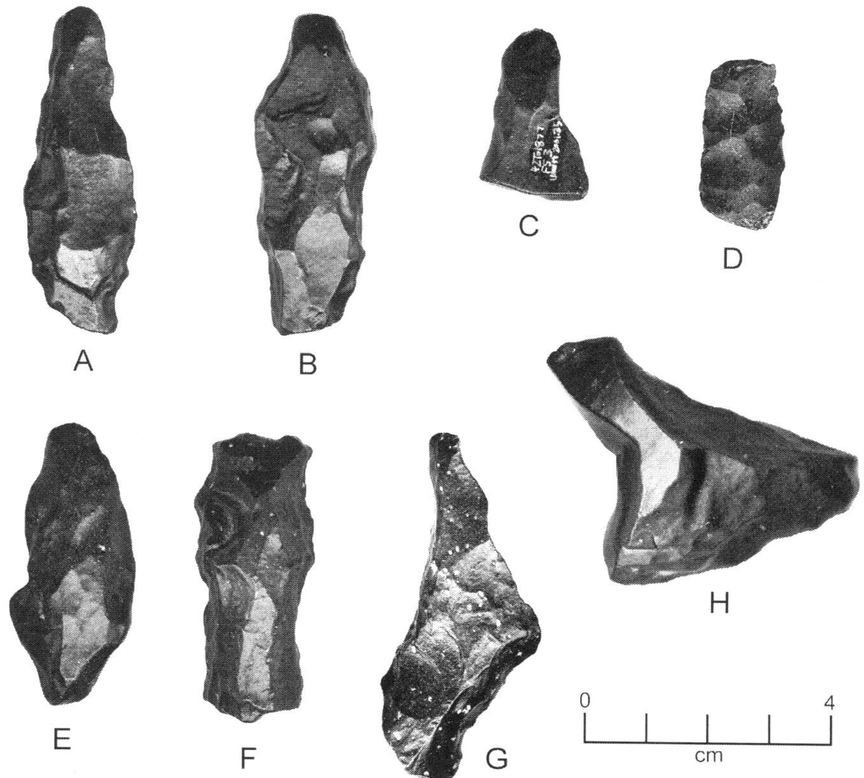

FIGURE 5.62. Chisels in the Old River Bed assemblage: (*A*) 42To1924, FS 103; (*B*) 42To1688, FS 55; (*C*) 42To1877, FS 3; (*D*) 42To1358, FS 4; (*E*) 42To1872, FS 50; (*F*) DPGIF 755; (*G*) 42To3235, FS 84; (*H*) 42To1685, FS 19. (Not shown: 42To1920, FS 25.)

One FGV core was found in association with a flake that had been struck from it (Figure 5.64E); the flake shows no retouch. Amorphous cores range considerably in size, weighing between 8.4 and 76.8 g (Table 5.69). They range widely in raw material as well.

Eight specimens are referred to as pebble cores (Figure 5.64A–C, G; Table 5.69). These are generally small and retain cortex. Four of the eight are chert, again suggesting the use of pebbles from a streambed. One specimen has been designated a disc core because of its round, relatively flat shape (Figure 5.64F). It is made from chert and is the largest of the cores.

The final three cores are prismatic blade cores; two are wedge-shaped (Figure 5.65A–B), and the third is conical (Figure 5.65C). All three are made from chert. For blade cores the ORB specimens are relatively small (as compared with Clovis blade cores, for example; see below), and the products are more appropriately referred to as "bladelets." Four bladelets have been recovered, three of which are shown in Figure 5.65D–F. The presence of blade cores as well as four bladelets in the ORB assemblage is intriguing. With the exception of those associated with Paleoindian assemblages, true blades are relatively rare in North America (see Parry 1994 for a discussion of North American blade technologies). Although the existence of prismatic blades in Paleoindian assemblages has been known for more than 60 years (e.g., Green 1963; Witthoft 1952), little attention has been paid to them until recently, except, perhaps, in the literature on the Northeastern record.

TABLE 5.67. Summary of Quantitative Data for Chisels in the Old River Bed Assemblage.

Variable	Statistic	Chisel Category	
		Chisel	Chisel/Scraper
Length (mm)	*n*	1	2
	Range	—	49.1–57.8
	Mean	52.6	53.45
	SD	—	6.152
Width (mm)	*n*	2	4
	Range	18.6–20.5	17.9–21.8
	Mean	19.55	20.33
	SD	1.344	1.733
Thickness (mm)	*n*	4	5
	Range	7.2–8.7	8.2–17.7
	Mean	8.09	11.58
	SD	.633	3.700

TABLE 5.68. Distribution of Old River Bed Chisels and Amorphous Unifaces Across Seven Qualitative Variables.

| | Tool Type | | | | | |
| | Chisel | | Chisel/Scraper | | Amorphous Uniface | |
Variable and State	n	%	n	%	n	%
Raw Material						
Chert					4	13.8
Obsidian	1	25.0			3	10.3
Fine-grained volcanics	3	75.0	5	100.0	21	72.4
Quartzite					1	3.4
Total	4	100.0	5	100.0	29	100.0
Size Category						
7					3	10.3
8					1	3.4
9			1	20.0	1	3.4
10	1	25.0	2	40.0		
11					2	6.9
12					1	3.4
Indeterminate	3	75.0	2	40.0	21	72.4
Total	4	100.0	5	100.0	29	100.0
Shape						
Linear	2	50.0	1	20.0	3	33.3
Oval			1	20.0	1	11.1
Irregular			2	40.0	2	22.2
Fragment	2	50.0	1	20.0	3	33.3
Total	4	100.0	5	100.0	9	100.0
Employable Units						
1	4	100.0	2	40.0	29	100.0
2			2	40.0		
4			1	20.0		
Total	4	100.0	5	100.0	29	100.0
Use-Wear						
Present					1	3.4
Indeterminate					28	96.6
Total					29	100.0
Maximum Angle (°)						
61–70			1	20.0		
81–90	3	75.0	4	80.0		
Not recorded	1	25.0			29	100.0
Total	4	100.0	5	100.0	29	100.00
Weathering Damage						
Medium	2	50.0	2	40.0	4	44.4
Heavy	2	50.0	2	40.0	4	44.4
Not recorded			1	20.0	1	11.1
Total	4	100.0	5	100.0	9	100.0

FIGURE 5.63. Examples of amorphous unifaces in the Old River Bed assemblage: (*A*) 42To3520, FS 116; (*B*) 42To3224, FS 13; (*C*) 42To3229, FS 22.

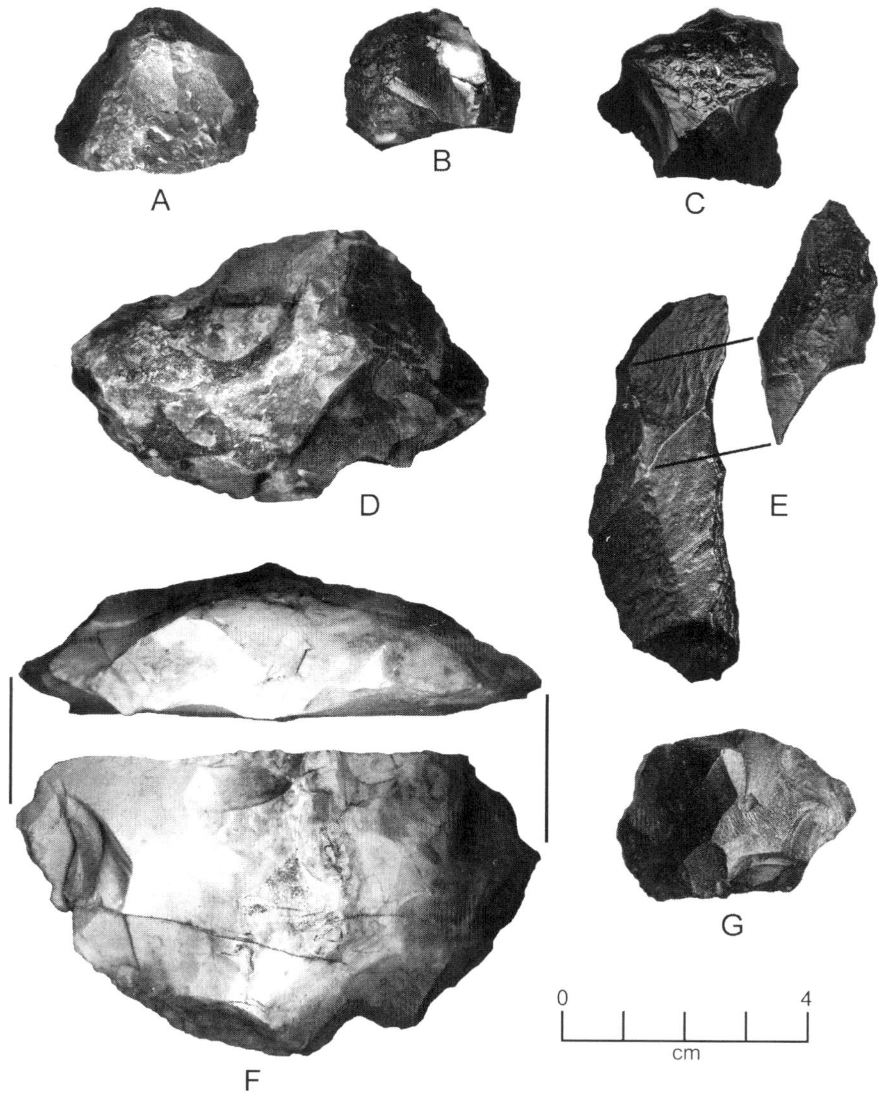

FIGURE 5.64. Examples of pebble (*A–C, G*), amorphous (*D–E*), and disc (*F*) cores in the Old River Bed assemblage: (*A*) 42To1173, FS 7; (*B*) 42To1173, FS 11; (*C*) 42To1182, FS 27; (*D*) 42To1358, FS 39; (*E*) 42To3237, FS 63a and b; (*F*) DPGIF 463; (*G*) 42To1173, FS 9.

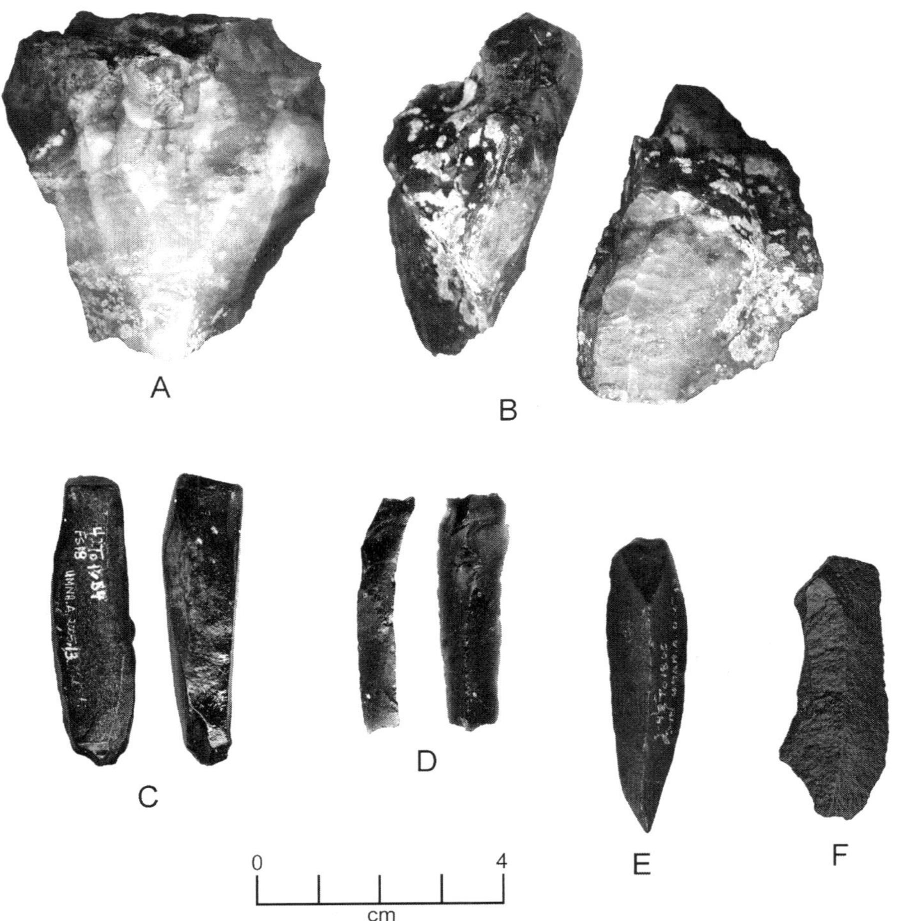

FIGURE 5.65. Examples of blade cores (*A–B*: wedge shaped; *C*: conical) and blades (*D–F*) in the Old River Bed assemblage: (*A*) 42To1358, FS 69; (*B*) 42To1671, FS 5; (*C*) 42To1684R, FS 18; (*D*) 42To2559, FS 56; (*E*) 42To1860, FS 18; (*F*) 04DM02, FS 1.

The importance of prismatic blades in Clovis assemblages has been stressed by Michael Collins (1999a, 1999b; Collins and Lohse 2004; see also Bradley et al. 2010). Collins describes Clovis blades as ranging between 50 and 160 mm in length and as having small platforms, flat bulbs, and marked curvature. Collins (1999a) outlined a set of specific criteria by which these blades could be identified, one of which is a curvature index. Collins and Lohse argue that "this constellation of traits is so distinctive that such blades are almost as diagnostic as are Clovis projectile points" (2004:176).

We have recently inventoried reported blades across North America, distinguishing between those that conform to all of Collins's criteria and those that may conform to some but not all (Beck and Jones 2010a; see also Beck and Jones 2015 for updated map of blade distribution). We found that those Collins would refer to as Clovis blades are most numerous in the southern Plains and parts of the Southeast, with hundreds reported from some quarry sites. Once out of those

areas, however, blades become much less prominent. It is notable that most of the reported blades from the Northeast are smaller than, and do not fit all of the criteria for, Clovis blades.

In the Great Basin very few blades have been reported, many of which do not fit all of Collins's criteria for Clovis blades. The ORB conical core, which is similar in form to Clovis conical cores (see Bradley et al. 2010), is much smaller than those cores. It is, in fact, more similar to many of those in the Northeast. This is true of the bladelets as well. We have argued that the paucity of prismatic blades in the Great Basin, especially those identifiable with Clovis, is evidence that Clovis technology is a relatively late arrival in the region (Beck and Jones 2010a). The non-Clovis blades probably represent a continuation for a short time of a technology that became too expensive relative to general core technology (Beck and Jones 2015). This may be true as well of the Northeastern blades since they are associated with post-Clovis fluted points. The Dugway cores and

TABLE 5.69. Quantitative and Qualitative Data and Site Locations for Old River Bed Cores and Blades.

Core/Blade Type and Site	Field Specimen No.	Raw Material	Actual Weight (g)	Size Category	Abrasion Damage
Amorphous Cores					
42To1172	7	FGV	37.50	11	Heavy
42To1172	10	Obsidian	8.40	8	Heavy
42To1182	27	FGV	14.39	9	Medium
42To1358	27	Obsidian	13.50	10	Heavy
42To1358	39	Quartzite	76.80	12	Minimal
42To1358	52	Chert	14.80	9	Medium
42To1371	52	FGV	29.38	10	Extreme
42To1666	6	Chert	56.56	12	Minimal
42To1666	7	Chert	19.76	9	Minimal
42To3237	63A	FGV	28.47	11	Minimal
Pebble Cores					
42To1173	7	Chert	23.7	10	Minimal
42To1173	9	Chert	18.0	10	Medium
42To1173	11	Chert	17.5	9	Medium
42To1358	71	Obsidian	26.30	10	Medium
42To1358	84	Obsidian	8.10	9	Medium
42To1359	5	Chert	13.30	9	Medium
42To3521	2	Obsidian	14.17	9	Heavy
DPGIF 198		Obsidian	5.0	8	Heavy
Disc Core					
DPGIF 463		Chert	54.76	Not recorded	Minimal
Wedge-Shaped Blade Cores					
42To1671	5	Chert	54.30	11	Minimal
42To1358	69	Chert	103.10	12	Minimal
Conical Blade Core					
42To1684R	18	Obsidian	12.88	9	Extreme
Prismatic Bladelets					
04DM02	1	FGV	3.48	9	Medium
2To1860	15	FGV	3.10	Not recorded	Heavy
42To1860	18	FGV	5.61	8	Medium
42To2559	56	Obsidian	2.67	—	Medium

Note: FGV = fine-grained volcanics.

blades are likely representative of the later, non-Clovis, blades given the lack of fluted points in the study area and the proposed early Holocene age of the assemblage.

DISCUSSION AND CONCLUSION

We have stated several times throughout our discussions that the Dugway ORB lithic tool assemblage is unusual in our experience. When comparisons are made with other Paleoarchaic assemblages throughout the Great Basin, the ORB assemblage is distinctive in several regards:

- It exhibits unusual patterns of raw material use.
- It contains projectile points of novel morphology, which often are very small.
- It evidences high levels of recycling and expedient tool manufacture.

We have suggested that these features are probably related. It seems that some of the occupants of the ORB delta were inadequately provisioned with raw material for tool manufacture. When their supplies of toolstone began to run out, they turned in part to the only source of raw material available to them in the marsh—the tools and tool fragments left by earlier visitors to the area. The size and condition of the tool fragments available for reuse gave rise to the diminutive tools, particularly projectile points, that are so common in the ORB assemblage.

Cross-dating and seriation combined with the chronology of landform development in the ORB delta demonstrate that human use of the area probably began sometime between 11,000 and 10,000 ^{14}C BP and continued into the early Holocene. As noted in Chapter 7, the most intense occupation likely occurred between ~10,500 and ~8800 ^{14}C BP. This is supported by the prevalence of Early Holocene projectile points in the assemblage and the presence of artifacts from the Eden and Cody complexes, as well as the presence of the non-Clovis blade cores and blades (see Beck and Jones 2010a). This occupation may have overlapped with the final phases of high-energy channel development associated with the end of the Gilbert lake phase. The principle period of occupation, however, probably coincides with the transgression and regression of the Gilbert episode and its immediate aftermath. With recession of the Gilbert lake, but with continued strong flow of the Old River, extensive marshlands developed in the ORB delta. After ~10,000 ^{14}C BP, when general drying throughout the eastern Great Basin caused smaller wetlands to disappear in many valleys, the delta maintained a vigorous wetland that lasted until ~8500 ^{14}C BP, which must have been attractive to Paleoarchaic foragers.

The seriation order of 28 ORB sites suggests that the period of occupation (at least at those 28 sites) occurred over a period of between 1,500 and 2,300 years. This is supported by the fact that both WST and EH points are present throughout the order. However, the peak of Silver Lake is near the earlier end, while the peak of Stubby occurs a little more than halfway up. The distributions of the remaining EH types are spotty, but in general, these points appear to be more abundant in the upper two-thirds of the order. These patterns suggest that use of WST points was more prevalent early on and then subsequently gave way to a greater reliance on EH types. Thus it appears that the earliest occupants of the ORB came with a typical WST tool kit, which consisted of unfluted concave-base and large stemmed projectile points, large leaf-shaped production bifaces, and other curated tools such as scrapers and crescents. These tools were made of obsidian from distant sources and FGV toolstone from sources nearer ORB, and a few were made from chert, the sources of which are unknown. With respect to stemmed points, which were manufactured primarily from obsidian and FGV toolstone, material selection was patterned in an expectable manner.

For unfluted concave-base points, crescents, scrapers, gravers, drills, and chisels, which are usually made from chert, however, the patterns are not typical.

Later occupants of the region appear to have practiced a more expedient and opportunistic approach, at least with respect to projectile points. Points, such as Stubbies, are smaller, often have irregular shapes, retain fracture facets and other obstructions on functional segments (stems and cutting edges), and are shaped by minimal retouch. These later assemblages appear to have developed from high levels of tool recycling—that is, from the conversion of broken and expended tools left by earlier occupants into new tool forms. While some of the projectile points we attribute to the Stubby and other short stemmed classes may have been manufactured by conventional biface reduction (i.e., using pressure retouch to shape a bifacially thinned preform into a finished point), the majority reflect less attention to achieving perfect longitudinal and cross-sectional symmetry. This may simply be a product of what the knapper had to work with and not any intent to produce "sloppy work." Starting from small, irregularly shaped pieces of toolstone appears to have limited the control of toolmakers in thinning preforms.

It seems to logically follow from the assemblage characteristics that occupants during the later period often operated with inadequate supplies of tool material. This may have been the result of longer residential time in the marshes, increased population size, economic changes resulting in the need for different types of tools (e.g., Stubbies), or some combination of these factors. Madsen et al. (Chapter 7) argue that toolstone was obtained logistically by a few individuals while the majority remained in the marsh. This is likely correct, but the degree of recycling indicates that raw material obtained during these trips had to be supplemented substantially through the use of old discarded tools as blanks for new ones. This might explain the low incidence of chert. The earliest occupants may have brought chert tools in with them, but the tools commonly made from this material, such as scrapers, do not generally break in such a way that a new, unrelated tool can be manufactured from the remnants. Thus, obsidian and FGV from old points and large bifaces were recycled into scrapers, crescents, gravers, and drills.

The early Holocene saw the shrinking of marshes throughout the Great Basin, and thus foragers dependent on these habitats had to travel increasingly greater distances between them. For example, people living at the Sunshine Locality in Long Valley of eastern Nevada during the later Pleistocene and the initial part of the early Holocene focused their efforts on an extensive marsh. In southern Butte Valley, just to the east, WST sites are scattered throughout the valley, perched along old lake terraces. However, sites containing Early Holocene and Archaic point types are limited to only one of these terraces, known as Hunter Point. This suggests that foragers were utilizing smaller patches of marshes rather than an extensive one as in Long Valley, and as these smaller habitats disappeared, they focused on the marsh remaining around Hunter Point.

By the mid-early Holocene marshes probably remained only in those areas where at an earlier time they had been extensive, such as the Sunshine Locality and the ORB delta. Areas in between, such as Butte Valley, where most of these habitats had disappeared, were almost completely dry. Thus, a foraging population utilizing the Sunshine Locality and ORB marshes probably remained longer within each habitat, expanding their diet to include more small seeds and other plant foods they had previously ignored. However, at the Sunshine Locality we do not see the recycling of toolstone that is so evident in the ORB delta. We believe that this relates to the nature of the marshes in these two areas, which differs considerably.

The Sunshine Locality is a 3-km-long, relatively narrow band of continuous artifact distribution along Sunshine Wash. The marsh that developed along the wash was easily entered by foragers on a daily basis from a base camp adjacent to it. Both chert and FGV were readily available within 10 to 15 km. Obsidian tools, made from distant sources, were heavily resharpened but eventually discarded. These tools were rarely recycled.

The marsh that developed in the ORB delta, however, covered a much larger area and consisted of a labyrinth of causeways, streams, and dry sand and gravel patches, all within the marsh habitat. Once deep inside the marsh, foragers would more likely have remained there rather than retreating to the edges of the wetland every day. Although most of the ORB artifacts occur as isolates or in small site assemblages, there are several sites with relatively large, diverse assemblages that could represent base camps (see Chapter 4). The downside of establishing base camps within the marsh is that toolstone would have been less accessible than in Long Valley. However, it *was* available in the form of previously deposited artifacts, some of which were quite large. Thus foragers probably collected any large pieces they came across and took them back to camp. It is interesting that production bifaces, which for the most part represent the largest tools and thus had the most potential for recycling, are made mostly from FGV (~85 percent), while the large majority of Stubbies, almost all of which were probably made from recycled material, are obsidian (~86 percent). This may reflect a preference for obsidian for recycling since an obsidian

biface would be easier to re-form than one made from FGV. If this is the case, it would account for the heavy representation of FGV among production bifaces because most of those made of obsidian would have been recycled. On the other hand, the latter may simply represent the replacement of large stemmed points, many of which are made from obsidian (see Table 5.10), in local material, as we have seen in eastern Nevada. In any event, the ORB assemblage appears to represent a unique situation.

Duke (2011; Duke and Young 2007) makes a similar argument for increasing residence time in the distal ORB throughout the early Holocene. He argues that highly mobile foragers, such as those in the Great Basin during the terminal Pleistocene and earliest early Holocene, needed a tool kit that was "transport efficient," meaning that they would have maximized utility relative to weight: Tools would have been larger (storage for raw material), and many objects were combination tools (reducing the overall number of objects). Tools would also have been made from high-quality raw material such as obsidian. As foragers began to remain in the marsh for longer periods, transport efficiency would have been sacrificed "for more cost-effective acquisition of toolstone to be used as needed according to increasingly predictable tool-use expectations" (Duke 2011:225–226). Thus tools would become smaller and more expedient, and use would have been made of lower-quality raw material such as FGV. We agree with Duke concerning increased residence in the marsh but part company with him regarding how later foragers acquired their toolstone.

Duke argues that although some scavenging did occur in the distal ORB, this approach to replenishing toolstone supplies would not have made up a regular part of the technology of these later residents because they would not have been able to plan ahead (i.e., could not know in advance where they would find old tools to scavenge) and thus the risk of running out of raw material would be high. Caching old artifacts would alleviate this problem, but upon their return foragers would have to relocate these caches, which could be difficult inside the marsh. Theoretically this is a sound argument, but it is not supported by the proximal ORB assemblage. As we stated earlier, Stubbies, which are mostly made on blades of broken WST points or production bifaces, constitute more than a quarter of the point assemblage. Also, there is evidence that other tools, such as scrapers, chisels, and gravers, were made from recycled material as well. With this much recycling so evident in the assemblage it is difficult to argue that scavenging did not play a strategic role in raw material acquisition during the later use of the proximal ORB marsh. We agree with

Duke concerning increased residence in the marsh but part company with him regarding how later foragers acquired their toolstone.

These conjectures do not imply that later Paleoarchaic groups were any less mobile than those who earlier visited the ORB, that their foraging range had grown smaller, or that they made semipermanent residence in the delta, although one or more of these changes may have been in process. It does appear to us, however, that residence time at sites throughout their territory increased, probably in concert with diet breadth expansion (Beck and Jones 2010b; Jones, Beck, and Kessler 2003). Unfortunately, these ideas are difficult to test. Based on provenance analysis mentioned earlier, we can say with some certainty that the earliest occupants traveled between the Topaz Mountain and Brown's Bench obsidian sources, since tools are made from both glasses (see discussion by Madsen et al. in Chapter 7). We cannot be equally certain of the meaning of source provenance data in the Stubby assemblages for the simple reason that they are, in large part, recycled and thus bear the provenance imprint of the older assemblages.

Duke (2011) was fortunate enough to be able to separate a set of relatively early sites from the very latest occupations in the distal ORB and demonstrate that the two separate areas contributed different relative frequencies of WST and EH points. Thus, he was able to demonstrate longer residence time for later foragers in that area. We have not been so fortunate, even though there is a slight trend in the use of WST and EH points from the earliest to latest channel, with the exception of the Light Blue channel. The proximal ORB channels are intermingled, crosscutting each other, and thus distinct differences in usage cannot be observed.

In sum, the early occupants of the area probably practiced a fairly high degree of mobility, carrying the typical WST tool kit, one that was easily and efficiently transported. As marshes across the region shrank, foragers began to spend more time in large marshes such as that in the ORB delta, and thus the lithic technology of these later occupants was not so clearly organized by concerns of long tool use-life and the production of efficient forms for transport. Theirs was not a transport-efficient tool kit but one quickly drafted into use and easily left behind. This last point seems particularly evident in the fact that such a high percentage of projectile points are complete, and though many are diminutive, these points appear to retain their functionality. We may find through additional studies in the region that diminutive points made their way across the landscape—Danger Cave does, in fact, contain a fair number of small, roughly made points in the early Holocene DII levels. However, we may learn that Stubbies and other short stemmed points were designed for somewhat different uses than other Paleoarchaic point forms. This may also be the case for Silver Lake points, which were found to be functionally different from Cougar Mountain, Lake Mohave, Parman, and Ovate (similar to Haskett) points in the Sunshine assemblage (see Beck and Jones 2009:188–192; but see Duke 2011 for a different argument).

To summarize, two lithic technological patterns are represented in the proximal ORB assemblages. The initial one follows a common and widespread pattern of material-use practices indicating high mobility and large territorial use, with tools designed for long use-lives and multiple uses. The other, later, pattern is one of recycling and expedient manufacture, occasioned by shortages of raw material and designed for immediate deployment. Alteration of the traditional pattern may have begun fairly early in the occupation sequence as foragers began to remain for increasingly longer periods, utilizing a continually widening range of resources. These ideas, of course, are difficult to evaluate, first, because the WST tool forms could date to any time during the approximately 2,300 or more years of human presence and, second, because of the impact that such extensive recycling had on the record of tool use.

6 Toolstone Sourcing, Lithic Resource Use, and Paleoarchaic Mobility in the Western Bonneville Basin

David Page and Daron G. Duke

Paleoarchaic forager movement in the Great Basin is often inferred from the distance to geologic sources of raw materials found in a site or sample of sites. In this study we focus on a large sample of obsidian and fine-grained volcanic (FGV) artifacts characterized using X-ray fluorescence (XRF) spectrometry to identify the geologic sources of toolstone used across assemblages of Paleoarchaic age from the Old River Bed (ORB) delta. We compare these results with sources of a sample of later materials from the surrounding region and then relate the overall results to baseline models of spatial patterning developed by other researchers for toolstone use in the Great Basin (Jones, Beck, Jones, and Hughes 2003; Jones et al. 2012; Madsen 2007; Newlander 2012; Smith 2007, 2010).

History of XRF Sourcing Research in the Bonneville Basin

Early investigations into obsidian use in the eastern Great Basin includes work by Condie and Blaxland (1970) at Danger and Hogup caves (Holmer 1997; Hughes and Bennyhoff 1986). At that time the characterization of regional source materials was in its infancy, and the resultant source attributions were generally unsuccessful, although they set the stage for decades of research and refinement of the regional source record. Nelson and Holmes (1979) conducted a large-scale sourcing study that included geologic source material from 10 obsidian source areas in Utah, Nevada, and southern Idaho. These were compared with more than 50 sites across southwestern Utah, including a number of sites located south of the study area in Juab and Mil-

lard counties. Their results indicated a general pattern of obsidian use from the nearest available geologic source. A synthesis of research efforts involving obsidian use in eastern Idaho and geologic obsidian sources in southern Idaho was presented by Holmer (1997 and references therein). He viewed usage between areas and noted that materials in the Snake River Plain were more frequently transported parallel to the Snake River while little material was transported perpendicular to this geographic barrier. There have been a number of large-scale obsidian characterization projects done in the Great Basin since the mid-1990s, with many in the last decade conducted by federal or state agencies through cultural resource management (CRM) projects (e.g., see Hull 1994; Jackson et al. 2009; Johnson and Haarklau 2005; Moore 2009; Seddon 2005). Efforts to further characterize obsidian sources in southern Idaho were continued by Plager (2001) and reexamined by Wilson (2007).

Initial sourcing in the ORB delta was completed by Hughes on 34 specimens from several sites in the distal ORB delta near Wildcat Mountain (Arkush and Pitblado 2000 and references therein). The earliest geochemical sourcing in the proximal delta area was done at Camels Back Cave, where 116 obsidian artifacts were sampled (Hughes 2001). This was followed by initial investigations on materials from the proximal ORB delta itself, where 45 specimens were sourced as part of a lithic analysis project report completed by Jones et al. (Hughes 2004; Jones, Beck, and Kessler 2003). Investigations to determine sources of regional FGV were started independently by Duke (Carter et al. 2004) and Page (2008; Page, Schmitt, Dalldorf, and Wazaney 2012) in 2003. Jones, Beck, Jones, and Hughes (2003) identify an Eastern Conveyance Zone (ECZ), defined largely by sourcing data that include the Bonneville basin and the ORB delta. However, they exclude the area in their more recent revision (Jones et al. 2012:363).

Eastern Great Basin Lithic Terrane

Four primary toolstone varieties dominate Great Basin prehistoric lithic assemblages to varying degrees: obsidian, FGV, chert (cf. Luedtke 1992) (i.e., cryptocrystalline silicates, chert, and chalcedony), and fine-grained quartzite (orthoquartzite/quartz arenite). Here, we consider primarily obsidian and FGV because of the limited presence of the other toolstone types in Paleoarchaic (PA) collections from the ORB delta.

Figure 6.1 shows the primary sources of obsidian and FGV in and around the eastern Great Basin and Snake River Plain. Obsidian was used extensively by prehistoric hunter-gatherers in the Great Basin and peripheral areas because its physical properties are conducive to toolmaking. It is widely distributed across the geologic landscape (but see Thomas 2012), and varieties can be distinguished geochemically using a variety of techniques. For these reasons, a disproportional effort has been expended by researchers to understand the use of this raw material type. Much has been learned about the use of various obsidian geochemical types, and a great deal has been written about these studies (e.g., Amick 1993, 1995, 1999; Arkush and Pitblado 2000; Basgall 1989; Beck and Jones 1990b, 1994, 1997; Beck et al. 2002; Duke 2011; Elston 1990, 2005; Hughes 1984, 1998; Jones, Beck, Jones, and Hughes 2003; Jones, Beck, and Kessler 2003; Jones et al. 1997; Madsen and Schmitt 2005; Schmitt and Madsen 2005; Schmitt et al. 2003). Until recently, outcomes of FGV characterization studies in the Basin have been less successful, and the resulting literature is still limited (Arkush and Pitblado 2000; Duke 2011; Duke and Young 2007; Graf 2002; Jones et al. 1997; Page 2008).

Chert

In contrast to the lithic terrane east of the Rocky Mountains, where bedded chert of marine origin can sometimes extend for many, even hundreds of, miles (e.g., Knife River, Edwards Plateau, Flint Hills, Alibates, Burlington), the Great Basin contains numerous localized volcanic and sedimentary rock precipitates, including chalcedony, jasper, opalite, or sinter (cf. Luedtke 1992). These materials originate in the various veins and joints of different rock types, with those from volcanic rocks containing high concentrations of silica. Where sedimentary geology dominates the mountain ranges, the valley alluvium can contain sporadic to abundant sources of chert stretching for miles. These sources vary considerably and are usually found in relatively small nodules. An exception is the extensive Tosawihi outcrop in north-central Nevada (e.g., Elston 1992). In tectonically active areas hydrothermal events have produced chert outcrops often associated with metallic ore bodies such as gold. For example, in eastern Nevada there are numerous small outcrops of good-quality chert that are sometimes identifiable by color and texture.

Archaeological sites usually contain a mix of chert from diverse sources. Thus, while archaeological assemblages often contain chert artifacts, they can be difficult to attribute to sources, partly as a result of little attention being paid to source geochemistry. This situation is changing as new approaches, including inductively coupled plasma mass spectrometry and the use of handheld portable XRF machines, are being applied to this important material.

Obsidian

Obsidian is well known as a preferred material for flint knapping. Its glassy nature is conducive to refined percussion and pressure techniques, and flaked edges can be exceedingly sharp. Thin obsidian flakes can possess edges only 3 nm thick, or about three hundred-thousandths (.00003) the thickness of a human hair (Buck 1982). Obsidian does vary in its quality but only to near negligible amounts in most cases, with some varieties having a somewhat sugary texture and others possessing extensive spherulites or bands of bubbly glass or pumice inclusions.

Cobble size represents the most distinctive limiting factor in obsidian selection. It is usually gathered from surface or near-surface deposits in both primary and secondary contexts. It can range from pea- to marble-sized pebbles at some sources where it originated as volcanic ejecta to automobile-sized blocks where massive flows occurred; however, fist- to cantaloupe-sized cobbles conducive to most flint-knapping needs can be found at many primary sources.

Sourcing studies have taken place intensively since the 1980s to provide a nearly complete view of obsidian distribution throughout the western United States and can be viewed online at the Northwest Research Obsidian Studies Laboratory (NWROSL) website (http://www.obsidianlab.com). A distinctive pattern has emerged from these studies, showing that obsidian largely surrounds the central Great Basin, where it is nearly absent (Thomas 2012). Obsidian is ubiquitous in Oregon and nearby parts of Nevada and California and then scattered along the Sierra Nevada into the Mojave Desert through southern Nevada and back north through western Utah and across southern Idaho.

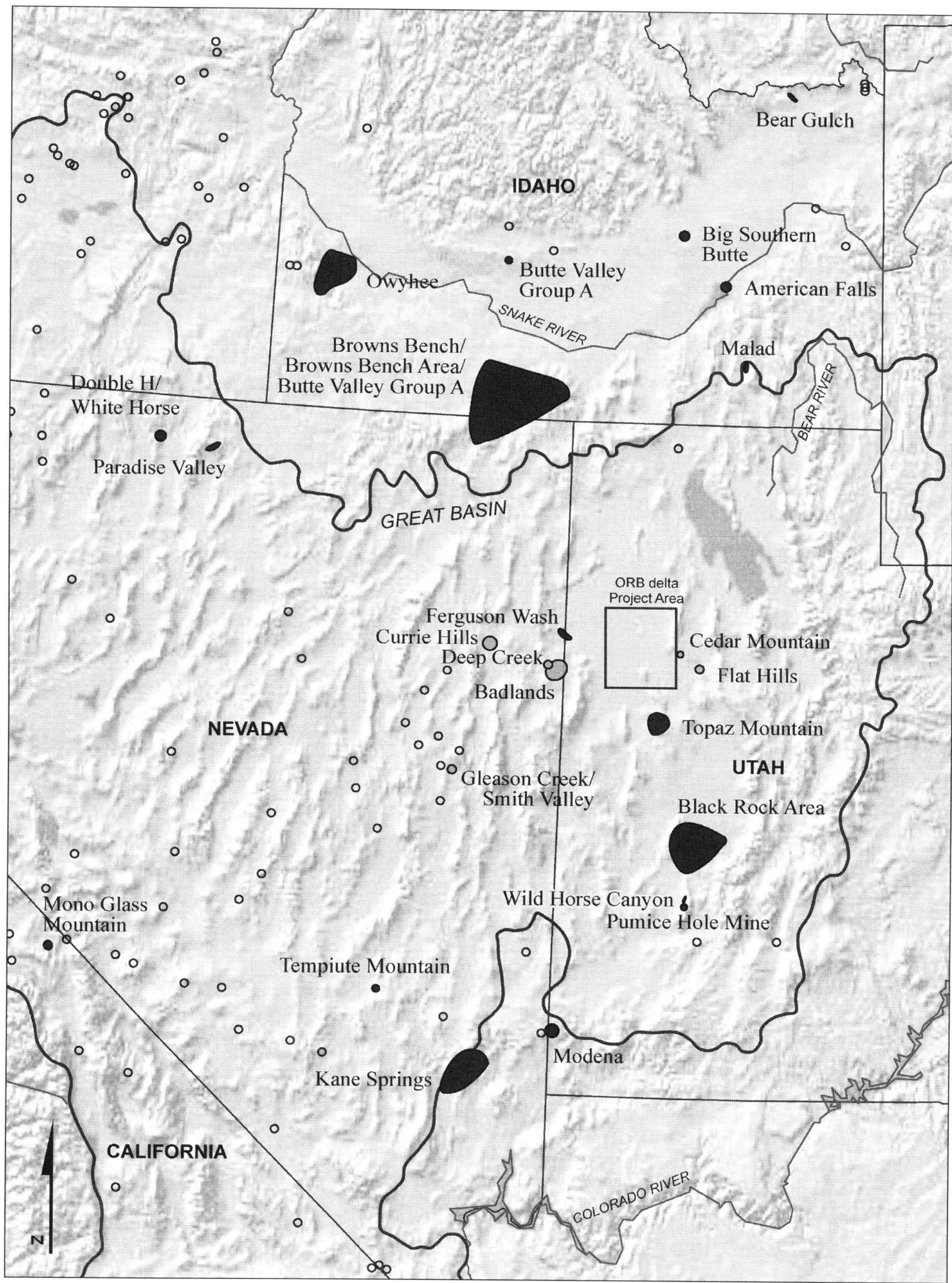

FIGURE 6.1. Geochemical source areas of obsidian (*dark gray polygons*) and fine-grained volcanic (*light gray polygons*) toolstone represented in the Old River Bed (ORB) delta (*rectangle*) and Dugway Proving Ground Archaic samples. Small open circles show select obsidian/fine-grained volcanic sources not represented in the sample analyzed (http://www.obsidianlab.com/) (base image from http://services.arcgisonline.com/ArcGIS/services/World_Shaded_Relief).

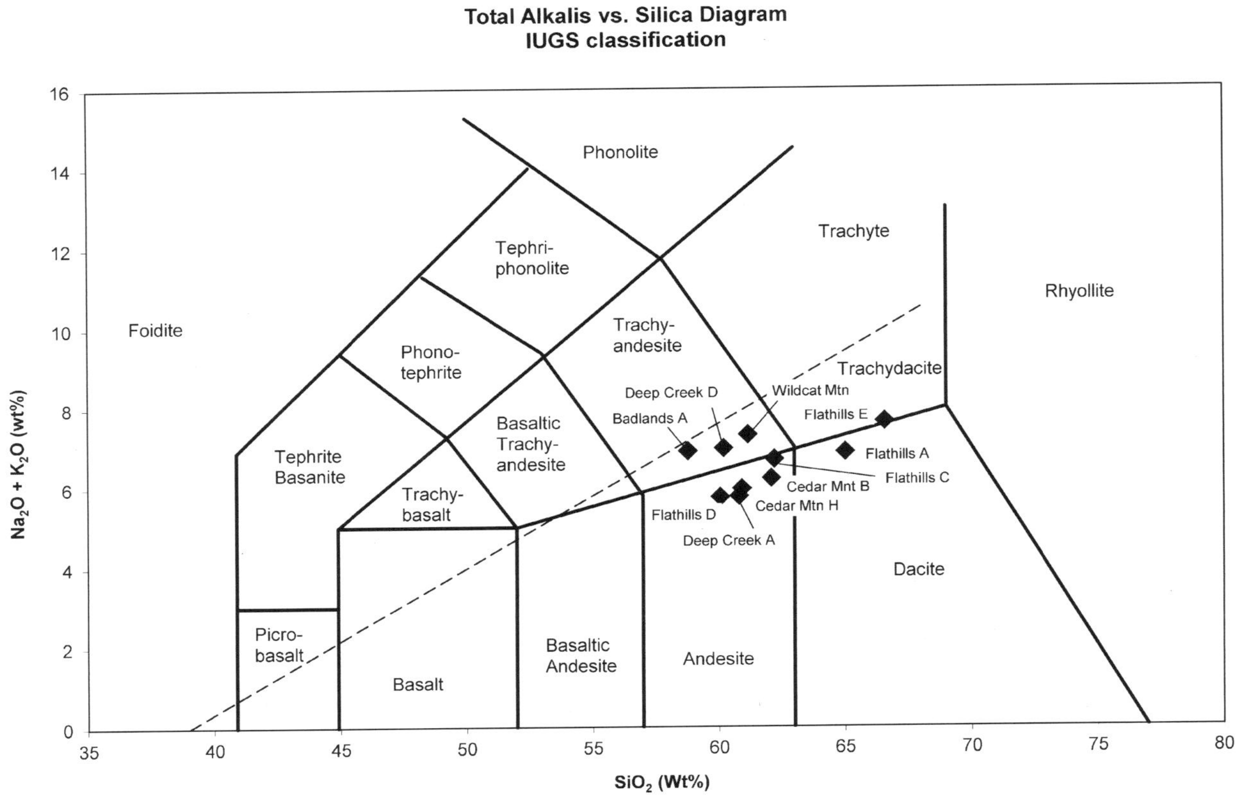

Total Alkalis vs. Silica Diagram
IUGS classification

FIGURE 6.2. International Union of Geological Sciences (IUGS) classification of igneous rock types for several eastern Great Basin fine-grained volcanic sources. Total alkali silica diagram shows weight percent of silica plotted against sodium and potassium oxides (after Le Bas et al. 1986). The data are single samples from individual rocks.

Fine-Grained Volcanics

The term *FGV* refers to the suite of fine-grained silica-rich nonglassy volcanic rocks commonly used for flaked-stone tool production (see Jones et al. 1997). This is often referred to generically as "basalt" in the archaeological literature, but this term now appears inaccurate based on geochemical analysis (Duke 2011; Page 2008). In these studies whole-rock XRF geochemical analysis was conducted on several notable varieties of FGV from the eastern Great Basin to identify concentrations of major elements/oxides, and these were plotted by International Union of Geological Sciences classification to identify specific igneous rock types (Figure 6.2). These materials contain higher amounts of silica and are not basalt at all but are classified as dacite, trachydacite, andesite, and trachyandesite. FGV toolstone usually exceeds ~58 percent SiO$_2$ by weight.

Bonneville Basin FGV Sources

Eight archaeologically significant FGV geochemical types were identified in the vicinity of the study area by Page (2008) and characterized by C. Skinner at NWROSL. Summary statistics of trace element composition and rock types for characterized geologic samples are presented in Table 6.1.

Two main source groups were identified on the eastern edge of the Bonneville basin and include two geographically distinct sources composed of a variety of geochemical variants. These eastern Cedar Mountain and Flat Hills sources are located along the southern edge of the Cedar Mountain Range and on the northern and eastern flanks of the Flat Hills, respectively (Page 2008). Fifteen chemical types were identified and geochemically characterized from these two sources. Wildcat Mountain FGV was previously identified in 2003 (Carter et al. 2004) but was misidentified as a major contributor to local archaeological assemblages due to an overlap in geochemical signatures.

Two main source groups were identified on the western edge of the Bonneville basin and include two geographically overlapping, but geochemically distinct, sources containing a number of geochemical variants. These western Deep Creek and Badlands sources are located in the Little Antelope Hills, across the heavily dis-

TABLE 6.1. Summary Statistics of Trace Element Composition and Rock Types for Fine-Grained Volcanic Geologic Samples.

Geochemical Source Group/Variant	Trace Element			Igneous Rock Type	Total
	Sr (ppm)	Rb (ppm)	Zr (ppm)		
Eastern Sources					
Flat Hills A				Dacite	12
Mean	657	99	294		
SD	18	9	11		
Flat Hills C				Andesite	7
Mean	759	100	277		
SD	35	11	6		
Flat Hills D				Andesite	3
Mean	370	102	248		
SD	21	6	10		
Flat Hills E				Trachydacite	2
Mean	287	217	249		
SD	15	1	0		
Cedar Mountain B				Andesite	16
Mean	513	131	319		
SD	14	4	7		
Western Sources					
Deep Creek A				Andesite	17
Mean	354	91	357		
SD	24	5	17		
Badlands A				Trachyandesite	14
Mean	484	178	201		
SD	37	15	13		
Currie Hills				Unknown	3
Mean	355	182	250		
SD	6	18	18		

Source: Modified from Page 2008.

sected piedmont surface between the Antelope Range and the Deep Creek Range, and entrained with the fluvial gravels/cobbles of Deep Creek (Page 2008). Three other archaeologically insignificant chemical types of FGV material were also identified in this western area and include Ferber Wash, Gold Hill Wash, and Little White Horse Badlands sources. Nine chemical types present in this western area were identified and geochemically characterized. One of these sources, Deep Creek A, was previously identified in 2003 (Carter et al. 2004) but was further explored and characterized by Page (2008).

ARCHAEOLOGICALLY SIGNIFICANT TOOLSTONE SOURCES

Figures 6.3 and 6.4 show the locations of obsidian and FGV toolstone source areas that are represented in PA sites in the delta and in Archaic sites across Dugway Proving Ground (DPG). The closest known sources of obsidian are found at Topaz Mountain and Ferguson Wash, while the closest known sources of FGV are Cedar Mountains and Flat Hills (Page 2008). These sources are located in opposite and cardinal directions from the study area—obsidian north–south, FGV east–west—making procurement directed at both material types problematic during any one collecting trip. During the Gilbert episode (~10,500 to ~10,000 ^{14}C BP; see Chapter 3), access to portions of the delta may have been geographically restricted by a large body of water to the north and west and deltaic wetlands to the south and east. Thus, movements to access toolstone from the ORB delta could have required substantial deviation from direct routes. For example, travel to Browns Bench would increase by ~100 km, while the distance to Topaz Mountain would remain much the same.

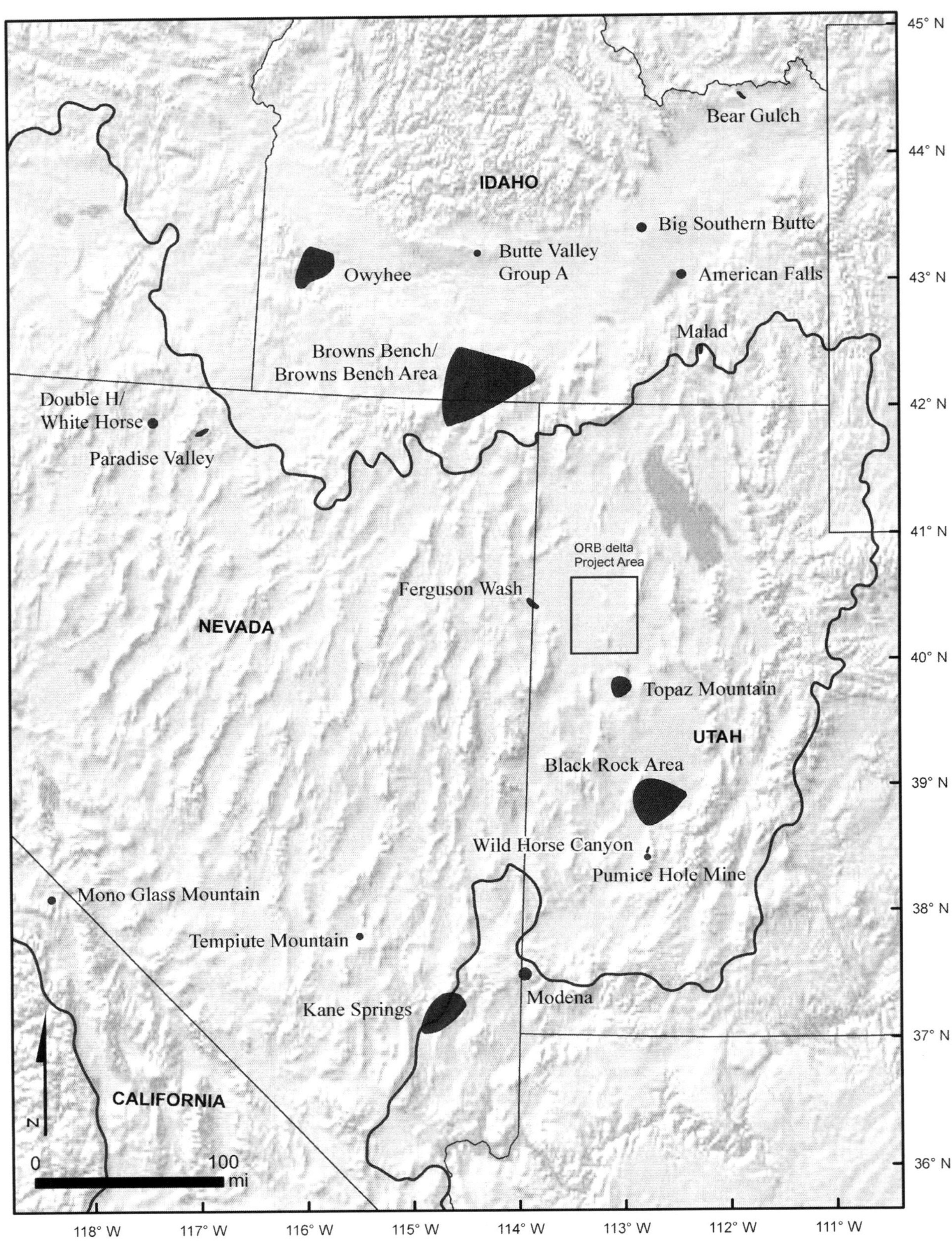

Figure 6.3. Obsidian toolstone sources (*dark gray polygons*) represented in the Old River Bed (ORB) delta and Dugway Proving Ground Archaic samples.

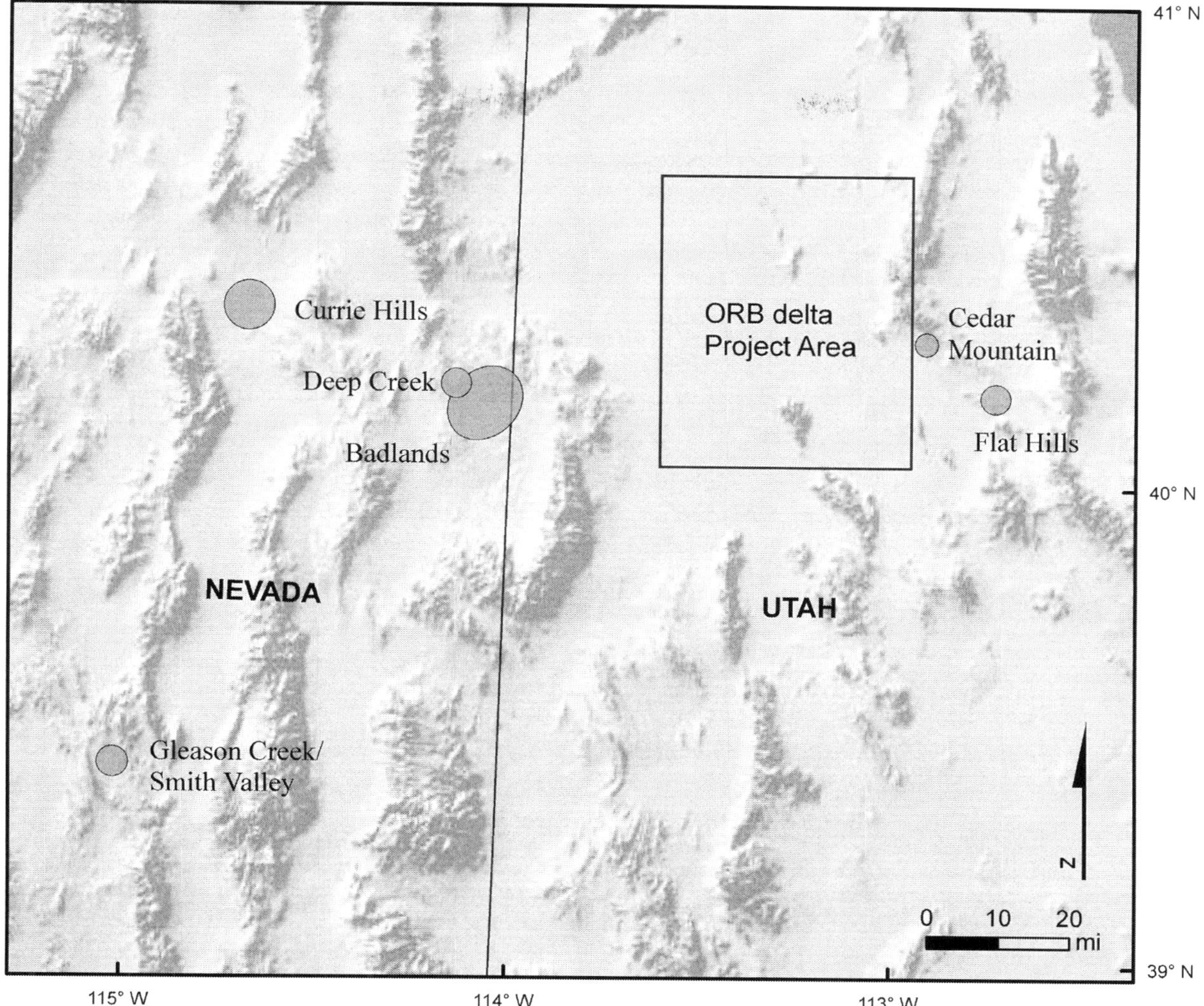

FIGURE 6.4. Fine-grained volcanic toolstone sources (*light gray polygons*) represented in the Old River Bed (ORB) delta sample.

Obsidian Sources

Detailed information about each archaeologically significant obsidian source/chemical type present in the sample is beyond the scope of this project. We refer the reader to regional studies by Jackson et al. (2009) and Moore (2009) on obsidian in Utah, Haarklau et al. (2005) for sources in southern Nevada, and Holmer (1997) and Plager (2001) for information on sources in Idaho.

Table 6.2 includes approximate distances and general direction to obsidian sources for the proximal and distal delta. Distances range from ~50 km up to more than 500 km. Sixteen obsidian sources/source areas are included. The distribution of the Browns Bench geochemical type(s) is widespread and chemically diverse,[1] and variants of Browns Bench Area and Butte Valley

Group A are included with the Browns Bench source group for the purposes of this study.[2] These obsidian sources are located primarily to the north and south of the ORB delta. No known sources are located to the east, and only one is located to the west.

FGV Sources

Detailed information about each archaeologically significant source of FGV present in the sample is also beyond the scope of this project. We refer the reader to studies by Page (2008) and Duke (2011) on FGV in Utah and Nevada. Table 6.3 includes approximate distances and general direction to represented FGV source areas for the proximal and distal delta. Six FGV source areas are included in the sample, and distances range from ~30 km to ~190 km. Several geochemically

TABLE 6.2. Distance and Direction to Represented Obsidian Sources.

Chemical Source Group	Distance to Proximal Old River Bed Delta/Dugway Proving Ground (km)	Distance to Distal Old River Bed Delta (km)	Direction
Topaz Mountain	53 [45]	82 [73]	S/SE
Ferguson Wash	69 [62]	54 [49]	W
Black Rock Area	159 [138]	187 [168]	S/SE
Wild Horse Canyon	200	228	SE
Pumice Hole Mine	206	234	SE
Browns Bench/Browns Bench Area/Butte Valley Group A	240 [202]	211 [174]	NW
Malad	256	235	NE
American Falls	316[a]	292	NE
Modena	314	338	SW
Tempiute Mountain	347	361[a]	SW
Butte Valley Group A (primary)	347	318	NW
Big Southern Butte	352[b] [350]	327[a]	N
Kane Springs	368 [348]	389 [369][a]	SW
Paradise Valley	389	367	NW
Owyhee	406 [393]	378 [366][a]	NW
Double H/White Horse	431[a]	410	NW
Bear Gulch	479[a]	456	NE
Mono Glass Mountain	535[b]	537[a]	SW

Note: Distances are approximate, rounded to the nearest kilometer, and measured as straight-line distances from source centroids to points north of Granite Peak and east of Wild Isle for the proximal delta and distal delta, respectively. The nearest available distance is provided in brackets for chemical types with broad (non–projectile point) geographic distributions.

[a] Not represented.
[b] Archaic sample only.

distinct variants are found within the Flat Hills source area. These FGV sources are located primarily to the east and west of the delta.

METHODS

Geochemical trace element characterization was conducted by means of X-ray fluorescence spectroscopy, an established method of geochemical characterization within geology and a method employed extensively in archaeology for obsidian and other toolstone investigations (Hughes 1984, 1998; Shackley 1998). The investigative technique of XRF analysis, in a simplified view, consists of exposing a sample to a source of radiation, causing it to react in a way that allows specific trace elements present in the sample to be identified as chemical signatures (Thomsen and Schatzlein 2002).

Our sample from the ORB delta consists of identifications using three different types of XRF. Nondestructive quantitative XRF analyses were conducted by R. Hughes at Geochemical Research Laboratory (GRL) following his standard laboratory protocols (http://

www.geochemicalresearch.com/about-xrf.html). Nondestructive quantitative XRF analyses of samples were also conducted by C. Skinner at the NWROSL following his standard laboratory protocols (http://www.obsidianlab.com/summary_xrf.html).

Nondestructive trace element analyses were conducted by David Page at Desert Research Institute using a Niton XL3t 900 handheld portable X-ray fluorescence (pXRF) elemental analyzer with a 50-kV X-ray tube, a Geometrically Optimized Large Area Drift Detector, 80-MHz real-time digital signal processing, and dual state-of-the-art embedded processors. An analysis protocol was established for commonly found obsidian and FGV types on DPG. Due to specifics of calibration, quantitative trace element measurements using pXRF are machine specific and not readily comparable to other XRF analyses from independent laboratories (Goodale et al. 2011; Shackley 2011). Resultant trace element values are internally precise but are not accurate according to recommended values for geologic standards.

TABLE 6.3. Distance and Direction to Represented Fine-Grained Volcanic Sources.

Chemical Source Group	Distance to Proximal Old River Bed Delta (km)	Distance to Distal Old River Bed Delta (km)	Direction
Cedar Mountains	29 [26]	45 [42][a]	E
Flat Hills	44 [41]	65 [61]	E/SE
Badlands	72 [63]	70 [61]	W
Deep Creek	78 [75]	73 [70]	W
Currie Hills	125 [120]	114 [108]	W
Gleason Creek/Smith Valley	180 [177]	186 [183]	SW

Note: Distances are approximate, rounded to the nearest kilometer, and measured as straight-line distances from source centroids to points north of Granite Peak and east of Wild Isle for the proximal delta and distal delta, respectively. The nearest available distance is provided in brackets for chemical types with broad (non–projectile point) geographic distributions.

[a] Not represented.

To cross-check these pXRF results, a sample of previously sourced obsidian artifacts with known source proveniences common to the region was reanalyzed using the pXRF unit and results of both sets of analyses were plotted on an XY scatter plot (Sr/Zr and Rb/Zr). After extensive experimentation with instrument calibration using the supplied Niton software package, analysis modes, filters, and count times, the spectrometer was run in Mining Cu/Zn Mode on the main filter at a short count time of 15 seconds using an 8-mm spot size. This short exposure time boosts productivity but still allows for replicable estimates in trace elements of interest, particularly mid-Z elements. Specimens from a number of recent CRM projects on the proximal delta were analyzed using pXRF and this analysis protocol (Page and Schmitt 2010, 2011a, 2011b; Page, Schmitt, Dalldorf, and Wazaney 2012; Page, Schmitt, Rhode, and Sapula 2012; Rhode et al. 2011; Schmitt and Page 2011; Schmitt et al. 2010; Schmitt et al. 2011). Artifact source assignments were based on element ratios (primarily Rb/Sr/Zr) but also included an assessment of other element ratios in concert with these elements—for example, the use of Y, Fe, and Ti levels in differentiating Topaz Mountain from Ferguson Wash obsidian and correspondence with the source database. Results and recommended values for the U.S. Geological Survey obsidian standard "RGM-2" were also obtained at each run for further reference. When samples did not match established geochemical types (i.e., identified within the known source universe) they were sent out for reanalysis at NWROSL.

In light of a debate about the reliability of portable XRF usage in archaeological sourcing studies (see Frahm 2012a, 2012b; Speakman and Shackley 2012), we should note that we agree with Frahm (2012a) and see

artifact sourcing as an archaeological problem that is addressed through geochemical methods (see discussion in Hughes 1998). While Speakman and Shackley (2012) question the "accuracy" of pXRF devices in determining the precise chemical makeup of a geological source, Frahm (2012a) contends that they are internally consistent and can reliably attribute a stone artifact to a particular source if both are analyzed using the same machine. Since the goal of archaeological toolstone identifications is to identify where artifacts come from, rather than to produce a precise chemical composition, the use of pXRF analyses is a valid archaeological research tool.

We consider our results with the handheld Niton spectrometer to be precise and internally consistent but numerically inaccurate. We made substantial effort, including attempted interaction with the instrument dealers, to recalibrate the pXRF to produce "true" values (i.e., recognized, published values), but we ultimately opted to rely on the inaccurate, yet sufficiently precise, values generated by the unit to differentiate between the chemical types found on Dugway. Accuracy (i.e., congruence with established values for source standards) of trace element values was not entirely necessary in this sample. We essentially knew what to expect (e.g., high use of specific obsidian and FGV types) prior to using the pXRF for any artifact sourcing. We had a large amount of existing sourcing data from both GRL and NWROSL that established a pattern of toolstone use for the region and limited the number of variables to a subset of principal sources from the larger known source universe. Our aim was not to work out the nuances of various unknown chemical sources but, rather, to supplement our existing database with data on sources that are well defined and readily distinguishable.

TABLE 6.4. Obsidian (Obs) and Fine-Grained Volcanic (FGV) Tools Included in This Study (Debitage Not Included).

Sample and Tool Type Group	No. of Sites/Isolates	No. of Non–Projectile Point Tools (Obs/FGV)	No. of Projectile Points (Obs/FGV)	Total
Proximal Old River Bed Delta	137/91			
EH		—	459 (321/138)	
WST		—	168 (75/93)	
PA		237 (54/183)	627 (396/231)	864
			308 (253/55)	
Distal Old River Bed Delta	145/87			
EH		—		
WST		—	291 (260/31)	
PA		257 (87/170)	599 (513/86)	856
Dugway Proving Ground Comparative Sample	110/40			
AR		—	287 (265/22)	287
Total	392/218	494 (141/353)	1,513 (1,174/339)	2,007

Note: EH = Early Holocene; WST = Western Stemmed Tradition; PA = Paleoarchaic; AR = Archaic.

The following toolstone types were identified with pXRF analyses: obsidian—Topaz Mountain, Ferguson Wash, Black Rock Area, Browns Bench, Browns Bench Area, Browns Bench/Butte Valley A, Wild Horse Canyon, and Malad; FGV—Cedar Mountains, Flat Hills variants, Deep Creek, Badlands, and Currie Hills. These chemical types constitute 98.4 percent of the stone tools in the ORB delta sample. Minor toolstone types were identified by C. Skinner and R. Hughes using laboratory XRF and include the following: obsidian—Pumice Hole Mine, American Falls, Modena, Tempiute Mountain, Big Southern Butte, Kane Springs, Paradise Valley, Double H/White Horse, Owyhee, Bear Gulch, and Mono Glass Mountain; FGV—Gleason Creek/Smith Valley.

SAMPLE ARRAY AND RESULTS

In total, 2007 obsidian and FGV tools were sampled and geochemically analyzed for this study (Table 6.4). These artifacts originate from some 392 sites and 218 isolated finds and include both temporally diagnostic projectile points and non–projectile point tool classes (e.g., bifaces, cores, drills/perforators, gravers/chisels, unifaces/scrapers, modified/utilized flakes, knives, etc.) from different-aged assemblages across the region (Figure 6.5).[3] Materials were included from both the proximal and distal ORB delta to investigate changes in source use across the delta. Paleoarchaic-aged projectile points were separated according to Beck and Jones (see Chapter 5) into named Western Stemmed Tradition (WST) types and Early Holocene (EH) types (Great Basin Stemmed Series, Pintos, and Butte Valley Corner-notched). Also a comparative sample of Archaic-aged projectile points from across DPG was analyzed to investigate temporal changes in both raw material use and geologic source use into the Archaic period.[4]

Results of XRF and pXRF analysis and data tables showing source assignments resulting from this study are available online in Chapter 6, Supplemental Digital Material (see link information at the end of Chapter 1). These attributions by material type include WST projectile points from the proximal delta, EH projectile points from the proximal delta, PA non–projectile point tools from the proximal delta, Archaic projectile points from across DPG, WST projectile points from the distal delta, EH projectile points from the distal delta, and PA non–projectile point tools from the distal delta. Data from these tables are presented graphically in the figures below.

GREAT BASIN TOOLSTONE AND MOBILITY STUDIES

The application of obsidian provenience data to the study of trade, settlement patterns, and population movements has been an integral part of archaeology in the Great Basin since the early 1970s (Hughes 1984). In several related works, Jones and Beck (Beck and Jones 2011; Jones 2008; Jones, Beck, Jones, and Hughes 2003; Jones et al. 2012) present iterations of regional-scale lithic conveyance zones that vary in extent based on the raw materials and artifact types included in the relevant study areas. Generally, these expansive zones are based primarily on patterns of obsidian and FGV distribution and to a lesser extent on reduction and use-wear data and are equated with "lithic foraging territories." Conveyance zones are also referenced in the litera-

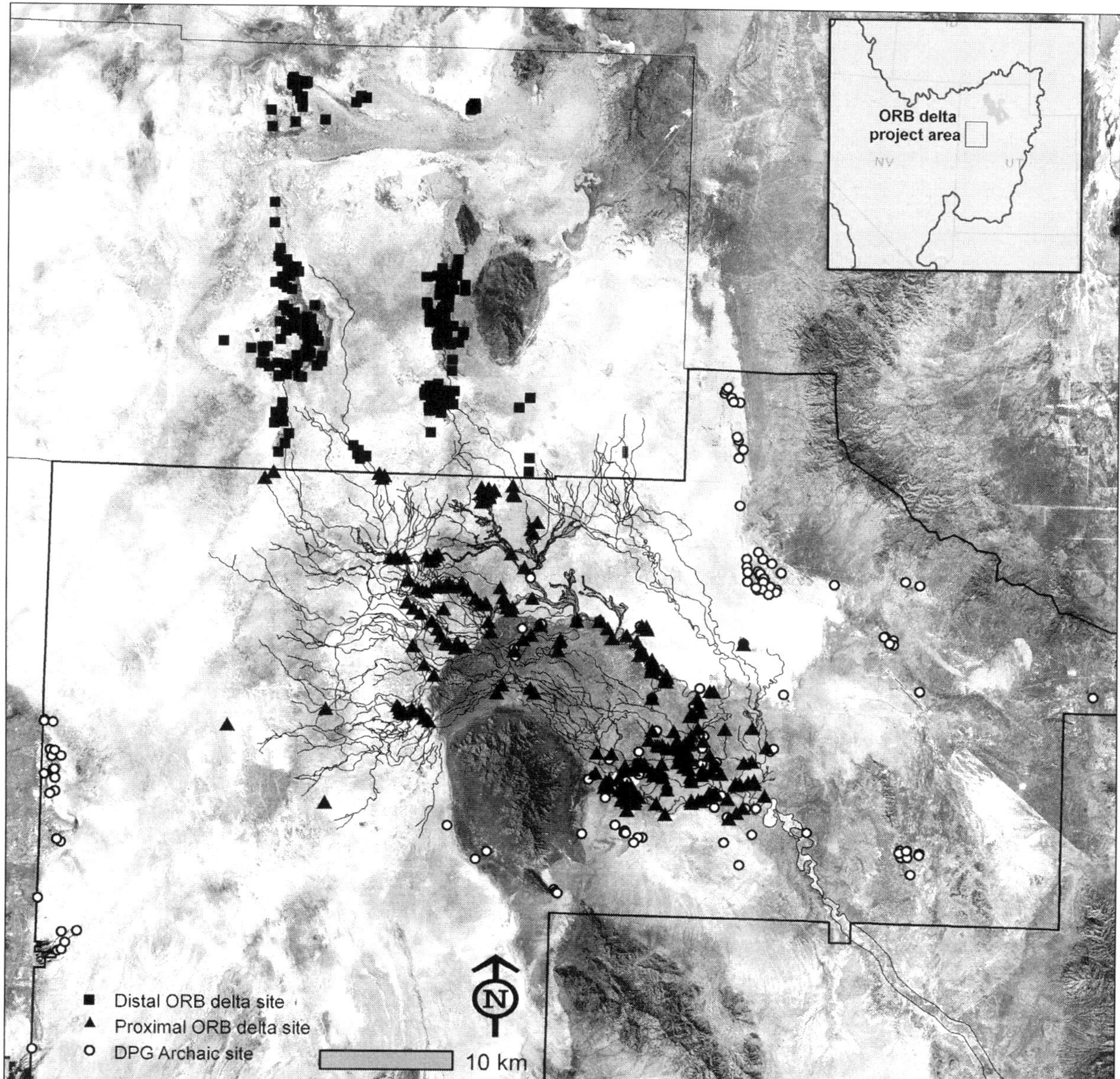

FIGURE 6.5. Assemblages sampled for this study. Squares show Paleoarchaic sites in the distal Old River Bed (ORB) delta on the Utah Test and Training Range; triangles show Paleoarchaic sites in the proximal ORB delta on Dugway Proving Ground (DPG); and circles show Archaic-aged sites sampled from across DPG.

ture as "a pattern of annual mobility" (Anderson and Hanson 1988:267), "geographic range for raw material procurement and use" (Beck and Jones 1990b:295), "annual subsistence cycle" (Buck et al. 1998:222), and "lithic supply zones" (Seeman 1994:275). According to Jones, Beck, Jones, and Hughes (2003; also see Jones et al. 2012), materials would have been directly procured within large established foraging territories or ranges. If trade and exchange are ruled out as primary means of acquisition, then it follows that the toolstone in a site accurately represents the area that the occupants of that

site "habitually occupied" as part of a seasonal round (Jones, Beck, Jones, and Hughes 2003:8) and that the material used to identify and demarcate these zones was directly procured from the source (Beck and Jones 2011). This characterization colors current debate about Paleoarchaic mobility (e.g., Duke and Young 2007; Madsen 2007; Newlander 2012; Simms 2008; Smith 2007, 2010), some of which appeals to behavioral factors that become conflated with toolstone distribution if the distinction is not made clear (see Hughes 2011 for a discussion of this problem).

ORB DELTA TOOLSTONE SOURCING

In light of the body of existing work for the eastern Great Basin, we seek here to expand the geochemical sourcing database for obsidian and FGV using a large number of sites across the Bonneville basin, with a focus on the ORB delta. The delta is unique in that it is located on active military land and is therefore protected from the impact of collectors. It contains a rich archaeological record composed almost entirely of obsidian and FGV lithic tools and debris. These mostly dark-colored materials are primarily surface finds and are easily identifiable against the white/tan ground surface. Moreover, most assemblages are single component or only minimally mixed. With the exception of some small chert gravels transported north from the Sevier basin within some of the higher-energy stream channels, there are few "local" (<10-km-distant) raw materials readily available for tool production within the delta. However, the delta is surrounded by geologic sources of obsidian and FGV that are well studied and geochemically characterized, with very few of our samples originating from "unknown" geologic provenience. Although a number of PA sites on the ORB delta are likely palimpsests, many appear to be discrete camps or single-episode use areas (see Chapter 4).

Toolstone-sourcing projects in the past have mainly focused on an examination of the toolstone of projectile points (Bamforth 2009; Jones, Beck, Jones, and Hughes 2003). In the western United States, this has largely centered on the acquisition and transport of obsidian, but to better understand behavior related to toolstone use we incorporate a more complete array of other lithic materials and tool types.

The Bonneville basin of western Utah and the adjacent valleys to the west in eastern Nevada have been host to a number of archaeological investigations into prehistoric mobility and lithic raw material use since the mid-1980s (Beck and Jones 1988, 1990b, 1997; Beck et al. 2002; Elston 2005; Hughes 1984; Jones, Beck, Jones, and Hughes 2003; Jones, Beck, and Kessler 2003; Jones et al. 1997; Madsen and Schmitt 2005). These studies suggest that the physiography and topography of the Basin and Range Province influenced the archaeological distribution of raw materials and constrained the flow of materials between and within regions of the Great Basin (Grayson 1993; Jones, Beck, Jones, and Hughes 2003). This archaeological distribution of toolstone has been used to produce a number of different models of Paleoarchaic mobility in the Great Basin (see review in Chapter 1). We do not repeat that review here, but in the sections that follow we employ ORB delta sourcing data to evaluate their relative explanatory utility.

RAW MATERIAL REPRESENTATION IN THE OLD RIVER BED DELTA

In this section, we use a series of stacked-column bar graphs to visually represent toolstone use across the delta. Materials included in the ORB delta WST/EH sample span an enormous area that measures approximately 1,350 km^2, while our Archaic sample spans an even larger area of approximately 2,050 km^2. In addition, the total sample includes many sites and isolated artifacts. Because sites included in the study area are distributed across such a large area, sites on one end of the delta are much closer to certain source areas.

Obsidian/FGV

Two simple plots compare the presence of obsidian and FGV toolstone across the delta and compare their presence among the WST, EH, and Archaic periods. In the proximal delta, there is an increasing presence of obsidian and a declining presence of FGV for projectile points through time when comparing the WST and EH periods (Figure 6.6). In the distal delta the trend is reversed. The differences are statistically significant (WST: $\chi^2 = 25.68$, $df = 1$, $p < .0001$; EH: $\chi^2 = 12.70$, $df = 1$, $p < .0004$). In the DPG Archaic sample there is also a high occurrence of obsidian projectile points, a trend that continues to increase through time within the Archaic period. Early in the Archaic sequence there is an elevated presence of FGV, especially in Northern Side-notched and Elko points, but these make up only 21 percent and 13 percent of the assemblage, respectively (Figure 6.6). Overall there is a slight increase in obsidian projectile points through time, but the difference between the WST and EH samples is not statistically significant ($\chi^2 = .980$; $df = 1$; $p = .3223$).

FGV dominates PA non–projectile point tools (e.g., bifaces, cores, scrapers, flake tools, etc.) across the delta but is higher in the proximal delta (Figure 6.7), although not at a statistically significant level ($\chi^2 = 2.371$; $df = 1$; $p = .1236$). In general, obsidian and FGV were used differently within the WST and EH periods across the entire delta, with more obsidian projectile points and more FGV non–projectile point tools (Figures 6.6 and 6.7). The presence of FGV in the Archaic period is greatly diminished, especially later in time (see Page 2008).

Obsidian Representation

The next two plots contrast the frequency of Topaz Mountain and Browns Bench projectile points during the WST and EH periods to that of other sources observed in the sample. This representation is then compared with the Archaic sample and with non–projectile point tools during the PA. Obsidian representation for

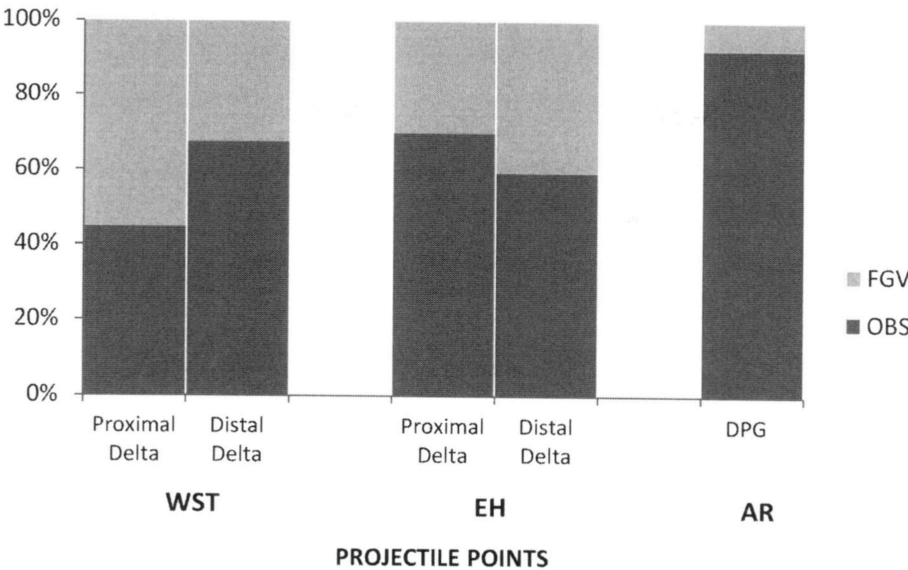

FIGURE 6.6. Frequency of obsidian (OBS) vs. fine-grained volcanic (FGV) use for points comparing the proximal and distal Old River Bed delta by temporal period and Archaic-aged sample. DPG = Dugway Proving Ground; WST = Western Stemmed Tradition; EH = Early Holocene; AR = Archaic.

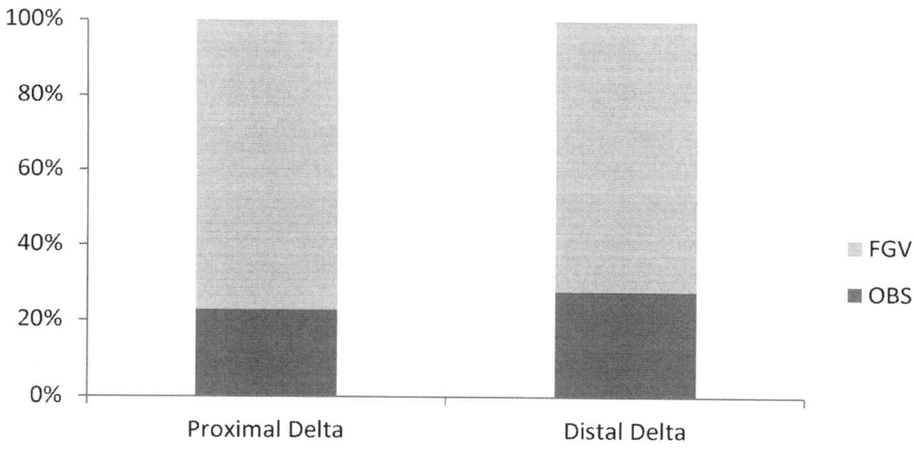

FIGURE 6.7. Frequency of obsidian (OBS) vs. fine-grained volcanic (FGV) use for Paleoarchaic non–projectile point tools in the proximal and distal Old River Bed delta.

projectile points shows a trend of increasing Topaz Mountain obsidian, decreasing Browns Bench, and fluctuations in other sources through time (Figure 6.8). This trend is apparent during the PA and into the Archaic. In viewing the entire assemblage by temporal periods (WST, EH, and Archaic) obsidian representation for pooled points and non–projectile point tool classes shows a very similar trend of increasing representation of Topaz Mountain through time and decreasing representation of both Browns Bench and

other sources when comparing the proximal and distal ORB delta.

In viewing obsidian representation in the manufacture of both WST and EH points, the proximal delta shows greater representation of Topaz Mountain and other sources than the distal delta, while the distal delta shows greater representation of Browns Bench. Topaz Mountain still dominates the distal assemblage. Both are likely due to proximity to each source. The representation of other sources during the EH remains quite

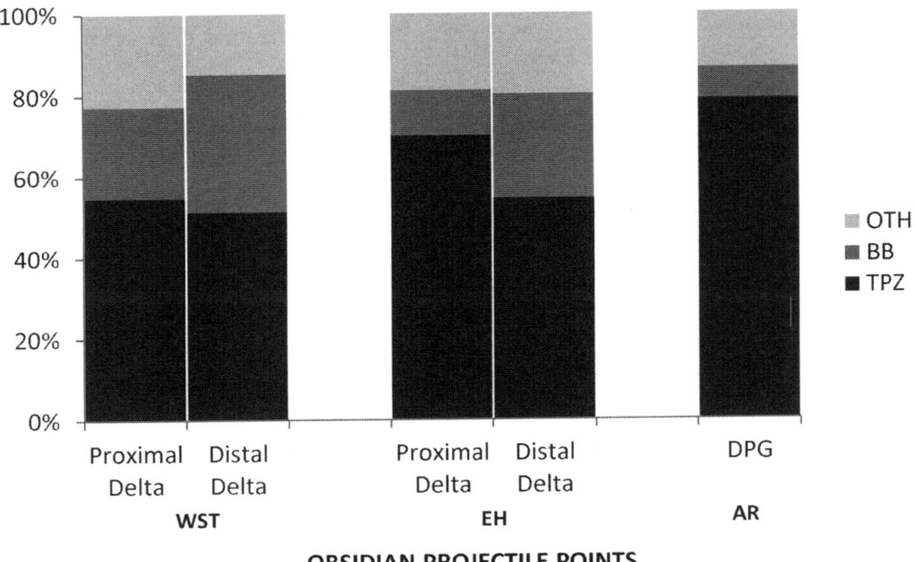

FIGURE 6.8. Frequency of obsidian source use for projectile points comparing the proximal and distal Old River Bed delta by temporal period and Archaic-aged sample. OTH = other; BB = Browns Bench; TPZ = Topaz Mountain; DPG = Dugway Proving Ground; WST = Western Stemmed Tradition; EH = Early Holocene; AR = Archaic.

similar across the delta in frequency, but composition varies, with more individually represented sources in the proximal delta (e.g., Owyhee, Pumice Hole Mine, and Tempiute Mountain) and more individually represented northern and western sources present in the distal delta (e.g., American Falls, Bear Gulch, Ferguson Wash, Modena, and Paradise Valley). Black Rock Area, Malad, and Wild Horse Canyon obsidians have a common representation in the assemblages. For obsidian represented in non–projectile point tool classes, there is elevated Topaz Mountain obsidian, while representation of both Browns Bench and other sources decreases (Figure 6.9).

FGV Representation

In comparing FGV source representation for projectile points in the proximal ORB delta assemblage, there is little change in the amount of west vs. east source representation, with only a slight decline in eastern sources through time (but not statistically significant [$p = .5066$]; Figure 6.10). In the distal ORB delta assemblage there is a similar, more pronounced trend (but still not statistically significant [$p = .0856$]). This trend continues across the entire sample of projectile points (WST, EH, and Archaic). In comparing representations of west and east FGV sources in non–projectile point tools for the proximal and distal ORB delta assem-

blages, there is greater representation of western FGV sources in the distal delta ($\chi^2 = 9.478$; $df = 1$; $p = .0021$; Figure 6.11). This may be because sites in the distal delta are farther (~20-km) away from eastern FGV sources (i.e., Flat Hills).

Example: Toolstone Representation at Site 42To3834

Site 42To3834 is a large surface assemblage from an open campsite (~300 × 200 m) located just west of a confluence of two ORB deltaic channels in an unexposed channel section of the ORB delta. The assemblage was recorded in detail and is composed of 483 flakes, 140 flaked-stone tools, a shaft straightener, what appears to be a mano, and two white quartz manuports of unknown function. All tools were collected for additional analyses. With the exception of a single Fremont-aged Rose Spring arrowpoint and perhaps the mano, all other tools appear to date to the late Paleoarchaic/early Holocene (Page, Schmitt, Dalldorf, and Wazaney 2012). Flaking debris is dominated by FGV, along with obsidian, chert, and some quartzite. The quartzite consists of butterscotch, white, red/rust, and tan/buff materials, and the chert includes brown, butterscotch, dark olive/black, red, pink, and red/brown. Twelve pieces of shatter (2.5 percent), 16 primary flakes (3.3 percent), and 282 secondary flakes (58.8 percent) reflect cobble/nodule

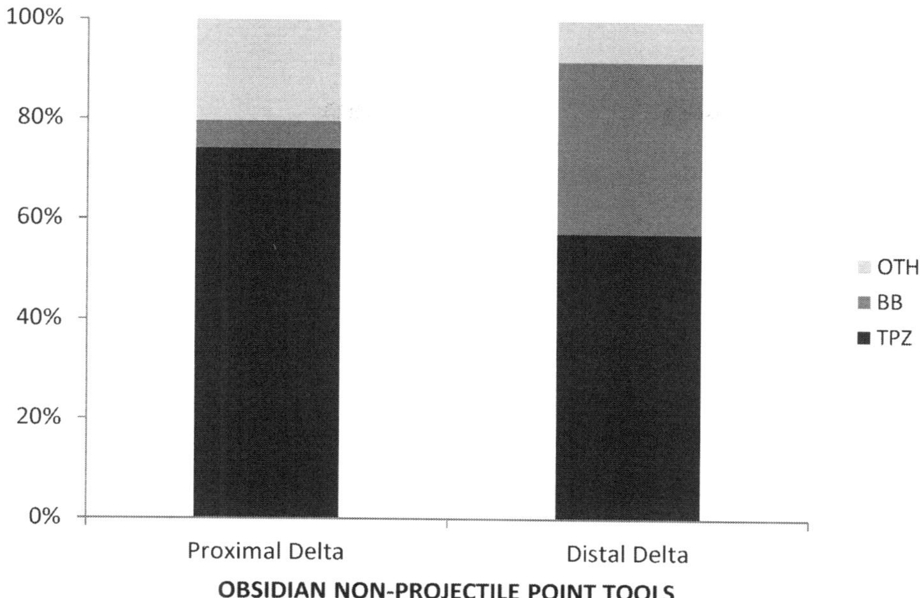

FIGURE 6.9. Frequency of obsidian source use for non–projectile point tools from the proximal and distal Old River Bed delta. OTH = other; BB = Browns Bench; TPZ = Topaz Mountain.

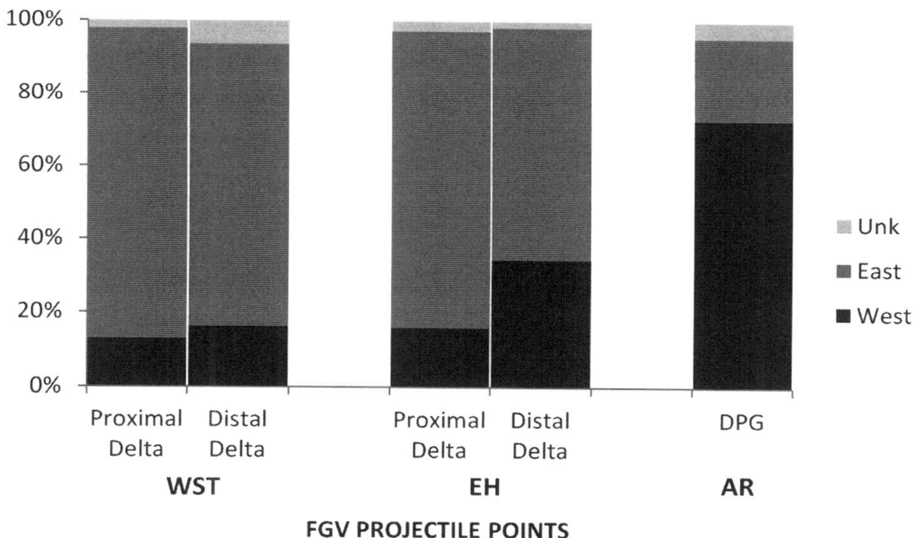

FIGURE 6.10. Frequency of fine-grained volcanic (FGV) source use for projectile points comparing the proximal and distal Old River Bed delta by temporal period and Archaic-aged sample. Unk = unknown; DPG = Dugway Proving Ground; WST = Western Stemmed Tradition; EH = Early Holocene; AR = Archaic.

and core reduction, and 170 tertiary flakes (35.4 percent) mark at least some on-site manufacture and/or maintenance of bifaces. Tools are diverse and include 11 cores (six chert, four obsidian, one quartzite), four scrapers (two obsidian, one FGV, one chert), 62 bifaces/biface fragments (11 chert, 14 obsidian, 35 FGV, two quartzite), five drills (one chert, four FGV), 25 edge-modified

flakes (17 FGV, five obsidian, one chert, two quartzite), one utilized chert flake, a pumice shaft straightener, a limestone mano fragment, nine FGV Great Basin Stemmed (GBS) points, 20 Pinto points (eight FGV, 11 obsidian, and one quartzite), and an obsidian Rosegate point. The Pinto and GBS point types reflect a period of early Holocene occupation, and a single obsidian

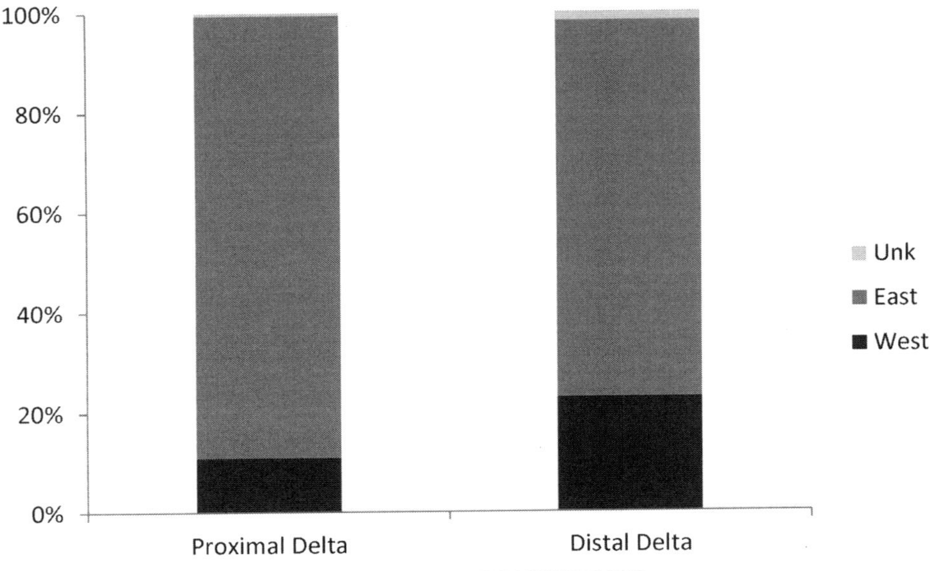

FIGURE 6.11. Frequency of fine-grained volcanic (FGV) source use for Paleoarchaic non–projectile point tools from the proximal and distal Old River Bed delta. Unk = unknown.

Rosegate point (and likely the mano fragment) marks at least one later visit during the Fremont period. Like most of the ORB delta sites in the uneroded (or minimally eroded) alluvial plain, most of the lithics exhibit limited weathering.

All tools were collected (142 tools: 22 chert [15.5 percent], 73 FGV [51.4 percent], 39 obsidian [27.5 percent], seven quartzite [5.0 percent], and one pumice [.7 percent]), and 111 obsidian and FGV tools were subjected to geochemical sourcing by pXRF. The sourced sample consisted of 33 obsidian tools (33 percent) and 74 FGV tools (67 percent). The obsidian originates from three sources, Topaz Mountain (75.7 percent), Wild Horse Canyon (18.9 percent), and Black Rock Area (5.4 percent). The FGV comes from four geographically distinct source areas: Cedar Mountain B (2.7 percent); Badlands A (9.5 percent); Flat Hills A (8.1 percent), D (75.7 percent), and E (2.7 percent); and one unknown source (1.4 percent).

The site is 25–35 km from the nearest FGV source and 40 km from the nearest obsidian source, and this differential proximity to toolstone may have had a significant effect on both the sources represented and the patterns of raw material usage. In the ratio of obsidian to FGV there is a higher representation of FGV for EH projectile points (60.6 percent, compared with 30 percent for other EH points in the proximal delta). There is a slightly elevated rate of obsidian representation for non–projectile point tools (29.9 percent, compared with 23 percent for PA non–projectile point tools in the proximal delta). In specific obsidian representation there is a slightly different pattern of representation compared with other EH points in the proximal delta. Topaz Mountain obsidian is represented more than expected in the point category (76.9 percent, compared with 69 percent for other EH points in the proximal delta). No points were made of Browns Bench obsidian, and other sources make up the remaining 23.1 percent (15.4 percent Wild Horse Canyon and 7.7 percent Black Rock Area), compared with 18 percent for other EH points in the proximal delta. For non–projectile point tools, obsidian representation follows a similar pattern, with a higher representation of Topaz Mountain than expected (73.9 percent, compared with 67 percent for PA non–projectile point tools in the proximal delta), no Browns Bench, and elevated representation of other obsidians (26.1 percent, compared with 19 percent for PA non–projectile point tools in the proximal delta). The representation of FGV compared with other points fits the pattern seen at other proximal delta sites, with slightly higher representation of eastern FGV sources and limited representation of western and unknown FGV sources. For non–projectile point tools, eastern FGV representation is also elevated more than expected (94.4 percent, compared with 89 percent for other EH points in the proximal delta).

Data from this site provide a useful example of results and interpretations that are generally beyond the reach of most sourcing studies. If we had sourced only a small sample of obsidian tools, as is typical, we

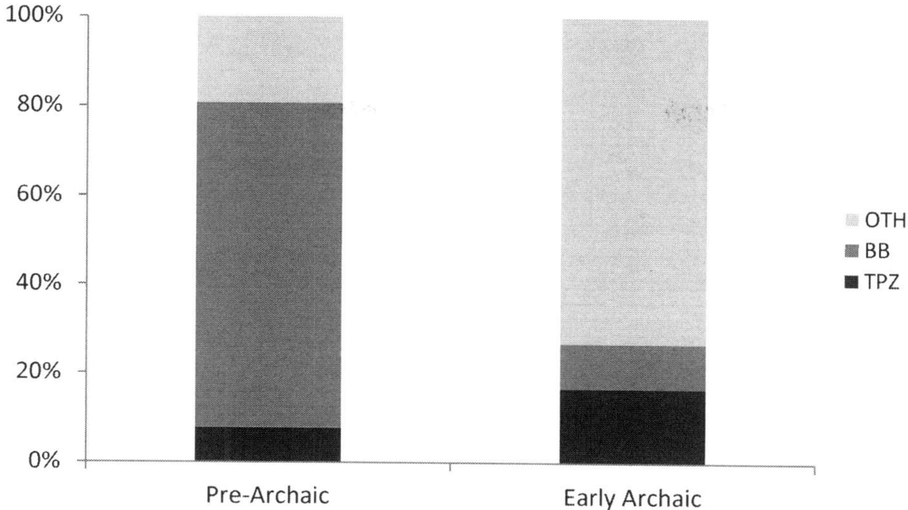

FIGURE 6.12. Frequency of obsidian source use in the Paleoarchaic and Early Archaic assemblages at Bonneville Estates Rockshelter. OTH = other; BB = Browns Bench; TPZ = Topaz Mountain.

likely would have produced different interpretations of toolstone representation depending on the number and kind of sourced items. In this site, obsidian and FGV account for ~80 percent of all tools but for only ~18 percent of the entire assemblage when debitage is included. Obsidian projectile points account for ~25 percent of flaked-stone tools but only 5.5 percent of the entire flaked-stone assemblage.

Toolstone Use at Bonneville Estates Rockshelter

Goebel (2007) suggests that Paleoarchaic foragers at Bonneville Estates Rockshelter, a site situated on the western edge of the Bonneville basin some 70 km northwest of the ORB wetlands, were highly mobile. This is based primarily on the sourcing of a small sample of obsidian tools ($n = 26$, Paleoarchaic; and $n = 30$, Early Archaic). Yet raw material diversity within the Paleoarchaic collection as a whole tells another story, as 44 percent is chert, 32 percent is FGV, and only 23 percent of the assemblage is obsidian. Based on the sourcing of a small sample of FGV tools ($n = 22$ [$n = 6$, PA; and $n = 16$, Early Archaic]) (Page 2008) and the presumed proximity of unknown chert sources (Goebel [2007:182] estimates the sources to be within 10 km of the rockshelter), most of the PA lithic material (>70 percent) originated from within 35 km of the site (Goebel 2007), a pattern not unlike that found in the ORB delta. More than 80 percent of the obsidian was sourced to either Topaz or Browns Bench (Figure 6.12), again as in the nearby ORB delta. Very little use of nearby Ferguson Wash obsidian was noted, but this is not surprising given the small

nodule size of this geochemical type. In later Early Archaic materials (63 percent chert, 17 percent FGV, and 19 percent obsidian) there is a prominent shift to more localized resource use and an increase in the use of Ferguson Wash obsidian that is consistent with a reduction in projectile point and tool size through time. Upward of 80 percent of the Early Archaic assemblage also originated from what is likely within 35 km of the site.

Overall, toolstone sourcing data from Bonneville Estates Rockshelter suggest, in our opinion, an acquisition pattern for the WST/EH periods not unlike that found in the ORB delta. Whether or not this pattern reflects "high mobility" as suggested by Goebel (2007) is another question entirely (see Chapter 7).

DISCUSSION

Gross trends indicate that the nearest raw materials are most often represented, with FGV from Flat Hills dominating assemblages roughly two to one and Topaz Mountain obsidian dominating the obsidian components by about the same proportion. Patterning among tool types within the ORB delta links toolstone preferences to spatially and temporally dynamic aspects of lithic resource use. Within the delta as a whole, obsidian was used more for projectile points, and FGV was used more for bifaces and flake tools/scrapers (Figure 6.13). The toolstone representation for bifaces follows the pattern observed in flake tools and scrapers and is statistically different from material usage patterns for projectile points ($\chi^2 = 299.7$; $df = 1$; $p < .0001$). Topaz Mountain obsidian accounts for ~60 percent of the

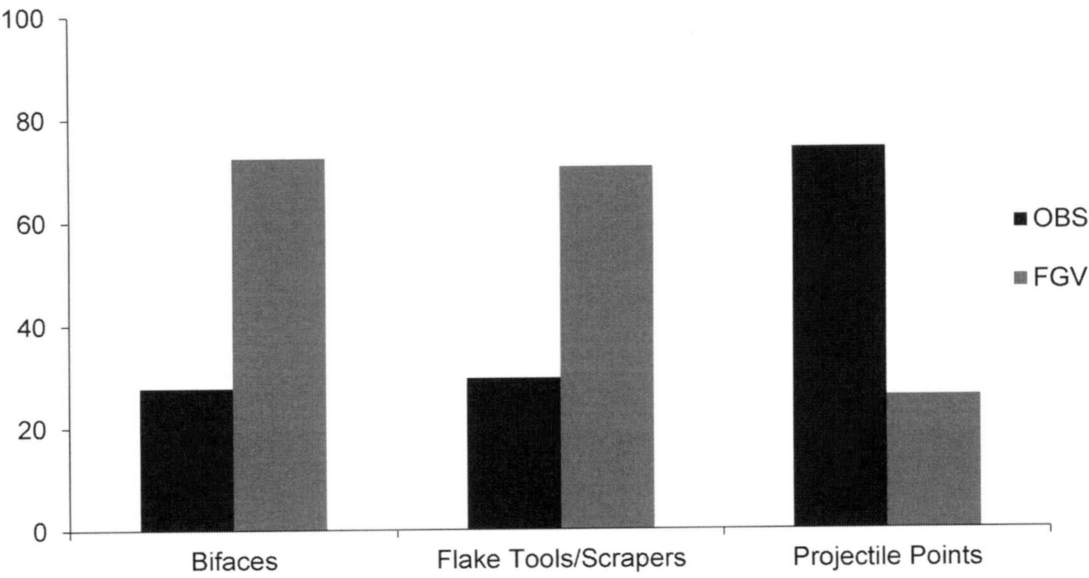

FIGURE 6.13. Frequency of obsidian (OBS) and fine-grained volcanic (FGV) use for bifaces, flake tools/scrapers, and projectile points in the Old River Bed delta.

obsidian used for tools (points and non–projectile point artifacts) across the delta. With the exception of Ferguson Wash obsidian, which is underrepresented in the PA regional record due to small cobble size, the closest viable source of obsidian, Topaz Mountain, is the one most represented. In viewing obsidian representation for three broad tool categories (projectile points, flake tools/scrapers, and bifaces) in the PA sample, the representation of obsidian from Topaz Mountain also dominates (Figure 6.14). Use of Topaz obsidian in biface production is elevated, yet representation in the flake tools/scrapers category is slightly depressed. Browns Bench accounts for ~22 percent of the obsidian used for tools (points and non–projectile point artifacts) in the delta, but its use also varies by tool type. Representation of Browns Bench is elevated in flake tool and scraper production but is slightly depressed for biface production. Representation of other obsidian sources accounts for ~18 percent of the obsidian used for tools (points and non–projectile point artifacts) delta-wide but also varies by tool type. Other obsidians are less frequently represented in flake tool and scraper production.

Andesite and dacite from Flat Hills were the preferred FGV toolstones in the delta. They account for ~80 percent of the FGV points and non–projectile point tools in both the proximal delta and distal delta. Page (2008) reports differences in the use of FGV types and concludes that silica content was likely a factor in the selection of one type over another in the production of points and bifaces because more silica is present in finer-grained material. Thus, dacite and trachydacite were used more in the production of projectile points, while andesite and trachyandesite were used more in the production of bifaces across the delta. A chi-square comparison shows a statistically significant difference between usage patterns at a 5 percent significance level. For the ORB delta assemblage at large, this comparison was repeated for Flat Hills D (andesite) and Flat Hills A (dacite; $\chi^2 = 46.27$; $df = 1$; $p < .0001$). This simple exercise suggests that silica content and properties effecting knapping resulted in the targeting of specific source localities in the production of specific artifact classes.

Distance Decay Model/Source Profiles

Toolstone representation in the delta fits a general distance decay model, with the closest viable toolstone sources contributing to more points and non–projectile point tools than more distant sources (Figures 6.15 and 6.16). Exceptions occur for those sources that only come in small packages (e.g., Ferguson Wash obsidian) or are of poor quality. Even in settings where all materials are quite distant, the most distant sources generally account for the smallest fraction of the overall assemblage. Sample sizes for distant sources are also low and usually are represented by projectile points and some bifaces. Occupants of the ORB delta drew on toolstone sources up to ~250 km away, supporting a pattern of convergence into the delta.

In viewing a compositional breakdown of the WST/ EH artifact sample by class from known geologic sources

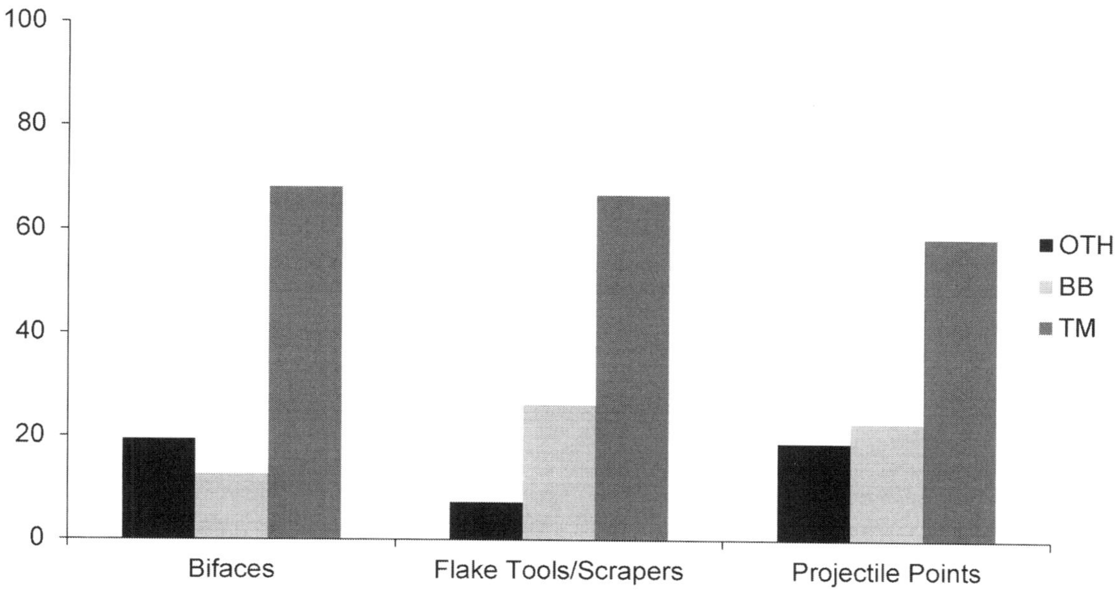

FIGURE 6.14. Frequency of obsidian types for bifaces, flake tools/scrapers, and projectile points delta-wide in the Old River Bed delta. OTH = other; BB = Browns Bench; TM = Topaz Mountain.

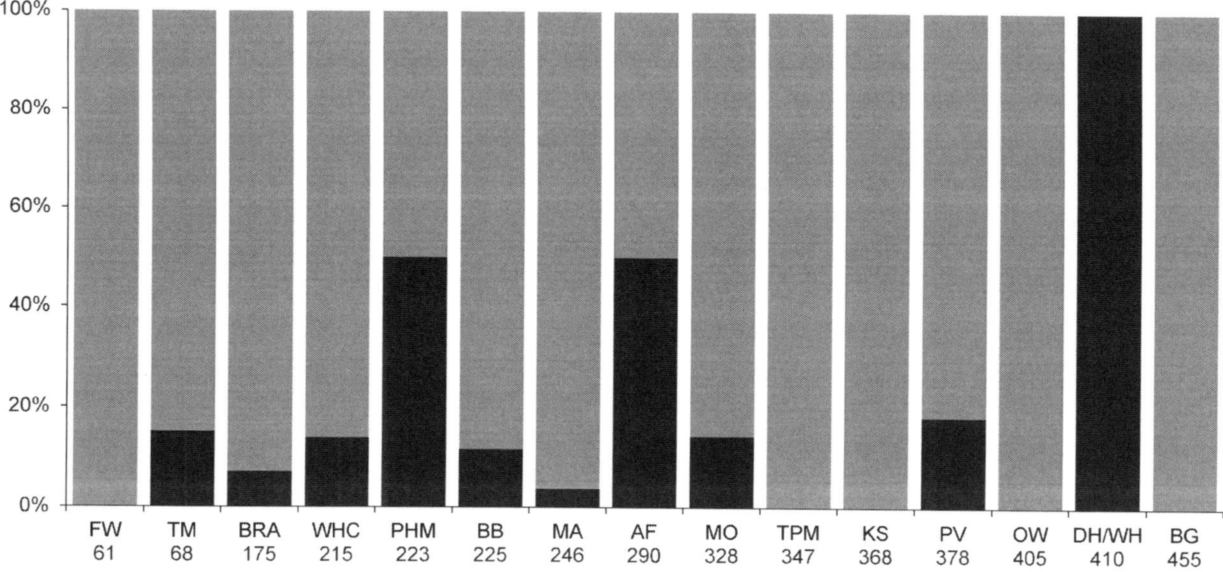

FIGURE 6.15. Obsidian use across the Old River Bed delta during the Paleoarchaic. Sources are sorted by distance (kilometers) from least to greatest (*left to right*). Light gray denotes the frequency of projectile points, and dark gray denotes the frequency of non–projectile point tools. FW = Ferguson Wash; TM = Topaz Mountain; BRA = Blackrock Area; WHC = Wild Horse Canyon; PHM = Pumice Hole Mine; BB = Browns Bench; MA = Malad; AF = American Falls; MO = Modena; TPM = Tempiute Mountain; KS = Kane Springs Wash Caldera Variety 2 (Kane Springs); PV = Paradise Valley; OW = Owyhee; DH/WH = Double H/White Horse; BG = Bear Gulch.

(n = 1,691, with no debitage included), 71.1 percent are projectile points (52.9 percent obsidian [n = 894] and 18.2 percent FGV [n = 308]) and 28.9 percent are non–projectile point tools (8.3 percent obsidian [n = 140] and 20.6 percent FGV [n = 349]). Obsidian accounts for

61.1 percent of the sample (86.5 percent projectile points and 13.5 percent non–projectile point tools), while FGV accounts for the remaining 38.9 percent (46.9 percent projectile points and 53.1 percent non–projectile point tools). Figure 6.17 is a compilation of Figures 6.15 and

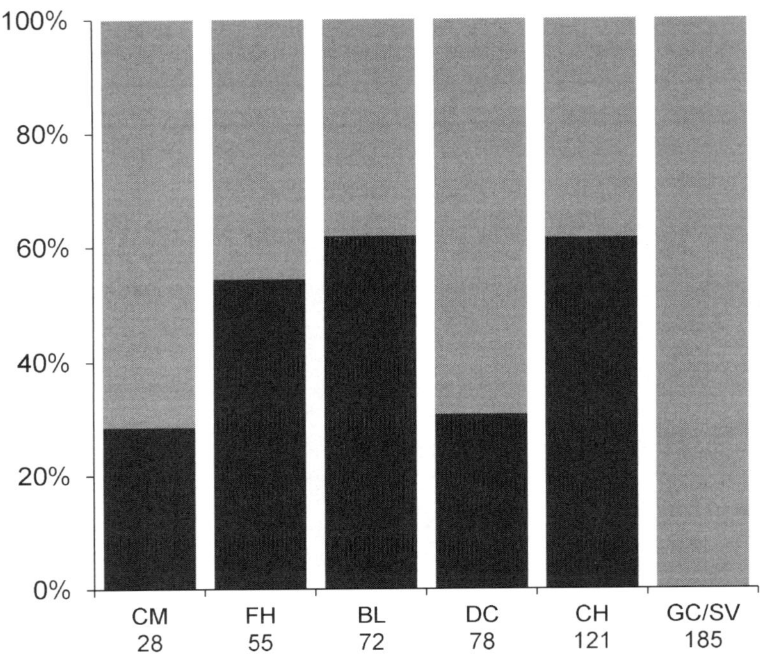

FIGURE 6.16. Fine-grained volcanic use across the Old River Bed delta during the Paleoarchaic. Sources are sorted by distance (kilometers) from least to greatest (*left to right*). Light gray denotes the frequency of projectile points, and dark gray denotes the frequency of non–projectile point tools. CM = Cedar Mountain; FH = Flat Hills; BL = Badlands; DC = Deep Creek; CH = Currie Hills; GC/SV = Gleason Creek/Smith Valley.

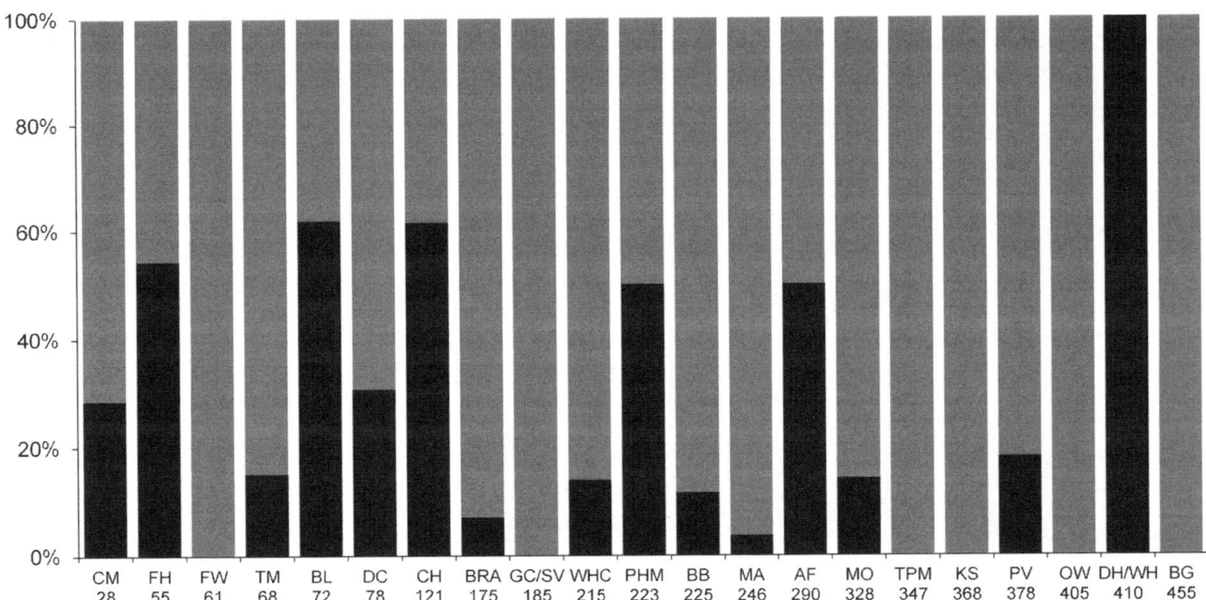

FIGURE 6.17. Frequency of obsidian and fine-grained volcanic source use across the Old River Bed delta during the Paleoarchaic. Sources are sorted by distance (kilometers) from least to greatest (*left to right*). Light gray denotes the frequency of projectile points, and dark gray denotes the frequency of non–projectile point tools. CM = Cedar Mountain; FH = Flat Hills; FW = Ferguson Wash; TM = Topaz Mountain; BL = Badlands; DC = Deep Creek; CH = Currie Hills; BRA = Blackrock Area; GC/SV = Gleason Creek/Smith Valley; WHC = Wild Horse Canyon; PHM = Pumice Hole Mine; BB = Browns Bench; MA = Malad; AF = American Falls; MO = Modena; TPM = Tempiute Mountain; KS = Kane Springs Wash Caldera Variety 2 (Kane Springs); PV = Paradise Valley; OW = Owyhee; DH/WH = Double H/White Horse; BG = Bear Gulch.

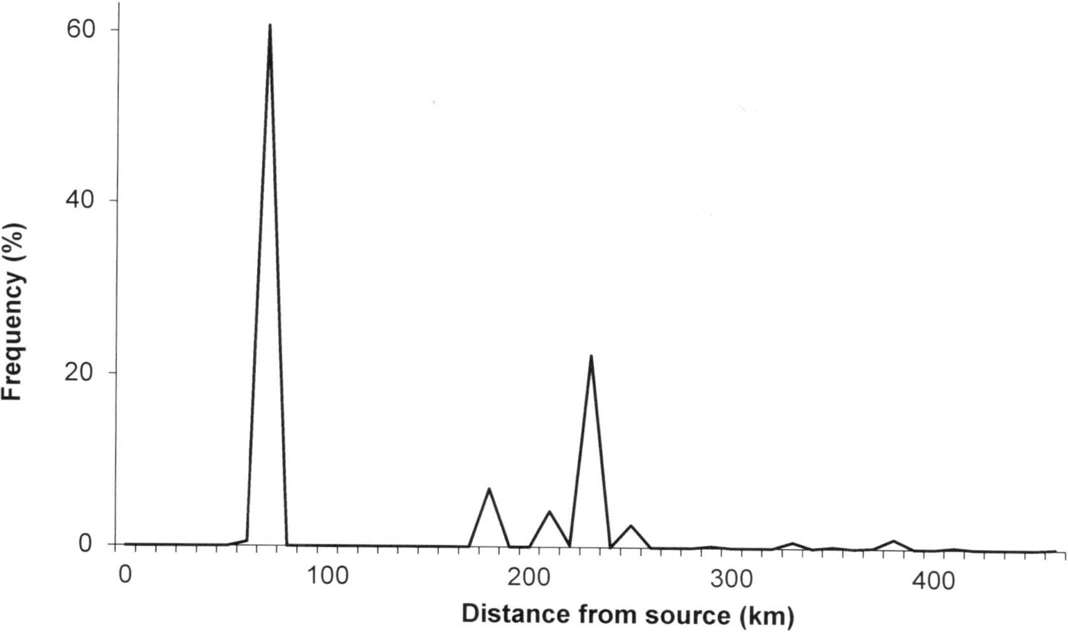

FIGURE 6.18. Distance decay graph showing the frequency of obsidian use in the Old River Bed delta. Prominent peaks show abundant use of Topaz Mountain (*left*) and, to a lesser extent, Browns Bench glass (*center*); points in between show limited use of Black Rock Area and Wild Horse Canyon, respectively (*left*) and Malad (*right*).

6.16 and shows 21 known sources/chemical types of obsidian and FGV sorted by distance and usage frequencies for projectile points and non–projectile point tools for each source. The bulk of artifacts were made on materials originating less than 250 km away, although some obsidian originates from more than 400 km away (Bear Gulch, at ~455 km). Approximately 97 percent of the obsidian used to produce tools originates <250 km from the delta, and ~61 percent of the obsidian originates <70 km from the delta. These spikes in usage are due to the high-frequency representation of the Topaz Mountain and Browns Bench sources (Figure 6.18). Approximately 98 percent of the FGV also originates <80 km from the delta (Figure 6.19).

Two anomalies stand out when viewing obsidian and FGV representation by projectile point type (Figure 6.20): elevated representation of FGV in Cougar Mountain/Haskett types and in WST point fragments (unidentifiable GBS stems/blades).[5] These two groups are also anomalous in patterns of obsidian source usage (Figure 6.21). There is abundant representation of other obsidians (dominated by Wild Horse Canyon and Black Rock Area) in Cougar Mountain/Haskett types and abundant representation of Browns Bench in WST point fragments. This seems to suggest a southerly oriented pattern of entry into the delta, with finished Cougar Mountain/Haskett points and replacement of these points with more-localized FGV. Both WST point frag-

ments and Parman points show elevated representation of Browns Bench, possibly suggesting a more northerly pattern of delta entry with points of these types. As Beck and Jones hypothesize in Chapter 5, this may also be the replacement of broken/exhausted/lost WST points with nearby FGV toolstone. If there is indeed a valid temporal distinction between WST points and EH GBS/Pinto/Butte Valley Corner-notched points (see Chapter 5), then sourcing data are consistent with this hypothesis, and we can infer that this change likely corresponded with increased use of a persistent ORB delta wetlands and a decline in alternative wetlands.

We use provenience data to plot the maximum extent of sourced materials from PA assemblages. The large dark gray area in Figure 6.22 measures ~850 km north–south × ~475 km east–west (>214,000 km²; >82,000 mi²) and extends from Bear Gulch in the north to Kane Springs in the south and then west to Paradise Valley. This area includes material from 15 obsidian source areas and six FGV source areas. Sources lie as far as 475 km from sites in the delta. This geographic zone covers an area more than five times larger than the area presented in the revised ECZs of Jones et al. (2012).

In an attempt to better identify where the majority of materials originated from in the ORB delta, two other polygons are plotted in Figure 6.21. The larger mid-tone gray oval area measures ~500 km north–south × ~165 km east–west (ca. 66,000 km²; >25,000 mi²) and accounts

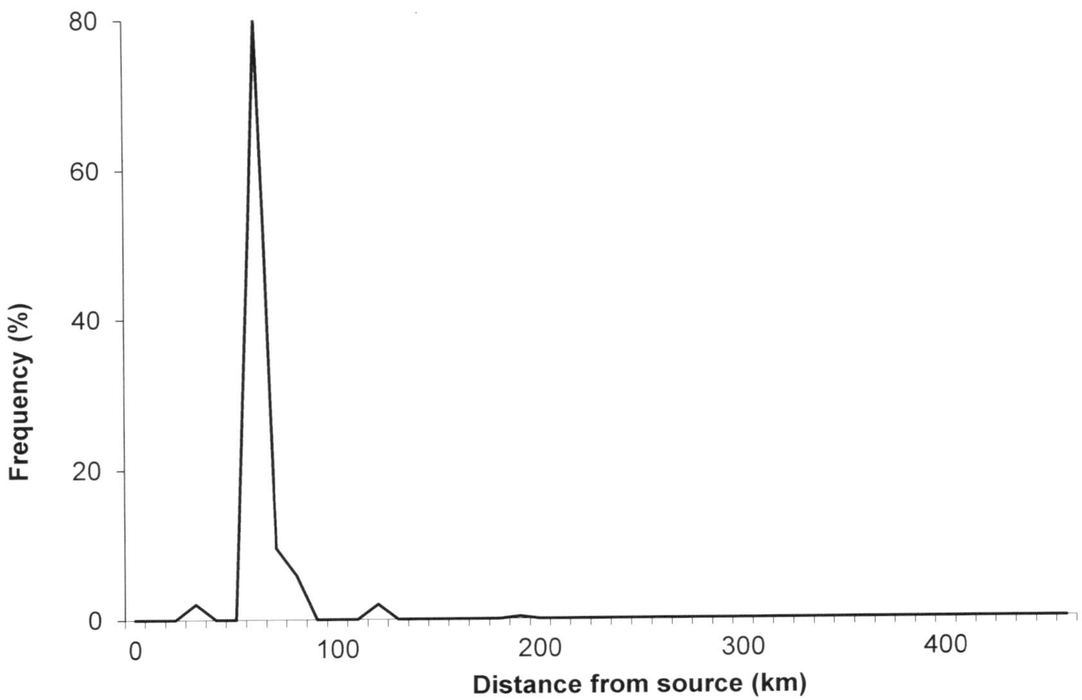

Figure 6.19. Distance decay graph showing the frequency of fine-grained volcanic use in the Old River Bed delta. Prominent peak shows abundant use of Flat Hills and, to a lesser extent, Cedar Mountain (*left*) and Badlands/Deep Creek chemicals (*right*).

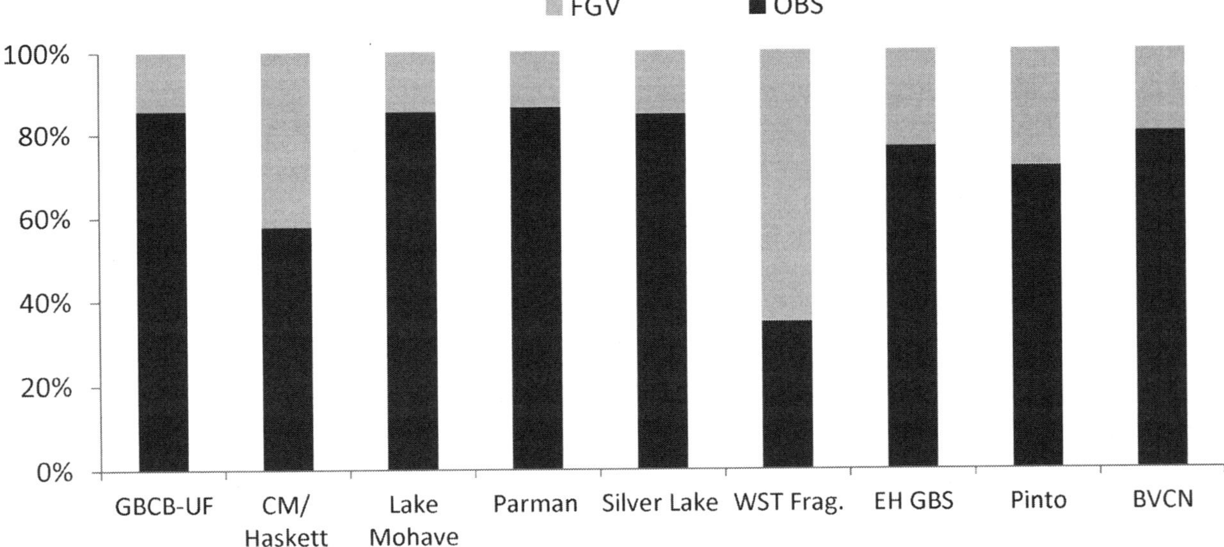

Figure 6.20. Frequency of obsidian (OBS) and fine-grained volcanic (FGV) use by Western Stemmed Tradition (WST) and Early Holocene (EH) point types. GBCB-UF = Great Basin Concave Base, Unfluted; CM = Cougar Mountain; Frag. = fragment; GBS = Great Basin Stemmed; BVCN = Butte Valley Corner-notched.

for 95.5 percent of the obsidian from known sources and 97.7 percent of the FGV from known sources. This locale includes material from six obsidian source areas and five FGV source areas and covers an area more than 1.5 times larger than the area presented in the revised ECZs

of Jones et al. (2012). The maximum distance between site and source within this primary core area is ~250 km. The third smaller light gray triangular polygon includes an estimated 73 percent of identified obsidian and FGV from known sources and measures ~130 km east–west ×

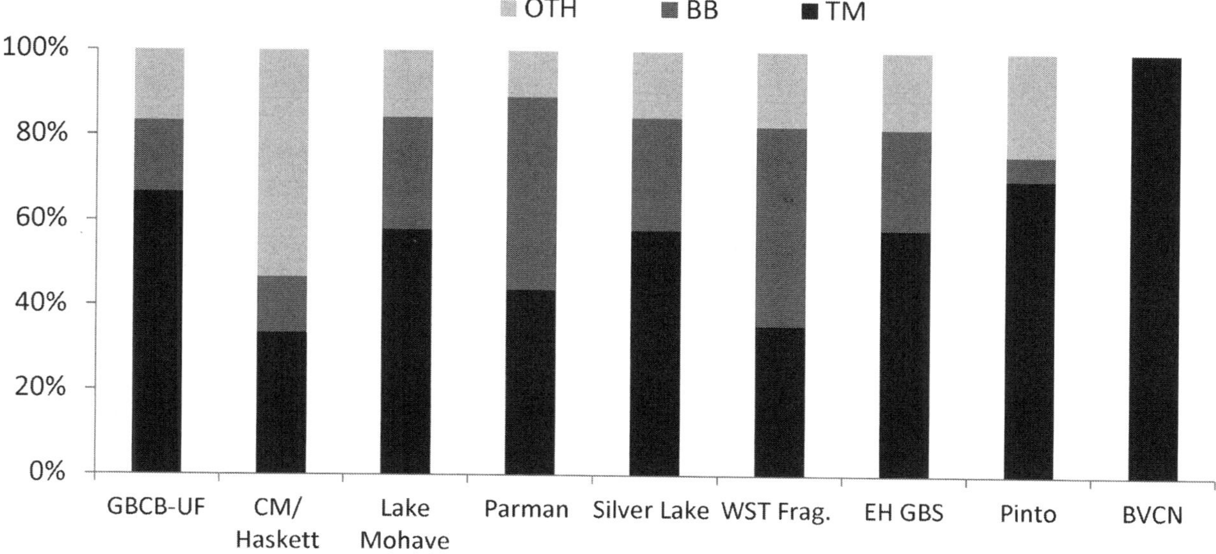

FIGURE 6.21. Frequency of general obsidian source use by Western Stemmed Tradition and Early Holocene point types. OTH = other; BB = Browns Bench; TM = Topaz Mountain; GBCB-UF = Great Basin Concave Base, Unfluted; CM = Cougar Mountain; Frag. = fragment; GBS = Great Basin Stemmed; BVCN = Butte Valley Corner-notched.

~60 km north–south (>5,000 km^2; ca. 1,950 mi^2). This inner core area includes obsidian from Topaz Mountain and FGV from Flat Hills, Deep Creek, and Badlands. The maximum distance between site and source in this inner core is <90 km.

For comparison, we use provenience data to plot the maximum extent of sourced materials from Archaic-aged assemblages from across the proximal delta. The maximum extent for these artifacts is ~700 km north–south and ~500 km east–west (>172,000 km^2; >66,000 mi^2) and includes material from up to 535 km away (Mono Glass Mountain). The acquisition area differs in extent and direction from the WST/EH area. It includes several sources not represented in the ORB delta, such as Big Southern Butte and Mono Glass Mountain, but excludes ones that are represented in the delta, such as Bear Gulch, Owyhee, Paradise Valley, Double H/White Horse, and Modena. However, the core area established for WST/EH sites in the delta still accounts for ~98 percent of sourced material in the Archaic sample. It is apparent that post-Paleoarchaic groups often scavenged the early-period artifacts found in surface sites, as evidenced by differential weathering/patination on artifacts, and this behavior may falsely extend the Archaic toolstone profile.

The Question of Exchange

We give little credence to the notions that toolstone from distant sources (outside the core area) was either a commodity to be traded or directly acquired through round-trip procurement. Without a complete view of both sides of a trade or exchange transaction (i.e., being able to track or identify the reciprocated items leaving the delta), there is no way to substantiate claims that toolstone or artifacts (even in low quantities) are products of trade/exchange. Both the remoteness of the ORB delta and the suspected low population densities during the Paleoarchaic argue against such behavior accounting for more than a very small fraction of the observed assemblage (cf. Beck and Jones 2011). A simple comparison of the presence/weight/frequency of the rare long-distance sources with the array of more local sources that make up the bulk of the lithic assemblage suggests that it is certainly possible that some level of exchange occurred. It is even possible that there was the occasional direct acquisition of these rarer materials by a splinter group, such as young men on long-range information-gathering trips. However, there is nothing in our data to suggest that an occupational group as a whole ever traveled beyond the core area for the sake of obsidian trade/acquisition. We agree with Madsen (2007), Newlander (2012), and Speth et al. (2013) that mobility by the whole group is unlikely to account for the most exotic of materials, but this could have come simply by replenishment of working tool kits by those on long-distance trips.

Differential source use in the ORB delta is most apparent when viewing the transport of obsidian from outside the Bonneville basin (Figure 6.23). While the majority of these toolstone types were used across many

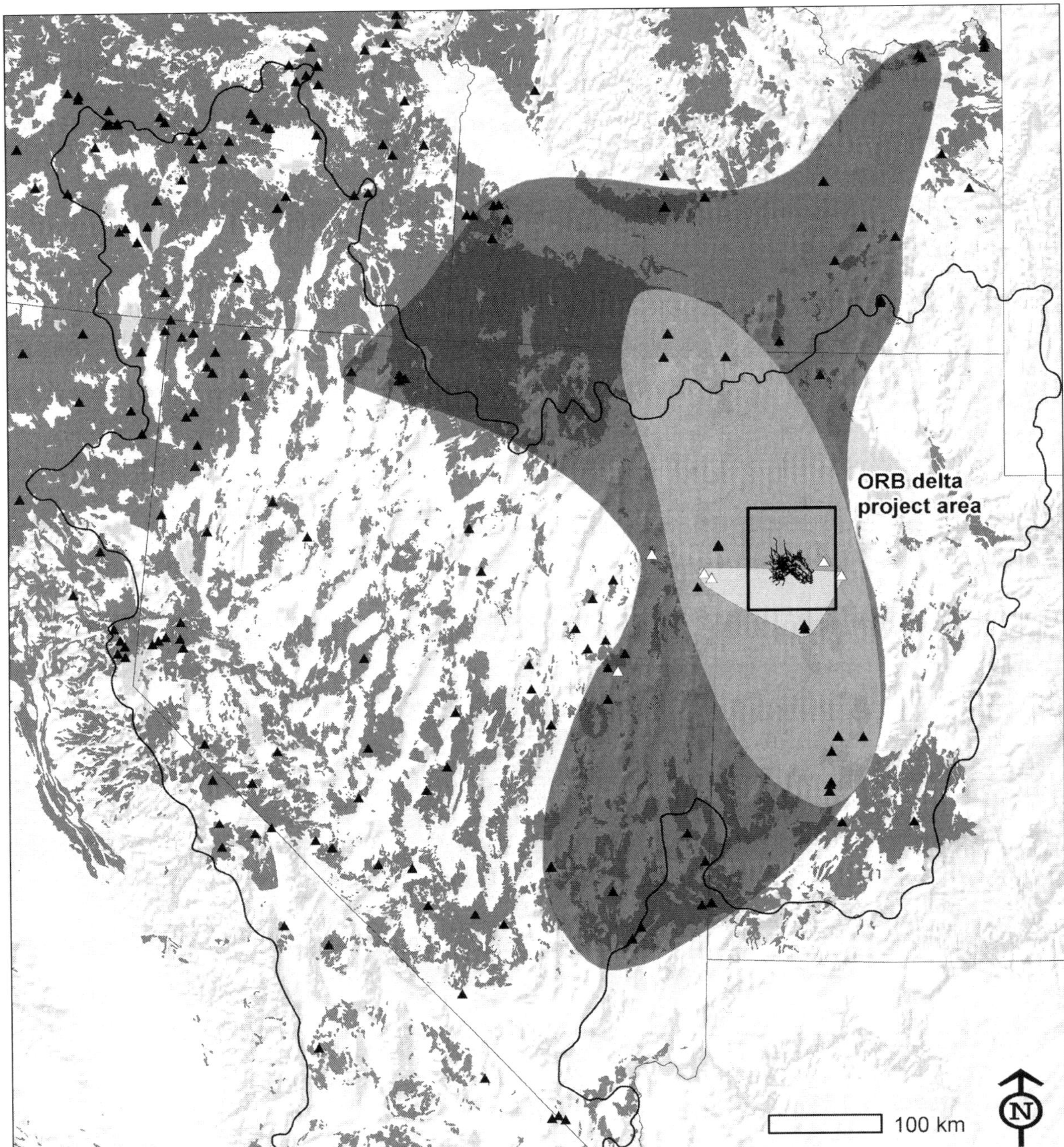

FIGURE 6.22. Maximum extent of all the identified obsidian and fine-grained volcanic (FGV) materials from Paleoarchaic assemblages from the Old River Bed (ORB) delta (*dark gray area*). The oval (*mid-tone gray*) includes an estimated 96 percent of identified obsidian and FGV, while the smaller triangular polygon (*light gray*) includes an estimated 73 percent of identified obsidian and FGV from the delta.

sites in the delta, some only show up in very limited quantities and occur in different subareas. This implies differential use of the wetlands by parties entering from different directions on the landscape. The distal delta shows limited use of two northern sources—Bear Gulch and American Falls—that are not represented in the proximal delta. The proximal delta shows limited use of another northern obsidian chemical type—Owyhee—that is not represented in the distal portion of the delta. Limited amounts of two southern obsidian types—Tempiute Mountain and Kane Springs—are also represented in the proximal delta but not seen in the distal delta. Obsidian use in the northern delta is spatially skewed to the north, covering an area that extends

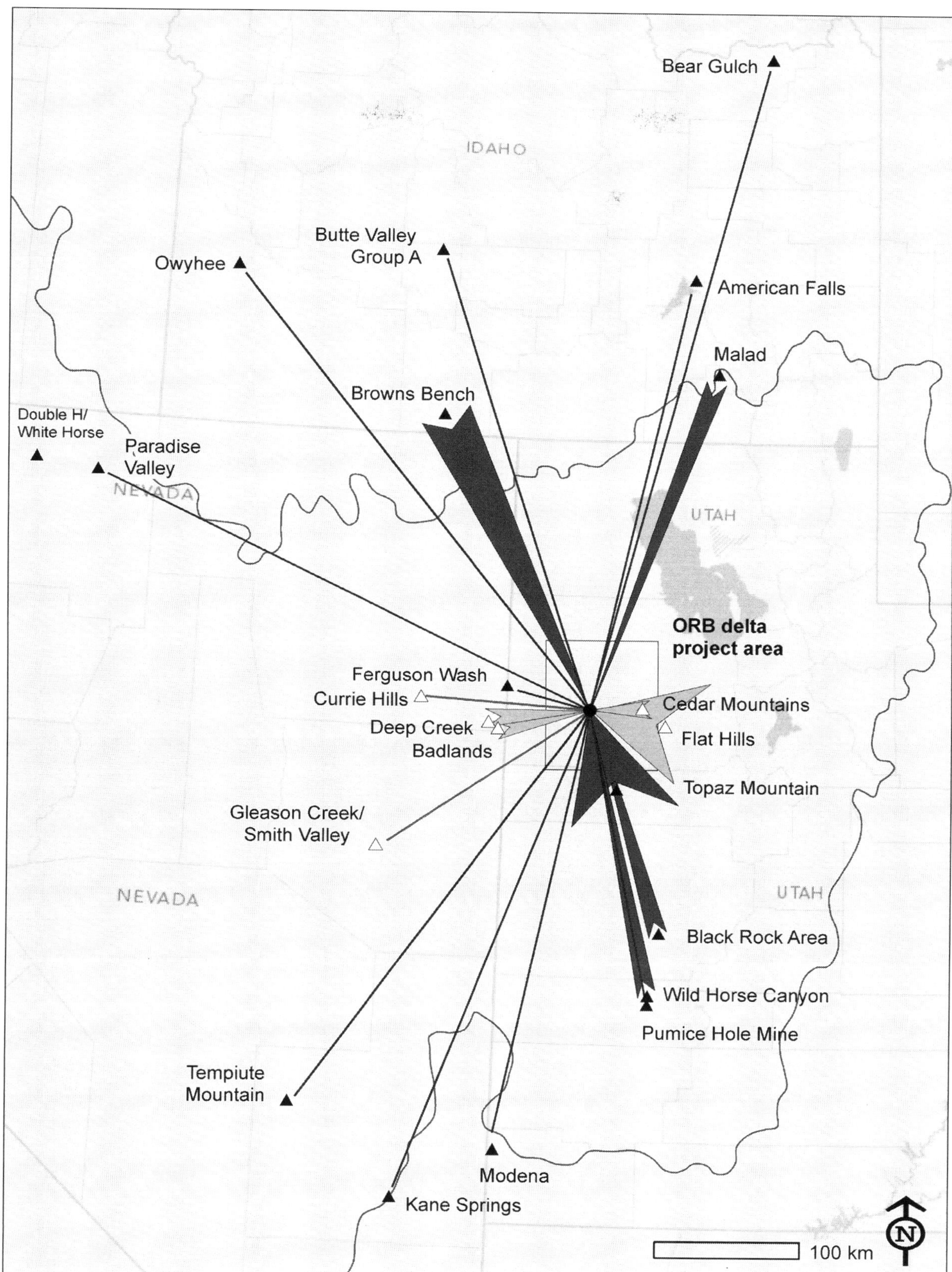

FIGURE 6.23A. Raw material source use within the Old River Bed (ORB) delta for projectile points. Arrow size indicates frequency by material type: large, >25 percent; intermediate, 15–25 percent; and small, 2–15 percent. Thin lines indicate <2 percent. Dark gray arrows/lines are obsidian sources, and lighter gray arrows/lines are fine-grained volcanic source areas.

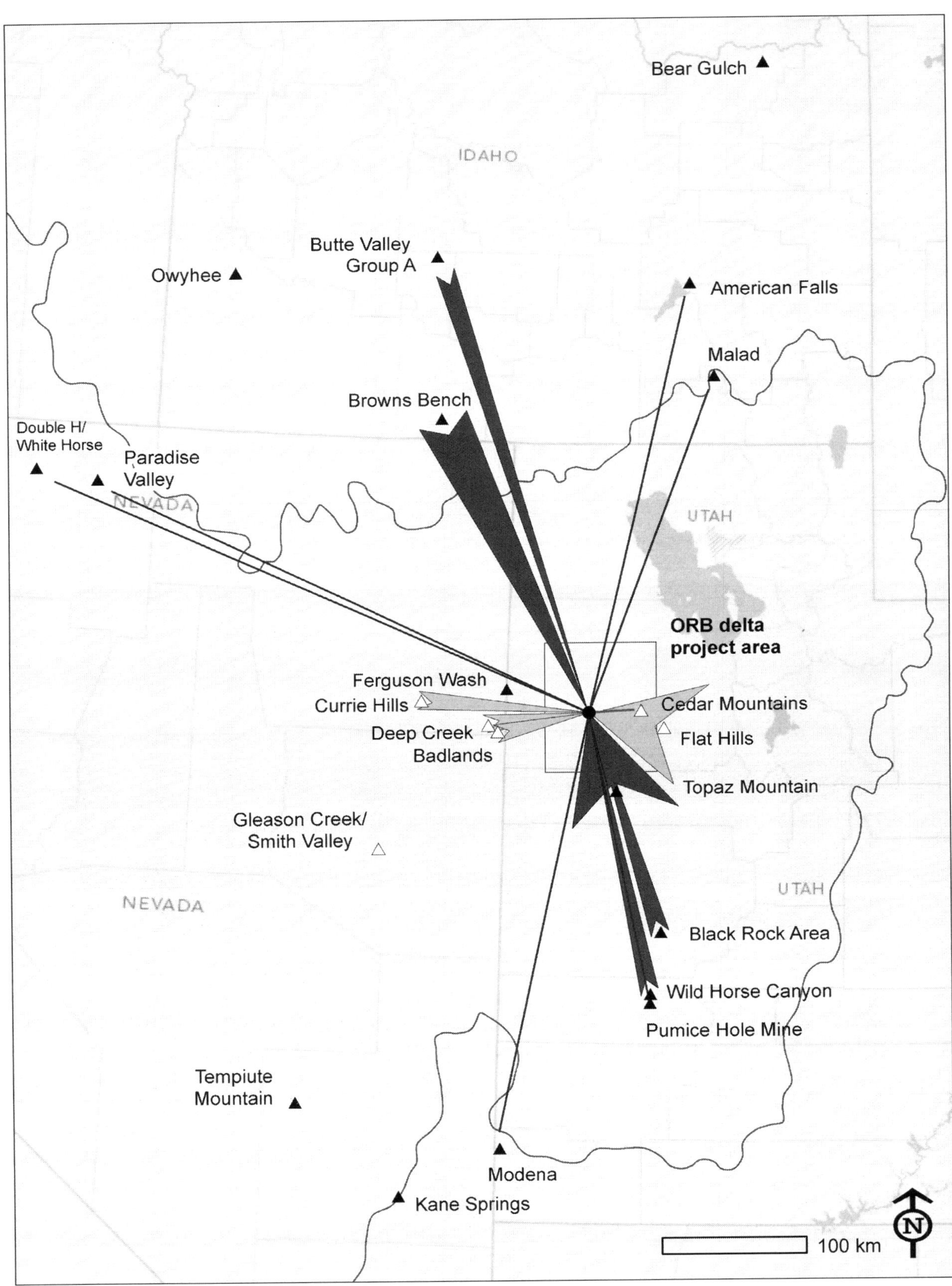

FIGURE 6.23B. Raw material source use within the Old River Bed (ORB) delta for non–projectile point tools. Arrow size indicates frequency by material type: large, >25 percent; intermediate, 15–25 percent; and small, 2–15 percent. Thin lines indicate <2 percent. Dark gray arrows/lines are obsidian sources, and lighter gray arrows/lines are fine-grained volcanic source areas.

some 135 km more northerly, while obsidian use in the southern delta is skewed to the south, covering an area that extends some 75–120 km more southerly.

Obsidian percentages can be viewed directionally by splitting the toolstone pool into three zones: 71 percent south (Utah), 26 percent north (Idaho), and less than 2 percent west (Nevada). Based on these obsidian data, there is not much of an eastern Nevada connection, as there was little use of Ferguson Wash, Paradise Valley, or Double H/White Horse obsidian. If we include FGV use, then there is a bit more of a western link with the inclusion of primarily Deep Creek and Badlands, but still only 17 percent of the FGV is from eastern Nevada.

There appears to be an association or affinity with southern Idaho and obsidian sources on/around the Snake River Plain. Twenty-six percent of obsidian overall comes from there, with 21 percent for non–projectile point tools and 26 percent for projectile points delta-wide. There is higher source diversity in the north, with seven source areas/chemical types, compared with five for the south and four for the west. Northern source use is skewed, with elevated use of source variants making up the Browns Bench geochemical group. A Snake River Plain/southern Idaho obsidian connection was also observed in eastern Nevada Paleoarchaic sites reported by Jones, Beck, Jones, and Hughes (2003). In a sample of 916 artifacts, there was moderate use of the Browns Bench source area ($n = 265$, 29 percent) and very limited use of Malad obsidian ($n = 6$, <1 percent).

Smaller basins/resource patches, such as Butte Valley, Jakes Valley, and Sunshine Locality, include many early WST tools, including fluted points, yet lack many later EH stem types and the abundant Pinto points that are found in the ORB (Beck and Jones 1988, 2009; Estes 2009). As environmental factors contributed to the rapid decline in productivity of these smaller resource patches (see Chapter 2), intensive occupation of these smaller patches would have correspondingly waned. If the ORB was indeed the place to be in this period of environmental flux (i.e., a sink), then we would also expect it to be a social hub or gathering place where there was exchange of various things, including genetics, ideas, raw materials, and so on.

Conclusions

This study of toolstone use and source provenience places the sites of the ORB delta into a larger context within the eastern Great Basin. The results and conclusions presented here are specific to the ORB and are tied directly to the context of a toolstone-poor, but otherwise resource-rich, refugium within a larger deteriorating environmental system at the Pleistocene–Ho-

locene transition. With different variables in place (e.g., more locally available toolstone, larger and less remote patches, greater and more sustained productivity in surrounding basins, etc.) foragers may have reacted much differently, producing a picture that is more in line with a system of lithic conveyance within a large-scale foraging territory. We do not doubt that some of the exotic toolstone found in low frequencies from distant sources is the result of exchange, but we do not envision formal trade as a primary mover. We think that the majority of stone in the assemblage was directly acquired, with distant materials representing population convergence into the delta over an extended period.

Our model is consistent with Madsen's (2007) patch choice–based model for Paleoarchaic mobility, but he also argues that male hunting parties may have driven much of the extended logistics of Paleoarchaic mobility. However, in traveling great distances from the ORB delta, any task group burdened with group-level hunting responsibilities could not produce a reasonable return rate, even if toolstone procurement was part of their responsibilities. Within the constraints of this economic model, people would have only traveled to neighboring uplands until the transport costs became excessive. Smith (2010) raises this concern, even in the smaller foraging territories he proposes. If we remove hunting as a primary task and simply place far-ranging, self-sufficient task groups on the landscape, there may be room to account for the less than 10–20 percent occurrence rates for exotic obsidians found in our assemblages. Given the low population density and large foraging territories implied, there are good sociocultural reasons to expect long trips to maintain group interactions, gather information about future moves, and acquire other types of useful resources that cannot be found in the wetlands.

Speth et al. (2013) focus on the possibility of exchange to consider that small parties may account for much of the toolstone profile in a given area and critique the common assumption of embedded procurement for Paleoindian groups. Yet there is nothing in the ORB delta data to indicate that people moved toolstone in any other than a straightforward economical manner. Our findings are predicted by formal land-use and technological models, and even direct procurement from residences could have occurred during lengthy occupations. If Speth and colleagues' (2013) "exotic materials" are taken to be the usual definition of nonlocal, then all of the material on the ORB delta is exotic. We certainly would not exclude long-range activities, social or otherwise, by certain individuals from being behind the presence of some of the most remote obsidians, but these need not be considered exchange if individuals simply visited distant sources. Furthermore, given the wide distribution of obsidian in the eastern Great Basin, we see no reason why it would need to be exchanged at all.

It is useful to remember that lithic conveyance zones are not behavioral models. There were no physical boundaries keeping foragers contained within one zone or out of another. Conveyance zones can be mapped any number of ways and will vary extensively according to what sites, tool types, material types, temporal period, and so on are sampled, and they also differ according to the spatial extent of surrounding geological sources of raw material. They are distributional overlays that allow us to frame geochemical sourcing data in such a way that they provide some constraints on how we conceptualize prehistoric behavior.

Finally, we reject the notion that governing aspects of Paleoarchaic mobility can be extrapolated from toolstone profiles alone, no matter how they are tracked across space (cf. Hughes 2011; Kelly 1992; Shackley 1996, 2005; Skinner et al. 2004). The analysis presented here centers on explaining distributions of lithic material in the ORB delta according to behavioral change through time within a particular ecological context. We favor an interpretation of lithic transport into the ORB delta based on a pattern of wetland settlement that increased in intensity into the early Holocene. In this sense, the ORB delta is more of a toolstone sink than an axis of movement. Contra the "conveyance zone" model used to describe distributions in other parts of the Great Basin (Jones, Beck, Jones, and Hughes 2003; Jones et al. 2012; Smith 2007, 2010), conceptualizing the entire ORB delta as a convergence zone yields a more satisfying explanation for local toolstone selection and patterning.

NOTES

1. Browns Bench obsidian is geographically extensive and has not yet been studied in detail. It is geochemically diverse and includes a number of varied chemical types referred to in the literature over the years as Browns Bench, Browns Bench Area, Butte Valley Group A, and Coal Bank Creek (Jones, Beck, Jones, and Hughes 2003); Brown's Bench, Hudson Ridge, Jackpot, Mahogany Butte, McMullen Basin, Monument Peak, Rock Creek, and Shoshone Creek (http://www.sourcecatalog.com/index.html); Brown's Bench Ranch and Twin Meadows Ranch (Hughes 1990); Goose Creek (Jackson et al. 2009); and Cedar Creek, Coal Bank Spring, Murphy Springs, Three Creeks Landfill, and Three Creeks 2 (Holmer 1997 and references therein). For simplicity these types, especially Browns Bench, Browns Bench Area, and Butte Valley Group A, are referred to under the general moniker "Browns Bench."

2. The chemical type "Butte Valley Group A" refers to unknown obsidian source "A" as reported by Jones and Beck (1990:89–91) in artifacts from Butte Valley, Nevada; also see discussion by Jones, Beck, Jones, and Hughes (2003:15) and http://www.sourcecatalog.com/id/s_id.html.

3. Several sites contained materials from Western Stemmed Tradition, Early Holocene, and/or Archaic temporal periods and are counted under multiple categories.

4. A comparative sample of 287 temporally diagnostic, Archaic-aged projectile points was selected from various sites across DPG. Items were chosen based on general availability from those artifacts curated by DPG. Only projectile points made of obsidian or FGV were sampled. The primary author was responsible for verifying typological determinations and did so according to Thomas (1981). Geochemical analysis was conducted by multiple analysts. pXRF ($n = 142$): Page and Schmitt 2010, 2011a; Page, Schmitt, Dalldorf, and Wazaney 2012; Page, Schmitt, Rhode, and Sapula 2012; Rhode et al. 2011; Schmitt and Page 2011; Schmitt et al. 2010; Schmitt et al. 2011. GRL ($n = 78$): Hughes 2004. And NWROSL ($n = 67$): Skinner 2010.

5. Typological identifications vary somewhat between the proximal delta and distal delta due to work by multiple lithic analysts. It is possible that some of the differences in source materials used in the two areas may be due to typological differences introduced by inconsistency in lithic analysis. Artifact classification has been conducted by multiple researchers, notably Duke (see Table 1.2 and references therein, especially Duke 2011) and Beck and Jones (Chapter 5). A large amount of the sourcing data was derived though thesis research by Page (2008) and dissertation and CRM research by Duke (2011). The remaining analyses were derived from various CRM projects conducted by the military on DPG (see Table 1.1, Chapter 1; also Page and Schmitt 2010, 2011a, 2011b; Page, Schmitt, Dalldorf, and Wazaney 2012; Page, Schmitt, Rhode, and Sapula 2012; Rhode et al. 2011; Schmitt and Page 2011; Schmitt et al. 2010; Schmitt et al. 2011) and the Utah Test and Training Range (Carter et al. 2004; Carter et al. 2005; Duke et al. 2008; Young et al. 2006). A small set of data published by Arkush and Pitblado (2000) represents the first XRF sourcing conducted in the area. Our large sample size likely suppresses bias to a great extent.

7 Integration and Synthesis

David B. Madsen, Dave N. Schmitt, and David Page

Geomorphological, Environmental, and Chronological Implications

In its heyday, between about 11,500 and 8800 ^{14}C BP, the Old River Bed (ORB) delta wetlands system occupied a roughly 2,600-km[2] area of western Utah on what is now the southwestern playa remnant of glacial Lake Bonneville. This wetland system was composed primarily of numerous distributary channels of the ORB river originating in the Sevier basin to the south but included a significant contribution by streams originating in the northeastern Deep Creek Mountains and, to a lesser extent, streams and groundwater-fed springs from the west flank of the Cedar Range (Figure 7.1). The ORB river probably began to flow at about 12,000 ^{14}C BP, as rapidly declining levels of Lake Bonneville eventually resulted in the separation of Lake Gunnison in the Sevier basin from the Great Salt Lake (GSL) basin to the north (see Chapter 3). In the GSL basin, this postglacial lake decline apparently continued in a rapid fashion, with lake levels in the basin falling to at or below modern levels by ~11,200 ^{14}C BP (Broughton et al. 2000; Madsen et al. 2001; Chapter 2), by which time the entire area later covered by the ORB delta was exposed. During the deposition of the ORB delta, the river was not constrained to a narrow valley, and bifurcating deltaic channels formed on the clay/silt sediments of what had been the bottom of Lake Bonneville. Wetlands and channels probably occupied a relatively narrow area between the end of the inset ORB river channel and the Wild Isle dune area to the northwest in the area where the oldest channels are dated to between ~11,500 and 11,000 ^{14}C BP.

As noted in Chapter 2, this drying trend was reversed during the Younger Dryas climatic interval from ~11,100 to 10,000 ^{14}C BP, as conditions in the Bonneville basin cooled. Oviatt et al. (2005) estimate that lake levels in the GSL basin began to transgress to higher levels during this period, reaching the Gilbert shoreline (probably about 1,295 m asl in the ORB delta area) sometime between about 10,500 and 10,000 ^{14}C BP. Although they suggest that the highest levels may have been maintained for only a short period, lake levels probably fluctuated above modern levels of the Great Salt Lake and below the Gilbert shoreline both before and after the lake reached its highest Younger Dryas peak. The ORB delta area was very likely constrained by ponding in the playa along its northwestern margins throughout this period and probably for as long as groundwater continued to flow in the ORB system. Currently, a shallow playa lake is present in a crescent-shaped depression on the western and northern margins of the ORB delta most of the time and only disappears during prolonged multiyear droughts. With increased water flow from both the ORB and Deep Creek Mountains distributary systems, this unnamed lake was sufficiently large that access to the ORB wetlands was restricted to southern or eastern entries even at lake levels below the Gilbert highstand (Figures 1.2 and 7.1).

After peaking during the Gilbert episode, surface water and groundwater feeding the ORB wetlands began to decrease between ~10,000 and 8300 ^{14}C BP, and by the end of that period both the ORB river and the wetlands it supported became extinct. Counterintuitively, the ORB delta may have supported a richer and

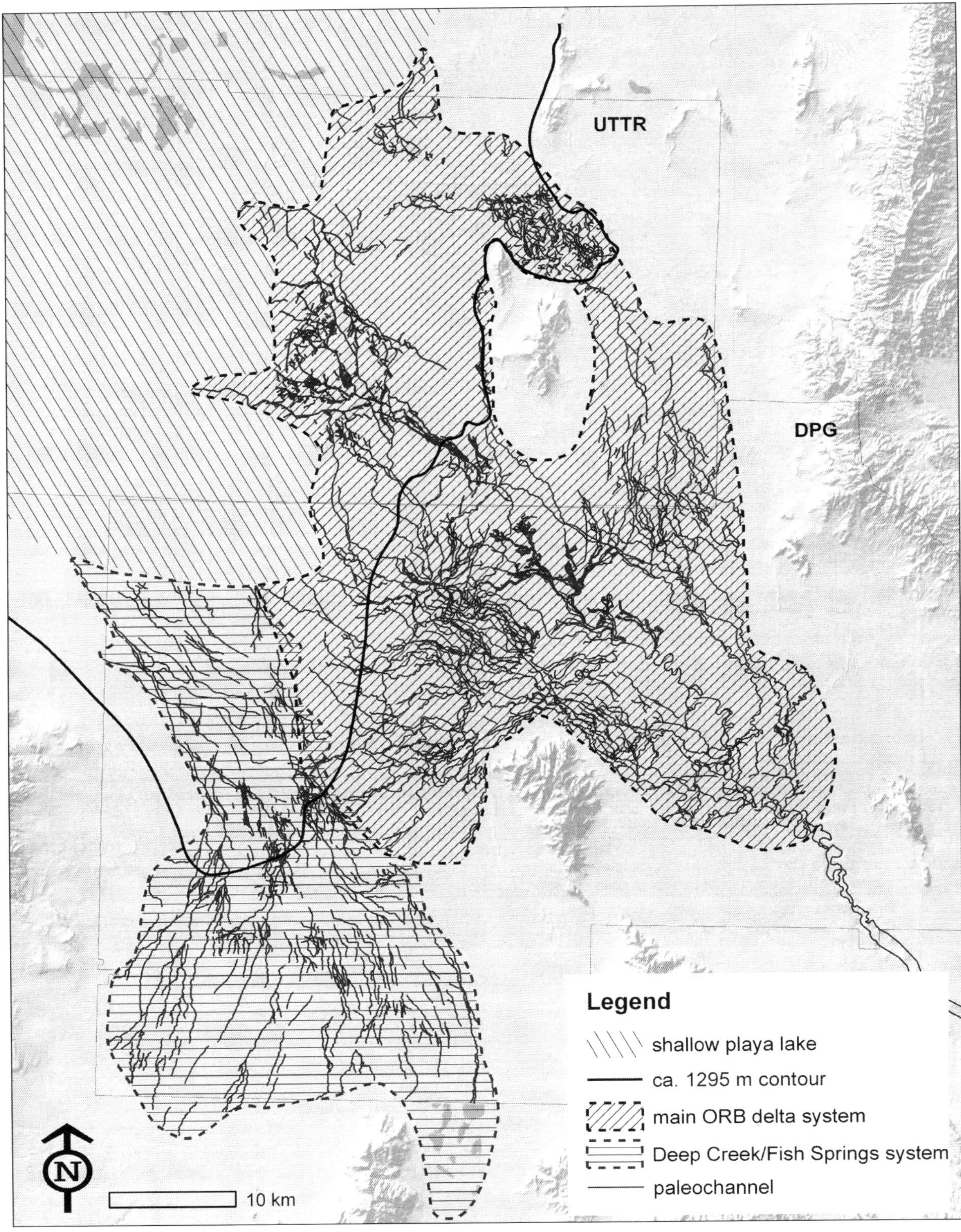

FIGURE 7.1. Schematic plan view of the Old River Bed (ORB) delta wetlands system showing contributory paleochannels originating from the Deep Creek Mountains and Fish Springs to the southwest and the main ORB delta area to the north. These were plotted from satellite imagery and not field checked in all cases. A number of these channels are recent (i.e., mid-Holocene or later), but the distribution does show the general pattern of distributary channels formed during the Paleoarchaic period. At present, a shallow playa lake seasonally covers the Great Salt Lake Desert north of the distributary system. The 1,295-m contour marks the approximate elevation below which a Gilbert lake formed in ~10,100 ^{14}C BP. The Utah Test and Training Range (UTTR) to the north is a U.S. Air Force facility. Dugway Proving Ground (DPG) to the south is managed by the U.S. Army.

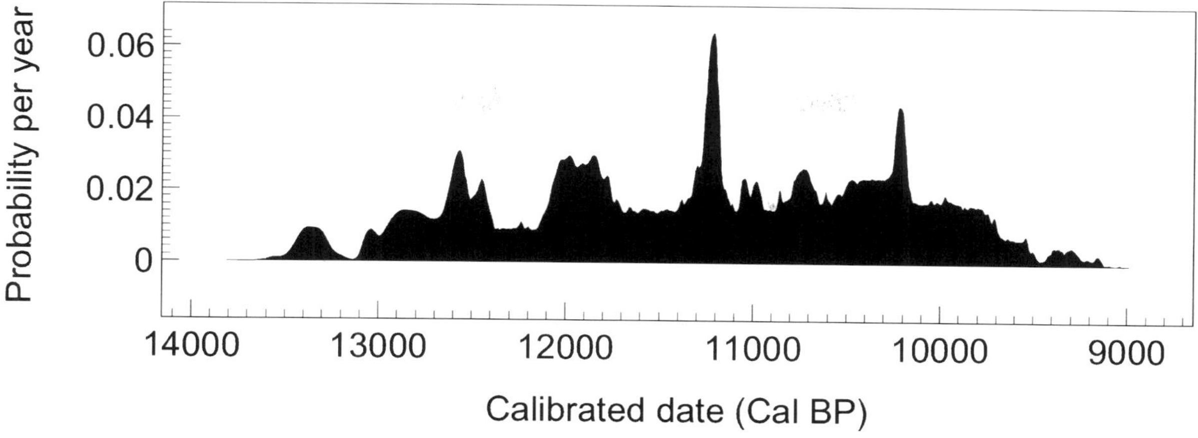

FIGURE 7.2. Summed probability distribution of 70 radiocarbon age estimates from Old River Bed delta distributary channels, produced using OxCal version 4.1.7 (Ramsey 2010). The dates are from Table 3.2, Chapter 3, with the addition of 14 age estimates on black mats located toward the distal end of the Green and Red channels that have yet to be formally reported (Daron Duke, personal communication 2012). With the possible exception of a brief interval at ~13,100 cal BP (~11,150 ^{14}C BP), this distribution suggests a generally continuous presence of wetlands in the delta between ~13,400 and 9300 cal BP (~11,500–8300 ^{14}C BP). Note that the modest general increase in wetlands dates to ~11,500–9800 cal BP (~10,000–8800 ^{14}C BP) as more and more channel systems were created. The marked decrease after ~9800 cal BP (8800 ^{14}C BP) suggests that the wetlands system was likely rapidly desiccating during this interval.

more abundant resource base for human foragers during this period rather than when water flow in the system was at its maximum. Water apparently overflowed from the Sevier basin until ~9900 ^{14}C BP, and until ~8800 ^{14}C BP groundwater flow was sufficient to maintain streams in the ORB delta. As noted in Chapter 3, the ORB delta is composed of hundreds of intersecting distributary channel systems that formed before, during, and after the lake reached its Gilbert highstand (Figure 7.2). Before and during the highstand, the number and extent of these distributary channels were relatively limited, but after this late Younger Dryas wet interval, as the delta distributary channels shifted back and forth across the delta from the southwest to the northeast, the number and overall length of these distributary channels increased almost exponentially. At any one time, the ORB wetland system contained channels with flowing water, abandoned groundwater-fed channels with open still water, and channels that had begun to silt in but still supported wetland vegetation rooted in water seeping along the sands and gravels of the channel fill. All of these channels were incised down into the relatively impermeable clays on the floor of Lake Bonneville, and water in the system likely fed all of the interlocking channels. As the number of channels increased, so too did the area supporting wetland floral communities and the fauna dependent on them. There are a limited number of channels dated to ~8800–8300 ^{14}C BP, suggesting that the system was drying out and only intermittently

supported by water flow, but in the centuries immediately prior to that time the ORB wetlands were likely at their largest and most productive size.

There is little direct evidence of the flora and fauna available to human foragers in the ORB delta wetlands, but it is possible to speculate using data from modern river and spring-fed marsh systems formed on Lake Bonneville lake-bottom muds, together with ethnographic data from historic foragers living in Great Basin wetlands. Certainly, it is clear from remains within the channels that both freshwater mussels (*Anodonta* sp.) and fish (Utah chub, *Gila atraria*) were common in the ORB wetlands.[1] A number of radiocarbon dates were run on sedges, such as *Schoenoplectus*, and reeds, such as *Phragmites*, but other marsh plants including cattails (*Typha* sp.) and rushes (*Juncus* sp.) were undoubtedly common, with the specific plant community in a particular channel varying along with differences in flow rates and water depths. A description of predevelopment modern vegetation in the Fish Springs Wildlife Refuge on the southern margin of what was the ORB wetlands area is likely representative:

The submerged-vegetation consists primarily of wigeon grass (*Ruppia maritime*), muskgrass (*Chara* spp.), and coontail (*Ceratophyllum demersum*) with wigeon predominating. The coontail is more abundant in the freshwater areas near the springs. Olney's three-square bulrush (*Scirpus olneyi*) is by far the dominated emergent species. It forms a band along nearly all shallow water areas and, in many cases, has completely covered the waterway. It is usually found in "dog hair" stands and averages five feet in height. There are a few patches of hardstem bulrush (*Scirpus acutus*) and cattail (*Typha* sp.) but these species are in the minority. Spikerush (*Eleocharis* spp.) is usually found on moist sites near the water. Cane (*Phragmites* spp.) is present in patches along the channels and also in sparse stands on the drier sites [Gueswel 1958].

Waterfowl, including a variety of ducks and geese, were available at least seasonally, and in such a large marsh area some waterfowl were likely year-round residents. The checklist of bird species present at Fish Springs includes 126 species, of which 34 are described as common or abundant and 26 of which are resident throughout the year (Madsen 1982a). Species that fall in both the abundant and the available year-round categories include black-crowned night heron, Canada goose, mallard, gadwall, pintail, green-winged teal, blue-winged teal, shoveler, ring-necked duck, Virginia rail, sora, American coot, and horned lark.

Small wetland game animals such as muskrats (*Ondata zibethicus*) were likely common, and, as noted in Chapter 2, rabbits (*Sylvilagus* sp.) were common at lower elevations in the Bonneville basin during the Pleistocene–Holocene transition and were likely common along the dryer, silted-in distributaries. Again at Fish Springs, a variety of small mammals are common, with several species of arvicoline rodents being particularly abundant (Madsen 1982a). Actual counts for the year 1977–1978 list ~12,000 muskrats, ~700 jackrabbits, and ~400 desert cottontail rabbits as present within the ~31-km^2 marsh area. Extrapolating those densities to the ~2,600-km^2 ORB delta wetlands produces estimated numbers of ~1,000,000, ~59,000, and ~34,000, respectively. Although these numbers are speculative, they do provide some insight into the size of the ORB wetlands resource base.

Pronghorn are common throughout the ORB delta area at present, but their numbers and distribution during the Paleoarchaic (PA) period are impossible to determine. Prior to extensive hunting in modern times, a herd of 60 was commonly in residence in the Fish Springs marsh area (Madsen 1982a). Again by extrapolation, the ORB wetlands would have contained a herd of more than 5,000 animals. We have observed extensive herds of both deer and pronghorn wintering in the watered areas of Dugway Proving Ground, and

similar winter occupations were likely present during the period of delta formation. Deer are also common residents at many wildlife refuges in the Great Basin, but as the refuges are managed primarily for birds, their numbers are rarely calculated and are hard to come by. However, a wildlife viewing guide for the state of Nevada is certainly suggestive. At the Sheldon National Wildlife Refuge, for example, there is an

excellent view of upland game, small mammals, pronghorn, mule deer and bighorn sheep spring through fall.... Hundreds of pronghorn winter at Big Spring and Gooch Tables, then pass spring and summer at Swan Lake. Mule deer winter on the eastern refuge, and by summer have moved up to higher elevations [Clark 1993:17].

Anecdotal reports for the Malheur Wildlife Refuge in the northwestern Great Basin describe as many as 1,000–1,500 deer as year-round marsh residents in the 757-km^2 wetlands area (Robert Elston, personal communication 2012).

SITE TYPES, DENSITY, AND DISTRIBUTION

As of this writing, 503 sites containing PA diagnostics, such as Great Basin Stemmed and Pinto points and other diagnostic artifacts, such as crescents, have been identified and recorded in the ORB delta wetlands. Of these, 238 are located in the proximal delta and are managed by the U.S. Army (six of these are actually in the Fish Springs/Callao areas south of the installation boundary), and 265 are located in the distal delta area managed by the U.S. Air Force. In addition, there are three fluted point sites and three isolated fluted point localities on or near the ORB delta containing a total of nine fluted points. All of these could be categorized as "Western Fluted," although two may be regarded as Folsom and may be of a younger age. Half of these are within the delta area itself, although one site, containing four of the fluted points, is actually near a quarry in the Cedar Mountains overlooking the wetlands, and an isolated fluted point was recovered from the slopes of Wig Mountain.

As noted in Chapters 1 and 4, sites in the two management areas were identified using different survey methods: block surveys in the distal ORB delta and primarily linear transects along channels in the proximal ORB. By using these surveys as samples of the ORB wetlands system as a whole, it is possible to estimate the total number of PA sites in the delta area in two different ways. Simply by extrapolating from the number of sites per square kilometer in the distal block surveys to the total ~2,600-km² area of the wetlands system, there may be more than 5,500 PA sites in the ORB delta (265 sites identified in 123.93 km² of surveyed area). It should be noted that in parts of these distal survey areas distributary channels are covered by dunes, thus reducing the number of sites per square kilometer by an unknown amount. On the other hand, ponding in some areas of the wetlands would have prevented occupation, so these factors may balance out in this crude estimation.

More importantly, except for areas in the northern mudflats where the boundaries of early sites are less distinctive, site locations are closely related to channels and channel margins, with even isolated artifacts rarely found between channels. Since this distribution makes a straightforward area-to-area extrapolation problematic, extrapolation from the number of sites along surveyed channel lengths to the total length of mapped channels in the ORB delta may provide a closer estimate of total PA sites in the wetlands. Based on these channel surveys, there is consistently about one site per kilometer of channel length. To date we have mapped 3,839 km of distributary channels in the ORB wetlands, suggesting that there are at least that many PA sites in the system. Since there are many additional unmapped channels, the number of PA sites in the ORB delta area is probably much higher and may even approach the numbers estimated from the area-based extrapolation.

By either estimation method, this is an astounding number of sites when compared with other areas of the Great Basin, but one must remember that these sites were created over a period of at least 2,500 years, with the most intense occupation occurring between ~10,500 and 8800 ^{14}C BP. That is, if one is to believe these rather speculative extrapolations, only one–two sites/year were created by PA foragers even in this resource-rich environment. On the other hand, as noted by Schmitt in Chapter 4, a number of the sites represent palimpsests created by multiple occupations, so the actual number of sites created per year was undoubtedly greater.

All sites were not created equal, of course, and a variety of site types are represented in our delta sample. Of the 100 sites investigated and categorized in Chapter 4, only 18 are considered to be base camps, with the remainder focused on diurnal tasks such as resource acquisition, processing, lithic reduction and fabrication, and general-purpose activities. Again extrapolating to the delta area as a whole, there may have been as many as 600–900 residential base camps occupied by PA foragers in the delta during the Pleistocene–Holocene transition. While this may seem to represent a high number of PA residential camps (and certainly is relative to other Great Basin localities), this estimated number, if accurate, still represents the occupation of only a single residential camp by one small group of PA foragers every two to three years during the time the ORB wetlands were in existence. As noted in the previous chapter, we think that these residential bases may have been occupied for months at a time, but even if they were occupied for much of the year, the number of foragers living in the marsh at any one time may have been rather small, and their corresponding impact on the resource base may have been equally limited.

Most of these residential base camps appear to be located on areas of higher ground such as point bars, natural levees, and the tops of older topographically inverted channels. In particular, the Black channels sit up to 4 m or more above the surrounding wetlands and, once those channels were abandoned, would have provided dry ground on which foragers could comfortably construct dwellings and conduct their day-to-day lives. These higher and dryer older channels would likely also have been less densely vegetated and would have provided elevated "highway" routes into and out of the ORB wetlands.

TOOLSTONE AS A MEASURE OF MOBILITY

In studies of Paleoindian mobility across North America, but particularly in the Great Basin, the use and distribution of artifacts from particular toolstone sources have been critical, if not primary, components of defining the way those foragers moved across the landscape. As we reviewed in Chapter 1 and as discussed in Chapter 6, estimates of how far PA foragers moved in the Great Basin, how often they moved, and even why they moved have been largely dependent on studies of toolstone sources and distribution and on hypotheses derived from those studies. Here we use the data we have obtained from studies of the ORB delta described in previous chapters to evaluate various aspects of these mobility hypotheses.

Points vs. Utilitarian Tools

Sourcing studies of PA toolstone distribution in the Great Basin have been almost entirely restricted to the identification of obsidian used to manufacture large formalized biface forms referred to generically as "points." We wonder if this bias may have influenced interpretations of Paleoarchaic foraging territories and ask: Do the types of toolstone reflected in different tool types suggest differently sized foraging territories? Further, does the toolstone of large, formal, well-made biface forms differ from that of more utilitarian tools such as scrapers? Finally, does such a difference, if there is one, indicate a differential mobility between male hunting parties and the rest of the PA peoples who occupied the ORB?

Page and Duke sourced 495 non–projectile point tools from sites in the proximal and distal ORB delta to compare with 1,226 "points" that could be categorized into formal point types (see Figure 6.23). Non–projectile point tools include artifacts such as choppers, scrapers, cores, drill/perforators, and bifaces. While we realize that many of these nonpoint tools may have been recycled from broken or worn-out points (or the reverse for that matter) and that "points" may have been used for tasks other than as projectiles, we anticipate that the group as a whole allows us to distinguish between toolstone used in hunting-related activities and that used for general utilitarian tasks around camp.

Assuming that the source identifications are accurate, there is a distinct bias toward the use of local toolstone vs. that from more distant sources, but that bias is much greater for nonpoint tools than it is for points. By "local" we mean sources in relatively close proximity to the ORB delta, including Topaz Mountain and Ferguson Wash obsidian and all the fine-grained volcanic (FGV) sources within 80 km of the central delta area (Figure 7.3). For "distant" we include all other obsidian toolstone (see Chapter 6 for actual distances from the ORB delta to these toolstone sources). The ratio of "distant" to "local" points is 1:2.17, while that of nonpoint tools is 1:7.52. Put another way, more than three out of every 10 projectile points (31.6 percent) were made from distant toolstone, while only about one out of every 10 nonpoint tools (11.7 percent) was manufactured from toolstone imported into the ORB wetlands from distant sources (Figure 7.4). Not only is the use of less-distant sources much rarer in the construction of nonpoint tools, but there are distinctly fewer distant sources so used. There are 14 different "distant" obsidian toolstone sources within the collection of ORB delta points, while only nine different distant sources were used to make nonpoint tools. Moreover, four of these nine sources are represented by a single biface each, which may have been imported as a point preform. Thus, in answer to one of our questions, there is a clear indication that the toolstone employed for the production of points differs significantly from the toolstone employed for the production of nonpoint tools. Not only is the proportion of locally acquired toolstone much higher in the nonpoint tools, but the number of distant sources is lower.

That said, there are some interpretive nuances hidden within these numbers. Many of the distant sources used in the manufacture of points are also rare within the collection, with several represented by a single point. This brings up the question of sample size, abundance, and diversity often found in discussions of faunal analyses (e.g., Grayson 1984). That is, the larger the sample size, the more likely it is for items of rare abundance to appear in a collection and, hence, for the diversity of the collection to appear to be greater. The situation is not quite the same for toolstone diversity since it is not the relative abundance of toolstone in the environment that is being measured but, rather, the frequency with which foragers made and/or traded for tools from the source locality and deposited them in the ORB delta. We discuss that issue below. Regardless, it is clear from proportional differences alone that the sourcing of only points produces a different view of conveyance zones, and hence, of foraging territories, than does the sourcing of non–projectile point tools. Sourcing only points creates the suggestion of larger foraging territories than does the sourcing of utilitarian tools. This is to be expected since, as Eerkens et al. (2007 and references therein) note, at sites occupied by mobile groups, waste flakes consist primarily of local materials, while expended and discarded tools disproportionally consist of distant toolstone.

Moreover, when sourcing only points and using the results to determine mobility patterns, chronological changes can be easily obscured. Beck and Jones in Chapter 5 indicate that there is a shift in the duration that foragers remained in the ORB delta during any one stay that is reflected by an increase in the way toolstone is husbanded. Yet this shift is poorly reflected in the use of local vs. nonlocal projectile points. The relative proportion of "local" to nonlocal toolstone used in the production of points is only slightly lower for PA/Western Stemmed Tradition (WST) points than it is for Early Holocene (EH)/Pinto points (64.5 percent and 70.8 percent, respectively). Perhaps, more importantly, however, the number and location of these sources change little between the earlier and later periods. The way points are made and the amount of wear they exhibit change through time, but their sources change only in slight relative proportions, not in number or kind.

The difference between toolstone use for points and that for nonpoint tools leads to the final question of whether or not it represents the different foraging territories of hunting parties and those of the remainder of the foraging groups in the ORB delta. The answer is not completely clear since, as is stressed in Chapter 5, many of the "points" in the formally recognized point categories may not be projectile points at all. But even assuming that most of the points did indeed tip hunting projectiles during at least a portion of their use-history, does that imply that hunters had larger foraging terri-

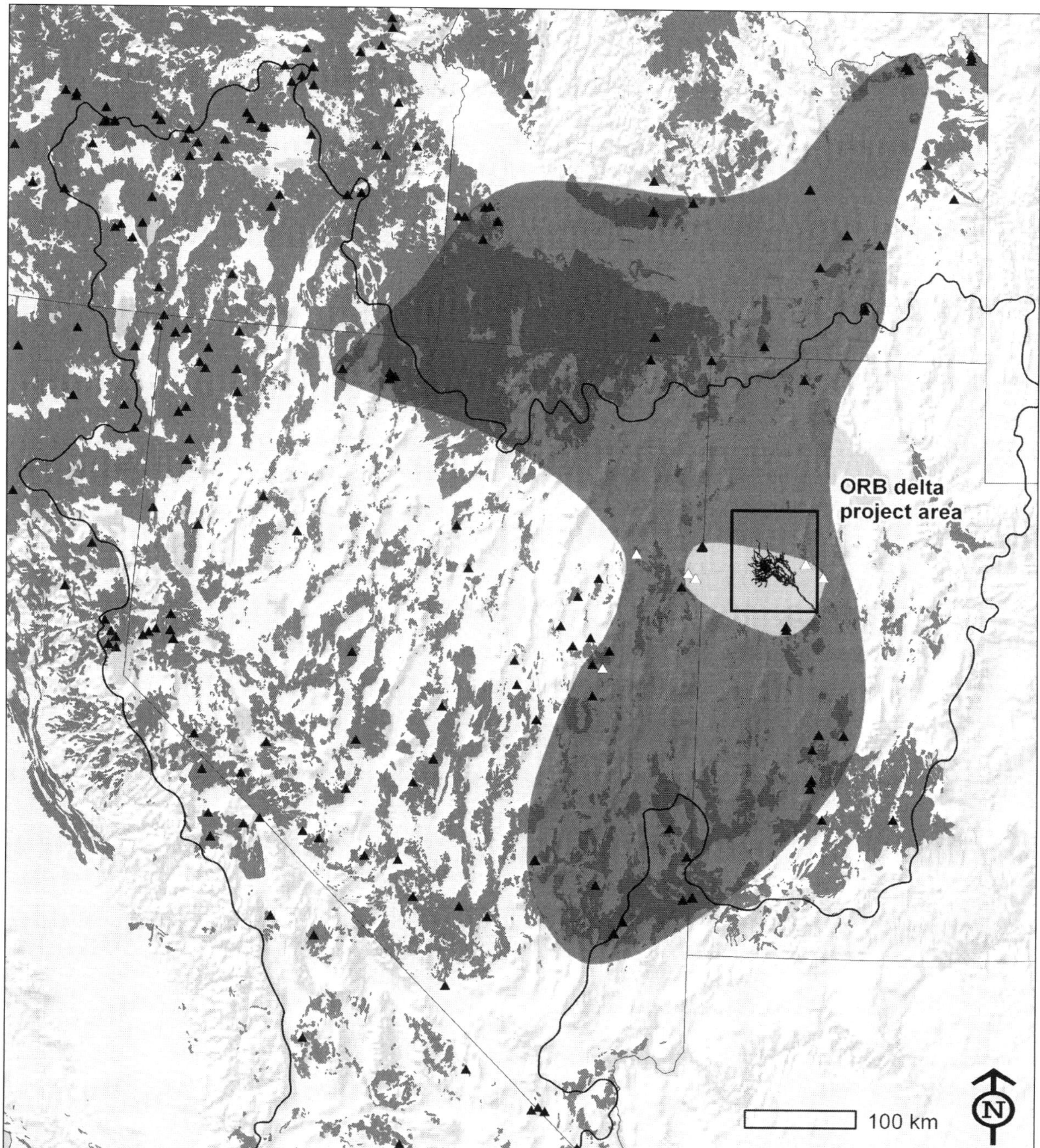

FIGURE 7.3. Area encompassing toolstone sources considered to be "local" (*light gray*) compared with the area of "distant" sources (*dark gray*) in the Old River Bed (ORB) delta.

tories than did the nonhunting members of the group? The sourcing data from the ORB delta certainly suggest that is the case, but it is also possible that finished "points" were more valuable as trade items than were utilitarian tools and that the differences in both proportions and numbers of sources are the product of a widespread and intensive trade network, as suggested by Newlander (2012). Such a case seems unlikely but cannot be ruled out without further study.

Page and Duke (Chapter 6), however, contend that the difference in the types of toolstone used for projectile points, as opposed to types used for utilitarian tasks, may be due to the functional requirements of particular tool types, with points requiring the sharpness

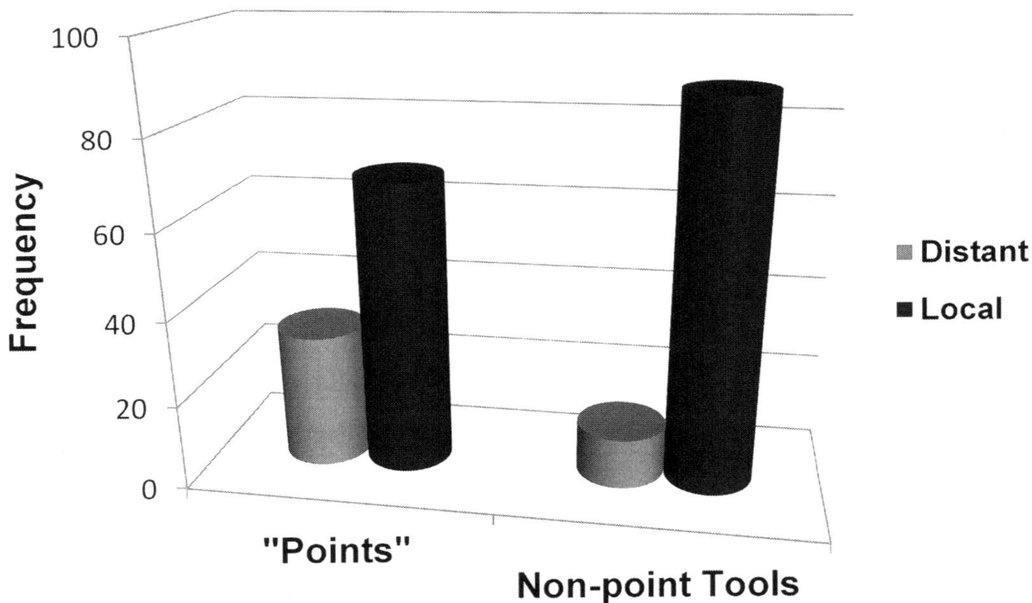

FIGURE 7.4. Relative proportions of "distant" (*gray*) and "local" (*black*) toolstone sources for points (*left*) and nonpoint tools (*right*).

of obsidian and many nonpoint tools requiring the strength and durability of FGV. The result is that proportionally more utilitarian tools are made locally than are projectile points, a conjecture that seems to be confirmed by the sourcing data. However, that implies a degree of experience-based planning by PA foragers traveling to the ORB delta wetlands, since they would have had to know that such lower-quality toolstone was available. For the present, the answer to this question must remain equivocal. The foraging territories of hunters *may* have been larger than that of the rest of the group, but substantially more support is needed to confirm that idea.

"Local" vs. Long-Distance Transport

As Page and Duke note in Chapter 6, more than 75 percent of all the sourced artifacts are made from what we consider to be local toolstone sources located less than 80 km from the central ORB delta area. But are those sources really "local"? Most lithic analysts consider only sources less than 20 km away to be local, meaning that all the toolstone in the ORB delta is therefore of "nonlocal" origin. However, there is really no consensus, that we can discover, on the distance demarcation between local and nonlocal toolstone. The <20-km distance was formalized by Surovell (2009), who, based on earlier work, considers it to be the distance a forager could travel in a day moving away from and returning to a base camp (40 km round-trip), and it is the distance used

subsequently in a number of toolstone-based Great Basin mobility studies (e.g., Smith 2011). However, as Smith notes,

ethnographic data show that hunter-gatherers also procure resources through multi-day logistical forays[, and] given that prehistoric groups could have procured toolstone either via daily foraging trips or longer logistical forays, it is problematic to use a single arbitrary distance to classify raw materials as either local or nonlocal, especially in cases where toolstone is unavailable <20 km from a site [2011:463].

We agree and, like Smith (2011), consider the nearest sources to be "local" regardless of the actual distances involved. In the case of the ORB delta we consider all the sources within 60–80 km to be local because, depending on where one is in the wetlands, they are all equidistant. The next-nearest major sources are 175–225 km removed from the marsh (disregarding a single rarely used FGV source at 120 km), at distances that represent a considerably greater investment in travel costs.

This unusual distinction between nonlocal toolstone sources and "local" toolstone sources that require more than a single day to reach and exploit leads to the question of whether the procurement of these local sources was of a logistical nature or was embedded in normal foraging activities. Virtually all toolstone-based

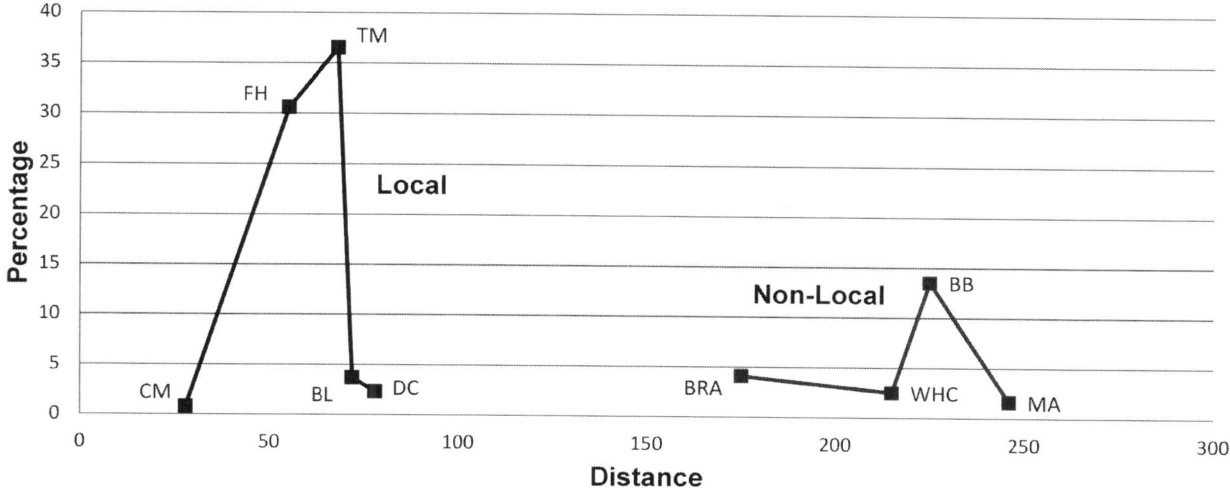

FIGURE 7.5. Relative amounts of toolstone sources in the Old River Bed delta collection vs. distance from the wetlands. Sources representing less than 1 percent of the collection are not shown. CM = Cedar Mountain; FH = Flat Hills; TM = Topaz Mountain; BL = Badlands; DC = Deep Creek; BRA = Black Rock Area; WHC = Wild Horse Canyon; BB = Browns Bench; MA = Malad.

studies of PA mobility in the Great Basin assume the latter (Jones, Beck, Jones, and Hughes 2003; Jones et al. 2012; Smith 2010, 2011), but that may not necessarily be the case. The local FGV and nearest obsidian quarries surrounding the ORB delta are all located in relatively barren foothill areas of low resource productivity that would be relatively unattractive to foragers. It is highly unlikely that PA foragers would have traveled to these quarries, or even areas surrounding the quarries, in order to pursue hunting and/or collecting activities.

It seems more likely that toolstone from these quarries was obtained in a logistical fashion, with its procurement being the primary purpose of the trip. This may have been accomplished by a small group of toolmakers who left the main foraging group behind in the wetlands for a few days while they traveled to the local quarries to replenish their toolstone supplies and then returned to the marsh. We cannot presently conceive of a way to test whether or not this was the case, but of the two possibilities it seems more parsimonious to assume that local toolstone was obtained logistically than that it was obtained as part of general foraging activities of the social group as a whole or even that of hunting parties.

More than 95 percent of the toolstone in the ORB wetlands comes from essentially three source areas (Figure 7.5). These are, in decreasing order of importance, (1) local FGV and obsidian sources found around the periphery of the ORB delta (76 percent), (2) sources within the Browns Bench area on the Idaho/Nevada/Utah border (15 percent), and (3) the Black Rock/Wild Horse Canyon area on the extreme southern margin of

the Bonneville basin (5 percent). These three source areas are spread along a narrow north–south axis more than 400 km long. If the conveyance zone represented by these three source areas is considered to be the equivalent of a foraging territory, then the PA foragers of the ORB delta were moving primarily along this narrow axis, camping within these three zones for unknown periods of time. We consider what this means in terms of mobility below.

The rest of the toolstone sources each makes up less than 1–2 percent of the ORB delta collection. Of the remaining 14 nonlocal toolstone types, seven are represented by only one or two artifacts, and two more are represented by only five and seven tools out of the 1,720 tested. Only the Malad obsidian and Currie Hills FGV sources, constituting 2.2 percent and 1.0 percent of the collection, respectively, could conceivably be considered as part of a conveyance zone that represents the foraging territories of the people who visited the ORB delta. Whether or not these minor components of the ORB collection represent direct procurement or exchange through a trade network of some kind cannot be determined. Smith suggests that the exchange of stone artifacts among Great Basin PA foragers is "unlikely to have occurred on a magnitude sufficient to move the substantial amounts of exotic toolstone represented in early assemblages" (2011:462), but these minor toolstone sources are not present in "substantial amounts" and could easily represent trade items. Beck and Jones, in a study of PA mobility and exchange in the east-central Great Basin, consider rare amounts of exotic obsidian from distant sources outside the "eastern conveyance zone" to be "evidence for informal reciprocal exchanges" (2011:76). We concur with this interpretation but find no way to confirm or reject it. However, recalling that the ORB delta collection comprises material deposited over the course of more than 2,500 years, it is difficult to conclude that one or two artifacts from a source can be reasonably used to argue for a high degree of residential mobility over a very large foraging territory.

But is this necessarily the case? Brantingham (2003) developed a neutral model of stone tool procurement that can be used to test this conclusion. A basic assumption of this model is that all stone raw materials are of equivalent utility and each unit weight of raw material, regardless of material type, is equivalent in terms of probabilities of procurement, consumption, and discard. These and a number of other model assumptions are unlikely to be met in real-world situations, but they do allow us to examine how those real-world situations differ, if at all, from modeled expectations.

In the case of the ORB delta, the collections are remarkably similar to model predictions, especially in that there is a rapid decline in proportional representation in the collection with distance to the toolstone source. If foragers are expected to have arrived at the ORB delta with a randomly encountered, used, and discarded collection of toolstone, then, as Brantingham suggests, "the quantity of material of a given type within the mobile toolkit generally follows an exponential decline with increasing distance from source" (2003:501). Moreover, sources located at great distances are expected to be extremely rare, or not present at all, despite being included in the tool kit at one time, exactly the situation found at the ORB delta. In short, if the amounts of various toolstone sources represented in the ORB delta collections are not greatly different from those that might be expected to have been deposited by foragers moving randomly about the Bonneville basin, then how are we to interpret the collection in terms of Paleoarchaic mobility? We do find that there are some clear differences between the expectations of the neutral model and the ORB delta collections, in that some relatively nearby toolstone sources are poorly represented while others at great distances, such as Brown Bench obsidian, are present in numbers that exceed expectations. However, computer simulations of the model suggest that even these variations are likely to occasionally occur by chance.

Such interpretive difficulties are compounded by our inability to precisely define what we mean by mobility, as discussed in Chapter 1. Most toolstone-based interpretations of Great Basin PA foragers conclude that they were "highly mobile," at least in comparison with later Archaic groups (e.g., Jones, Beck, Jones, and Hughes 2003; Smith 2011). But what exactly does that mean? Although it is not explicitly defined, from the context we conclude that the phrase is meant to convey the notion that PA foragers moved often within rather large foraging ranges, although neither the number of moves nor the average distance of each is ever made clear. Moreover, the implication is that such a mobility pattern was *the* pattern followed by Great Basin foragers, with little variation between groups. As more, and larger, toolstone collections are subjected to chemical analyses, estimates of the size of these foraging ranges are beginning to drop (Jones et al. 2012; Smith 2010). Even the degree of residence time at various locations within those ranges is now conceded to be variable (Jones et al. 2012).

These changing interpretations of mobility still do not, in our opinion, capture the degree of variance we think may be represented in the adaptations of PA foragers across western North America—or even only within the Great Basin. We think, for example, that two different groups may have had equally large foraging territories, moved around within those territories exactly the same number of times, and still had vastly different patterns of residential mobility. One group may have moved once a month for 12 months, while the other group may have moved once a week for 11 weeks and stayed the rest of the year in one place. We hypothesize that the relative attractiveness of resource patches within those foraging territories may explain these differences.

A Gravity Model of Great Basin Paleoarchaic Mobility

Using patch choice models, noting the apparent focus of PA foragers on wetland settings, and utilizing a model of differential men's and women's foraging strategies proposed by Elston and Zeanah (2002), Madsen (2007) suggests that the foraging radius and residence time of Great Basin foragers should vary with the relative size of available wetland habitats. In regions where marsh environments are large and closely spaced, moves should be limited, and foraging ranges should be small. Where such habitats are small and widely spaced, moves should be frequent and of greater distance. As a corollary, Madsen also envisions that the foraging territories of hunting parties and those of the rest of the social group were likely different, as the hunters may have been moving logistically from long-term residential base camps in the wetlands occupied by the larger group. Such a foraging pattern is based, essentially, on that found ethnographically among the "Cattail-Eaters" of the Carson Sink in the western Great Basin (Fowler 1992; Zeanah 2004). This model has found some support in studies of lithic conveyance zones (Jones et al. 2012), but we think that it can be usefully expanded by including a notion of "megapatches" in the model's formulation.

The ORB delta wetlands area, at more than 2,600 km², is essentially what we term a "megapatch." We consider a megapatch to be a large geographical area containing a relatively uniform ecosystem with an extensive number of smaller, but similar and contiguous, resource patches within that geographical area. Foragers in such a megapatch can move around within the large geographical area from one patch to an adjacent one that is essentially similar to the one they left behind. All of the contiguous patches require the same adaptive strategies and tools needed to survive, and all of the patches have overall return rates that are basically similar. In the particular case of the ORB delta wetlands, foragers would have been able to move from one distributary channel system to another while absorbing little in the way of transport costs, the production of alternative tools, or shifting of their resource-procurement strategies. One ORB distributary channel was much like any other.

Patch choice models fail to fully incorporate the differential implications of moving around from patch to patch within a megapatch like the ORB delta, and moving between patches in markedly different ecosystems such as between a small spring-fed valley marsh and a canyon mouth location in a pinyon-juniper woodland, because they fail to incorporate the notion of patch recovery rates. Patch choice models suggest that foragers will move from one patch to another when the return rate within the patch they currently occupy falls below the average return rate for all patches (MacArthur and Pianka 1966). As a basic rule, the utility of such models remains unchanged when considering their application to moves within a megapatch vs. moves between patches in different ecosystems.

Foragers still move from one patch to another when their resource return rate falls below what they would, on average, obtain by moving. If, however, the overall return rate of the megapatch never falls below the average for all patches within their foraging territory, then patch choice models predict that the foragers would never, or rarely, leave the megapatch. As a result, to understand when foragers would leave a megapatch, one must understand how many similar patches there are within a particular megapatch and how quickly those individual patches recover once the foragers have moved on. If there are a sufficient number of similar patches

within a megapatch, and if those patches quickly recover to a level where return rates within them reach or exceed the return rate of the megapatch as a whole, it then follows that foragers would only leave the megapatch when the average return rate for the megapatch as a whole falls below that of all patches. If that overall return rate continually exceeds the average return rate for all patches in the broader environment, then foragers would never, or infrequently, leave the megapatch. Foragers may leave due to behavioral requirements (such as the need to find mating partners), due to the need to obtain nonfood resources unavailable in the megapatch (e.g., toolstone), or because of a dramatic increase in population size caused by the arrival of additional foraging groups (and a concomitant increase in the rate of resource depression) but not because food acquisition became too costly. Given the small group size that we envision occupied the ORB delta (one residential base created every one–three years), such a group may not have been motivated to leave the megapatch for food procurement reasons.

When added to Madsen's patch choice–based model of Great Basin Paleoarchaic foraging strategies, the concept of a megapatch alters it sufficiently that we can think of variations in the relative attractiveness of wetland habitats, or indeed all habitats, in the Great Basin in terms of what we call a gravity model of PA foraging behavior. Gravity models are widely used in geography to explain the movement of people from place to place and in economics to explain the flow of trade. Generally, they refer to the relative attractiveness of population centers, with larger population centers being more "attractive" than smaller centers. Here, we use the phrase to explain the movement of people living in small-scale foraging societies but base attractiveness not on relative population size but on the relative productivity of resource patches and megapatches.

In such a gravity model, the relative attractiveness of a spring-fed marsh, a river delta, or a subalpine meadow can be ranked according to the amount and rate of resource productivity, each of which can be considered as a separate patch. For Great Basin Paleoarchaic foragers those habitat patches consisted primarily of wetland habitat patches, since evidence of the utilization of other habitats is rare. Such upland patches were undoubtedly utilized from time to time, but it appears from site settlement data that the PA substance focus was on marsh resources.

Ranking habitat patches on the basis of "attractiveness" rather than on average resource return rates incorporates recovery rates into the equation. In the simple patch choice model of Paleoarchaic mobility envisioned by Madsen, the size of a marsh habitat is important because there are simply more resources in larger patches and it therefore takes longer for foragers to depress return rates to the average found in all marsh patches. What is missing in this simple model, however, is that as patch sizes increase and become sufficiently large, depleted resources can start to regenerate before foragers have reduced overall return rates to the environmental average. As a result, this partial recovery extends the foragers' residence time in that particular patch. Because moving is an expensive endeavor, foragers are more often drawn to such patches relative to smaller resource patches. Thus, size matters not only because there are more resources to deplete but also because resource patches above a certain size have the ability to regenerate even while being subject to depredation. The larger the patch, then, the more it attracts foragers.

Where resource patches become sufficiently large and approach the size of a megapatch, as does the ORB delta area, they become somewhat analogous to a gravity sink in that foragers never, or rarely, leave. Again, they may move around in that megapatch, but it is sufficiently large that resources have a chance to fully recover as foragers move from patch to patch within the megapatch. This notion of movement within a megapatch, combined with the concept of prolonged stays, produces different expectations of what we should see in toolstone distributions than do models of high residential mobility among Great Basin PA foragers.

Internal Mobility

As an example of these differential expectations, we earlier hypothesized that toolmakers in the ORB delta may have left the marsh on multiday logistical trips to local toolstone quarries on visits that were not embedded in other foraging activities. Movement from patch to patch within the greater ORB delta wetlands may have mitigated the costs of even those trips. The local toolstone sources utilized by delta foragers are found around the periphery of the wetlands. As the ORB delta is a relatively large place, sites on one edge of the wetlands are much closer to one or more of these source areas than they are to others. As foragers moved from place to place within the ORB wetlands megapatch, other foraging costs, beyond those involved in obtaining food, fuel, and potable water, may have come into play. That is, when toolstone needs increased, it was possible to move closer to local toolstone sources without increasing the difficulty in obtaining other resources. Such decision making may not even have involved extensive planning. Foragers need only to have recognized that they happened to be closer to one or another toolstone

source and decided to gear up when acquisition costs were at their minimum.

The travel distances to these local sources given above are only approximate distances to central points in the delta, and actual distances from any one spot vary greatly. For example, the Flat Hills FGV source is only ~25 km from the nearest ORB distributary channel, while Topaz Mountain obsidian is available in stream channels draining the northwest side of the Thomas Range less than 20 km from Fish Springs on the southern end of the ORB wetlands. These distances approach or are below the 20-km limit for day trips suggested by Surovell (2009) and may not have required overnight logistical forays. We would therefore expect that as foragers moved around in the marsh from one patch to another, they would have retrieved toolstone from the nearest source and that this should be reflected in the relative proportions of different local toolstone at sites in the delta.

Most of our work, unfortunately, has been focused on areas well out into the wetlands area, and we have conducted little work on the edges of the wetlands that would allow us to test this proposition. Even at these more distant locations, however, this bias is reflected in the toolstone of debitage, as shown in Table 4.11, Chapter 4. Sites reported in this table show a significant degree of variability in the relative proportions of FGV and obsidian flakes and, further, that this variability is relatively consistent from one channel to the next. While it was not possible to source these flakes in the field, we do know from the material that was sourced that the very large majority of FGV toolstone comes from the Flat Hills source, while much of the obsidian comes from the Topaz Mountain source, and we can assume that the crude field identifications reflect these sources. Flakes on sites along the Light Blue and Black channels, located an average of 35 and 44.4 km from the Flat Hills source area, respectively, and 42.5 and 49 km from the nearest margin of the Topaz Mountain source area, respectively, are 80–90 percent FGV. On the other extreme, flakes on sites along the Lime and Lavender channels, located 50.7 and 48.9 km from the Flat Hills source, respectively, and 46 and 40 km from the nearest margin of the Topaz Mountain source, respectively, are 70–80 percent obsidian. Sites on the Blue and Yellow channels, located more equidistant from the two toolstone source areas, have roughly proportional amounts of FGV and obsidian.

Assuming that there are no chronological issues involved here (both the Light Blue and the Lavender channels are about the same age, for example), this distribution suggests that the savings of round-trip travel costs of as little as 8–10 km may have been important in the decision-making process. This fits theoretical expectations that debitage should reflect the closest local toolstone source (e.g., Eerkens et al. 2007) but also illustrates the fact that foragers within the delta wetlands were able to differentially exploit the environment around them. There is not a single pattern of toolstone exploitation within the wetlands but, rather, many, reflecting the movement of foragers from place to place within the ORB delta. Put another way, the "conveyance zone" or "foraging territory" reflected at any one site in the wetlands can differ significantly from those reflected at another.

Megapatches and Great Basin Mobility

In a gravity model of PA mobility, if the attractiveness of a megapatch exceeds that of surrounding patches, it will draw in foragers, even if there are many patches whose initial overall return rates are equal to that of the megapatch. This is simply because the smaller surrounding patches degrade faster than does the megapatch. This seems contrary to the notion that PA foragers became substantially less mobile as the number and size of Great Basin wetlands gradually decreased as the transition to a more arid Holocene climate progressed (e.g., Smith 2011). While this may be true to some extent, we suggest that the mobility patterns of both early and late PA foragers were highly variable, dependent on and affected by the presence or absence of megapatches such as the ORB delta wetlands (e.g., Smith et al. 2013).

As Madsen (2007) and others have noted, where available wetland patches were small and widely spaced, PA foragers would likely have been highly mobile, as suggested by Jones, Beck, Jones, and Hughes (2003) for eastern Nevada. In such regions the degree of mobility may have remained high, and even increased, as the size and number of these wetland patches decreased with the transition to the Holocene. Where wetland patches were larger, especially when they were much larger, foragers would have been less mobile even during the earliest periods of PA occupation (again, discounting movement within the wetland patch). With increasing aridity wetland patches may have become relatively even more attractive as surrounding areas dried up, but they would always have been more attractive to a greater or lesser degree. If this were true, however, the relative mobility patterns of PA foragers in the ORB delta should not have changed greatly over time. We return to this issue by revisiting the relative proportions of local and distant toolstone in the ORB delta through time.

Based on Great Basin chronologies as a whole, Beck and Jones in Chapter 5 have separated projectile point forms found on ORB delta sites into an early PA/WST phase and a later EH phase. The earlier phase includes Cougar Mountain, Haskett, Silver Lake, Lake Mohave, and generalized WST forms, while the EH phase consists of Eden, various Great Basin Stemmed varieties, Dugway Stubbies, Butte Valley Corner-notched, and Pinto forms. The relative proportions of local vs. distant toolstone for these early and later forms can be found in Chapter 6 but are shown visually in Figure 7.6. As noted earlier, 64.5 percent of the earlier-phase projectile points are made from local toolstone sources, while 70.8 percent of the later-phase points are of local toolstone. The relative proportions vary slightly, but, while the small difference is real and not likely due to chance, during both periods toolstone was derived only from the same three places: Local stone was acquired from the Flat Hills and Topaz Mountain sources, and distant toolstone was acquired from the Browns Bench area and from the Black Rock/Wild Horse Canyon area. This finding suggests a couple of interpretive options: (1) The use of toolstone sources as a guide to forager mobility must be seriously questioned, or (2) patterns of mobility may appear relatively similar from one period to another (or from one area to another) but contain subtle variations that alter the degree to which stone tools were utilized.

There are two interpretive subtleties that may be important for understanding toolstone use in the ORB wetlands. First, the timing of the shift to a more intensive use of toolstone and the husbanding of toolstone appears to have happened rather late in the ORB delta occupational sequence. In Chapter 5, Beck and Jones distinguish a later period of intensified toolstone use from an earlier one, and such a distinction certainly seems to be a reasonable one (see also Duke 2011). What was not made clear, however, is that this later phase was actually quite late in the overall history of the ORB delta. The "eastern" area used by Duke (2011) dates to well after 9000 ^{14}C BP, during the last several centuries of the ~2,500-year occupational span. Similarly for the proximal delta material, many of the heavily reworked and/or unique small stubby tool forms come from an area on the extreme southwest of the distributary system where channels, such as the Lavender channel, date to only about 9000 ^{14}C BP. These last few centuries of the ORB delta's existence were characterized by a marked reduction in the productivity of the wetlands. After ~9500 ^{14}C BP the Sevier basin stopped overflowing, and the distributary channels were supplied only by groundwater. By ~9000 ^{14}C BP even this groundwater supply began to dwindle, and the few remaining distributary channels were small, low-energy streams that, seasonally, may not have flowed at all. Relative to surrounding areas, the ORB wetlands may have been attractive during this transition to the drier middle Holocene, but even the wetlands themselves were much different than they had been throughout much of their history.

Second, if foragers during this last short occupational phase were scavenging and reworking older material, then the material they reworked would have been in the same toolstone proportions as that used by earlier foragers. These last, late foragers may have employed different foraging patterns than their predecessors and may even have imported toolstone from different sources and/or in different proportions. Determining any such differences using toolstone sourcing alone may be impossible, however, both because of the short time span involved and because of the reworking of older material. Subtle toolstone source variations may be statistically swamped and invisible to us. Regardless, it appears either that the mobility patterns of earlier and later foragers in the ORB wetlands, as revealed by toolstone sourcing, were essentially similar, which on technological grounds described in Chapter 5 appears to be unlikely, or that the use of toolstone sourcing to investigate prehistoric mobility patterns is of limited utility. It then follows that such technological features, such as evidence of scavenging and others described by Beck and Jones in Chapter 5, may be more important in understanding shifts in forager mobility.

Conjectural Mobility Patterns of ORB Delta Foragers

We can envision any number of ways that PA foragers may have been occupying the ORB delta. We present two here that could conceivably have been employed at various times in the ORB wetlands megapatch, or, for that matter, at different places in the Great Basin, but there are many more. An important factor in devising these hypothetical mobility patterns is that we really do not know what was going on outside the wetlands and contend that the possible variations in these external factors may have been critical in the patterns of mobility reflected internally within the ORB wetlands.

We start by contemplating how long people may have actually stayed in the wetlands during any one year, something we really do not know. Did they stay continuously for six months at a time? Nine months? Year-round? Obviously, the latter possibility is remote because of the presence of significant amounts of toolstone from the areas far to the north and south, but when did they leave, what time of the year did they do it, and for how long? Was it once for two or three months at a time? Two or three times during which they were away for only short periods? It is impossible to say, but however long the PA foragers were gone, they probably did not leave because of resource depression that reduced return rates in the wetlands.

We have noted that despite the astounding number of estimated PA sites in the delta area, that number still represents the creation of only one residential base

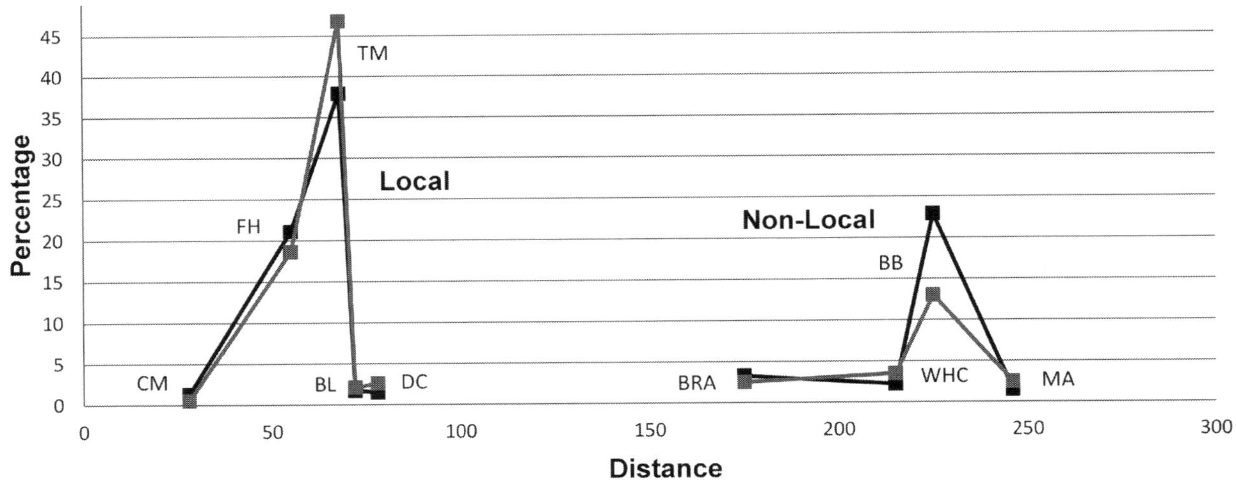

FIGURE 7.6. Relative amounts of toolstone sources in the Old River Bed delta collection vs. distance from the wetlands for the early Paleoarchaic/Western Stemmed Tradition period (*black*) and the later early Holocene/Pinto period (*gray*). Sources representing less than 1 percent of the collection are not included. CM = Cedar Mountain; FH = Flat Hills; TM = Topaz Mountain; BL = Badlands; DC = Deep Creek; BRA = Black Rock Area; WHC = Wild Horse Canyon; BB = Browns Bench; MA = Malad.

every one–three years. While there may have been some loss in the number of sites through taphonomic processes, the sites along the ORB distributaries appear to be relatively intact despite being deflated. Certainly, burial of sites does not appear to be an issue. Even assuming that there are actually more than three times the number of estimated sites in the ORB delta, and assuming that a single forager group was occupying each residential base, that still means that only one residential base per year was occupied each year.

Given the size of the wetlands, large enough that we consider it to be a "megapatch," it seems unlikely that a single small to medium group of foragers could reduce overall return rates to the point where they would prefer to move elsewhere. They were more likely to simply move elsewhere in the marsh, as we have discussed. It also seems unlikely that there was some seasonal resource depression that necessitated their leaving the wetlands. As Madsen (1982b, 1988, 2002; Madsen and Kelly 2010) has frequently noted, one of the great advantages of Great Basin wetlands is the year-round availability of resources that throughout the year sequentially reach the height of their productivity. While the toolstone sourcing data make it clear that PA foragers did leave the ORB wetlands (or, less likely, were resupplied by a population influx from outside the delta), they likely did so either because there were seasonally available resources with much higher overall return rates located elsewhere or because of non-resource-related reasons. For example, in the ethnographic Great Basin the main group of Cattail-Eaters only left the terminal marshes of the Carson River during the late fall and early winter of years when pinyon pine nuts were especially abundant (Fowler 1992).

Although environmental conditions were obviously somewhat different during the Pleistocene–Holocene transition (pinyon had not yet replaced limber pine, for instance), we speculate that the PA foragers in the ORB delta were behaving much as were the Northern Paiute of the Stillwater marshes and were only leaving for a few months at a time. This was likely in the late summer to early winter, since during other times of the year the productivity and return rates for resources in the wetlands, especially women's resources (Zeanah 2004), tend to outweigh those available elsewhere (Madsen 2002), at least under modern conditions. In short, we suggest that PA foragers in the ORB delta may have stayed in the wetlands for as long as six–nine months at a time from late fall through midsummer.

There may be a difference, however, in how long earlier and later foragers stayed in the wetlands. The higher degree of husbanding of toolstone during later periods

as described in Chapter 5 suggests that later stays were, on average, longer than earlier stays. In addition (following Smith 2011), the slight, but statistically significant, differences in the relative amount of local vs. nonlocal toolstone suggest that earlier and later PA foragers were residing in the ORB wetlands for relatively different amounts of time. Unfortunately, we do not know how great that difference was. The production of tool forms such as "Dugway Stubbies" during both periods suggests that stays were prolonged throughout the occupation of the ORB delta; it is just that they were even more prolonged during the later period as alternative patch choices diminished.

But what of the rest of the year? Again, we can only speculate, but the toolstone data suggest that the foragers were primarily only visiting two other places: (1) an uplands area on the Idaho/Nevada/Utah border where Browns Bench obsidian could be obtained and (2) a wetlands area not unlike the ORB delta on the southern end of the Sevier basin where they obtained Black Rock/Wild Horse Canyon obsidian (Figure 7.7). This holds for both the earlier and later periods of occupation. The relative proportion of the distant Browns Bench obsidian vs. the local Topaz Mountain obsidian is slightly higher in the earlier period, but the locations that earlier and later foragers were visiting (according to the toolstone) remained unchanged. If the toolstone sourcing data can tell us anything, it is that the mobility pattern represented in the ORB delta was persistent for a very long period of time.

What the PA foragers were doing in these locations and how long they stayed are unknown. In the Bonneville basin during the Pleistocene–Holocene transition, wetlands akin to those in the ORB delta extended from Fish Springs and the end of the entrenched ORB river south to the Sevier and Beaver River delta areas. There are only about 50 km of upland terrain between the two large marsh areas, and they are connected by the riverine wetlands of the ORB river valley itself. The delta area of the Sevier and Beaver rivers is almost as large as the ORB wetlands system, and the relatively broad riverine wetlands along the lower Beaver River, where it flows through the floor of the Bonneville basin, are almost as extensive. The Black Rock and Wild Horse Canyon sources are located at the southern end of this chain of wetlands, and foragers could have moved along this corridor from the ORB delta to the vicinity of the Black Rock and Wild Horse Canyon obsidian sources without having to shift their collecting strategies in the least. But if the ORB wetlands were indeed a megapatch, as we have speculated, why would foragers leave one such patch for another? We are not sure that they

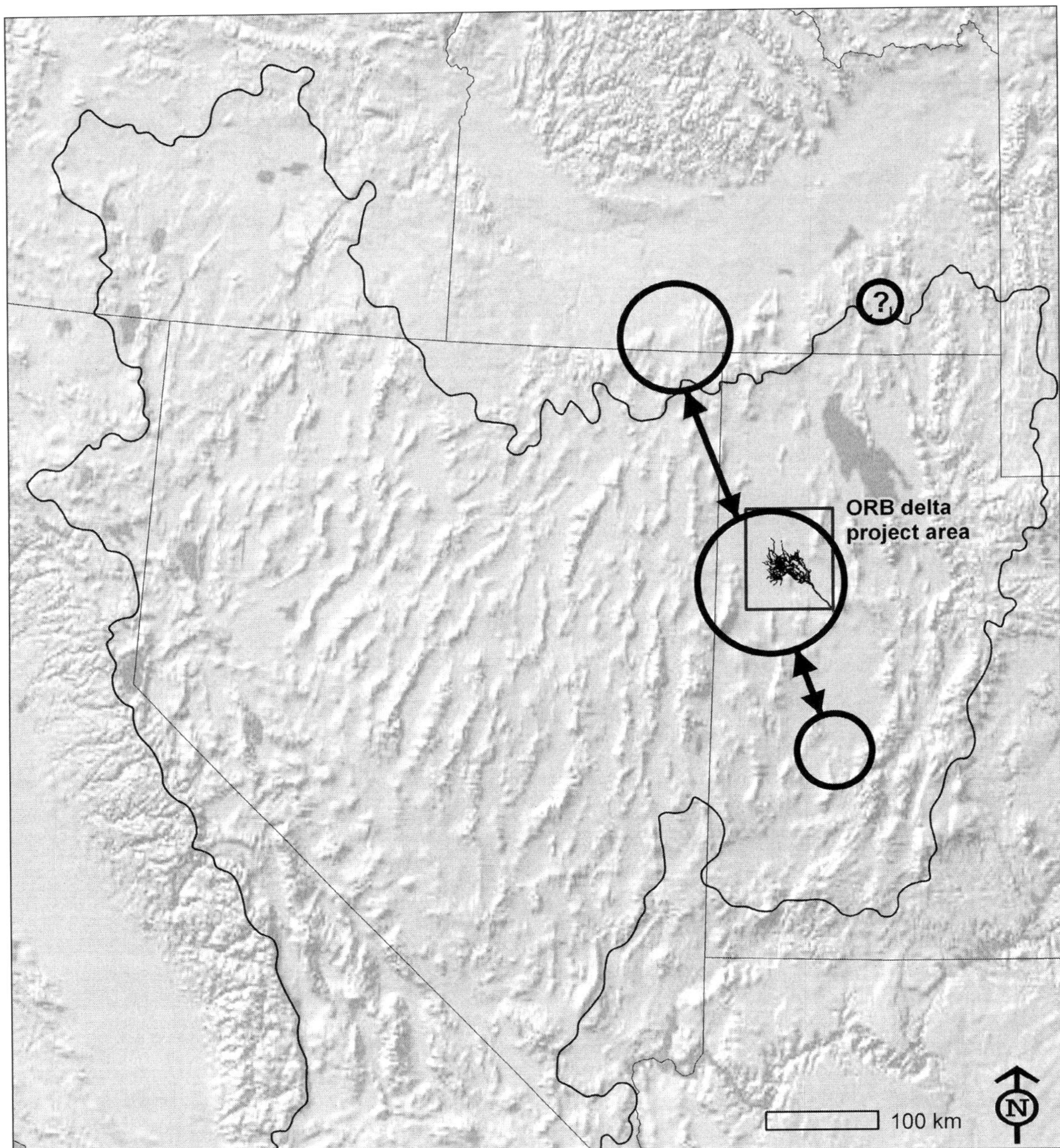

FIGURE 7.7. Three toolstone source areas likely visited by Old River Bed (ORB) foragers. The travel pattern was apparently consistent and equally limited during both the early and late phases of occupation in the delta wetlands. The Malad obsidian source (*small circle with the question mark*), representing ~2 percent of the ORB delta collections, may be a product of either trade or rare direct procurement over the ~2,500 years of occupation.

did. The Black Rock and Wild Horse Canyon obsidian sources together constitute <10 percent of both the early and later ORB delta collections. Whether or not this is a "significant" proportion of the toolstone, significant enough to represent direct procurement rather than trade, is an open and presently unanswerable question.

If there were related PA groups scattered along this north–south wetlands corridor, it is probable that there would have been some trade up and down the line.

If the toolstone does represent direct procurement, the question of why PA foragers would leave the ORB wetlands still remains. The answer may simply be that there is no singular mobility pattern for Great Basin PA foragers. If, as we think, the mobility patterns of PA groups varied widely across the landscape as well as through time, then it is impossible to define *the* Paleoarchaic mobility pattern as "highly mobile" or "relatively sedentary" or something in between. The ORB wetlands were occupied over the course of some 2,500 years, and during that time there were undoubtedly periods of extended aridity when the productivity of the wetlands was greatly reduced. During such times foragers who visited the delta may have had to move around the landscape much more often. Our evidence for the reworking of older material and the production of small ill-shaped tool forms suggests that, for the most part, foragers in the ORB wetlands were staying for prolonged periods of time, although we do not know for precisely how long. At times, however, even that may not have been possible, with the end result that toolstone from other nearby wetlands may occasionally have been imported more often.

Why ORB foragers were venturing to the uplands around the Browns Bench obsidian source is even less clear. The environment around this toolstone source is very unlike the wetland habitats in and around which the very large majority of Great Basin PA sites are found. It presently consists of sagebrush- and woodland-covered hills, and while the species of trees that composed those woodlands during the Pleistocene/Holocene were different, the environment then was not greatly different than it is today. What would have been particularly attractive in terms of food resources, as opposed to that available elsewhere, is not evident. Perhaps the foragers in the ORB delta really liked the qualities of Browns Bench obsidian and merely went there to get it as quickly as they could, living on what was available along the way and returning just as quickly. Perhaps they did so often. Perhaps, alternatively, that area was common ground for many Paleoarchaic groups who met there for social events, and the ORB foragers went there to obtain marriage partners, exchange information, and have a good time. If obsidian from the Malad source, which occurs in ~2 percent of the sample, was obtained directly rather than through trade, then this northern foraging area was much larger and may have included the Snake River Plain, and it is even possible that the PA foragers were traveling north for salmon and camas roots. It is impossible to say.

What we can say is that during most of the time the ORB was occupied, foragers were likely spending most of the year in the ORB wetlands. They may or may not have been relatively sedentary while there, but it is possible, indeed likely, that they were moving around in the marsh without leaving its boundaries. When they left, they went to only one or two distant places that we know of, one more than 200 km to the north and one more than 200 km to the south. They may have gone to more places, but we have no evidence for that. When they did leave they may have gone straight to the northern or southern locality and quickly returned, or they may have spent a lot of time moving around to places of which we have no record. They may have left to procure resources of which we have neither a record nor even a suggestion of what they might have been. They may have gone to visit relatives at a jamboree and/or to trade. There is at least some suggestion that toolstone trade may have occurred in very minor amounts, but the nature of that trade, for economic, social, spiritual, or other reasons, is also unknown. In short, there is a lot more we do not know about the mobility of foragers who lived in the ORB delta than there is that we do know. It is clear, however, that the mobility patterns of the people who lived there were not like those found in many other areas of the Great Basin and that it is impossible to define a single mobility pattern for Great Basin Paleoarchaic foragers even within a single "conveyance zone."

ITERATION AND SUMMARY

From around 11,500 ^{14}C BP until shortly after about 8800 ^{14}C BP a major northward-flowing river, the ORB river, connecting the Sevier and Great Salt Lake basins, created a complex network of braided distributary channels on the floor of what is now the Great Salt Lake Desert. Together with contributory distributary channels originating in the Deep Creek Mountains and Fish Springs, this network of both flowing and abandoned channels formed an extensive wetland habitat covering more than 2,600 km^2. This enormous wetland was likely the home to a wide array of plants and animals known to have been utilized ethnographically by Great Basin foragers, but we have direct knowledge only of fish (*Gila atraria*) and freshwater clams (*Anodonta* sp.). From what we know of modern Great Basin wetlands, there were aquatic mammals such as muskrats and beaver, small terrestrial mammals such as rabbits, and large herds of antelope, as well as a great variety of ducks, geese, and other aquatic birds. Extensive fields of cattails, bulrush, and other semiaquatic plants with edible rhizomes and pollen likely covered the wetlands.

Between about 10,500 and 9000 ^{14}C BP the wetlands were probably at their most productive, as more and more distributary channels fed by groundwater were constructed and abandoned but continued to support marsh habitats. We have mapped more than 3,800 km of distributary channels in the wetlands, and it is evident that there were likely four–five times that many that we have not been able to map. Water flow in the channels probably reached its peak shortly before about 10,000 ^{14}C BP when the Great Salt Lake rose to the Gilbert level and partially flooded the extreme distal end of the wetlands. After about 9900 ^{14}C BP overflow from Lake Gunnison in the Sevier basin probably ended, but the distributary channels continued to be fed by groundwater. By about 8800–8300 ^{14}C BP even this groundwater had dried up, and the wetlands ceased to exist. Until very recently, when the area was reoccupied by the Department of Defense, the area was essentially a mudflat unattractive for human occupation.

For approximately 2,500 years the ORB wetlands were the home to Great Basin Paleoarchaic foragers who lived along the distributary channels in the marsh. More than 600 sites once occupied by these forager groups have been identified and recorded in surveys of only a small portion of the total wetlands area. Extrapolating these numbers to the wetlands as a whole, we estimate that as many as 4,000–5,000 Paleoarchaic sites may be present in what was once the greater ORB wetland area. These sites are quite visible and easily recognizable because, after the wetlands dried up, the area has been subject to extensive deflation and wind erosion that exposed the sites and left the distributary channels topographically inverted. This exposure has prevented the preservation of organic materials and removed very small flakes from most sites, leaving an archaeological record composed principally of larger lithic artifacts. Apart from wind ablation, these larger stone artifacts remain largely in place, and site boundaries are largely intact and coherent. Due to the area's isolation and general unattractiveness, the sites in what is now a mostly barren mudflat desert were apparently not subject to scavenging by later Holocene foragers, and the designation of the area as a military reservation has discouraged modern artifact collecting as well. As a result, the sites contain a remarkable array of stone tools.

Based on the diversity of these stone tools, we have been able to classify these sites into different categories based on how they were likely used. These break down into basically two major divisions: residential sites that were probably the focus of long-term occupation and special-purpose sites that represent diurnal resource-procurement activities. In a subsample of 115 of the 600+ sites that have been identified (see Chapter 4), 18 sites could be classified as residential sites, with the remainder being the result of daily activities that took place away from the main residential campsite. Again extrapo-

lating to the wetlands as a whole, there may be 700–900 residential sites in the wetlands, or one for every two–three years the ORB wetlands were occupied. The sites are located both in and along the distributary channels, and there is virtually no evidence of occupations between channels. Sites along the channels appear to have been formed on natural levees and on point bars along flowing channels, where higher ground provided a suitable occupation area. Many of the diurnal-activity sites occur within channel margins, indicating that they were formed after the channel was abandoned and had been largely filled in. Black organic mats in these abandoned channels indicate that groundwater flowing along the channels still supported wetland plant communities. The dates from these black mats, together with the dates from mollusk shells from channel fill, provide the only means of dating these archaeological sites, though indirect. The channels appear to have formed and been abandoned rather quickly, probably on the order of only a few hundred years at most, and dating of the channels provides a reasonable estimate for the age of sites associated with them. We have obtained more than 70 radiocarbon dates on channel deposits, and these suggest that the earliest occupations date to ~11,000 ^{14}C BP, with the latest dating to ~8500 ^{14}C BP. The distributary channels appear to have formed continuously throughout the 3,000 years of the wetlands' existence, and human occupation of the channel margins may have been relatively continuous as well. Periods of extended drought may have temporarily reduced marsh productivity, and hence, human occupation, but we have no record of this happening. Particularly wet periods, such as during the height of the Gilbert episode, may also have made it a difficult proposition to live in some parts of the delta.

Formal stone tools on the ORB delta sites consist primarily of projectile point forms associated with the Western Stemmed Tradition and with Early Holocene technologies. These stemmed points range from larger Haskett and Cougar Mountain forms to smaller Silver Lake and Lake Mohave forms and include forms unique to the ORB delta that we have called "Dugway Stubbies." Many of these points were probably not used as projectiles at all and may have been used as knives or multipurpose tools. Crescents are also present on the ORB delta sites; it remains unclear whether or not these were used as projectiles in waterfowl hunting, but it appears that they were used as cutting tools. Western Fluted points occur in very small numbers. Early Holocene projectile point forms such as Butte Valley Corner-notched and Pinto occur on sites associated with the younger ORB delta distributary channels. Pinto points, commonly associated with Archaic occupations in areas outside the ORB delta, occur as early as about 9500 ^{14}C BP and appear to be a transitional point form used by both Paleoarchaic and Archaic groups.

Many of the stemmed points appear to have been reworked down to near stubs, to a point well beyond where they would have been discarded elsewhere. Broken and abandoned tools also appear to have been recycled into attenuated stemmed point forms rarely found in locations outside the ORB delta. These reworked and recycled forms suggest prolonged stays in the wetlands and reflect dwindling toolstone supplies necessitating the husbanding of available lithic material. Reworking and recycling may have been more intense during the later portion of the ORB delta occupation, suggesting even longer resident times in the wetlands, but when this shift occurred is unknown. Based on the association of sites with channels, the shift appears to have occurred sometime after about 9500 ^{14}C BP and more intensively after about 9000 ^{14}C BP, perhaps due to a greater decline in resource productivity elsewhere. The shift in resident times is only relative, however, as prolonged stays are evident throughout the occupational period.

Toolstone sourcing, based on X-ray fluorescence determinations conducted by two different laboratories and a portable X-ray fluorescence elemental analyzer, is consistent across multiple data sets and allows some limited assessment of the movement of the foragers who occupied the ORB delta. More than 1,700 artifacts from the wetland sites were sourced and suggest a surprisingly consistent use of toolstone sources through time. More than 75 percent of the artifacts we sampled are made from what we term "local" sources. These local sources require round-trips of 50 to 200 km depending on where one starts from in the wetlands and are far beyond the ~40-km-round-trip distance normally thought of as local. They are, however, the closest available sources, as no suitable toolstone occurs in the wetlands. The two principal local sources are fine-grained volcanic toolstone variants from the Flat Hills east of the delta and Topaz Mountain obsidian from south of the wetlands area. Minor amounts of local materials also come from sources in the northern Deep Creek Mountains west of the delta area. The remainder of the toolstone comes primarily from only two other areas and consists of Browns Bench obsidian from the Idaho/Nevada/Utah border area and Black Rock and Wild Horse Canyon obsidian from the southern Bonneville basin. Both of these localities require round-trips of 400–500 km. A number of other distant obsidian sources from as far away as the Snake River Plain and the western Great

Basin are represented by a very few tools, most often only one or two points, which appear to be present as trade items. Certainly they do not represent consistent travel to these distant areas.

Based on these toolstone sourcing data, a limited assessment of the mobility patterns favored by the forgers of the ORB delta is possible. Essentially, the foragers who occupied the delta visited only two other distant places, the Browns Bench obsidian area to the north and the Black Rock/Wild Horse Canyon obsidian area to the south. Stays in the ORB wetlands appear to have been prolonged, possibly lasting many months at a time, although there may have been, and likely was, movement within the wetlands themselves. The lengths of stays in the other two locations are not known, nor do we know what the foragers were doing or how they were surviving in these localities. The environment of the southern location is much like the ORB wetlands, and presumably subsistence activities were much the same. The Browns Bench area is in the uplands, however, and would require a different kind of subsistence focus. We also do not know how swiftly foragers traveled among the three areas or exactly what time(s) of the year they may have done so. It is highly likely that such long-distance travel was part of the seasonal round involving stops at a number of localities, but we have no idea where these might have been, how many stops there were, or how frequently the moves between them occurred. We do know that the mobility pattern, at least as exhibited by the toolstone sourcing data, was consistent through time, with the basic pattern of obsidian use from three source areas along a north–south axis persistent throughout the length of the ORB delta's existence. The relative proportion of the northern Browns Bench obsidian to local Topaz Mountain obsidian decreased slightly over time, but the basic pattern remained. Overall, the mobility pattern of the Paleoarchaic foragers appears markedly different from that found elsewhere at the same time in the Great Basin, and it is apparent there was no single mobility pattern that characterized foragers during this early time period. Likely there were many such patterns depending on the nature, availability, and distribution of resources available to different groups of people.

By ~8500–8300 ^{14}C BP the ORB delta had dried up, and the vast wetlands were no longer available to local foragers. As in other areas of the Great Basin, foragers shifted to a different mode of subsistence and settlement, one that we recognize as "Archaic." Exactly how different is a matter of debate. The principal difference seems to be the appearance of intensive seed processing involving seed grinding. While the Paleoarchaic foragers of the ORB delta were likely subsisting on a broad array of resources much like their Archaic descendants, we have no evidence from the huge number of ORB sites created over a prolonged occupational span that any seed grinding occurred. They may have collected and eaten seeds, but they did not use manos and metates to process them. The change in chipped-stone tools appears to have been gradual, with Pinto points and corner-notched forms such as Butte Valley Corner-notched appearing during the late Paleoarchaic period and persisting well into the early Archaic. This new Archaic mode of subsistence and settlement may have been more difficult simply because the ORB wetlands were no longer in existence. For ~2,500 years or more the wetlands supported a highly successful and consistent lifestyle that persisted relatively unchanged throughout that entire time period. With the passing of the Old River Bed delta, and the wetlands it supported, that lifestyle passed into history as well.

NOTE

1. Fish bones in the ORB distributary channels were identified as Utah chub (*Gila atraria*) by Jack Broughton, Department of Anthropology, University of Utah. The bones suggest fish that were generally small, averaging less than 20 cm long. Utah chub remains were recovered in channels dating to virtually the entire period of delta formation, including those formed after 9000 ^{14}C BP. Their presence indicates continuous flow and open water in the distributary system well after the channels became disconnected from the Great Salt Lake following the Gilbert episode.

References Cited

Aikens, C. M.
1970   *Hogup Cave.* University of Utah Anthropological
        Papers 93. Salt Lake City.
Alley, R. B., and P. U. Clark
1999   The Deglaciation of the Northern Hemisphere:
        A Global Perspective. *Annual Reviews of Earth and
        Planetary Sciences* 27:149–182.
Amick, D. S.
1993   Toolstone Use and Distribution Patterns Among
        Western Pluvial Lakes Tradition Points from South-
        ern Nevada. *Current Research in the Pleistocene* 10:
        49–51.
1995   Raw Material Selection Patterns Among Paleo-
        indian Tools from the Black Rock Desert, Nevada.
        Current Research in the Pleistocene 12:55–57.
1997   Geochemical Source Analysis of Obsidian Paleo-
        indian Points from the Black Rock Desert, Nevada.
        Current Research in the Pleistocene 14:97–99.
1999   Using Lithic Artifacts to Explain Past Behavior. In
        *Models for the Millennium. Great Basin Anthropology
        Today,* edited by C. Beck, pp. 161–170. University of
        Utah Press, Salt Lake City.
Amsden, C. A.
1937   The Lake Mojave Artifacts. In *The Archaeology of
        Pleistocene Lake Mohave: A Symposium,* by E. W. C.
        Campbell, W. H. Campbell, E. Antevs, C. A.
        Amsden, J. A. Barbieri, and F. D. Bode, pp. 51–98.
        Southwest Museum Papers No. 11. Los Angeles.
Anderson, D. G., and G. T. Hanson
1988   Early Archaic Settlement in the Southeastern
        United States: A Case Study from the Savannah
        River Valley. *American Antiquity* 53:262–286.
Arkush, B. S., and B. L. Pitblado
2000   Paleoarchaic Surface Assemblages in the Great Salt
        Lake Desert, Northwestern Utah. *Journal of Califor-
        nia and Great Basin Anthropology* 22:12–42.
Bamforth, D. B.
2009   Projectile Points, People, and Plains Paleoindian
        Perambulations. *Journal of Anthropological Archaeol-
        ogy* 28:142–157.
Bartram, L. E., E. M. Kroll, and H. T. Bunn
1991   Variability in Camp Structure and Bone Food Re-
        fuse Patterning at Kua San Hunter-Gatherer Camps.
        In *The Interpretation of Archaeological Spatial Pattern-
        ing,* edited by E. M. Kroll and T. D. Price, pp. 77–148.
        Plenum Press, New York.
Basgall, M. E.
1989   Obsidian Acquisition and Use in Prehistoric Central-
        Eastern California: A Preliminary Assessment. In
        Obsidian Studies in California, edited by R. E. Hughes,
        pp. 111–126. Contributions of the University of Cali-
        fornia Archaeological Research Facility 48. Berkeley.
Basgall, M. E., and M. C. Hall
2000   Morphological and Temporal Variation in Bifurcate-
        Stemmed Dart Points of the Western Great Basin.
        Journal of California and Great Basin Anthropology
        22:237–276.
Beck, C.
1984   *Steens Mountain Surface Archaeology: The Sites.* Ph.D.
        dissertation, University of Washington, Seattle.
        University Microfilms, Ann Arbor.
1998   Projectile Point Types as Valid Chronological Units.
        In *Measuring Time, Space, and Material: Unit Issues
        in Archaeology,* edited by A. F. Ramenofsky and A.
        Steffen, pp. 21–40. University of Utah Press, Salt
        Lake City.
1999   Dating the Archaeological Record and Modeling
        Chronology. In *Models for the Millennium. Great
        Basin Anthropology Today,* edited by C. Beck,
        pp. 171–181. University of Utah Press, Salt Lake City.
Beck, C., and G. T. Jones
1988   Western Pluvial Lakes Tradition Occupation in
        Butte Valley, Eastern Nevada. In *Early Human Oc-
        cupation in Far Western North America: The Clovis–
        Archaic Interface,* edited by J. A. Willig, C. M. Aikens,
        and J. L. Fagan, pp. 273–302. Nevada State Museum
        Anthropological Papers No. 21. Carson City.
1990a   The Late Pleistocene/Early Holocene Archaeol-
        ogy of Butte Valley, Nevada: Three Season's Work.
        Journal of California and Great Basin Anthropology
        12:231–261.
1990b   Toolstone Selection and Lithic Technology in Early
        Great Basin Prehistory. *Journal of Field Archaeology*
        17:283–299.
1994   Dating Suface Assemblages Using Obsidian Hy-
        dration. In *Dating in Exposed and Surface Contexts,*
        edited by C. Beck, pp. 47–76. University of New
        Mexico Press, Albuquerque.

1997   The Terminal Pleistocene/Early Holocene Archaeology of the Great Basin. *Journal of World Prehistory* 11:161–236.

2007   Early Paleoarchaic Point Morphology and Chronology. In *Paleoarchaic or Paleoindian? Great Basin Human Ecology at the Pleistocene–Holocene Transition*, edited by K. E. Graf and D. N. Schmitt, pp. 23–41. University of Utah Press, Salt Lake City.

2009   *The Archaeology of the Eastern Nevada Paleoarchaic, Pt. I: The Sunshine Locality.* University of Utah Anthropological Papers 126. Salt Lake City.

2010a   Clovis and Western Stemmed: Population Migration and the Meeting of Two Technologies in the Intermountain West. *American Antiquity* 75:81–116.

2010b   *Analysis of Post-2003 Artifacts from the Old River Bed Delta, Dugway Proving Ground, Utah.* Report on file, Desert Research Institute, Reno.

2011   The Role of Mobility and Exchange in the Conveyance of Toolstone During the Great Basin Paleoarchaic. In *Perspectives on Prehistoric Trade and Exchange in California and the Great Basin*, edited by R. E. Hughes, pp. 55–82. University of Utah Press, Salt Lake City.

2012   The Clovis-Last Hypothesis: Investigating Early Lithic Technology in the Intermountain West. In *Meetings at the Margins: Prehistoric Cultural Interactions in the Intermountain West*, edited by D. Rhode, pp. 23–46. University of Utah Press, Salt Lake City.

2015   A Case of Extinction in Paleoindian Technology. In *Lithic Technological Systems: Stone, Human Behavior, Evolution*, edited by N. Goodale and W. Andrefsky, Jr., pp. 83–99. Cambridge University Press, Cambridge.

Beck, C., A. Taylor, G. T. Jones, C. M. Fadem, C. R. Cook, and S. A. Millward

2002   Rocks Are Heavy: Transport Costs and Paleoarchaic Quarry Behavior in the Great Basin. *Journal of Anthropological Archaeology* 21:481–507.

Bedwell, S. F.

1973   *Fort Rock Basin: Prehistory and Environment.* University of Oregon Books, Eugene.

Behle, W. H., E. D. Sorenson, and C. M. White

1985   *Utah Birds—A Revised Checklist.* Utah Museum of Natural History Occasional Publication 4. Salt Lake City.

Beiswenger, J. M.

1991   Late Quaternary Vegetational History of Grays Lake, Idaho. *Ecological Monographs* 61:165–182.

Bennett, K., and G. T. Jones

2012   Is Western Stemmed Biface Technology Derived from Clovis? Poster presented at the 77th Annual Meeting of the Society for American Archaeology, Memphis, Tennessee.

Benson, L. V., S. P. Lund, J. P. Smoot, D. Rhode, R. J. Spencer, K. L. Verosub, L. A. Louderback, C. A. Johnson, R. O. Rye, and R. M. Negrini

2011   The Rise and Fall of Lake Bonneville Between 45 and 10.5 ka. *Quaternary International* 235:57–69.

Benson, L. V., J. P. Smoot, S. P. Lund, S. A. Mensing, F. F. Foit, Jr., and R. O. Rye

2013   Insights from a Synthesis of Old and New Climate-Proxy Data from the Pyramid and Winnemucca Lake Basins for the Period 48 to 11.5 cal ka. *Quaternary International* 310:62–82.

Bever, M. R.

2006   Too Little, Too Late? The Radiocarbon Chronology of Alaska and the Peopling of the New World. *American Antiquity* 71:595–620.

Beyries, S.

1988   Functional Variability of Lithic Sets in the Middle Paleo-Lithic. In *Upper Pleistocene Prehistory of Western Eurasia*, edited by H. Dibble and A. Montet-White, pp. 213–223. University Museum, Philadelphia.

Bradley, B. A., M. B. Collins, and A. Hemmings

2010   *Clovis Technology.* International Monographs in Prehistory 17. University of Michigan Press, Ann Arbor.

Bradley, B. A., and G. C. Frison

1987   Projectile Points and Specialized Bifaces from the Horner Site. In *The Horner Site. The Type Site of the Cody Cultural Complex*, edited by G. C. Frison and L. C. Todd, pp. 199–231. Academic Press, New York.

Brantingham, P. J.

2003   A Neutral Model of Stone Raw Material Procurement. *American Antiquity* 68:487–509.

Bright, R. C.

1966   Pollen and Seed Stratigraphy of Swan Lake, Southeastern Idaho: Its Relation to Regional Vegetational History and to Lake Bonneville History. *Tebiwa* 9(1):1–47.

Broughton, J. M.

2000   The Homestead Cave Ichthyofauna. In *Late Quaternary Paleoecology in the Bonneville Basin*, by D. B. Madsen, pp. 103–122. Utah Geological Survey Bulletin 130. Salt Lake City.

Broughton, J. M., D. A. Byers, R. A. Bryson, W. Eckerle, and D. B. Madsen

2008   Did Climatic Seasonality Control Late Quaternary Artiodactyl Densities in Western North America? *Quaternary Science Reviews* 27:1916–1937.

Broughton, J. M., D. B. Madsen, and J. Quade

2000   Fish Remains from Homestead Cave and Lake Levels of the Past 13,000 Years in the Bonneville Basin. *Quaternary Research* 53:392–401.

Bryan, A. L.

1979   Smith Creek Cave. In *The Archaeology of Smith Creek Canyon, Eastern Nevada*, edited by D. R. Tuohy and D. L. Rendall, pp. 162–251. Nevada State Museum Anthropological Papers No. 17. Carson City.

1980   The Stemmed Point Tradition: An Early Technological Tradition in Western North America. In *Anthropological Papers in Memory of Earl H. Swanson*, edited by L. B. Harten, C. N. Warren, and D. R. Tuohy, pp. 77–107. Idaho State Museum of Natural History, Pocatello.

1986   Paleoamerican Prehistory as Seen from South America. In *New Evidence for the Pleistocene Peopling of the Americas*, edited by A. L. Bryan, pp. 1–4. Center for the Study of Early Man, Orono, Maine.

1988   The Relationship of the Stemmed Point and Fluted Point Traditions in the Great Basin. In *Early Human Occupation in Far Western North America: The Clovis–Archaic Interface*, edited by J. A. Willig, C. M. Aikens, and J. L. Fagan, pp. 53–74. Nevada State Museum Anthropological Papers 21. Carson City.

Buck, B. A.

1982   Ancient Technology in Contemporary Surgery. *Western Journal of Medicine* 136:265–269.

Buck, P. E., W. T. Hartwell, and G. Haynes

1998   *Archaeological Investigations at Two Early Holocene Sites near Yucca Mountain, Nye County Nevada.* Topics in Yucca Mountain Archaeology No. 2. Desert Research Institute, Las Vegas. Submitted to the U.S. Department of Energy, Washington, D.C.

Callahan, E.

1979   The Basics of Biface Knapping in the Eastern Fluted Point Tradition: A Manual for Flint-Knappers and Lithic Analysts. *Archaeology of Eastern North America* 7:1–180.

Campbell, E. W. C., and W. H. Campbell

1935   *The Pinto Basin Site*. Southwest Museum Papers 9. Los Angeles.

Campbell, E. W. C., W. H. Campbell, E. Antevs, C. A. Amsden, J. A. Barbieri, and F. D. Bode

1937   *The Archaeology of Pleistocene Lake Mohave*. Southwest Museum Papers 11. Los Angeles.

Cannon, M. D., and D. J. Meltzer

2004   Early Paleoindian Foraging: Examining the Faunal Evidence for Large Mammal Specialization and Regional Variability in Prey Choice. *Quaternary Science Reviews* 23:1955–1987.

2008   Explaining Variability in Early Paleoindian Foraging. *Quaternary International* 191:5–17.

Carlson, R. L., and M. P. R. Magne

2007   *Projectile Point Sequences in Northwestern North America*. Archaeology Press, Simon Fraser University, Burnaby.

Carter, J. A.

1999   *TS-5 Target Complex Cultural Resource Inventory, Wendover Air Force Range, Tooele County, Utah*. Historical Research Associates, Inc., Seattle. Submitted to the Hill Air Force Base, Utah, and National Park Service, Midwest Archaeological Center, Lincoln, Nebraska.

Carter, J. A., D. Duke, and D. C. Young

2004   *Limited Site Testing at 22 Archaeological Sites in the Wild Isle Project Area, Utah Test and Training Range, Tooele County, Utah*. Historical Research Associates, Inc., Seattle. Submitted to the Hill Air Force Base, Utah.

Carter, J. A., D. Duke, D. C. Young, and C. Bialas

2005   *Wildcat Mountain Cultural Resource Inventory, Utah Test and Training Range, Tooele County, Utah*. Historical Research Associates, Inc., Seattle. Report submitted to the Hill Air Force Base, Utah.

Carter, J. A., and D. C. Young

2002   *TS-5 Central Area and Craners Cultural Resources Inventory, Utah Test and Training Range, Tooele and Box Elder Counties, Utah*. Historical Research Associates, Inc., Seattle. Submitted to the Hill Air Force Base, Utah, and National Park Service, Midwest Archaeological Center, Lincoln, Nebraska.

Clague, J. J., R. W. Mathewes, and T. A. Ager

2004   Environments of Northwestern North America Before the Last Glacial Maximum. In *Entering America: Northeast Asia and Beringia Before the Last Glacial Maximum*, edited by D. B. Madsen, pp. 63–94. University of Utah Press, Salt Lake City.

Clark, D. L., C. G. Oviatt, and D. Page

2008   *Interim Geologic Map of Dugway Proving Ground and Adjacent Areas—Parts of the Wildcat Mountain, Rush Valley, and Fish Springs 30′ × 60′ Quadrangles, Tooele County, Utah (Year 2 of 2)*. Utah Geologic Survey Open-File Report 532. Salt Lake City.

Clark, J. D.

1958   Some Stone Age Woodworking Tools in Southern Africa. *South African Archaeological Bulletin* 13:144–152.

Clark, J. L.

1993   *Nevada Wildlife Viewing Guide*. Falcon Press, Helena.

Clark, P. U., J. D. Shakun, P. A. Baker, P. J. Bartlein, S. Brewer, E. Brook, A. E. Carlson, H. Cheng, D. S. Kaufman, Z. Liu, T. M. Marchitto, A. C. Mix, C. Morrill, B. L. Otto-Bliesner, K. Pahnke, J. M. Russell, C. Whitlock, J. F. Adkins, J. L. Blois, J. Clark, S. M. Colman, W. B. Curry, B. P. Flower, F. He, T. C. Johnson, J. Lynch-Stieglitz, V. Markgraf, J. McManus, J. X. TMitrovica, P. I. Moreno, and J. C. Williams

2012   Global Climate Evolution During the Last Deglaciation. *Proceedings of the National Academy of Sciences* 109(19):E1134–E1142.

Clewlow, C. W., Jr.
1968   Surface Archaeology of the Black Rock Desert, Nevada. *University of California Archaeological Survey Reports* 73:1–94. Berkeley.

Collins, M. B.
1999a   *Clovis Blade Technology*. University of Texas Press, Austin.
1999b   Clovis and Folsom Lithic Technology on and near the Southern Plains. In *Folsom Lithic Technology. Explorations in Structure and Variation*, edited by D. S. Amick, pp. 12–38. International Monographs in Prehistory, Archaeological Series 12. University of Michigan Press, Ann Arbor.
2014   Initial Peopling of the Americas: Context, Findings, and Issues. In *The Cambridge World Prehistory*, edited by C. Renfrew and P. Bahn, pp. 893–912. Cambridge University Press, Cambridge.

Collins, M. B., and J. C. Lohse
2004   The Nature of Clovis Blades and Blade Cores. In *Entering America. Northeast Asia and Beringia Before the Last Glacial Maximum*, edited by D. B. Madsen, pp. 159–183. University of Utah Press, Salt Lake City.

Condie, K. C., and A. B. Blaxland
1970   Sources of Obsidian in Hogup and Danger Caves. In *Hogup Cave*, by C. M. Aikens, pp. 275–281. University of Utah Anthropological Papers 93. Salt Lake City.

Connolly, T. J.
1999   *Newberry Crater. A Ten-Thousand-Year Record of Human Occupation and Environmental Change in the Basin–Plateau Borderlands*. University of Utah Anthropological Papers No. 121. Salt Lake City.

Connolly, T. J., and D. L. Jenkins
1999   The Paulina Lake Site (35DS34). In *Newberry Crater. A Ten-Thousand-Year Record of Human Occupation and Environmental Change in the Basin–Plateau Borderlands*, by T. J. Connolly, pp. 86–127. University of Utah Anthropological Paper No. 121. Salt Lake City.

Crabtree, D. E.
1972   *An Introduction to Flintknapping*. Occasional Papers of the Idaho State Museum, Pocatello.

Cressman, L. S.
1966   Man in Association with Extinct Fauna in the Great Basin. *American Antiquity* 31:866–867.

Cressman, L. S., F. C. Baker, P. S. Conger, H. P. Hansen, and R. F. Heizer
1942   *Archaeological Researches in the Northern Great Basin*. Carnegie Institution of Washington Publication 538. Washington, D.C.

Currey, D. R.
1980   Radiocarbon Dates and Their Stratigraphic Implications from Selected Localities in Utah and Wyoming. *Encyclia: Journal of the Utah Academy of Sciences, Arts, and Letters* 57:110–115.
1982   *Lake Bonneville: Selected Features of Relevance to Neotectonic Analysis*. U.S. Geological Survey Open-File Report 82-1070. Boulder.
1990   Quaternary Paleolakes in the Evolution of Semidesert Basins, with Special Emphasis on Lake Bonneville and the Great Basin, U.S.A. *Palaeogeography, Palaeoclimatology, Palaeoecology* 76:189–214.

Currey, D. R., G. Atwood, and D. R. Mabey
1984   Major Levels of Great Salt Lake and Lake Bonneville. Map 73, Scale 1:750,000. Utah Geological and Mineralogical Survey, Salt Lake City.

Danzeglocke, U., O. Jöris, and B. Weninger
2012   CalPal-2007[online]. Electronic document, http://www.calpal-online.de/, accessed February 24, 2012.

Daugherty, R. D.
1956   Archaeology of the Lind Coulee Site, Washington. *Proceedings of the American Philosophical Society* 100:223–278.

Davis, E. L.
1973   The Hord Site, a Paleoindian Camp. *Pacific Coast Archaeological Society Quarterly* 10:1–6.
1978   *The Ancient Californials: Rancholabrean Hunters of the Mojave Lakes Country*. Natural History Museum of Los Angeles County Science Series 29. Los Angeles.

Davis, E. L., C. W. Brott, and R. Shutler, Jr.
1969   *The Western Lithic Co-Tradition*. San Diego Museum Papers 6. San Diego.

Davis, L. G., and C. E. Schweger
2004   Geoarchaeological Context of Late Pleistocene and Early Holocene Occupation at the Cooper's Ferry Site, Western Idaho, USA. *Geoarchaeology* 19:685–687.

Davis, L. G., S. C. Willis, and S. J. Macfarlan
2012   Lithic Technology, Cultural Transmission, and the Nature of the Far Western Paleoarchaic/Paleoindian Co-Tradition. In *Meetings at the Margins: Prehistoric Cultural Interactions in the Intermountain West*, edited by D. Rhode, pp. 48–64. University of Utah Press, Salt Lake City.

Davis, O. K., and D. S. Shafer
1992   An Early-Holocene Maximum for the Arizona Monsoon Recorded at Montezuma Well, Central Arizona. *Palaeogeography, Palaeoclimatology, Palaeoecology* 92:107–119.

Deller, D. B.
1980   Parkhill (Brophey) Site, AhHk-49: Analyses of Surface Materials. Unpublished Master's thesis, Wilfrid Laurier University, Waterloo, Ontario.

Deller, D. B., and C. J. Ellis
1992   *Thedford II. A Paleo-Indian Site in the Ausable River Watershed of Southwestern Ontario*. Museum of Anthropology Memoirs 24. University of Michigan, Ann Arbor.

Denton, G. H., R. F. Anderson, J. R. Toggweiler, R. L. Edwards, J. M. Schaefer, and A. E. Putnam
2010   The Last Glacial Termination. *Science* 328:1652–1656.

Des Lauriers, M. R.
2006   Terminal Pleistocene and Early Holocene Occupants of Isla de Cedros, Baja, California. *Journal of Island and Coastal Archaeology* 1:255–270.

Dixon, E. J.
2007   Bifaces from On Your Knees Cave, Southeast Alaska. In *Projectile Point Sequences in Northwestern North America*, edited by R. L. Carlson and M. P. R. Magne, pp. 11–18. Archaeology Press, Simon Fraser University, Burnaby.
2013   Late Pleistocene Colonization of North America from Northeast Asia: New Insights from Large-Scale Paleogeographic Reconstructions. *Quaternary International* 285:57–67.

Duke, D.
2003   *Cultural Resources Inventory of the South Route to Wild Isle and TS-5, Utah Test and Training Range, Tooele County, Utah*. Geo-Marine, Inc., Las Vegas. Submitted to the U.S. Army Corps of Engineers, Fort Worth, and U.S. Air Force, Hill Air Force Base, Ogden.
2011   If the Desert Blooms: A Technological Perspective on Paleoindian Ecology in the Great Basin from the Old River Bed, Utah. Unpublished Ph.D. dissertation, University of Nevada, Reno.
2015   *Cultural Resources Inventory of 17,000 Acres at Knolls Dunes, Utah Test and Training Range, Tooele County, Utah*. Far Western Anthropological Research Group, Inc., Las Vegas. Report for the Hill Air Force Base, Utah, in preparation.

Duke, D., T. Carpenter, and D. Page
2007   New Obsidian Hydration Findings Suggest the Use of Split-Stem Points by Great Basin Paleoindians. *Current Research in the Pleistocene* 24:80–82.

Duke, D., and D. C. Young
2007   Episodic Permanence in Paleoarchaic Basin Selection and Settlement. In *Paleoindian or Paleoarchaic? Great Basin Human Ecology at the Pleistocene–Holocene Transition*, edited by K. E. Graf and D. N. Schmitt, pp. 123–138. University of Utah Press, Salt Lake City.

Duke, D., D. C. Young, and J. A. Carter
2004   A Unique Example of Early Technology and Land Use in the Eastern Great Basin. *Current Research in the Pleistocene* 21:32–34.

Duke, D., D. C. Young, and A. R. Garner
2008   *Cultural Resources Inventory of High Probability Lands near Wildcat Mountain: The 2006 Field Season on the U.S. Air Force Utah Test and Training Range-South, Tooele County, Utah*. Far Western Anthropological Research Group, Inc., Virginia City, Nevada. Submitted to the Hill Air Force Base, Utah.

Dunnell, R. C.
1970   Seriation Method and Its Evaluation. *American Antiquity* 35:305–319.
1978   Archaeological Potential of Anthropological and Scientific Models of Function. In *Archaeological Essays in Honor of Irving B. Rouse*, edited by R. C. Dunnell and E. S. Hall, Jr., pp. 41–73. Mouton Publishers, New York.

Durrant, S. D.
1952   *Mammals of Utah: Taxonomy and Distribution*. Museum of Natural History, University of Kansas Publications 6. Lawrence.

Eerkens, J. W., J. R. Ferguson, M. D. Glascock, C. E. Skinner, and S. A. Waechter
2007   Reduction Strategies and Geochemical Characterization of Lithic Assemblages: A Comparison of Three Case Studies from Western North America. *American Antiquity* 72:585–597.

Eiselt, B. S.
1997   Fish Remains from the Spirit Cave Paleofecal Material: 9,400 Year Old Evidence for Great Basin Utilization of Small Fishes. *Nevada Historical Quarterly* 40:117–139.

Elias, S. A.
1997   The Mutual Climatic Range Method of Palaeoclimate Reconstruction Based on Insect Fossils: New Applications and Interhemispheric Comparisons. *Quaternary Science Reviews* 16:1217–1225.
2007   Beetle Records/Late Pleistocene of North America. In *Encyclopedia of Quaternary Science*, edited by S. A. Elias, pp. 222–236. Elsevier, New York.

Ellis, C. J., and D. B. Deller
1982   Hi-Lo Materials from Southwestern Ontario. *Ontario Archaeology* 38:3–22.

Elston, R. G.
1982   Good Times, Hard Times: Prehistoric Culture Change in the Western Great Basin. In *Man and Environment in the Great Basin*, edited by D. B. Madsen and J. F. O'Connell, pp. 186–206. Society for American Archaeology Papers 2. Washington, D.C.
1990   Lithic Raw Materials: Sources and Utility. In *The Archaeology of James Creek Shelter*, edited by R. G. Elston and E. E. Budy, pp. 165–174. University of Utah Anthropological Papers 115. Salt Lake City.
1992   The Lithic Terrane of Tosawihi. In *Archaeological Investigations at Tosawihi, a Great Basin Quarry, Pt. 1: The Periphery*, edited by R. G. Elston and C. Raven, pp. 31–47. Intermountain Research, Silver City, Nevada. Submitted to the Bureau of Land Management, Elko District, Elko, Nevada.
2005   Lithic Assemblage Variability. In *Camels Back Cave*, by D. N. Schmitt and D. B. Madsen, pp. 120–135. University of Utah Anthropological Papers 125. Salt Lake City.

Elston, R. G., and D. W. Zeanah
2002   Thinking Outside the Box: A New Perspective on Diet Breadth and Sexual Division of Labor in the Prearchaic Great Basin. *World Archaeology* 34:103–130.

Erlandson, J. M., and T. J. Braje
2011   From Asia to the Americas by Boat? Paleogeography, Paleoecology, and Stemmed Points of the Northwest Pacific. *Quaternary International* 239:28–37.

Erlandson, J. M., T. J. Braje, and G. J. Snitker
2008   Two Chipped-Stone Crescents from CA-SMI-680, Cardwell Bluff, San Miguel Island, California. *Current Research in the Pleistocene* 25:81–83.

Erlandson, J. M., T. C. Rick, T. J. Braje, M. Casperson, B. J. Culleton, B. Fulrost, T. Garcia, D. A. Guthrie, N. Jew, D. J. Kennett, M. L. Moss, L. Reeder, C. Skinner, J. Watts, and L. Willis
2011   Paleoindian Seafaring, Maritime Technologies, and Coastal Foraging on California's Channel Islands. *Science* 331:1181–1885.

Estes, M. B.
2009   Paleoindian Occupations in the Great Basin: A Comparative Study of Lithic Technological Organization, Mobility, and Landscape Use from Jakes Valley, Nevada. Unpublished Master's thesis, University of Nevada, Reno.

Fagan, J. L.
1988   Clovis and Pluvial Lakes Tradition Lithic Technologies at the Dietz Site in South-Central Oregon. In *Early Human Occupation in Far Western North America: The Clovis–Archaic Interface*, edited by J. A. Willig, C. M. Aikens, and J. L. Fagan, pp. 389–416. Nevada State Museum Anthropological Papers 21. Carson City.

Faith, J. T., and T. A. Surovell
2009   Synchronous Extinction of North America's Pleistocene Mammals. *Proceedings of the National Academy of Science* 106:20641–20645.

Fawcett, W. B., and S. R. Simms
1993   *Archaeological Test Excavations in the Great Salt Lake Wetlands and Associated Analyses.* Utah State University Contributions to Anthropology No. 14. Logan.

Fedje, D. W., Q. Mackie, E. J. Dixon, and T. H. Heaton
2004   Late Wisconsin Environments and Archaeological Visibility on the Northern Northwest Coast. In *Entering America: Northeast Asia and Beringia Before the Last Glacial Maximum*, edited by D. B. Madsen, pp. 97–138. University of Utah Press, Salt Lake City.

Fedje, D. W., Q. Mackie, T. Lacourse, and D. McLaren
2011   Younger Dryas Environments and Archaeology on the Northwest Coast of North America. *Quaternary International* 242:452–462.

Fedje, D. W., and R. W. Mathewes
2005   *HaidaGwaii: Human History and Environment from the Time of Loon to the Time of the Iron People.* University of British Columbia Press, Vancouver.

Fladmark, K. R.
1979   Routes: Alternative Migration Corridors for Early Man in North America. *American Antiquity* 44:55–69.

Fowler, C. S.
1992   *In the Shadow of Fox Peak: An Ethnography of the Cattail-Eater Northern Paiute People of Stillwater Marsh.* U.S. Fish and Wildlife Service Cultural Resource Series No. 5. Portland, Oregon.

Frahm, E.
2012a   Is Obsidian Sourcing About Geochemistry or Archaeology? A Reply to Speakman and Shackley. *Journal of Archaeological Science* 40:1444–1448.

2012b   Validity of "Off-the-Shelf" Handheld Portable XRF for Sourcing Near Eastern Obsidian Chip Debris. *Journal of Archaeological Science* 40:1080–1092.

Frison, G. C.
1988   Paleoindian Subsistence and Settlement During Post-Clovis Times on the Northwestern Plains, the Adjacent Mountain Ranges, and Intermontane Basins. In *America Before Columbus: Ice Age Origins*, edited by R. C. Carlisle, pp. 83–106. University of Pittsburgh Ethnology Monographs 12. Pittsburgh.

1991   *Prehistoric Hunters of the High Plains.* 2nd ed. Academic Press, New York.

1992   The Foothills–Mountain and the Open Plains: The Dichotomy in Paleoindian Subsistence Strategies Between Two Ecosystems. In *Ice Age Hunters of the Rockies*, edited by D. J. Stanford and J. S. Day, pp. 323–342. Denver Museum of Natural History, Denver.

1998   Paleoindian Large Mammal Hunters on the Plains of North America. *Proceedings of the National Academy of Sciences* 95:14576–14583.

Gibbon, G. E., and K. M. Ames
1998   *Archaeology of Prehistoric North America: An Encyclopedia.* Garland Reference Library of the Humanities, Vol. 1537. Garland Press, New York.

Gifford, E. W., and W. E. Schenck
1926   Archaeology of the Southern San Joaquin Valley. *University of California Publications in American Archaeology and Ethnology* 25:1–122. Berkeley.

Gilbert, G. K.
1890   *Lake Bonneville.* U.S. Geological Survey Monograph 1. Washington, D.C.

Gilbert, M. T. P., D. L. Jenkins, A. Götherstrom, N. Naveran, J. J. Sanchez, M. Hofreiter, P. F. Thomsen, J. Binladen, T. F. G. Higham, R. M. Yohe, II, R. Parr, R. Scott Cummings, and E. Willerslev
2008   DNA from Pre-Clovis Human Coprolites in Oregon, North America. *Science* 320:786–789.

Gillette, D. D., and D. B. Madsen
1992   The Short-Faced Bear *Arctodussimus* from the Late

Quaternary in the Wasatch Mountains of Central Utah. *Journal of Vertebrate Paleontology* 12:107–112.

1993   The Columbian Mammoth, *Mammuthus columbi*, from the Wasatch Mountains of Central Utah. *Journal of Paleontology* 67(4):669–680.

Godsey, H. S., D. R. Currey, and M. A. Chan

2005   New Evidence for an Extended Occupation of the Provo Shoreline and Implications for Regional Climate Change, Pleistocene Lake Bonneville, Utah, USA. *Quaternary Research* 63:212–223.

Godsey, H. S., C. G. Oviatt, D. M. Miller, and M. A. Chan

2011   Stratigraphy and Chronology of Offshore to Nearshore Deposits Associated with the Provo Shoreline, Pleistocene Lake Bonneville, Utah, USA. *Palaeogeography, Palaeoclimatology, Palaeoecology* 310:442–450.

Goebel, T.

2007   Pre-Archaic and Early Archaic Technological Activities at Bonneville Estates Rockshelter: A First Look at the Lithic Artifact Record. In *Paleoarchaic or Paleoindian? Great Basin Human Ecology at the Pleistocene–Holocene Transition*, edited by K. E. Graf and D. N. Schmitt, pp. 156–184. University of Utah Press, Salt Lake City.

Goebel, T., K. Graf, B. Hockett, and D. Rhode

2007   The Paleoindian Occupations at Bonneville Estates Rockshelter, Danger Cave, and Smith Creek Cave (Eastern Great Basin, U.S.A): Interpreting Their Radiocarbon Chronologies. In *On Shelter's Ledge: Histories, Theories and Methods of Rockshelter Research*, edited by M. Kornfield, S. Vasi'ev, and L. Miotti, pp. 147–161. British Archaeological Reports, Oxford.

Goebel, T., B. Hockett, K. D. Adams, D. Rhode, and K. Graf

2011   Climate, Environment, and Humans in North America's Great Basin During the Younger Dryas, 12,900–11,600 Calendar Years Ago. *Quaternary International* 242:479–501.

Goebel, T., and J. L. Keene

2014   Are Great Basin Stemmed Points as Old as Clovis in the Intermountain West? A Review of the Geochronological Evidence. In *Archaeology in the Great Basin and Southwest: Papers in Honor of Don D. Fowler*, edited by N. J. Parezo and J. C. Janetski, pp. 35–60. University of Utah Press, Salt Lake City.

Goodale, N., D. G. Bailey, G. T. Jones, C. Prescott, E. Scholz, N. Stagliano, and C. Lewis

2011   pXRF: A Study of Inter-Instrument Performance. *Journal of Archaeological Science* 39:875–883.

Graf, K. E.

2001   Paleoindian Technological Provisioning in the Western Great Basin. Unpublished Master's thesis, University of Nevada, Reno.

2002   Paleoindian Procurement and Mobility in the Western Great Basin. *Current Research in the Pleistocene* 19:87–89.

2007   Stratigraphy and Chronology of the Pleistocene to Holocene Transition at Bonneville Estates Rockshelter, Eastern Great Basin. In *Paleoarchaic or Paleoindian? Great Basin Human Ecology at the Pleistocene–Holocene Transition*, edited by K. E. Graf and D. N. Schmitt, pp. 82–104. University of Utah Press, Salt Lake City.

Graf, K. E., and D. N. Schmitt (editors)

2007   *Paleoarchaic or Paleoindian? Great Basin Human Ecology at the Pleistocene–Holocene Transition.* University of Utah Press, Salt Lake City.

Gramly, R. M.

1982   *The Vail Site: A Palaeo-Indian Encampment in Maine.* Bulletin of the Buffalo Society of Natural Sciences 30. Buffalo.

1990   *Guide to the Palaeo-Indian Artifacts of North America.* Persimmon Press Monographs in Archaeology, Buffalo.

Grayson, D. K.

1984   *Quantitative Zooarchaeology: Topics in the Analysis of Archaeological Faunas.* Academic Press, Orlando.

1988   *Danger Cave, Last Supper Cave, and Hanging Rock Shelter: The Faunas.* American Museum of Natural History Anthropological Papers 66(1). New York.

1993   *The Desert's Past: A Natural Prehistory of the Great Basin.* Smithsonian Institution Press, Washington, D.C.

1998   Moisture History and Small Mammal Community Richness During the Latest Pleistocene and Holocene, Northern Bonneville Basin, Utah. *Quaternary Research* 49:330–334.

2000a   Mammalian Responses to Middle Holocene Climatic Change in the Great Basin of the Western United States. *Journal of Biogeography* 27:181–192.

2000b   The Homestead Cave Mammals. In *Late Quaternary Paleoecology in the Great Basin*, by D. B. Madsen, pp. 67–89. Utah Geological Survey Bulletin 130. Salt Lake City.

2006   The Late Quaternary Biogeographic Histories of Some Great Basin Mammals (Western USA). *Quaternary Science Reviews* 25:2964–2991.

2007   Deciphering North American Pleistocene Extinctions. *Journal of Anthropological Research* 63:185–214.

2011   *The Great Basin: A Natural History.* University of California Press, Berkeley.

Green, F. E.

1963   The Clovis Blades: An Important Addition to the Llano Complex. *American Antiquity* 29:145–165.

Green, T. J., B. Cochran, T. W. Fenton, J. C. Woods, G. L. Titmus, L. Tieszen, M. A. Davis, and S. J. Miller

1998   The Buhl Burial: A Paleoindian Woman from Southern Idaho. *American Antiquity* 63:437–456.

Grimes, J. R., and B. G. Grimes
1985 Flakeshavers: Morphometric, Functional and Life-Cycle Analyses of a Paleoindian Unifacial Tool Class. *Archaeology of Eastern North America* 13:35–57.

Gueswel, W. E.
1958 *Report on Fish Springs*. U.S. Department of the Interior, Fish and Wildlife Service. Manuscript on file, Fish Springs Wildlife Refuge, Dugway, Utah.

Haarklau, L., L. Johnson, and D. L. Wagner (editors)
2005 *Fingerprints in the Great Basin: The Nellis Air Force Base Regional Obsidian Sourcing Study*. Prewitt and Associates, Inc., Austin. Submitted to Nellis Air Force Base, Nevada.

Harrington, M. R.
1933 *Gypsum Cave, Nevada*. Southwest Museum Papers 8. Los Angeles.

Hart, W. S., J. Quade, D. B. Madsen, D. S. Kaufman, and C. G. Oviatt
2004 The ^{87}Sr/^{86}Sr Ratios of Lacustrine Carbonates and Lake-Level History of the Bonneville Paleolake System. *Geological Society of America Bulletin* 16:1107–1119.

Hayden, B.
1985 Use and Misuse: The Analysis of Endscrapers. *Lithic Technology* 15:65–70.

Haynes, C. V., Jr.
2008 Younger Dryas "Black Mats" and the Racholabrean Termination in North America. *Proceedings of the National Academy of Sciences* 105:6520–6525.

Haynes, G.
2007 Paleoindian or Paleoarchaic? In *Paleoarchaic or Paleoindian? Great Basin Human Ecology at the Pleistocene–Holocene Transition*, edited by K. E. Graf and D. N. Schmitt, pp. 251–258. University of Utah Press, Salt Lake City.

Haynes, G. (editor)
2009 *American Megafaunal Extinctions at the End of the Pleistocene*. Springer, New York.

Hayward, C. L., C. Cottam, A. M. Woodbury, and H. H. Frost
1976 *Birds of Utah*. Great Basin Naturalist Memoirs No. 1. Brigham Young University Press, Provo.

Heaton, T.
1999 Late Quaternary Vertebrate History of the Great Basin. In *Vertebrate Paleontology in Utah*, edited by D. D. Gillette, pp. 501–507. Utah Geological Survey Miscellaneous Publication 99-1. Salt Lake City.

Heizer, R. F., and M. A. Baumhoff
1970 Big Game Hunters in the Great Basin: A Critical Review of the Evidence. In *Papers on Anthropology of the Western Great Basin*, pp. 1–12. Contributions of the University of California Archaeological Research Facility 7. Berkeley.

Heizer, R. F., and C. R. Berger
1970 Radiocarbon Age of the Gypsum Cave Culture. In *Papers on Anthropology of the Western Great Basin*, pp. 13–18. Contributions of the University of California Archaeological Research Facility 7. Berkeley.

Hill, M. E., Jr.
2008 Variation in Paleoindian Fauna Use on the Great Plains and Rocky Mountains of North America. *Quaternary International* 191:34–52.

Hintze, L. F.
1975 *Geologic History of Utah*. Brigham Young University Geology Studies 20(3). Brigham Young University Press, Provo.
1988 *Geologic History of Utah*. Brigham Young University Geology Studies Special Publication 7. Provo.

Hockett, B.
2007 Nutritional Ecology of Late Pleistocene to Middle Holocene Subsistence in the Great Basin: Zooarchaeological Evidence from Bonneville Estates Rockshelter. In *Paleoindian or Paleoarchaic? Great Basin Human Ecology at the Pleistocene–Holocene Transition*, edited by K. E. Graf and D. N. Schmitt, pp. 204–230. University of Utah Press, Salt Lake City.

Holliday, V. T.
2000 The Evolution of Paleoindian Geochronology and Typology on the Great Plains. *Geoarchaeology: An International Journal* 15:227–290.

Holmer, R. N.
1980 Projectile Points. In *Sudden Shelter*, by J. D. Jennings, A. R. Schroedl, and R. N. Holmer, pp. 63–83. University of Utah Anthropological Papers 103. Salt Lake City.
1986 Common Projectile Points of the Intermountain West. In *Anthropology of the Desert West: Essays in Honor of Jesse D. Jennings*, edited by C. J. Condie and D. D. Fowler, pp. 89–115. University of Utah Anthropological Papers No. 110. Salt Lake City.
1997 Volcanic Glass Utilization in Eastern Idaho. *Tebiwa* 26(2):186–204.

Hughes, R. E.
1984 Obsidian Sourcing Studies in the Great Basin: Problems and Prospects. In *Obsidian Studies in the Great Basin*, edited by R. E. Hughes, pp. 1–20. Contributions of the University of California Archaeological Research Facility 45. Berkeley.
1990 Obsidian Sources at James Creek Shelter, and Trace Element Geochemistry of Some Northeastern Nevada Volcanic Glasses. In *The Archaeology of James Creek Shelter*, edited by R. G. Elston and E. E. Budy, pp. 297–305. University of Utah Anthropological Papers 115. Salt Lake City.
1998 On Reliability, Validity, and Scale in Obsidian Sourcing Research. In *Unit Issues in Archaeology*, edited by A. F. Ramenofsky and A. Steffen, pp. 103–127. University of Utah Press, Salt Lake City.
2001 Determination of the Geologic Sources for Obsidian Artifacts from Camels Back Cave, and Trace Element Analysis of Some Western Utah and

Eastern Nevada Volcanic Glasses. In *The Archaeology of Camels Back Cave*, by D. N. Schmitt and D. B. Madsen, pp. 387–406. Utah Geological Survey, Salt Lake City. Submitted to the Directorate of Environmental Programs, U.S. Army Dugway Proving Ground, Utah.

2004   *1997–2004 Letter Reports on Results of XRF Studies on Artifacts from Dugway Proving Ground*. Geochemical Research Laboratory, Portola Valley, California. On file at Environmental Programs, U.S. Army Dugway Proving Ground, Utah, and Desert Research Institute, Reno.

2011   Sources of Inspiration for Studies of Prehistoric Resource Acquisition and Materials Conveyance in California and the Great Basin. In *Perspectives on Prehistoric Trade and Exchange in California and the Great Basin*, edited by R. E. Hughes, pp. 1–21. University of Utah Press, Salt Lake City.

Hughes, R. E., and J. A. Bennyhoff

1986   Early Trade. In *Great Basin*, edited by W. L. d'Azevedo, pp. 238–255. Handbook of North American Indians, Vol. 11, W. C. Sturtevant, general editor, Smithsonian Institution, Washington, D.C.

Hull, K. L.

1994   Obsidian Studies. In *Kern River Pipeline Cultural Resources Data Recovery Report: Utah, Vol. I: Research Context and Data Analysis*, by Dames and Moore, pp. 7-1–7-63. Report submitted to the Federal Energy Regulatory Commission, Washington, D.C.

Hunt, J. M., D. Rhode, and D. B. Madsen

2000   Homestead Cave Flora and Non-Vertebrate Fauna. In *Late Quaternary Paleoecology in the Bonneville Basin*, by D. B. Madsen, pp. 47–58. Utah Geological Survey Bulletin 130. Salt Lake City.

Ives, R. L.

1946   Desert Ripples. *American Journal of Science* 244: 492–501.

Jackson, R. L., A. Kovak, J. Spidell, and D. Kennelly-Spidell

2009   *A Historic Context for Native American Procurement of Obsidian in the State of Utah*. Pacific Legacy, Inc., Cameron Park, California. Submitted to Logan Simpson Design, Inc., Salt Lake City.

Jahren, A. H., R. Amundson, C. Kendall, and P. Wigand

2001   Paleoclimatic Reconstruction Using the Correlation in $\delta^{18}O$ of Hackberry Carbonate and Environmental Water, North America. *Quaternary Research* 56: 252–263.

Janetski, J. C., M. L. Bodily, B. A. Newbold, and D. T. Yoder

2012   The Paleoarchaic to Early Archaic Transition on the Colorado Plateau: The Archaeology of North Creek Shelter. *American Antiquity* 77:125–159.

Janetski, J. C., and D. B. Madsen (editors)

1990   *Wetland Adaptations in the Great Basin*. Museum of Peoples and Cultures Occasional Papers No. 1. Brigham Young University, Provo.

Jenkins, D. L.

2007   Distribution and Dating of Cultural and Paleontological Remains in the Paisley Five Mile Point Caves in the Northern Great Basin: An Early Assessment. In *Paleoindian or Paleoarchaic? Great Basin Human Ecology at the Pleistocene–Holocene Transition*, edited by K. E. Graf and D. N. Schmitt, pp. 57–81. University of Utah Press, Salt Lake City.

Jenkins, D. L., L. G. Davis, T. W. Stafford, Jr., P. F. Campos, B. Hockett, G. T. Jones, L. S. Cummings, C. Yost, T. J. Connolly, R. M. Yohe, II, S. C. Gibbons, M. Raghavan, M. Rasmussen, J. L. A. Paijmans, M. Hofreiter, B. M. Kemp, J. L. Barta, C. Monroe, M. T. P. Gilbert, and E. Willerslev

2012   Clovis Age Western Stemmed Projectile Points and Human Coprolites from the Paisley Caves. *Science* 337:223–228.

Jennings, J. D.

1957   *Danger Cave*. University of Utah Anthropological Papers 27. Salt Lake City.

Jertberg, P. M.

1986   The Eccentric Crescent: Summary Analysis. *Pacific Coast Archaeological Quarterly* 22:35–64.

Jodry, M. A.

1999   Paleoindian Stage. In *Colorado Prehistory: A Context for the Rio Grande Basin*, edited by M. A. Martorano, T. Hoefer, III, M. A. Jodry, V. Spero, and M. L. Taylor, pp. 45–114. Colorado Council of Professional Archaeologists, Denver.

Johnson, J. K.

1981   *Yellow Creek Archaeological Project, Vol. 2*. Tennessee Valley Authority Publications in Anthropology 2. Knoxville.

Johnson, J. R., T. W. Stafford, Jr., H. O. Aije, and D. P. Morris

2002   Arlington Springs Revisited. In *Proceedings of the Fifth California Islands Symposium*, edited by D. R. Brooks, K. C. Mitchell, and H. W. Chaney, pp. 541–545. Santa Barbara Museum of Natural History, Santa Barbara.

Johnson, L., and L. Haarklau

2005   Results of the Regional Obsidian Projectile Point Sourcing Study. In *Fingerprints in the Great Basin: The Nellis Air Force Base Regional Sourcing Study*, edited by L. Haarklau, L. Johnson, and D. L. Wagner, pp. 115–150. Prewitt and Associates, Inc., Austin. Submitted to Nellis Air Force Base, Nevada.

Jones, G. T.

2008   *Archaeological Field Studies in East-Central Nevada*. Department of Anthropology, Hamilton College, Clinton, New York. Bureau of Land Management, Ely Field Office, Ely, Nevada.

Jones, G. T., D. G. Bailey, and C. Beck

1997   Source Provenance of Andesite Artifacts Using Non-Destructive XRF Analysis. *Journal of Archaeological Science* 24:929–943.

Jones, G. T., and C. Beck

1990 An Obsidian Hydration Chronology of Late Pleistocene/Early Holocene Surface Assemblages from Butte Valley, Nevada. *Journal of California and Great Basin Anthropology* 12:84–100.

1999 Paleoarchaic Archaeology in the Great Basin. In *Models for the Millennium: Great Basin Anthropology Today*, edited by C. Beck, pp. 83–95. University of Utah Press, Salt Lake City.

2012 The Emergence of the Desert Archaic in the Great Basin. In *From the Pleistocene to the Holocene: Human Organization and Cultural Transformations in Prehistoric North America*, edited by B. C. Bousman and B. J. Vierra, pp. 105–124. Texas A&M Anthropology Series No. 17. College Station.

2014 Moving into the Mid-Holocene: The Paleoarchaic/Archaic Transition in the Intermountain West. In *Archaeology in the Great Basin and Southwest: Papers in Honor of Don D. Fowler*, edited by N. J. Parezo and J. C. Janetski, pp. 61–84. University of Utah Press, Salt Lake City.

Jones, G. T., C. Beck, E. E. Jones, and R. E. Hughes

2003 Lithic Source Use and Paleoarchaic Foraging Territories in the Great Basin. *American Antiquity* 68:5–38.

Jones, G. T., C. Beck, and R. A. Kessler

2003 *Analysis of Artifacts from the Old River Bed Delta, Dugway Proving Ground, Utah.* Department of Anthropology, Hamilton College, Clinton, New York. Submitted to the Directorate of Environmental Programs, U.S. Army Dugway Proving Ground, Utah.

Jones, G. T., L. M. Fontes, R. A. Horowitz, C. Beck, and D. G. Bailey

2012 Reconsidering Paleoarchaic Mobility in the Central Great Basin. *American Antiquity* 77:351–367.

Jones, G. T., D. K. Grayson, and C. Beck

1983 Sample Size and Functional Diversity in Archaeological Assemblages. In *Lulu Linear Punctated: Essays in Honor of George Irving Quimby*, edited by R. C. Dunnell and D. K. Grayson, pp. 55–73. University of Michigan, Museum of Anthropology Papers No. 72. Ann Arbor.

Kashani, B. H., U. A. Perego, A. Olivieri, N. Angerhofer, F. Gandini, V. Carossa, H. Lancioni, O. Semino, S. R. Woodward, A. Achilli, and A. Torroni

2011 Mitochondrial Haplogroup C4c: A Rare Lineage Entering America Through the Ice-Free Corridor? *American Journal of Physical Anthropology* 147:35–39.

Kelly, R. L.

1992 Mobility/Sedentism: Concepts, Archaeological Measures, and Effects. *Annual Review of Anthropology* 21:43–66.

Kiahtipes, C. A.

2009 Fire in the Desert: Holocene Paleoenvironments in the Bonneville Basin. Unpublished Master's thesis, Washington State University, Pullman.

Kirner, D. L., R. Burky, K. Selsor, D. George, R. E. Taylor, and J. R. Southon

1997 Dating the Spirit Cave Mummy: The Value of Re-examination. *Nevada Historical Society Quarterly* 40:54–56.

Knudson, R.

1973 Organizational Variability in Late Paleoindian Assemblages. Unpublished Ph.D. dissertation, Washington State University, Pullman.

1979 Inference and Imposition in Lithic Analysis. In *Lithic Use-Wear Analysis*, edited by B. Hayden, pp. 269–281. Academic Press, New York.

Kornfeld, M., and G. C. Frison

2000 Paleoindian Occupation of the High Country: The Case of Middle Park, Colorado. *Plains Anthropologist* 45:129–153.

Kornfeld, M., M. L. Larson, D. J. Rapson, and G. C. Frison

2001 10,000 Years on the Rocky Mountains: The Helen Lookingbill Site. *Journal of Field Archaeology* 28:307–324.

Krieger, A. D.

1944 The Typological Concept. *American Antiquity* 9:271–288.

Kutzbach, J. E.

1981 Monsoon Climate of the Early Holocene: Climate Experiment with the Earth's Orbital Parameters for 9000 Years Ago. *Science* 214:59–61.

Kutzbach, J. E., and P. J. Guetter

1986 The Influence of Changing Orbital Parameters and Surface Boundary Conditions on Climate Simulations for the Past 18,000 Years. *Journal of Atmospheric Sciences* 43:1726–1759.

Laabs, B. J. C., D. Marchetti, J. S. Munroe, K. A. Refsnider, J. C. Gosse, E. W. Lips, R. A. Becker, D. M. Mickelson, and B. S. Singer

2011 Chronology of Latest Pleistocene Mountain Glaciation in the Western Wasatch Mountains, Utah, USA. *Quaternary Research* 76:272–284.

Laabs, B. J. C., K. A. Refsnider, J. S. Munroe, D. M. Mickelson, P. J. Applegate, B. S. Singer, and M. W. Caffee

2009 Latest Pleistocene Glacial Chronology of the Unita Mountains: Support for Moisture Driven Asynchrony of the Last Deglaciation. *Quaternary Science Reviews* 28:1171–1187.

Le Bas, M. J., R. W. Le Maitre, A. Streckeisen, and B. Zanettin

1986 A Chemical Classification of Volcanic Rocks Based on the Total Alkali-Silica Diagram. *Journal of Petrology* 27(3):745–750.

Light, A.

1996 Amino Acid Paleotemperature Reconstruction and Radiocarbon Shoreline Chronology of the Lake Bonneville Basin, USA. Unpublished Master's thesis, University of Colorado, Boulder.

Livingston, S. D.

2000   The Homestead Cave Avifauna. In *Late Quaternary Paleoecology in the Bonneville Basin*, by D. B. Madsen, pp. 91–102. Utah Geological Survey Bulletin 130. Salt Lake City.

Louderback, L. A., D. K. Grayson, and M. Llobera

2011   Middle Holocene Climates and Human Population Densities in the Great Basin, Western USA. *Holocene* 21:366–373.

Louderback, L. A., and D. Rhode

2009   15,000 Years of Vegetation Change in the Bonneville Basin: The Blue Lake Record. *Quaternary Science Reviews* 28:308–326.

Luedtke, B. E.

1978   Chert Sources and Trace Element Analysis. *American Antiquity* 43:413–423.

1992   *An Archaeologist's Guide to Chert and Flint*. Institute of Archaeology, Archaeological Research Tools 7. University of California, Los Angeles.

MacArthur, R. H., and E. R. Pianka

1966   On Optimal Use of a Patchy Environment. *American Naturalist* 100:603–609.

Madsen, D. B.

1982a   Prehistoric Occupation Patterns, Subsistence Adaptations, and Chronology in the Fish Springs Area, Utah. In *Archaeological Investigations in Utah*, edited by D. B. Madsen and R. E. Fike, pp. 1–59. U.S. Bureau of Land Management Utah Cultural Resource Series 12. Salt Lake City.

1982b   Get It Where the Getting's Good: A Variable Model of Great Basin Subsistence and Settlement Based on Data from the Eastern Great Basin. In *Man and Environment in the Great Basin*, edited by D. B. Madsen and J. F. O'Connell, pp. 207–226. Society for American Archaeology Papers No. 2. Washington, D.C.

1988   The Prehistoric Use of Great Basin Marshes. In *Preliminary Investigations in Stillwater Marsh: Human Prehistory and Geoarchaeology*, edited by C. Raven and R. G. Elston, pp. 414–419. U.S. Fish and Wildlife Service Cultural Resource Series No. 1. Portland, Oregon.

2000a   *Late Quaternary Paleoecology in the Bonneville Basin*. Utah Geological Survey Bulletin 130. Salt Lake City.

2000b   A High Elevation Allerød–Younger Dryas Megafauna from the West-Central Rocky Mountains. In *Intermountain Archaeology*, edited by D. B. Madsen and M. D. Metcalf, pp. 100–115. University of Utah Anthropological Papers 122. Salt Lake City.

2002   Great Basin Peoples and Late Quaternary Aquatic History. In *Great Basin Aquatic Systems History*, edited by R. Hershler, D. B. Madsen, and D. R. Currey, pp. 387–405. Smithsonian Contributions to Earth Sciences No. 33. Smithsonian Institution Press, Washington, D.C.

2007   The Paleoarchaic to Archaic Transition in the Great Basin. In *Paleoarchaic or Paleoindian? Great Basin Human Ecology at the Pleistocene–Holocene Transition*, edited by K. E. Graf and D. N. Schmitt, pp. 3–20. University of Utah Press, Salt Lake City.

2012   Archaeological Perspectives on the Great Basin Culture Area. In *Meetings at the Margins: Prehistoric Cultural Interactions in the Intermountain West*, edited by D. Rhode, pp. 271–287. University of Utah Press, Salt Lake City.

Madsen, D. B., and D. R. Currey

1979   Late Quaternary Glacial and Vegetation Changes, Little Cottonwood Canyon Area, Wasatch Mountains, Utah. *Quaternary Research* 12:254–270.

Madsen, D. B., and R. L. Kelly

2010   The "Good Sweet Water" of Great Basin Marshes. In *The Great Basin: People and Place in Ancient Times*, edited by C. S. Fowler and D. D. Fowler, pp. 79–86. School for Advanced Research Press, Santa Fe, New Mexico.

Madsen, D. B., and D. Page

2008   *Year 2008 Follow-Up Archaeological and Geomorphic Studies of the Old River Bed Delta, Dugway Proving Ground, Utah*. Desert Research Institute, Reno. Submitted to the Directorate of Environmental Programs, U.S. Army Dugway Proving Ground, Utah.

2009   *Year 2008 Follow-Up Archaeological and Geomorphic Studies of the Old River Bed Delta, Dugway Proving Ground, Utah: Phase 2*. Desert Research Institute, Reno. Submitted to the Directorate of Environmental Programs, U.S. Army Dugway Proving Ground, Utah.

Madsen, D. B., D. Page, and D. N. Schmitt

2006   *Archaeological Survey of Additional Channel Remnants in the Old River Bed Delta, Dugway Proving Ground, Western Utah*. Desert Research Institute, Reno. Submitted to the Directorate of Environmental Programs, U.S. Army Dugway Proving Ground, Utah.

Madsen, D. B., and D. Rhode

1990   Early Holocene Pinyon (*Pinus monophylla*) in the Northeastern Great Basin. *Quaternary Research* 33: 94–101.

Madsen, D. B., D. Rhode, D. K. Grayson, J. M. Broughton, S. D. Livingston, J. M. Hunt, J. Quade, D. N. Schmitt, and M. W. Shaver, III

2001   Late Quaternary Environmental Change in the Bonneville Basin, Western U.S.A. *Palaeogeography, Palaeoclimatology, Palaeoecology* 167:243–271.

Madsen, D. B., and D. N. Schmitt

2005   *Buzz-Cut Dune and Fremont Foraging at the Margin of Horticulture*. University of Utah Anthropological Papers No. 124. Salt Lake City.

Madsen, D. B., D. N. Schmitt, and J. M. Hunt

2000   *Archaeological Evaluation of Areas Associated with the Gilbert Shoreline and Old River Bed Delta, Dugway Proving Ground, Utah*. Utah Geological Survey, Salt

Lake City. Submitted to the Directorate of Environmental Programs, Dugway Proving Ground, Utah.

Madsen, D. B., D. N. Schmitt, R. Quist, and D. Rhode
2004 *2003 Test Excavations at Two Paleoarchaic Sites in the Old River Bed Delta, Dugway Proving Ground, Utah.* Utah Geological Survey, Salt Lake City. Submitted to the Directorate of Environmental Programs, U.S. Army Dugway Proving Ground, Utah.

Marquardt, W. H.
1978 Advances in Archaeological Seriation. In *Advances in Archaeological Method and Theory*, Vol. 1, edited by M. B. Schiffer, pp. 257–314. Academic Press, New York.

Mason, R. J.
1981 *Great Lakes Archaeology*. Academic Press, New York.

Mayer, J., W. Eckerle, S. Taddie, and A. Geery
2010 *Geoarchaeology of the Milford Wind Corridor Project Area, Millard and Beaver Counties, Utah*. Report on file at SWCA Environmental Consultants, Salt Lake City.

McDonald, H. G.
2002 *Platygonus compressus* from Franklin County, Idaho, and a Review of the Genus in Idaho. In *And Whereas…: Papers on the Vertebrate Paleontology of Idaho Honoring John A. White*, Vol. 2, edited by W. A. Akersten, M. E. Thompson, D. J. Meldrum, R. A. Rapp, and H. G. McDonald, pp. 141–149. Idaho Museum of Natural History Occasional Paper 37. Pocatello.

McGee, D., J. Quade, R. L. Edwards, W. S. Broecker, H. Cheng, P. W. Reiners, and N. Evenson
2012 Lacustrine Cave Carbonates: Novel Archives of Paleohydrologic Change in the Bonneville Basin (Utah, USA). *Earth and Planetary Science Letters* 351–352:182–194.

Mead, J. I., R. S. Thompson, and T. R. Van Devender
1982 Late Wisconsinan and Holocene Fauna from Smith Creek Canyon, Snake Range, Nevada. *Transactions of the San Diego Society of Natural History* 20:1–26.

Mehringer, P. J., Jr.
1985 Late-Quaternary Pollen Records from the Interior Pacific Northwest and Northern Great Basin of the United States. In *Pollen Records of Late Quaternary North American Sediments*, edited by V. M. Bryant, Jr., and R. Holloway, pp. 167–189. American Association of Stratigraphic Palynologists Contribution Series 16. Dallas.

Meltzer, D. J.
2009 *First Peoples in a New World: Colonizing Ice Age America*. University of California Press, Berkeley.

Meltzer, D. J., and V. T. Holliday
2010 Would North American Paleoindians Have Noticed Younger Dryas Age Climate Changes? *Journal of World Prehistory* 23:1–41.

Miller, D. M., C. G. Oviatt, and J. P. McGeehin
2013 Stratigraphy and Chronology of Provo Shoreline Deposits and Lake-Level Implications, Late Pleistocene Lake Bonneville, Eastern Great Basin, USA. *Boreas* 42:342–361.

Miller, D. M., K. M. Schmidt, S. A. Mahan, J. P. McGeehin, L. A. Owen, J. A. Barron, F. Lehmkuhl, and R. Löhrer
2010 Holocene Landscape Response to Seasonality of Storms in the Mojave Desert. *Quaternary International* 215:45–61.

Miller, W. E.
1976 Late Pleistocene Vertebrates of the Silver Creek Local Fauna from North Central Utah. *Great Basin Naturalist* 36:387–424.
1987 *Mammut americanum*, Utah's First Record of the American Mastodon. *Journal of Paleontology* 61: 168–183.
2002 Quaternary Vertebrates of the Northeastern Bonneville Basin and Vicinity of Utah. In *Great Salt Lake: An Overview of Change*, edited by J. W. Gwynn, pp. 54–69. Utah Department of Natural Resources Special Publication, Salt Lake City.

Mock, C. J., and P. J. Bartlein
1995 Spatial Variability of Late-Quaternary Paleoclimates in the Western United States. *Quaternary Research* 44:425–433.

Moore, J.
2009 *Great Basin Tool-Stone Sources*. NDOT Obsidian and Tools-Stone Sourcing Project: 2002 Progress Report. Report on file, Nevada Department of Transportation, Carson City.

Morrow, J. E.
1997 End Scraper Morphology and Use-Life: An Approach for Studying Lithic Technology and Mobility. *Lithic Technology* 22:70–85.

Munroe, J. S.
2001 Late Quaternary History of the Northern Uinta Mountains, Northeastern Utah. Unpublished Ph.D. dissertation, University of Wisconsin, Madison.
2005 Glacial Geology of the Northern Uinta Mountains. In *Uinta Mountain Geology*, edited by C. M. Dehler, J. L. Pederson, D. A. Sprinkel, and B. J. Kowallis, pp. 215–234. Utah Geological Association Publication 33. Salt Lake City.

Munroe, J. S., B. J. C. Labbs, J. D. Shakun, B. S. Singer, D. M. Mickelson, K. A. Refsnider, and M. W. Caffee
2006 Latest Pleistocene Advance of Alpine Glaciers in the Southwestern Uinta Mountains, Utah, USA: Evidence for the Influence of Local Moisture Sources. *Geology* 34:841–844.

Munyikwa, K., J. K. Feathers, T. M. Rittenour, and H. K. Shrimpton
2011 Constraining the Late Wisconsin Retreat of the Laurentide Ice Sheet from Western Canada Using

Luminescence Ages from Postglacial Aeolian Dunes. *Quaternary Geochronology* 6:407–422.

Murchison, S. B.

1989   Fluctuation History of Great Salt Lake, Utah, During the Last 13,000 Years. Unpublished Ph.D. dissertation, University of Utah, Salt Lake City.

Muto, G. R.

1971   A Stage Analysis of the Manufacture of Stone Tools. In *Great Basin Anthropological Conference 1970: Selected Papers*, edited by M. E. Aikens, pp. 109–118. University of Oregon Anthropological Papers No. 1. Eugene.

Napton, L. K.

1997   The Spirit Cave Mummy: Coprolite Investigations. *Nevada Historical Society Quarterly* 40:97–104.

Nelson, F. W., and R. D. Holmes

1979   Trace Element Analysis of Obsidian Sources from Western Utah. *Utah State Historical Society, Antiquities Section Selected Papers* 6(15):65–80. Salt Lake City.

Nelson, M. E., and J. H. Madsen, Jr.

1987   A Review of Lake Bonneville Shoreline Faunas (Late Pleistocene) of Northern Utah. In *Cenozoic Geology of Western Utah—Sites for Precious Metal and Hydrocarbon Accumulations*, edited by R. S. Kopp and R. E. Cohenour, pp. 319–334. Utah Geological Association Publication 16. Salt Lake City.

Newlander, K. S.

2012   Exchange, Embedded Procurement, and Hunter-Gatherer Mobility: A Case Study from the North American Great Basin. Unpublished Ph.D. dissertation, University of Michigan, Ann Arbor.

O'Brien, P. J.

1984   The Tim Adrian Site (14NT604): A Hell Gap Quarry Site in Norton County, Kansas. *Plains Anthropologist* 29:41–55.

O'Connell, J. F.

1987   Alyawara Site Structure and Its Archaeological Implications. *American Antiquity* 52:74–108.

O'Connell, J. F., K. Hawkes, and N. B. Jones

1991   Distribution of Refuse-Producing Activities at Hadza Base Camps: Implications for Analyses of Site Structure. In *The Interpretation of Archaeological Spatial Patterning*, edited by E. M. Kroll and T. D. Price, pp. 61–76. Plenum Press, New York.

Oetting, A. C.

1994   Chronology and Time Markers in the Northwestern Great Basin: The Chewaucan Cultural Chronology. In *Archaeological Researches in the Northern Great Basin: Fort Rock Archaeology Since Cressman*, edited by C. M. Aikens and D. L. Jenkins, pp. 41–62. University of Oregon Anthropological Papers 50. Eugene.

Oviatt, C. G.

1988   Late Pleistocene and Holocene Lacustrine Fluctuations in the Sevier Lake Basin, Utah, U.S.A. *Journal of Paleolimnology* 1:9–21.

1989   *Quaternary Geology of Part of the Sevier Desert, Millard County, Utah*. Utah Geological and Mineral Survey Special Studies 70. Salt Lake City.

1999   *Surficial Deposits and Late Quaternary History of Dugway Proving Ground, Utah: Preliminary Notes.* Utah Geological Survey, Salt Lake City. Submitted to the Directorate of Environmental Programs, Dugway Proving Ground, Utah.

2014   *The Gilbert Episode in the Great Salt Lake Basin, Utah.* Miscellaneous Publication 14-3. Utah Geological Survey, Salt Lake City.

Oviatt, C. G., D. R. Currey, and D. Sack

1992   Radiocarbon Chronology of Lake Bonneville, Eastern Great Basin, USA. *Palaeogeography, Palaeoclimatology, Palaeoecology* 99:225–241.

Oviatt, C. G., and D. B. Madsen

2000   *Surficial Deposits and Late Quaternary History of Dugway Proving Ground, Utah: Progress Report, 1999.* Utah Geological Survey, Salt Lake City. Submitted to the Directorate of Environmental Programs, Dugway Proving Ground, Utah.

2001   *Surficial Deposits and Late Quaternary History of Dugway Proving Ground, Utah: Progress Report, 2000.* Manuscript on file, Environmental Directorate, Dugway Proving Ground, Utah.

Oviatt, C. G., D. B. Madsen, and D. N. Schmitt

2003   Late Pleistocene and Early Holocene Rivers and Wetlands in the Bonneville Basin of Western North America. *Quaternary Research* 60:200–210.

Oviatt, C. D., D. M. Miller, J. P. McGeehin, C. Zachary, and S. Mahan

2005   The Younger Dryas Phase of Great Salt Lake, USA. *Palaeogeography, Palaeoclimatology, Palaeoecology* 219:263–284.

Page, D.

2008   Fine-Grained Volcanic Toolstone Sources and Early Use in the Bonneville Basin of Western Utah and Eastern Nevada. Unpublished Master's thesis, University of Nevada, Reno.

Page, D., D. B. Madsen, and D. N. Schmitt

2008   *Additional Archaeological Surveys in the Old River Bed Delta, Dugway Proving Ground, Utah: The Southwestern Channels, Phase 2.* Desert Research Institute, Reno. Submitted to the Directorate of Environmental Programs, U.S. Army Dugway Proving Ground, Utah.

Page, D., and D. N. Schmitt

2010   *A Class III Cultural Resource Inventory of the Northwest All Purpose Grid Area and RHR Parcel, U.S. Army Dugway Proving Ground, Tooele County, Utah.* Desert Research Institute, Reno. Submitted to the Directorate of Environmental Programs, U.S. Army Dugway Proving Ground, Utah.

2011a   *A Class III Cultural Resource Inventory of 2,693 Acres in Support of Training Area Expansion, U.S. Army*

Dugway Proving Ground, Tooele County, Utah. Desert Research Institute, Reno. Submitted to the Division of Environmental Programs, U.S. Army Dugway Proving Ground, Utah.

2011b *An Intuitive Archaeological Reconnaissance and Recordation of Two Toolstone Quarries in the Southern Cedar Mountains, U.S. Army Dugway Proving Ground, Tooele County, Utah*. Desert Research Institute, Reno. Submitted to the Division of Environmental Programs, U.S. Army Dugway Proving Ground, Utah.

Page, D., D. N. Schmitt, G. Dalldorf, and B. Wazaney

2012 *A Class III Cultural Resource Inventory in Support of Phase II of the PD-Tess Range Modernization Project, U.S. Army Dugway Proving Ground, Tooele County, Utah*. Desert Research Institute, Reno. Submitted to the Division of Environmental Programs, U.S. Army Dugway Proving Ground, Utah.

Page, D., D. N. Schmitt, D. Rhode, and A. Sapula

2012 *A Class III Cultural Resource Inventory in Support of Phase 3 of the PD-Tess Range Modernization Project, U.S. Army Dugway Proving Ground, Tooele County, Utah*. Desert Research Institute, Reno. Submitted to the Division of Environmental Programs, U.S. Army Dugway Proving Ground, Utah.

Parry, W. J.

1994 Prismatic Blade Technologies in North America. In *The Organization of North American Prehistoric Chipped Stone Technologies*, edited by P. J. Carr, pp. 87–98. International Monographs in Prehistory No. 7. University of Michigan, Ann Arbor.

Pendleton, L. S.

1979 Lithic Technology in Early Nevada Assemblages. Unpublished Master's thesis, California State University, Long Beach.

Perego, U. A., A. Achilli, N. Angerhofer, M. Accetturo, M. Pala, A. Olivieri, B. H. Kashani, K. H. Ritchie, R. Scozzari, Q.-P. Kong, N. M. Myres, A. Salas, O. Semino, H.-J. Dandelt, S. R. Woodward, and A. Torroni

2008 Distinctive Paleo-Indian Migration Routes from Beringia Marked by Two Rare mtDNA Haplogroups. *Current Biology* 19:1–8.

Pinson, A. O.

1999 Foraging in Uncertain Times: The Effects of Risk on Subsistence Behavior During the Pleistocene–Holocene Transition in the Oregon Great Basin. Unpublished Ph.D. dissertation, University of New Mexico, Albuquerque.

2007 Artiodactyl Use and Adaptive Discontinuity Across the Paleoarchaic/Archaic Transition in the Northern Great Basin. In *Paleoarchaic or Paleoindian? Great Basin Human Ecology at the Pleistocene–Holocene Transition*, edited by K. E. Graf and D. N.

Schmitt, pp. 187–203. University of Utah Press, Salt Lake City.

Pitblado, B. L.

1998 Peak to Peak in Paleoindian Time: Occupation of Southwest Colorado. *Plains Anthropologist* 43:333–348.

2003 *Late Paleoindian Occupation of the Southern Rocky Mountains: Early Holocene Projectile Points and Land Use in the High Country*. University Press of Colorado, Boulder.

Plager, S. R.

2001 Patterns in the Distribution of Volcanic Glass Across Southern Idaho. Unpublished Master's thesis, Idaho State University, Pocatello.

Polyak, V. J., Y. Asmeron, S. J. Burns, and M. S. Lachniet

2012 Climatic Backdrop to the Terminal Pleistocene Extinction of North American Mammals. *Geology* 40:1023–1026.

Polyak, V. J., J. B. T. Rasmussen, and Y. Asmeron

2004 Prolonged Wet Period in the Southwestern United States Through the Younger Dryas. *Geology* 32:5–8.

Quade, J.

2000 Strontium Ratios and the Origin of Early Homestead Cave Biota. In *Late Quaternary Paleoecology in the Bonneville Basin*, by D. B. Madsen, pp. 44–46. Utah Geological Survey Bulletin 130. Salt Lake City.

Ramsey, C. B.

2010 OxCal Program, V. 4.1.7. Radiocarbon Accelerator Unit, University of Oxford. Electronic document, http://c14.arch.ox.ac.uk/embed.php?File=oxcal .html, accessed January 2013.

Rasmussen, S. O., K. K. Andersen, A. M. Svensson, J. P. Steffensen, B. M. Vinther, H. B. Calusen, M. L. Siggaard-Andersen, S. J. Johnsen, L. B. Larsen, D. Dahl-Jensen, M. Bigler, R. Röthlisberger, H. Fischer, K. Goto-Azuma, M. E. Hansson, and U. Ruth

2006 A New Greenland Ice Core Chronology for the Last Glacial Termination. *Journal of Geophysical Research* 111. doi:06110.01029/02005JD006079.

Raven, C.

1990 *Prehistoric Human Geography in the Carson Desert, pt. II: Archaeological Field Tests of Model Predictions*. U.S. Fish and Wildlife Service Cultural Resource Series No. 4. Portland, Oregon.

Raven, C., and R. G. Elston (editors)

1988 *Preliminary Investigations in Stillwater Marsh: Human Prehistory and Geoarchaeology*. U.S. Fish and Wildlife Service Cultural Resource Series No. 1. Portland, Oregon.

Raymond, A. W.

1984 Prehistoric Reduction and Curation of Topaz Mountain, Utah, Obsidian: A Technological Analysis of Two Lithic Scatters. Unpublished Master's thesis, Washington State University, Pullman.

Reeder, L. A., J. M. Erlandson, and C. R. Torben

2011   Younger Dryas Environments and Human Adaptation on the West Coast of the United States and Baja California. *Quaternary International* 242:463–478.

Reimer, P. J., M. G. L. Baillie, E. Bard, A. Bayliss, J. W. Beck, P. G. Blackwell, C. Bronk Ramsey, C. E. Buck, G. S. Burr, R. L. Edwards, M. Friedrich, P. M. Grootes, T. P. Guilderson, I. Hajdas, T. J. Heaton, A. G. Hogg, K. A. Hughen, K. F. Kaiser, B. Kromer, F. G. McCormac, S. W. Manning, R. W. Reimer, D. A. Richards, J. R. Southon, S. Talamo, C. S. M. Turney, J. van der Plicht, and C. E. Weyhenmeyer

2009   IntCal09 and Marine09 Radiocarbon Age Calibration Curves, 0–50,000 Years cal BP. *Radiocarbon* 51:1111–1150.

Reynolds, R. L., J. C. Yount, M. Reheis, H. Goldstein, P. Chavez, Jr., R. Fulton, J. Whitney, C. Fuller, and R. M. Forester

2007   Dust Emission from Wet and Dry Playas in the Mojave Desert, USA. *Earth Surface Processes and Landforms* 32:1811–1827.

Rhode, D.

2000a   Middle and Late Wisconsin Vegetation in the Bonneville Basin. In *Late Quaternary Paleoecology in the Bonneville Basin*, by D. B. Madsen, pp. 137–147. Utah Geological Survey Bulletin 130. Salt Lake City.

2000b   Holocene Vegetation History in the Bonneville Basin. In *Late Quaternary Paleoecology in the Bonneville Basin*, by D. B. Madsen, pp. 149–163. Utah Geological Survey Bulletin 130. Salt Lake City.

2008   Dietary Plant Use by Middle Holocene Foragers in the Bonneville Basin, Western North America. *Before Farming* 2008/3: article 2. Electronic document, http://www.waspress.co.uk/journals/beforefarming/journal_20083/abstracts/index.php.

Rhode, D., T. Goebel, K. Graf, B. Hockett, K. T. Jones, D. B. Madsen, C. G. Oviatt, and D. N. Schmitt

2005   Latest Pleistocene–Early Holocene Human Occupation and Paleoenvironmental Change in the Bonneville Basin, Utah–Nevada. In *Interior Western United States*, edited by J. Pederson and C. Dehler, pp. 211–230. Geological Society of America Field Guide 6. Boulder.

Rhode, D., and L. A. Louderback

2007   Dietary Plant Use in the Bonneville Basin During the Terminal Pleistocene/Early Holocene Transition. In *Paleoindian or Paleoarchaic? Great Basin Human Ecology at the Pleistocene–Holocene Transition*, edited by K. E. Graf and D. N. Schmitt, pp. 231–247. University of Utah Press, Salt Lake City.

Rhode, D., and D. B. Madsen

1995   Late Wisconsin/Early Holocene Vegetation Change in the Bonneville Basin. *Quaternary Research* 44:246–256.

Rhode, D., D. B. Madsen, and K. T. Jones

2006   Antiquity of Early Holocene Small Seed Consumption and Processing at Danger Cave, Utah, USA. *Antiquity* 80:328–329.

Rhode, D., D. Page, and D. N. Schmitt

2011   *Archaeological Data Recovery at Site 42To0567 in the Wig Mountain Training Area, U.S. Army Dugway Proving Ground, Tooele County, Utah.* Desert Research Institute, Reno. Submitted to the Directorate of Environmental Programs, U.S. Army Dugway Proving Ground, Utah.

Rogers, R. A.

1986   Spurred End Scrapers as Diagnostic Paleoindian Artifacts: A Distributional Analysis of Stream Terraces. *American Antiquity* 51:338–341.

Schmitt, D. N.

2004   Ecological Change in Western Utah: Comparisons Between a Late Holocene Archaeological Fauna and Modern Small-Mammal Surveys. In *Zooarchaeology and Conservation Biology*, edited by R. L. Lyman and K. P. Cannon, pp. 178–192. University of Utah Press, Salt Lake City.

Schmitt, D. N., and K. D. Lupo

2005   The Camels Back Cave Mammalian Fauna. In *Camels Back Cave*, by D. N. Schmitt and D. B. Madsen, pp. 136–176. University of Utah Anthropological Papers 125. Salt Lake City.

2012   The Bonneville Estates Rockshelter Rodent Fauna and Changes in Late Pleistocene–Middle Holocene Climates and Biogeography in the Northern Bonneville Basin, USA. *Quaternary Research* 78:95–102.

Schmitt, D. N., and D. B. Madsen

2005   *Camels Back Cave.* University of Utah Anthropological Papers No. 125. Salt Lake City.

Schmitt, D. N., D. B. Madsen, K. E. Callister, and R. Quist

2003   *Archaeological Survey of Areas Associated with Lake Öferneet and Additional Old River Bed Deltaic Channel Features, Dugway Proving Ground, Utah.* Utah Geological Survey, Salt Lake City. Submitted to the Directorate of Environmental Programs, U.S. Army Dugway Proving Ground, Utah.

Schmitt, D. N., D. B. Madsen, J. M. Hunt, K. Callister, K. Jensen, and R. Quist

2002   *Archaeological Inventories of Areas Associated with West Baker Dunes in the Old River Bed Delta at U.S. Army Dugway Proving Ground, Utah.* Utah Geological Survey, Salt Lake City. Submitted to the Directorate of Environmental Programs, U.S. Army Dugway Proving Ground, Utah.

Schmitt, D. N., D. B. Madsen, and K. D. Lupo

2002   Small-Mammal Data on Early and Middle Holocene Climates and Biotic Communities in the Bonneville Basin, U.S.A. *Quaternary Research* 58:255–260.

2004   The Worst of Times, the Best of Times: Jackrabbit Hunting by Middle Holocene Human Foragers in the Bonneville Basin of Western North America. In *Colonization, Migration and Marginal Areas: A Zooarchaeological Approach*, edited by M. Mondini, S. Muñoz, and S. Wickler, pp. 86–95. Oxbow Books, Oxford.

Schmitt, D. N., D. B. Madsen, C. G. Oviatt, and R. Quist

2007   Late Pleistocene/Early Holocene Geomorphology and Human Occupation of the Old River Bed Delta, Western Utah. In *Paleoindian or Paleoarchaic? Great Basin Human Ecology at the Pleistocene–Holocene Transition*, edited by K. E. Graf and D. N. Schmitt, pp. 105–119. University of Utah Press, Salt Lake City.

Schmitt, D. N., D. B. Madsen, D. Page, and G. M. Smith

2007   *Additional Archaeological Surveys in the Old River Bed Delta, Dugway Proving Ground, Utah: The Southwestern Channels.* Desert Research Institute, Reno. Submitted to the Directorate of Environmental Programs, U.S. Army Dugway Proving Ground, Utah.

Schmitt, D. N., and D. Page

2011   *A Class III Cultural Resource Inventory in Support of Phase 5 of the PD-Tess Range Modernization Project, U.S. Army Dugway Proving Ground, Tooele County, Utah.* Desert Research Institute, Reno. Submitted to the Directorate of Environmental Programs, U.S. Army Dugway Proving Ground, Utah.

Schmitt, D. N., D. Page, and M. Memmott

2010   *A Class III Cultural Resource Inventory for Phase I of the PD-TESS Project, U.S. Army Dugway Proving Ground, Tooele County, Utah.* Desert Research Institute, Reno. Submitted to the Directorate of Environmental Programs, U.S. Army Dugway Proving Ground, Utah.

Schmitt, D. N., D. Page, B. Wazaney, and G. Dalldorf

2011   *A Class III Cultural Resource Inventory of the UAS Loiter Area North of MAAF, U.S. Army Dugway Proving Ground, Tooele County, Utah.* Desert Research Institute, Reno. Submitted to the Division of Environmental Programs, U.S. Army Dugway Proving Ground, Utah.

Schubert, B. W.

2010   Late Quaternary Chronology and Extinction of North American Giant Short-Faced Bears (*Arctodussimus*). *Quaternary International* 217:188–194.

Seddon, M. T.

2005   A Revised Relative Obsidian Hydration Chronology for Wild Horse Canyon, Black Rock Area, and Panaca Summit/Modena Obsidian. In *Kern River 2003 Expansion Project: Utah, Vol. IV: Prehistoric Synthesis—Examination of Data by Archaeological Unit*, edited by A. Reed, M. Seddon, and H. Stettler, pp. 447–524. SWCA Environmental Consultants, Salt Lake City. Submitted to the Federal Energy Regulatory Commission, Washington, D.C.

Seeman, M. F.

1994   Intercluster Lithic Patterning at Nobles Pond: A Case for "Disembedded" Procurement Among Early Paleoindian Societies. *American Antiquity* 59:273–288.

Shackley, M. S.

1996   Range and Mobility in the Early Hunter-Gatherer Southwest. In *Early Formative Adaptations in the Southern Southwest*, edited by B. Roth, pp. 5–16. Monographs in World Prehistory 25. Prehistory Press, Madison, Wisconsin.

2005   *Obsidian: Geology and Archaeology in the North American Southwest.* University of Arizona Press, Tucson.

2011   *X-Ray Fluorescence Spectrometry (XRF) Geoarchaeology.* Springer, New York.

Shackley, M. S. (editor)

1998   *Archaeological Obsidian Studies: Method and Theory.* Advances in Archaeological and Museum Science, Vol. 3. Plenum Press, New York.

Shakun, J. D., and A. E. Carlson

2010   A Global Perspective of Last Glacial Maximum to Holocene Climate Change. *Quaternary Science Reviews* 29:1801–1816.

Shakun, J. D., P. U. Clark, F. He, S. A. Marcott, A. C. Mix, Z. Liu, B. Otto-Bliesner, A. Schmittner, and E. Bard

2012   Global Warming Preceded by Increasing Carbon Dioxide Concentrations During the Last Deglaciation. *Nature* 484:49–55.

Shott, M. J.

1994   Size and Form in the Analysis of Flake Debris: Review and Recent Approaches. *Journal of Archaeological Method and Theory* 1:69–110.

1995   How Much Is a Scraper? Curation, Use Rates, and the Formation of Scraper Assemblages. *Lithic Technology* 20:53–72.

1996   An Exegesis of the Curation Concept. *Journal of Anthropological Research* 52:259–280.

Simms, S. R.

2008   *Ancient Peoples of the Great Basin and Colorado Plateau.* Left Coast Press, Walnut Creek.

Simms, S. R., and M. C. Isgreen

1984   *Archaeological Investigations in the Sevier and Escalante Deserts, Western Utah.* University of Utah Archaeological Center Reports of Investigations 83-12. Salt Lake City.

Simms, S. R., and L. W. Lindsay

1989   42MD300, an Early Holocene Site in the Sevier Desert. *Utah Archaeology* 2:56–66.

Simms, S. R., D. N. Schmitt, and K. Jensen

1999   Playa View Dune: A Mid-Holocene Campsite in the Great Salt Lake Desert, Dugway Utah. *Utah Archaeology* 12:31–49.

Skinner, C. E.

2010   *2006–2010 Letter Reports on Results of XRF Studies*

on *Artifacts from Sites in the Great Salt Lake Desert and Vicinity, and Obsidian and Fine-Grained Volcanic Materials from Regional Source Areas.* Northwest Research Obsidian Studies Laboratory, Corvallis. On file at Environmental Programs, U.S. Army Dugway Proving Ground, Utah, and Desert Research Institute, Reno.

Skinner, C. E., J. E. Thatcher, D. L. Jenkins, and A. C. Oetting
2004   X-Rays, Artifacts, and Procurement Ranges: A Mid-Project Snapshot of Prehistoric Procurement Patterns in the Fort Rock Basin of Oregon. In *Early and Middle Holocene Archaeology of the Northern Great Basin*, edited by D. L. Jenkins, T. J. Connolly, and C. M. Aikens, pp. 221–232. University of Oregon Anthropological Papers No. 62. Eugene.

Slingerland, R., and N. D. Smith
2004   River Avulsions and Their Deposits. *Annual Review of Earth and Planetary Science* 32:257–285.

Smith, G. M.
2007   Pre-Archaic Mobility and Technological Activities at the Parman Localities, Humboldt County, Nevada. In *Paleoindian or Paleoarchaic? Great Basin Human Ecology at the Pleistocene–Holocene Transition*, edited by K. E. Graf and D. N. Schmitt, pp. 139–155. University of Utah Press, Salt Lake City.
2010   Footprints Across the Black Rock: Temporal Variability in Prehistoric Foraging Territories and Toolstone Procurement Strategies in the Western Great Basin. *American Antiquity* 75:865–885.
2011   Shifting Stones and Changing Homes: Using Toolstone Ratios to Consider Relative Occupation Span in the Northwestern Great Basin. *Journal of Archaeological Science* 38:461–469.

Smith, G. M., and J. Kielhofer
2011   Through the High Rock and Beyond: Placing Last Supper Cave and Parman Paleoindian Lithic Assemblages into a Regional Context. *Journal of Archaeological Science* 38:3568–3576.

Smith, G. M., E. S. Middleton, and P. A. Carey
2013   Paleoindian Technological Provisioning Strategies in the Northwestern Great Basin. *Journal of Archaeological Science* 40:4180–4188.

Spaulding, W. G., and L. J. Graumlich
1986   The Last Pluvial Climatic Episodes in the Deserts of Southwestern North America. *Nature* 320:441–444.

Speakman, R. J., and M. S. Shackley
2012   Silo Science and Portable XRF in Archaeology: A Response to Frahm. *Journal of Archaeological Science* 40:1435–1443.

Spencer, R. J., M. Baedecker, H. P. Eugster, R. M. Forester, M. B. Goldhaber, B. F. Jones, K. Kelts, J. MacKenzie, D. B. Madsen, S. L. Rettig, M. Rubin, and C. J. Browser
1984   Great Salt Lake and Precursors, Utah: The Last 30,000 Years. *Contributions to Mineralogy and Petrology* 86:321–334.

Speth, J. D., K. Newlander, A. A. White, A. K. Lemke, and L. E. Anderson
2013   Early Paleoindian Big-Game Hunting in North America: Provisioning or Politics? *Quaternary International* 285:111–139.

Steffensen, J. P., K. K. Andersen, M. Bigler, H. B. Clausen, D. Dahl-Jenzen, H. Fischer, K. Goto-Azuma, M. Hansson, S. J. Johnsen, J. Jouzel, V. Masson-Delmotte, T. Popp, S. O. Rasmussen, R. Rothlisberger, U. Ruth, B. Stauffer, M. L. Siggaard-Andersen, A. E. Sveinbjörnsdóttir, A. Svensson, and J. W. C. White
2008   High-Resolution Greenland Ice Core Data Show Abrupt Climate Change Happens in Few Years. *Science* 312:680–684.

Steiger, J. I., and G. W. Freethy
2001   *Ground-Water Hydrology of Dugway Proving Ground and Adjoining Areas, Tooele and Juab Counties, Utah.* U.S. Geological Survey Water-Resources Investigations Report 00-4240. Boulder.

Stephens, J. C., and C. T. Sumsion
1978   *Hydrologic Reconnaissance of the Dugway Valley–Government Creek Area, West-Central Utah.* Division of Water Rights Technical Publication No. 59. Utah Department of Natural Resources, Salt Lake City.

Storck, P. L.
1997   *The Fisher Site. Archaeological, Geological and Paleobotanical Studies at an Early Paleo-Indian Site in Southern Ontario, Canada.* Museum of Anthropology Memoirs 30. University of Michigan, Ann Arbor.

Stuiver, M., P. J. Reimer, and R. W. Reimer
2013   CALIB 7.0. Electronic document, http://calib.qub.ac.uk/calib/.

Surovell, T. A.
2009   *Toward a Behavioral Ecology of Lithic Technology: Cases from Paleoindian Archaeology.* University of Arizona Press, Tucson.

Tadlock, W. L.
1966   Certain Crescentic Stone Objects as a Time Marker in the Western United States. *American Antiquity* 31:662–675.

Thomas, D. H.
1981   How to Classify the Projectile Points from Monitor Valley, Nevada. *Journal of California and Great Basin Anthropology* 3:7–43.
1988a   Pattern Recognition: Quantitative Variability in Assemblage Size and Diversity. In *The Archaeology of Monitor Valley, 3: Survey and Additional Excavations*, by D. H. Thomas, pp. 380–393. American Museum of Natural History Anthropological Papers 66(2). New York.
1988b   Pattern Recognition: Variability in Elevation. In *The Archaeology of Monitor Valley, 3: Survey and Additional Excavations*, by D. H. Thomas, pp. 415–436. American Museum of Natural History Anthropological Papers 66(2). New York.

1989   Diversity in Hunter-Gatherer Cultural Geography. In *Quantifying Diversity in Archaeology*, edited by R. D. Leonard and G. T. Jones, pp. 85–91. Cambridge University Press, Cambridge.

2012   The Chert Core and the Obsidian Rim: Some Long-Term Implications for the Central Great Basin. In *Meetings at the Margins: Prehistoric Cultural Interactions in the Intermountain West*, edited by D. Rhode, pp. 254–270. University of Utah Press, Salt Lake City.

Thomas, S., M. Rondeau, and P. O'Grady

2013   Filling the Void (2000–2013): Clovis Spear Points and Other Diagnostic Artifacts in the Far Northern Great Basin (Southeastern Oregon). Poster presented at the Paleoamerican Odyssey Conference, Santa Fe, New Mexico.

Thompson, R. S.

1990   Late Quaternary Vegetation and Climate in the Great Basin. In *Packrat Middens: The Last 40,000 Years of Biotic Change*, edited by J. L. Betancourt, T. R. Van Devender, and P. S. Martin, pp. 201–239. University of Arizona Press, Tucson.

1992   Late Quaternary Environments in Ruby Valley, Nevada. *Quaternary Research* 37:1–15.

Thompson, R. S., and K. H. Anderson

2000   Biomes of Western North America at 18,000, 6000 and 0 ^{14}C yrbp Reconstructed from Pollen and Packrat Midden Data. *Journal of Biogeography* 27:555–584.

Thompson, R. S., C. Whitlock, P. J. Bartlein, S. P. Harrison, and W. G. Spaulding

1993   Climate Changes in the Western United States Since 18,000 yr BP. In *Global Climates Since the Last Glacial Maximum*, edited by H. E. Wright, Jr., J. E. Kutzbach, T. Webb, III, W. F. Ruddiman, F. A. Street-Perrott, and P. J. Bartlein, pp. 468–513. University of Minnesota Press, Minneapolis.

Thomsen, V., and D. Schatzlein

2002   Advances in Field-Portable XRF. *Spectroscopy* 17(7):14–21.

Tuohy, D. R.

1969   Breakage, Burin Facets, and the Probable Linkage Among Lake Mohave, Silver Lake, and Other Varieties of Paleo-Indian Projectile Points in the Desert West. *Nevada State Museum Anthropological Papers* 14:132–152.

Tuohy, D. R., and T. N. Layton

1977   Toward the Establishment of a New Series of Great Basin Projectile Points. *Nevada Archaeological Survey Reporter* 10(6):1–5.

Vest, D. E.

1962   The Plant Communities and Associated Fauna of Dugway Valley in Western Utah. Unpublished Ph.D. dissertation, University of Utah, Salt Lake City.

Wakelin-King, G. A.

1999   Banded Mosaic ("Tiger Bush") and Sheetflow Plains: A Regional Mapping Approach. *Australian Journal of Earth Sciences* 46:53–60.

Wanner, H., O. Solomina, M. Grosjean, S. P. Ritz, and M. Jetel

2011   Structure and Origin of Holocene Cold Events. *Quaternary Science Reviews* 30:3109–3123.

Wardle, H. N.

1913   Some Implements of Surgery from San Miguel Island, California. *American Anthropologist* 15:656–660.

Warren, C. N.

1967   The San Dieguito Complex: A Review and Hypothesis. *American Antiquity* 32:168–185.

Waters, M. R., and T. W. Stafford, Jr.

2007   Redefining the Age of Clovis: Implications for the Peopling of the Americas. *Science* 315:1122–1126.

Waters, M. R., T. W. Stafford, Jr., H. G. McDonald, C. Gustafson, M. Rasmussen, E. Cappellini, J. V. Olsen, D. Szklarczyk, L. J. Jensen, M. T. P. Gilbert, and E. Willerslev

2011   Pre-Clovis Mastodon Hunting 13,800 Years Ago at the Manis Site, Washington. *Science* 334:351–353.

Weng, C., and S. T. Jackson

1999   Late-Glacial and Holocene Vegetation and Paleoclimate of the Kaibab Plateau, Arizona. *Palaeogeography, Palaeoclimatology, Palaeoecology* 153:179–201.

Wigand, P., and D. Rhode

2002   Great Basin Vegetation History and Aquatic Systems: The Last 150,000 Years. In *Great Basin Aquatic Systems History*, edited by R. Hershler, D. B. Madsen, and D. R. Currey, pp. 309–368. Smithsonian Contributions to the Earth Sciences 33. Smithsonian Institution Press, Washington, D.C.

Willig, J. A.

1988   Paleo-Archaic Adaptations and Lakeside Settlement Patterns in the Northern Alkali Basin, Oregon. In *Early Human Occupation in Far Western North America: The Clovis–Archaic Interface*, edited by J. A. Willig, C. M. Aikens, and J. L. Fagan, pp. 417–482. Nevada State Museum Anthropological Papers 21. Carson City.

1989   Paleo-Archaic Broad Spectrum Adaptations at the Pleistocene–Holocene Boundary in Far Western North America. Unpublished Ph.D. dissertation, University of Oregon, Eugene.

Willig, J. A., and C. M. Aikens

1988   The Clovis–Archaic Interface in Far Western North America. In *Early Human Occupation in Far Western North America: The Clovis–Archaic Interface*, edited by J. A. Willig, C. M. Aikens, and J. L. Fagan, pp. 1–40. Nevada State Museum Anthropological Papers 21. Carson City.

Willig, J. A., C. M. Aikens, and J. L. Fagan (editors)
1988   *Early Human Occupation in Far Western North America: The Clovis–Archaic Interface.* Nevada State Museum Anthropological Papers 21. Carson City.

Wilson, C. A.
2007   A Re-Evaluation on X-Ray Fluorescence Data from Idaho and Southeastern Oregon. *Idaho Archaeologist* 30:17–26.

Witthoft, J.
1952   A Paleo-Indian Site in Eastern Pennsylvania: An Early Hunting Culture. *Proceedings of the American Philosophical Society* 96:464–495.

Wormington, H. M.
1957   *Ancient Man in North America.* Denver Museum of Natural History, Popular Series No. 4. Denver.

Young, D. C.
2008   *The Archaeology of Shifting Environments in the Great Salt Lake Desert: A Geoarchaeological Sensitivity Model and Relative Chronology for the Cultural Resources of the U.S. Air Force Utah Test and Training Range.* Far Western Anthropological Research Group, Inc., Davis. Submitted to the Hill Air Force Base, Utah.

Young, D. C., D. Duke, and T. Wriston
2006   *Cultural Resources Inventory of High Probability Lands near Wildcat Mountain: The 2005 Field Season on the US Air Force Utah Test and Training Range-South, Tooele County, Utah.* Far Western Anthropological Research Group, Inc., Virginia City, Nevada. Submitted to the Hill Air Force Base, Utah.

Zeanah, D. W.
2004   Sexual Division of Labor and Central Place Foraging: A Model for the Carson Desert of Western Nevada. *Journal of Anthropological Archaeology* 23:1–32.

Contributors

DAVID B. MADSEN
Texas Archaeological Research Laboratory
University of Texas at Austin
and
Department of Anthropology
Texas State University
San Marcos, Texas

DAVE N. SCHMITT
Desert Research Institute
Division of Earth and Ecosystem Sciences
Reno, Nevada
and
Department of Anthropology
Southern Methodist University
Dallas, Texas

DAVID PAGE
Desert Research Institute
Division of Earth and Ecosystem Sciences
Reno, Nevada

CHARLOTTE BECK
Department of Anthropology (Professor Emeritus)
Hamilton College
Clinton, New York

DARON G. DUKE
Far Western Anthropological Research Group
Henderson, Nevada

GEORGE T. JONES
Department of Anthropology
Hamilton College
Clinton, New York

LISBETH A. LOUDERBACK
Department of Anthropology
University of Utah
Salt Lake City, Utah

CHARLES G. OVIATT
Department of Geology
Kansas State University
Manhattan, Kansas

DAVID RHODE
Desert Research Institute
Division of Earth and Ecosystem Sciences
Reno, Nevada

D. CRAIG YOUNG
Far Western Anthropological Research Group
Carson City, Nevada

Index

Page numbers in *italics* indicate illustrations.

acquisition loci, and classification of Paleoarchaic sites in ORB, 82, 95

aeolian dunes, 5, 6, 36, 38–40

aerial photography, and mapping of distributary channels, 42–46

Aikens, C. M., 10

Air Force, U.S., 3, 8, 240

American Falls, 232

Amick, D. S., 172, 175–76

amorphous bifaces: definition of, *99*; examples of, *172*; and lithic analysis of ORB assemblage, 170–71; site locations of, *159*

amorphous cores, *203, 205*

amorphous unifaces: definition of, *99*; examples of, *203*; and lithic analysis of ORB assemblage, 198, *202*; site locations of, *200*

andesite, 226. *See also* fine-grained volcanics

animals. *See* mammals; megafauna

Antevs, Ernst, 9

Archaeological Resources Protection Act, 63

Archaic period, and shift in subsistence adaptations and settlement patterns in ORB, 255–56

Archaic projectile point types: definition of, *112*; examples of, *149*; and lithic analysis of ORB assemblage, 145–47, *148*; site locations of, *150*

ArcMap 9.x, 42

Arkush, B. S., 236n5

Arlington Springs (Channel Islands), 15

Army, U.S., 3, 240

Army Regulation 200–204, 63

"attractiveness," ranking of habitat patches on basis of in gravity model of mobility, 247–48

Badlands, as fine-grained volcanics source, 224

Baja Peninsula, and Western Stemmed Tradition tool forms, 12, 15

basalt. *See* fine-grained volcanics

base camps: lithic assemblages and identification of in ORB, 207; and models of mobility, 246, 250–51; and site types in ORB, 95, 241, 254

Basgall, M. E., 142–43

Beaked scrapers, *181, 194, 195, 197*

Bear Gulch, 232

Beaver River, 35, 251

Beck, Charlotte, 13, 17, 20, 97, 100, 105, 108–10, 125–27, 132, 134, 137–38, 150–52, 160–61, 167, 180, 210, 218–19, 229, 235, 236n2, 242, 245–46, 249–50

Benson, L. V., 25

beveling: of Archaic projectile point types, *148*; of Butte Valley Corner-notched points, *144*; and descriptions of projectile points, 107; of Dugway Stubby points, *141*; of Eden points, *144*; of Elko points, *144*; of Expanding Stem points, *135*; of Lake Mohave points, *122*; of Pinto points, *144*; of Short Stem points, *144*; of Square Stem points, *129*; of Western Stemmed Tradition points, *118*

biface(s): and amorphous bifaces, 170–71; and drills, 170; and indeterminate bifaces, 171; and knives, 169–70; production bifaces and biface reduction, 157–68; and unique bifaces, 170

biface-reduction flakes (BRFs), 164, 166, 168

bighorn sheep (*Ovis canadensis*), 6

Big Southern Butte, 231

biotic communities, and environment of ORB, 5–7. *See also* birds; mammals; plants

birds: and biotic communities of Great Basin, 6–7, 240, 253; and Early Holocene in Bonneville Basin, 29; species in modern Fish Springs Wildlife Refuge as model for ORB in Paleoarchaic, 240

Black Channel: chronology and geomorphology of, 54–55; and distributary channel forms, *40, 41*; mapping and coding of, *43*; and projectile points, 153, *154*, 156; site descriptions and classifications, 66–73, 87, 91, 93, 94, 95, 241; and sources of toolstone, 248

black mats. *See* "organic sediments"

Black Rock Area, as obsidian source, 222, 224, 251–52, 255

Black Rock Desert, 172

black-tailed rabbits (*Lepus californicus*), 6

"bladelets," 201

Blaxland, A. B., 209

Blue Channel: chronology and geomorphology of, 56; mapping and coding of, *43, 45*; and projectile points, *155*, 156; and site descriptions and classifications, 68, 78, 79–83, *91, 92, 93, 94, 95*; and sources of toolstone, 248

Blue Lake, 25, 26–27, 29

Bonneville, Lake, 22, 23, 32, 34

Bonneville Basin: fine-grained volcanic sources in, 212–13;